Fetal Pig Version

Human Anatomy and Physiology Laboratory Manual

SECOND EDITION

Elaine Nicpon Marieb, R.N., Ph.D.
Holyoke Community College

The Benjamin/Cummings Publishing Company, Inc.
Menlo Park, California • Reading, Massachusetts
Don Mills, Ontario • Wokingham, U.K. • Amsterdam • Sydney
Singapore • Tokyo • Mexico City • Bogota • Santiago • San Juan

Sponsoring Editor: *Andrew Crowley*
Production Supervisor: *Mimi Hills*
Interior Design: *John Edeen*
Cover Design: *Gary Head*
Page Layout: *Betty Duncan-Todd*
Photo Research: *Lisa Stanziano*
Back cover photograph: *Robert Aller*

Artists: *Marjorie Garlin, Cathleen Jackson, Neesh Wallace, Barbara Haynes, Fran Milner, Jill Leland, Doreen Davis Masterson, Sara Lee Steigerwald, and Carol Verbeeck. Cat musculature art by Sibyl Graber-Gerig. Art to accompany photomicrographs by Linda McVay.*

Photomicrographs 3.5, 6.3b, 6.5a, 6.6b, c, 14.1a and inside front and back covers: *Ed Reschke*

Photomicrographs 6.3a, c-f; 6.5 b-e; 6.6a, 6.7, 14.2, 17.1, 17.5: *Victor B. Eichler*

The cover shows a cross section of myelinated nerve fibers. Photomicrograph by Ed Reschke.

Of related Interest from the Benjamin/ Cummings Series in the Life Sciences

J. B. Jenkins
Human Genetics (1983)

S. W. Langjahr and R. A. Brister
Human Anatomy: A Computerized Review and Coloring Atlas (1985)

E. N. Marieb
Human Anatomy and Physiology Lab Manual: Brief Version (1983)

E. N. Marieb
Human Anatomy and Physiology Lab Manual: Fetal Pig Version, second edition (1985)

E. B. Mason
Human Physiology (1983)

A. P. Spence
Basic Human Anatomy (1982)

A. P. Spence and E. B. Mason
Human Anatomy and Physiology, second edition (1983)

Library of Congress Cataloging in Publication Data

Marieb, Elaine Nicpon, 1936–
  Human anatomy and physiology laboratory manual.

  Includes index.
  1. Human physiology—Laboratory manuals. 2. Anatomy, Human—Laboratory manuals. 3. Cats–Physiology–Laboratory manuals. 4. Cats—Anatomy—Laboratory manuals.
1. Title. [DNLM: 1. Anatomy—laboratory manuals.
2. Anatomy, Comparative—laboratory manuals.
3. Physiology—laboratory manuals. 4. Swine—anatomy and histology—laboratory manuals. 5. Swine—physiology—laboratory manuals. QS 25 M334h]
QP44.N54   1985   612'.0028   84-20338

ISBN 0-8053-6726-8

    DEFGHIJ-SE-89876

The Benjamin/Cummings Publishing Company, Inc.
2727 Sand Hill Road
Menlo Park, California 94025

PREFACE

Human Anatomy and Physiology Laboratory Manual, Second Edition, is a self-contained learning aid designed for introductory courses in human anatomy and physiology. It presents a wide range of laboratory experiences for students concentrating in nursing, physical therapy, occupational therapy, respiratory therapy, dental hygiene, pharmacology, health and physical education, as well as biology and pre-medical programs.

ORGANIZATION

The organization and scope of this manual lend themselves to use in both one- and two-term courses. The variety of both anatomical studies and physiological experiments enables instructors to gear their courses to specific academic programs, or to their own teaching preferences. The manual is independent of any textbook, and contains all the background discussion and terminology necessary to perform all experiments. This eliminates the need for students to bring a textbook into the laboratory.

Many anatomy and physiology textbooks were consulted during the writing of this manual to ensure consistency of terminology. Spence and Mason's *Human Anatomy and Physiology* was particularly helpful in this regard, and I gratefully acknowledge the permission of the publisher and authors to borrow and adapt many of the outstanding illustrations from that book.

Each of the 47 exercises leads students toward a coherent understanding of the structure and function of the human body. The manual begins with anatomic terminology and an orientation to the body, which provides the necessary tools for studying the various body systems. The exercises which follow reflect the dual focus of the manual—both anatomic and physiologic aspects receive considerable attention. As the various organ systems of the body are introduced the initial exercises focus on organization, from the cellular to the organ level. The anatomic exercises are followed by physiologic experiments to familiarize students with the scientific method, important laboratory skills, and an appreciation of the critical fact that function follows structure. Homeostasis of the body as a whole, and of its subdivisions, is continually emphasized. This holistic approach ultimately gives students an integrated understanding of the human body.

Most exercises can be completed in a two-hour lab session, although some exercises, such as those on blood, cardiovascular physiology, and muscle physiology, may require more time. In such exercises, generally requiring animal dissection or elaborate equipment setups, students can work in teams or can conduct selected portions of the exercises and share their results with the others. Instructor demonstrations have been kept to a minimum in the interest of maximizing student involvement in the learning process.

NEW IN THIS EDITION

The strengths of the previous edition have been retained; the manual is comprehensive and balanced enough to be flexible, and carefully written so that students can successfully complete exercises with little supervision. Changes made in the second edition of *Human Anatomy and Physiology Laboratory Manual* respond to the consonstructive suggestions and requests of readers and reviewers. Among the specific changes implemented are the following:

1. One entirely new exercise (47), "Principles of Heredity," has been added. This exercise provides an overview of the basic principles and terminology of genetics. Explanations, examples, and student activities emphasize human traits to inspire the students' interest in this challenging topic.
2. An additional experiment has been added to Exercise 28, "Experiments on Hormonal Action." Many reviewers requested that a laboratory experience dealing with the topic of metabolism be included in this edition. Such an experiment, "Effect of Thyroxine on Metabolic Rate," is now included.
3. The lab review section has been partially reformatted. More matching, multiple-choice, and labeling exercises have been added; fewer of the questions requiring short essay-type student explanations have been retained. Whereas such exercises offer valuable learning experiences, they are time consuming both to complete and correct. Every attempt has been made to achieve an acceptable balance between explanatory and recognition type questions in the second edition.
4. The original line drawings of pig musculature have been replaced by photographs accompanied by corresponding line art to maximize the students' abilities to identify pig muscles.
5. More photographs (and roentograms) have been

included throughout the manual.

6. Twelve full-color photomicrographs of various body tissues and organs are keyed to the corresponding exercises and have been placed inside the front and back covers for student reference.

7. A second color has been added to the line art to provide emphasis and improve effectiveness.

8. The descriptive text has been streamlined, and explanations and experimental procedures have been updated or revised to increase clarity and to incorporate current knowledge.

9. At the request of instructors, a comprehensive list of laboratory materials has been added to facilitate the task of ordering materials for the experiments.

THE DISSECTION ANIMAL

Although this manual emphasizes the study of the human body, the fetal pig is the major dissection animal. Market research indicates that the majority of laboratory courses in anatomy and physiology use cats as dissection animals, but due to the ever increasing price of cats, the fetal pig is gaining popularity in laboratory classrooms. For courses using the cat as a dissection animal, an alternate Cat Version of this manual is available (ISBN 0-8053-6726-8). The only differences between this manual and the Cat Version are in the instructions and artwork for the dissections.

SPECIAL FEATURES

The numerous pedagogical features of the first edition appreciated by the users of the manual have been retained in the present edition. These special features include:

- Learning objectives and a materials list precede each exercise. In addition, a complete materials list is now included in the front matter for easy reference and to facilitate the ordering of necessary apparatus.
- Key terminology appears in boldface print, and each term is defined when introduced.
- Illustrations are plentiful, large, and of exceptional quality. Structures are consistently shown in their natural anatomical positions within the body.

- Exercises are balanced between histological, gross anatomical, and physiological topics.
- There are numerous physiological experiments for each organ system, ranging from simple experiments that can be performed without specialized tools to more complex experiments using laboratory equipment or instrumentation techniques.
- Tear-out Laboratory Review-Study Sections, keyed to the exercises, require students to label diagrams and answer multiple-choice, short-answer, and review questions. Space is provided for recording and interpreting experimental results.
- A Solutions Manual is available to instructors free, on request from the publisher.

ACKNOWLEDGMENTS

I wish to thank the following people for their contributions to this edition of *Human Anatomy and Physiology Laboratory Manual:* Donna S. Klaaren (Miami University, Ohio), Robert Thomas (College of the Redwoods, California), Glenn Yoshida (Los Angeles Southwest College), and Janice Young, Ph.D. (McHenry County College, Illinois). These instructors reviewed the manuscript and provided invaluable help.

I also wish to acknowledge my indebtedness to Jim Behnke, Editor-in-Chief, for his constant encouragement; Andy Crowley, Acquisitions Editor; Mimi Hills, Production Supervisor; Betty Duncan-Todd, Lisa Stanziano, and the many others at Benjamin/Cummings who have contributed to this effort.

My continuing thanks also go to those who contributed to the first edition of this manual—the reviewers, the production staff at Benjamin/Cummings, my mentor, Dr. Curtis Smith, and as always, my husband Joseph.

Finally, since preparations for the next edition begin well in advance of its publication, I invite users of this edition to send me their comments and suggestions for subsequent editions.

Elaine Nicpon Marieb
Department of Biology
Holyoke Community College
303 Homestead Avenue
Holyoke, MA 01040

CONTENTS

List of Laboratory Materials

Exercise 1
Human torso model (dissectible)
Human skeleton
Demonstration: sectioned and labeled kidneys (three separate kidneys uncut or cut so that a. entire, b. transverse section, and c. longitudinal sectional views are visible)

Exercise 2
Freshly killed or preserved rat for dissection (one for every two to four students), or dissected human cadaver
Dissecting pans and pins
Scissors
Forceps
Human torso model (dissectible)
Related film: *Man: The Incredible Machine* (color, sound, 16 mm, 28 minutes, National Geographic Educational Services)

Excroioc 3
Compound microscope
Millimeter ruler
Prepared slides of the letter *e* or newsprint
Immersion oil
Lens paper
Prepared slide of grid ruled in millimeters (grid slide)
Prepared slide of crossed colored threads
Clean microscope slide and cover slip
Toothpicks (flat tipped)
Physiological saline in a dropper bottle
Methylene blue stain (dilute) in a dropper bottle
Forceps

Exercise 4
Three-dimensional model of the "composite" animal cell or laboratory chart of cell anatomy
Prepared slides of simple squamous epithelium ($AgNO_3$ stain), teased smooth muscle, human blood cell smear, and sperm
Compound microscope
Clean slide and cover slip
Physiologic saline

Flat-tipped toothpick
Janus green stain and medicine dropper
Filter paper or some other type of porous paper
Forceps
Prepared slides of whitefish blastulae
Three-dimensional models of mitotic stages
Related films: *Cell Biology: Life Functions* (color, sound, 16 mm, 19 minutes, Coronet Films); *Mitosis* (color, sound, 16 mm, 24 minutes, Encyclopaedia Britannica Films); *Cell Biology: Structure and Composition* (color, sound, 16 mm, 13 minutes, Coronet Films)

Exercise 5
For Passive Transport Experiments:
Clean slides and cover slips
Forceps, glass stirring rods, 15 ml graduated cylinders
Compound microscopes
Brownian motion: milk in dropper bottles, hot plate
Diffusion:
Demonstration: potassium permanganate crystals in distilled water; set up 24 hours before in a 500 cc graduated cylinder
Four dialysis sacs
Large beakers (500 ml)
40% glucose solution
Fine twine
10% NaCl solution, boiled starch solution
Laboratory balance
Benedict's solution in dropper bottle
Test tubes in racks, test tube holder
Wax marker
Silver nitrate ($AgNO_3$) in dropper bottle
Lugol's iodine solution in dropper bottle
Lancets, alcohol swabs
Mammalian Ringer's solution in dropper bottle, physiologic saline in dropper bottle
Yeast suspension
Congo red dye in dropper bottle
filtration:
Ring stand, ring, clamp, filter paper, funnel, solution of uncooked starch, powdered charcoal, and copper sulfate ($CuSO_4$)

For Active Transport:
Culture of starved ameba (*Amoeba proteus*)
Depression slide
Tetrahymena pyriformis culture
Compound microscope

Exercise 6
Compound microscope
Prepared slides of simple squamous, simple cuboidal, simple columnar, stratified squamous (nonkeratinizing), pseudostratified ciliated, and transitional epithelium
Prepared slides of adipose, areolar, and dense fibrous connective tissue (tendon); hyaline cartilage; and bone (cross section)
Prepared slides of skeletal, cardiac, and smooth muscle (longitudinal sections)
Prepared slide of nervous tissue (spinal cord smear)
Related film: *Nature of Life: Cells, Tissues, and Organs* (color, sound, 16 mm, 15 minutes, Coronet Films)

Exercise 7
Skin model (three-dimensional, if available)
Compound microscope
Prepared slide of human skin with hair follicles
Related film: *Human Skin* (color, 8 mm film loop, Encyclopaedia Britannica Films)

Exercise 8
Compound microscope
Prepared slides of trachea (cross section) and small intestine (cross section)
Prepared slide of serous membrane
Longitudinally cut fresh beef joint (if available)

Exercise 9
Numbered disarticulated bones that demonstrate classic examples of the four bone classifications (long, short, flat, and irregular)
Long bone sawed longitudinally (beef bone from a slaughterhouse,

if possible, or prepared laboratory specimen)
Compound microscope
Prepared slide of ground bone (cross section)
Long bone soaked in 10% nitric acid until flexible
Long bone baked at 250°F for more than 2 hours

Exercise 10

Intact skull and Beauchene skull
Colored pencils
X-rays of individuals with scoliosis, lordosis, and kyphosis (if available)
Articulated skeleton
Isolated cervical, thoracic, and lumbar vertebrae, sacrum, and coccyx
Disarticulated bones of the thorax

Exercise 11

Articulated skeletons
Disarticulated skeletons (complete)
Articulated pelves (male and female for comparative study)

Exercise 12

Isolated fetal skull
Fetal skeleton
Adult skeleton

Exercise 13

Articulated skeleton, skull
Diarthrotic beef joint (fresh)
Anatomical chart of joint types (if available)
X-rays of normal and arthritic joints (if available)

Exercise 14

Three-dimensional model of skeletal muscle cells (if available)
Forceps
Dissecting needles
Microscope slides and cover slips
Frog Ringer's solution in dropper bottles
Pithed frog, thigh muscle exposed
Compound microscope
Histological slides of skeletal muscle (longitudinal and cross section), skeletal muscle showing myoneural junctions, smooth muscle and cardiac muscle (all longitudinal section)
Three-dimensional models of smooth and cardiac muscle (if available)
Three dimensional model of skeletal muscle showing myoneural junction (if available)
Related film: *Muscles* (color, sound, 16 mm, 30 minutes, CRM Educational Films)

Exercise 15

Disposable gloves or protective skin cream
Preserved and injected pig (one for every two to four students)
Dissecting trays and instruments
Name tag and large plastic bag
Paper towels
Embalming fluid (formalin)
Human torso model or large anatomical chart showing human musculature

Exercise 16

ATP muscle kits (glycerinated rabbit psoas muscle; ATP and salt solutions. Obtainable from Carolina Biological Supply, Item #20-3525)
Petri dishes
Microscope slides
Cover glasses
Millimeter scale
Compound microscope
Dissecting microscope
Small beaker (50 ml)
Frog Ringer's solution
Scissors
Metal needle probes
Pointed glass probes
Medicine dropper
Cotton thread
Forceps
Glass or porcelain plate
Live bullfrog
Apparatus A or B
A: kymograph, kymograph paper (smoking stand, burner, and glazing fluid if using a smoke-writing apparatus), skeletal muscle lever, signal magnet, laboratory stand, clamp, electronic stimulators
B: physiograph (polygraph), polygraph paper and ink, myograph, pin and clip electrodes, stimulator output extension cable, transducer cable, straight pins, frog board, laboratory stand, clamp

Exercise 17

Model of a "typical" neuron (if available)
Compound microscope
Histological slides of an ox spinal cord smear and teased myelinated nerve fibers
Prepared slides of Purkinje cells (cerebellum), pyramidal cells (cerebrum), and a dorsal root ganglion
Prepared slide of a nerve (cross section)

Exercise 18

Human brain model (dissectible)
Preserved human brain (if available)
Coronally sectioned human brain slice (if available)
Preserved sheep brain (meninges and cranial nerves intact)
Dissecting tray and instruments

Exercise 19

Spinal cord model (cross section)
Laboratory charts of the spinal cord and spinal nerves and sympathetic chain
Red and blue pencils
Preserved cow spinal cords with meninges and nerve roots intact (or spinal cord segment saved from the brain dissection in Exercise 18)
Dissecting tray and instruments
Dissecting microscope
Histological slide of spinal cord (cross section)
Compound microscope
Pig specimen from previous dissections

Exercise 20

Supply area 1:
Bullfrogs
Dissecting instruments and tray
Ringer's solution
Thread
Glass rods or probes
Glass plates or slides
Ring stand and clamp
Stimulator; platinum electrodes
Filter paper
0.01% hydrochloric acid (HCl) solution
Sodium chloride (NaCl) crystals
Heat-resistant mitts
Bunsen burner
Absorbent cotton
Ether
Pipettes
1-cc syringe with small-gauge needle
0.5% tubocurarine solution

Supply area 2:
Indifferent (plate) electrode (or EKG electrode)
Rubber straps
Insulated copper wire to be used as electrode lead or exploring electrode
Electrode paste
Penknife
Saline solution

Exercise 21

Compound microscope
Histological slides of pacinian corpuscles (longitudinal section), Meissner's corpuscles, Golgi tendon organs, and muscle spindles
Temperature cylinders (or pointed

metal rods of approximately 1-in.
diameter and 2- to 3-in. length,
with insulated blunt end
Large beaker of ice water; chipped
ice
Hot water bath set at 45°C
Thermometer
Fine-point, felt-tipped markers
(black, red, and blue)
Small metric ruler
Von Frey's hairs (or 1-in. lengths of
horsehair glued to a matchstick)
or sharp pencil
Towel
Caliper or esthesiometer
Four coins (nickels or quarters)
Three large finger bowls or 1000-ml
beakers

Exercise 22
Reflex hammer
Lancets
Alcohol swabs
Cot (if available)
Absorbent cotton
Tongue depressor
Metric ruler
Flashlight
100- or 250-ml beaker
10- or 25-ml graduated cylinder
Lemon juice
Wide-range pH paper
12-in. ruler

Exercise 23
Oscilloscope and EEG lead-selector
box or polygraph and high-gain
preamplifier
Cot
Electrode gel
EEG electrodes and leads
Collodion gel

Exercise 24
Dissectible eye model
Chart of eye anatomy
Preserved cow or sheep eye
Dissecting pan and instruments
Laboratory lamp
Histological section of an eye show-
ing retinal layers
Snellen eye chart (Floor marked
with chalk to indicate 20 ft. dis-
tance from posted Snellen chart.)
Ishihara's color blindness plate
Ophthalmoscope
Compound microscope
1-inch diameter disks of colored
paper (white, red, blue, green)
White, red, blue, and green chalk
Metric ruler
Common straight pins
Related film: *The Eyes and Seeing*
(color, sound, 16 mm, 20 minutes,
Encyclopaedia Britannica Films)

Exercise 25
Three-dimensional dissectible
ear model and/or chart of ear
anatomy
Histological slides of the cochlea of
the ear
Compound microscope
Tuning forks (range of frequencies)
Rubber mallet
Absorbent cotton
Otoscope (if available)
Alcohol swabs
12-in. ruler
Related film: *The Ears and Hearing*
(black and white, sound, 16 mm,
10 minutes, Encyclopaedia
Britannica Films)

Exercise 26
Prepared histological slides: the
tongue showing taste buds; the
nasal epithelium (longitudinal
section)
Compound microscope
Paper towels
Small mirror
Granulated sugar
Cotton-tipped swabs
Absorbent cotton
Paper cups
Prepared vials of 10% NaCl, 0.1%
quinine or Epsom salt solution,
5% sucrose solution, and 1% acetic
acid
Prepared dropper bottles of oil of
cloves, oil of peppermint, and oil
of wintergreen
Like-sized food cubes of cheese,
apple, raw potato, dried prunes,
banana, raw carrot, and hard-
cooked egg white
Chipped ice

Exercise 27
Human torso model
Anatomical chart of the human endocrine system
Compound microscopes
Colored pencils
Histological slides of the anterior-
posterior pituitary (differential
staining), thyroid gland, parathyroid glands, adrenal gland, pancreas tissue
Dissection animal, dissection tray,
and instruments (if desired by the
instructor)
Related film: *The Hormones: Small
but Mighty* (black and white,
sound, 16 mm, 29 minutes, Indiana University, Audio-Visual
Center, Bloomington, IN 47401)

Exercise 28
Female frog (*Rana pipiens*)
Syringe
20- to 25-gauge needle

Pituitary extract
Physiologic saline
Battery jar
Wax marking pencils

500- or 600- ml beakers
10% glucose solution
Commercial insulin in dropper
bottle
Small fresh water fish (bluegill,
sunfish, or goldfish)

1:10,000 epinephrine (Adrenalin)
solution in dropper bottles
Dissecting pan, instruments, and
pins
Frog Ringer's solution in dropper
bottle

Glass dessicator; manometer, 20 ml
glass syringe, two-hole cork, and
T-joint (1 each/ 3-4 students)
Soda lime (dessicant)
Hardware cloth squares
Petrolatum
Rubber tubing; tubing clamps; scis-
sors
3-in. pieces of glass tubing
Animal balances
Young rats of the same box, obtained
two weeks prior to the laboratory
session and treated as follows for
14 days:
Group 1: control group—fed normal
rat chow and water
Group 2: experimental group
A—fed normal rat chow and
drinking water containing 0.02%
propylthiouracil
Group 3: experimental group
B—fed rat chow containing dessicated thyroid (2% by weight) and
normal drinking water
Chart set up on chalkboard to allow
each student group to record its
computed metabolic rate figures
under the appropriate heading

Exercise 29
Compound microscope
Immersion oil
Clean microscope slides
Lancets
Alcohol swabs
Wright's stain in dropper bottle
Distilled water in dropper bottle
Three-dimensional model of blood
cells (if available)
Plasma (obtained from a clinical
agency or prepared by centrifug-
ing whole blood)
Wide-range pH paper
Test tubes
Test tube racks
Because many blood tests are to be
conducted in this exercise, it
seems advisable to set up a
number of appropriately labeled

supply areas for the various tests. These are designated below.

General supply area:
Heparinized blood (obtained from a clinical agency) if desired by the instructor
Blood lancets
Alcohol swabs
Absorbent cotton balls
Millimeter ruler
Heparinized capillary tubes
Pipette cleaning solutions—(1) 10% household bleach solution, (2) distilled water, (3) 70% ethyl alcohol, (4) acetone

Blood cell count supply area:
Hemacytometer
RBC and WBC dilution fluids (Hayem's solution for RBCs; 1% acetic acid for WBCs)
RBC and WBC dilution pipettes and tubing
Mechanical hand counters

Hematocrit supply area:
Heparinized capillary tubes
Microhematocrit centrifuge and reading gauge (if the reading gauge is not available, millimeter ruler may be used)
Seal-ease (Clay Adams Co.) or modeling clay

Hemoglobin determination supply area:
Tallquist hemoglobin scales and test paper
Sahli hemoglobin determination kit (or another type of hemoglobino-meter such as Spencer or American Optical)
1% HCl
Slender glass stirring rod

Sedimentation rate supply area:
Landau Sed-rate pipettes with tubing and rack
Wide-mouthed bottle of 5% sodium citrate
Mechanical suction device

Coagulation time supply area:
Capillary tubes (nonheparinized)
Triangular file

Blood typing supply area:
Blood typing sera (anti-A, anti-B, and anti-Rh [D])
Rh typing box
Wax marker
Toothpicks
Clean microscope slides

Exercise 30
Torso model or laboratory chart showing heart anatomy

Red and blue pencils
Preserved sheep heart, pericardial sacs intact (if possible)
Dissecting pan and instruments
Pointed glass rods for probes
Three-dimensional model of cardiac muscle
Heart model (three-dimensional)
X-ray of the human thorax for observation of the position of the heart in situ

Exercise 31
ECG recording apparatus
Electrode paste; alcohol swabs
Cot

Exercise 32
Anatomical charts of human arteries and veins (or a three-dimensional model of the human circulatory system)
Anatomical charts of the following specialized circulations: pulmonary circulation, hepatic portal circulation, arterial supply and circle of Willis of the brain (or a brain model showing this circulation), fetal circulation
Compound microscope
Prepared microscope slides showing cross sections of an artery and vein
Dissection animals
Dissecting pans and instruments

Exercise 33
Stethoscope
Sphygmomanometer
Watch (or clock) with a second hand
Step stool
Alcohol swabs
Millimeter ruler
Ice
Small basin suitable for the immersion of one hand
Laboratory thermometer (°C)
Record or audiotape: "Interpreting Heart Sounds" (available on free loan from the local chapters of the American Heart Association)
Phonograph or tape deck

Exercise 34
Frog
Dissecting pan and instruments
Petri dishes
Medicine dropper
Millimeter ruler
Frog Ringer's solutions (at room temperature, 5°C, and 32°C)
Kymograph recording apparatus and paper
Two heart levers
Signal marker
Stimulator and platinum electrodes

Thread, rubber bands, fine common pins
Frog board
Ring stand; right angle clamps (three)
Cotton balls
Freshly prepared solutions of the following in dropper bottles:
5% atropine sulfate
1% epinephrine
5% potassium chloride (KCl)
1% calcium chloride (CaCl$_2$)
0.7% sodium chloride (NaCl)
0.01% histamine
0.01 N HCl

Exercise 35
Large anatomical chart of the human lymphatic system
Prepared microscope slides of lymph nodes
Compound microscope
Dissection animal, tray, and instruments

Exercise 36
Human torso model
Respiratory organ model and/or chart of the respiratory system
Larynx model (if available)
Sheep pluck (preserved or fresh from the slaughterhouse)
Dissection animals, trays, and instruments
Source of compressed air
2-foot length of laboratory rubber tubing
Alcohol swabs
Histological slides of the following (if available): trachea (cross section), lung tissue, pathological specimens such as section taken from lung tissues exhibiting bronchitis, pneumonia, emphysema, atelectasis, or other conditions
Compound and dissecting microscopes
Bronchogram (if available)
Suggested film: *Respiration in Man* (color, sound, 16 mm, 26 minutes, Encyclopaedia Britannica Films)

Exercise 37
Model lung (bell jar demonstrator)
Tape measure
Spirometer
Disposable mouthpieces
Nose clips
Alcohol swabs
Paper bag
Pneumograph and recording attachments
Recording apparatus (kymograph or physiograph)
Stethoscope

pH meter (standardized with buffer of pH 7)
Buffer solution (pH 7)
Concentrated HCl and NaOH
250-milliliter beakers
Glass stirring rod
Stethoscope
Frog
Dissecting pan (wax surface)
Dissecting instruments and fine dissecting pins
Frog Ringer's solution (room temperature and 15°C) in dropper bottles
Cork

Exercise 38

Dissectible torso model
Anatomical chart of the human digestive system
Dissection animal, trays, instruments, and bone cutters
Model of a villus and liver (if available)
Jaw model or human skull
Prepared microscope slides of the liver, pancreas, longitudinal sections of the esophageal-cardiac junction and tooth, and cross sections of the stomach, duodenum, and ileum
Hand lens
Compound microscope

Exercise 39

Supply area 1:
Water pitcher
Paper cups
Stethoscope

Supply area 2:
Rats or guinea pigs that have fasted for 24 hours (The rats should be force-fed cream and meat extract 2 hours before laboratory experiment.)
Large bell jar
Ether
Absorbent cotton
Dissecting instruments and pan
Hand lens
Physiological saline

Supply area 3:
materials in supply area 2, plus physiograph or kymograph recording apparatus (as employed in Unit 7, Exercise 16) with muscle transducers or isotonic muscle levers for recording muscle activity.
L-shaped glass tubing
Rubber hose length
Tyrode's solution
Water bath set at 37°C
250-ml beakers
Laboratory thermometer (°C)
Metric ruler

Alcohol burner or Bunsen burner
Medicine dropper
Thread
0.5-ml pipettes or 1-cc prepackaged syringes with needles
Ice
Epinephrine 1:10,000 and acetylcholine 1:10,000 solutions
Solutions in dropper bottles: 7.5% KCl, 2% NaOH, 20% $CaCl_2$, and 2% HCl

Exercise 40

Supply area 1:
50-ml or 100-ml graduated cylinders
Test tubes and rack
Paraffin
Dropper bottles with Lugol's IKI (Lugol's iodine) and Benedict's solution
0.1% starch solution
1.0% maltose solution
1 N NaOH and 1 N HCl in dropper bottles
Medicine dropper
Glass stirring rod
Wide-range pH paper
Wax markers
Hot plate and 500-ml beakers
Water bath at 37°C
Ice bath

Supply area 2:
5% pepsin solution
Fibrin or finely minced egg white; 1% albumin solution or egg white
0.8% HCl, 0.5% NaOH, 10% NaOH, 1% $CuSO_4$
Evaporating dishes; test tubes and test tube rack
10-ml graduated cylinder; glass stirring rod
Freshly prepared 0.1% Ninhydrin solution
0.1% alcoholic solution of glycine or alanine

Supply area 3:
1% pancreatin solution in 0.2% Na_2CO_3
Litmus cream (fresh cream to which powdered litmus has been added to achieve a blue color)
0.1 N HCl: glass stirring rod
Bile salts (sodium taurocholate)
Test tubes and test tube rack
10-ml graduated cylinder
Vegetable oil in dropper bottle

Exercise 41

Human dissectible torso model and/or anatomical chart of the human urinary system
3-dimensional model of the cut kidney and of the nephron (if available)

Dissection animal, tray, and instruments
Pig or sheep kidney, doubly or triply injected
Prepared histological slides of a longitudinal section of kidney and cross sections of the bladder
Compound microscope

Exercise 42

Urine samples (collected in clean bottles by the students before the laboratory begins)
Pathological urine specimens obtained from a health facility by the instructor and numbered
Urinometer
Hot plate
1000-ml beakers
Wide-range pH paper
Dip sticks (Clinistix, Ketostix, Albustix, Hemastix)
Ictotest reagent
Test reagents for sulfates: 10% barium chloride solution, dilute HCl (hydrochloric acid)
Test reagent for glucose (Benedict's reagent)
Test reagent for phosphates (dilute nitric acid, dilute ammonium molybdate)
Test reagent for proteins and urea (concentrated nitric acid in dropper bottle)
Test reagent for Cl: 3.0% silver nitrate solution ($AgNO_3$)
Test tubes, test tube rack, and test tube holders
10-cc graduated cylinders
Glass stirring rods
Medicine droppers
Wax marking pencils
Microscope slides
Cover slips
Compound microscope
Centrifuge and centrifuge tubes
"Sedi-stain"

Exercise 43

Distilled water
0.9% and 2.0% saline (NaCl) solutions, 0.1% $NaHCO_3$
Wide-range pH paper
Urinometer
Wax marking pencils
Colored pencils
20% potassium chromate in dropper bottle
2.9% $AgNO_3$ in dropper bottle ($AgNO_3$ should be made fresh daily)
500-ml and 1000-ml beakers
100-ml and 500-ml graduated cylinders (meticulously cleaned)
Test tubes and rack
Test tube holders

Exercise 44

Models or large laboratory charts of the male and female reproductive tracts

Prepared microscope slides of cross sections of the penis, epididymis, ovary, and uterus (showing endometrium)

Dissection animal, tray, and instruments

Bone clippers

Related film: *Human Body: Reproductive System* (color, sound, 16 mm, Coronet Films)

Exercise 45

Prepared microscope slides of testis, ovary, human sperm, and uterine endometrium (showing menses, proliferative, and secretory stages)

Three-dimensional models illustrating meiosis, spermatogenesis, and oogenesis

Compound microscope

Demonstration Area: microscopes set up to demonstrate the following stages of oogenesis in *Ascaris megalocephala:*

1. primary oocyte with fertilization membrane, sperm nucleus, and aligned tetrads apparent
2. formation of the first polar body
3. secondary oocyte with dyads aligned
4. second polar body formation and ovum
5. fusion of the male and female pronuclei to form the fertilized egg

Exercise 46

Prepared slides of sea urchin development (zygote through larval stages)

Compound microscope

Human development models or plaques (if available)

Life Before Birth, Educational Reprint #27, available from Time-Life Educational Materials, Box 834, Radio City P.O., New York, N.Y. 10019

Pregnant cat, rat, or pig uterus (one per laboratory session) with uterine wall dissected to allow student examination

Dissecting instruments

Model of pregnant human torso

Preserved human embryos (if available)

Fresh or formalin-preserved placenta (obtained from a clinical agency)

Microscope slide of placenta tissue

Exercise 47

Pennies (for coin tossing)

PTC (phenylthiocarbamide) taste strips

Sodium benzoate taste strips

Chart drawn on chalkboard for tabulation of class results of human phenotype/genotype determinations

Blood typing sera: Anti-A and Anti-B; slides, toothpicks, wax pencils, sterile lancets, alcohol swabs

UNIT 1

THE HUMAN BODY: AN ORIENTATION

The Language of Anatomy

EXERCISE

1

OBJECTIVES

1. To describe the anatomic position verbally or by demonstration.

2. To use proper anatomic terminology to describe body directions, planes, and surfaces.

3. To name the body cavities and note the important organs in each.

MATERIALS

Human torso model (dissectible)

Human skeleton

Demonstration: sectioned and labeled kidneys (three separate kidneys uncut or cut so that a. entire, b. transverse section, and c. longitudinal sectional views are visible)

Most of us have a natural curiosity about our bodies. This fact is amply demonstrated by infants, who early in life become fascinated with their own waving hands or their mother's nose. The study of the gross anatomy of the human body elaborates on this fascination. Unlike the infant, however, the student of anatomy must learn to identify and observe the dissectible body structures formally. The purpose of any gross-anatomy experience is to examine the three-dimensional relationships of body structures—a goal that can never completely be achieved by using illustrations and models, regardless of their excellence.

When beginning the study of any science, the student is often initially overcome by the jargon unique to the subject. The study of anatomy is no exception. But without this special-

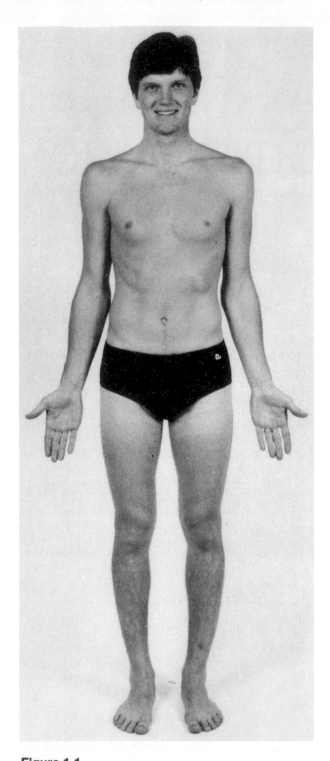

Figure 1.1

Anatomic position.

This unit presents some of the most important anatomic terminology used to describe the body and introduces you to basic concepts of **gross anatomy,** the study of body structures visible to the naked eye.

ANATOMIC POSITION

When anatomists or doctors discuss the human body, they refer to specific areas in relation to other body parts. This is done in accordance with a universally accepted standard position called the **anatomic position**. It is essential to understand this position, because all body terminology employed in this book refers to this body positioning, regardless of the position the body happens to be in. In the anatomic position the human body is erect, with feet together, head and toes pointed forward, and arms hanging at the sides with palms facing forward (Figure 1.1).

Assume the anatomic position, and note that it is not particularly comfortable. The hands are held unnaturally forward rather than hanging partially cupped toward the thighs.

BODY ORIENTATION AND DIRECTION

Study the following terms, referring to Figure 1.2. Note that certain terms have a different connotation for a four-legged animal than they do for a human.

Superior/inferior (*above/below*): These terms refer to placement of a body structure along the long axis of the body. Superior structures always appear above other structures. For example, the nose is superior to the mouth, and the abdomen is inferior to the chest region.

Cephalad/caudad (*Toward the head/toward the tail*): In humans these terms are used interchangeably with *superior* and *inferior*. But in four-legged animals they are synonymous with *anterior* and *posterior*.

Anterior/posterior (*front/back*): In humans the most anterior surfaces are those that are most forward—the face, chest, and abdomen. Posterior surfaces are those on the backside of the body—the back and buttocks.

Dorsal/ventral (*backside/belly side*): These terms are used chiefly in discussing the comparative anatomy of animals, assuming the animal is on all fours. *Dorsum* is a Latin word meaning "back"; thus *dorsal* refers to the backside of the animal's body or of any other structures. For instance, the posterior surface of the leg is its dorsal surface. The term *ventral* derives from the Latin term *venter*, meaning "belly," and thus always refers to the belly side of animals. In humans the terms ventral and dorsal may be used

ized terminology, confusion is inevitable. For example, what do *over, on top of, superficial to, above,* and *behind* mean in reference to the human body? Anatomists have an accepted set of reference terms that are universally understood. These allow body structures to be located and identified with a minimum of words and a high degree of clarity.

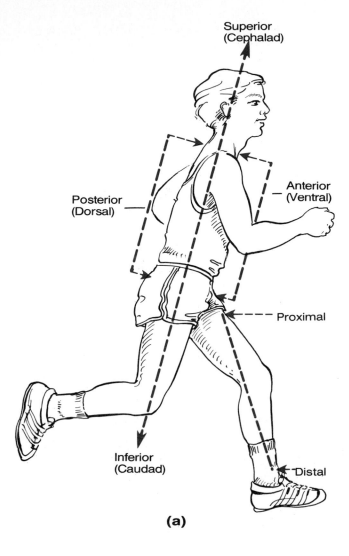

(a)

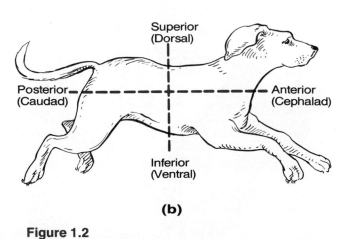

(b)

Figure 1.2

Anatomic terminology describing body orientation and direction; (a) with reference to a human and (b) with reference to a four-legged animal.

interchangeably with the terms anterior and posterior, but in four-legged animals ventral and dorsal are synonymous with inferior and superior.

Medial/lateral (*toward the midline/away from the midline or median plane*): The ear is lateral to the eye; the breastbone is medial to the ribs.

Proximal/distal (*nearer the trunk or attached end/farther from the trunk or point of attachment*): These terms are used primarily to locate various areas of the body limbs. For example, the fingers are distal to the elbow; the knee is proximal to the toes.

Before continuing, use a human torso model, a skeleton, or your own body to specify the relationship between the following structures. Use the correct anatomic terminology:

The wrist is _____ to the hand.

The trachea is _____ to the spine.

The brain is _____ to the spinal cord.

The kidneys are _____ to the liver.

The bridge of the nose is _____ to the eyes.

BODY PLANES AND SECTIONS

To observe internal structures, it is often helpful and necessary to make use of a **section**, or cut. When the section is made through the body wall or through an organ, it is made along an imaginary line called a **plane**. Anatomists commonly refer to three planes (Figure 1.3), or sections, which lie at right angles to one another:

Sagittal section: When the cut is made along a longitudinal plane, dividing the body into right and left parts, it is referred to as a sagittal section. If it divides the body into equal parts, right down the median plane of the body, it is called a **midsagittal section**.

Frontal section: Sometimes called a **coronal section**, this cut is made along a longitudinal plane that divides the body into anterior and posterior parts.

Transverse section: A cut made along a transverse or horizontal plane, dividing the body into superior and inferior parts, may also be called a **cross section**.

These terms can also be used to describe the way organs are cut for observation. An organ cut along the sagittal plane provides quite a different view from one cut along the transverse plane as shown in Figure 1.4.

Go to the demonstration area and observe the transversely and longitudinally cut organ specimens.

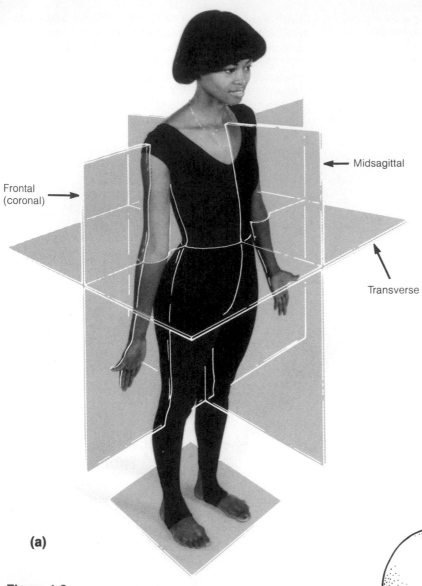

(a)

Figure 1.3

Planes of the body.

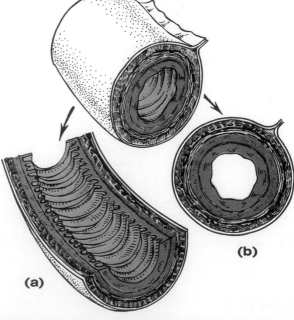

(a)

(b)

Figure 1.4

Segment of the small intestine (a) cut longitudinally and (b) cut transversely.

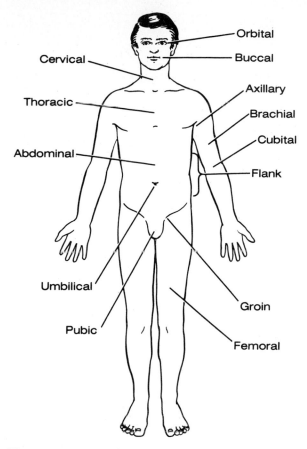

Figure 1.5

Anterior body landmarks.

SURFACE ANATOMY

Body surfaces provide visible landmarks for study of the body.

Anterior Body Landmarks

Note the following regions in Figure 1.5:

Orbital: pertaining to the eye
Buccal: pertaining to the mouth
Cervical: pertaining to the neck region
Thoracic: pertaining to the chest
Axillary: pertaining to the armpit
Brachial: pertaining to the arm
Umbilical: pertaining to the navel
Abdominal: pertaining to the anterior body trunk
Cubital: pertaining to the anterior surface of the elbow
Groin: pertaining to the area where the thigh meets the body trunk
Femoral: pertaining to the thigh
Flank: pertaining to the lateral surface of the body from the rib cage to the hip
Pubic: pertaining to the genital region

Since the abdominal surface covers a large area, it is helpful to divide it into smaller areas for study. Two medical division schemes exist for this purpose. One scheme divides the abdominal surface (and the abdominopelvic cavity deep to it) into four more or less equal regions called **quadrants**; these quadrants arc then named according to their relative position; that is, right upper quadrant, right lower quadrant, left upper quadrant, and left lower quadrant (Figure 1.6(a)). The second medical division scheme divides the abdominal surface into nine separate regions by four planes, as shown in Figure 1.6(b) and described as follows:

Umbilical region: the centermost region, which includes the umbilicus
Epigastric region: immediately above the umbilical region; overlies most of the stomach
Hypogastric region: immediately below the umbilical region; encompasses the pubic area
Iliac regions: lateral to the hypogastric region and overlying the inferior portion of the hip bones
Lumbar regions: overlying the flaring portions of the hip bones
Hypochondriac regions: flanking the epigastric region and overlying the lower ribs

Locate the anterior body landmarks, including the regions of the abdominal surface, on a torso model and on yourself before continuing.

Posterior Body Landmarks

Note the following body surface regions in Figure 1.7:

Scapular: pertaining to the scapula or shoulder blade area
Lumbar: pertaining to the area of the back between the ribs and hips
Gluteal: pertaining to the buttocks or rump
Popliteal: pertaining to the knee region
Calf: pertaining to the posterior surface of the lower leg
Occipital: pertaining to the posterior surface of the head
Deltoid: pertaining to the curve of the shoulder formed by the large deltoid muscle

BODY CAVITIES

The body has two sets of cavities, which provide quite different degrees of protection to the organs within them (Figure 1.8).

Dorsal Body Cavity

The dorsal body cavity can be subdivided into the **cranial cavity**, in which the brain is enclosed within the rigid skull, and the **spinal cavity**, in which the delicate spinal cord is protected within a bony vertebral column. Because the cord is a

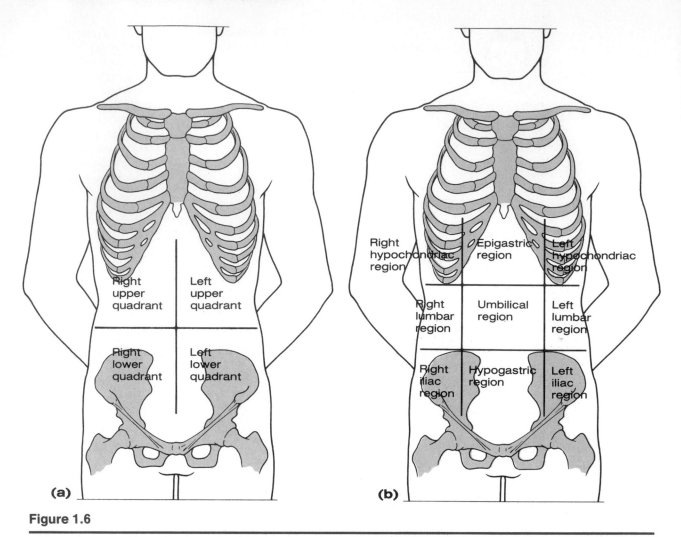

Figure 1.6

Abdominopelvic cavity. (a) Four quadrants; (b) nine regions delineated by four planes. The superior horizontal plane is just inferior to the ribs; the inferior horizontal plane is just superior to the hipbones. Vertical planes are just medial to the nipples.

continuation of the brain, these cavities are continuous with each other.

Ventral Body Cavity

Like the dorsal cavity, the ventral body cavity is subdivided. The superior **thoracic cavity** is separated from the rest of the ventral cavity by the dome-shaped diaphragm. The heart and lungs, located in the thoracic cavity, are afforded some measure of protection by the bony rib cage. The cavity below the diaphragm is often referred to as the **abdominopelvic cavity**, since there is no further physical separation of the ventral cavity. Some prefer to subdivide the abdominopelvic cavity into a superior **abdominal cavity**, which houses the stomach, intestines, liver, and other organs, and an inferior **pelvic cavity**, containing the reproductive organs, bladder, and rectum. Note in Figure 1.8 that

the abdominal and pelvic cavities are not continuous with each other in a straight plane but that the pelvic cavity is tipped away from the perpendicular.

The inner body wall of the ventral cavity is lined with a smooth serous membrane, the **parietal serosa**, which is continuous with a similar membrane, the **visceral serosa**, covering the external surfaces of the organs within the cavity (Figure 1.9). These membranes produce a thin lubricating fluid that allows the organs to slide over one another or to rub against the body wall without friction.

The specific names of the serous membranes depend on the structures they envelop. Thus the serosa lining the abdominal cavity and covering its organs is the **peritoneum**, that enclosing the lungs is the **pleura**, and that around the heart is the **pericardium**.

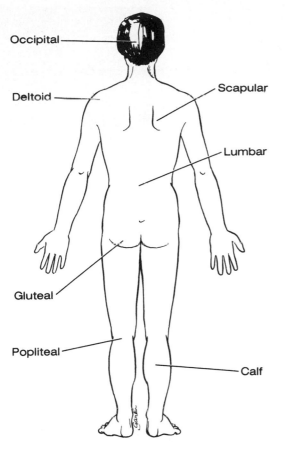

Figure 1.7

Posterior body landmarks.

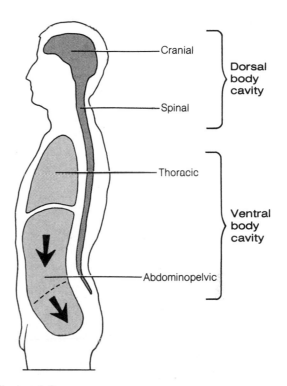

Figure 1.8

Body cavities; angle of the relationship between the abdominal and pelvic cavities shown by arrows.

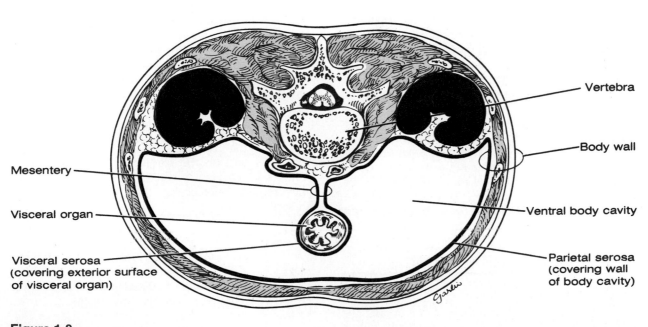

Figure 1.9

Parietal and visceral serosa (transverse section through body wall).

Organan Systems Overview

OBJECTIVES

1. To name the major human organ systems and state the major functions of each.

2. To list two or three organs of each system, and categorize the various organs by organ system.

3. To identify these organs in a dissected rat or on a dissectible torso.

4. To identify the correct organ system for each organ when presented with a list of organs (as studied in the laboratory).

MATERIALS

Freshly killed or preserved rat for dissection (one for every two to four students), or dissected human cadaver

Dissecting pans and pins

Scissors

Forceps

Human torso model (dissectible)

Related film: *Man: The Incredible Machine* (color, sound, 16 mm, 28 minutes, National Geographic Educational Services)

The basic unit or building block of all living things is the **cell.** Cells fall into four different categories according to their functions. Each of these corresponds to one of the four tissue types: epithelial, muscular, nervous, and connective. An **organ** is a structure composed of two or more tissue types that performs a specific function for the body. For example, the small intestine, which digests and absorbs nutrients, is composed of all four tissue types. An **organ system** is a group of organs that act together to perform a particular body function. For example, the organs of the digestive system work together to assure that food moving through the digestive system is properly broken down and that the end products are absorbed into the bloodstream to provide nutrients and fuel for all the body's cells. In all, there are 10 organ systems, which are described in Table 2.1. Read through this summary before beginning the rat dissection.

RAT DISSECTION

Now you will have a chance to observe the size, shape, location, and distribution of the organs and organ systems. Many of the external and internal structures of the rat are quite similar in structure and function to those of the human, so a study of

the gross anatomy of the rat should help you understand your own physical structure.

The following instructions have been written to complement and direct the student's dissection and observation of a rat, but the descriptions for organ observations from procedure 4 (p. 10) apply as well to superficial observations of a previously dissected human cadaver. In addition, the general instructions for observation of external structures can easily be extrapolated to serve human cadaver observations.

Note that four of the organ systems listed in Table 2.1 will not be studied at this time (integumentary, nervous, skeletal, and muscular), as they require microscopic study or more detailed dissection.

External Structures

1. Obtain a preserved or freshly killed rat (one for every two to four students), a dissecting pan, dissecting pins, scissors, and forceps.

2. Observe the major divisions of the animal's body—head, trunk, and extremities. Compare these divisions to those of humans.

TABLE 2.1   Overview of Organ Systems of the Body

Organ system	Major component organs	Function
Integumentary (Skin)	Epidermal and dermal regions; cutaneous sense organs and glands	• Protects deeper organs from mechanical, chemical, and bacterial injury, and dessication (drying out) • Excretion of salts and urea • Aids in regulation of body temperature
Skeletal	Bones, cartilages, tendons, ligaments, and joints	• Body support and protection of internal organs • Provides levers for muscular action • Cavities provide a site for blood cell formation
Muscular	Muscles attached to the skeleton (heart, and smooth muscles in walls of hollow organs)	• Primarily function to contract or shorten; in doing so, skeletal muscles allow locomotion (running, walking, etc.), grasping and manipulation of the environment, and facial expression; contraction of the heart propels blood through the blood vessels; contraction of smooth muscles changes the shape of a hollow organ acting to propel substances (blood, urine, food) along predetermined passageways
Nervous	Brain, spinal cord, nerves, and sensory receptors	• Allows body to detect changes in its internal and external environment and to respond to such information by activating appropriate muscles or glands • Maintains homeostasis of the body
Endocrine	Pituitary, thyroid, parathyroid, adrenal, and pineal glands; ovaries, testes, and pancreas	• Maintains body homeostasis, growth, and development; produces chemical "messengers" (hormones) that travel in the blood to exert their effect(s) on various "target organs" of the body
Circulatory	Heart, blood vessels, blood, lymphatic vessels, and nodes	• Primarily a transport system that carries blood containing oxygen, carbon dioxide, nutrients, wastes, ions, hormones, and other substances to and from the tissue cells where exchanges are made; blood is propelled through the blood vessels by the pumping action of the heart • White blood cells and antibodies in the blood act to protect the body
Respiratory	Nasal passages, pharynx, larynx, trachea, bronchi, and lungs	• Keeps the blood continuously supplied with oxygen while removing carbon dioxide
Digestive	Oral cavity, esophagus, stomach, small and large intestines, rectum, and accessory structures (teeth, salivary glands, liver, and pancreas)	• Acts to break down ingested foods to minute particles, which can be absorbed into the blood for delivery to the body cells • Undigested residue removed from the body as feces
Urinary	Kidneys, ureters, bladder, and urethra	• Rids the body of nitrogen-containing wastes (urea, uric acid, and ammonia), which result from the breakdown of proteins and nucleic acids by body cells • Maintains water and ionic balance of blood
Reproductive	Male: testes, scrotum, penis, and duct system, which carries sperm to the body exterior Female: ovaries, fallopian tubes, uterus, and vagina	• Provides germ cells (sperm and eggs) for perpetuation of the species • Female uterus houses the developing fetus until birth

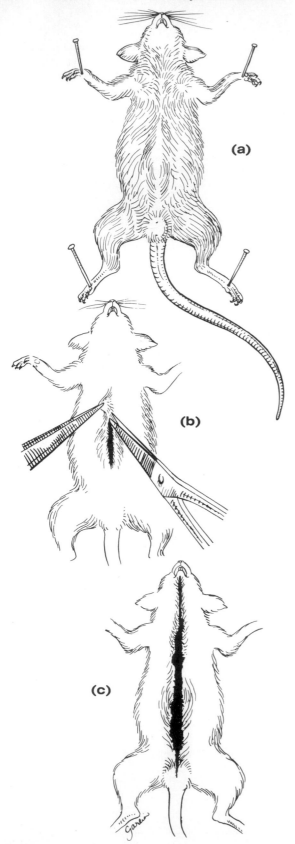

Figure 2.1

Rat dissection: pinning and the initial incision.
(a) Securing the rat to the dissection tray; (b) using scissors to make the incision on the medial line of the abdominal region; (c) completed incision from the pelvic region to the lower jaw.

3. Examine the structures of the oral cavity. Identify the teeth and tongue. Observe the extent of the hard palate (the portion underlain by bone) and the soft palate (immediately posterior to the hard palate, with no bony support). Note that the posterior end of the oral cavity leads into the throat, or pharynx. The pharynx is a passageway used by both the digestive and respiratory systems.

Ventral Body Cavity

1. Pin the animal to the wax of the dissecting pan by placing its dorsal side down and securing its extremities to wax as shown in Figure 2.1(a). (If the dissecting pan is not waxed, secure the animal with twine. Make a loop knot around one upper limb, pass the twine under the pan, and secure the opposing limb. Repeat for the lower extremities.)

2. Lift the abdominal skin with a forceps, and cut through it with the scissors (Figure 2.1(b)). Close the scissor blades and insert them under the cut skin. Moving in a superior direction, open and close the blades to loosen the skin from the underlying connective tissue and muscle. Once this skin-freeing procedure has been completed, cut the skin on the medial line, from the pubic region to the lower jaw (Figure 2.1(c)). Make a lateral cut about halfway down the ventral surface of each limb. Complete the job of freeing the skin with the scissor tips, and pin the flaps to the tray (Figure 2.2(a)). The underlying tissue that is now exposed is the skeletal musculature of the body wall and limbs. It allows voluntary body movement. Note that the muscles are packaged in sheets of pearly white connective tissue (fascia), which protects the muscles and binds them together.

3. Carefully cut through the muscles of the abdominal wall in the pubic region, avoiding the underlying organs. Remember, to *dissect* means "to separate"—not mutilate! Now, hold and lift the muscle layer with a forceps and cut through the muscle layer from the pubic region to the bottom of the rib cage. Make two lateral cuts through the rib cage (Figure 2.2(b)). A thin membrane attached to the inferior boundary of the rib cage should be obvious; this is the **diaphragm,** which separates the thoracic and abdominal cavities. Cut the diaphragm away to loosen the rib cage. You can now lift the ribs to view the contents of the thoracic cavity.

4. Examine the structures of the thoracic cavity, starting with the most superficial structures and working deeper. As you work, refer to Figure 2.3 (p. 12), which shows the superficial organs.

Thymus: an irregular mass of glandular tissue overlying the heart

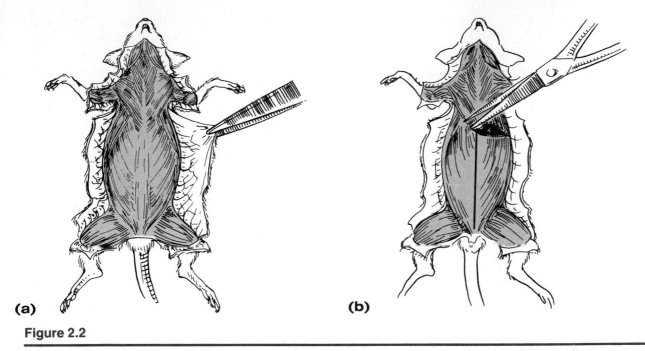

(a)   **(b)**

Figure 2.2

Rat dissection. (a) Reflection (folding back) of the skin to expose the underlying muscles; (b) making lateral cuts at the base of the rib cage.

Push the thymus to the side to view the heart.

Heart: median oval structure enclosed within the pericardium (serous membrane sac)
Lungs: flanking the heart on either side

Now observe the throat region

Trachea: tubelike "windpipe" running medially down the throat; part of the respiratory system

Follow the trachea into the thoracic cavity; note where it divides. These are the bronchi.

Bronchi: two passageways that plunge laterally into the tissue of the two lungs.

Now push the trachea to one side to expose the esophagus.

Esophagus: literally a food chute; the part of the digestive system that transports food from the throat to the stomach

Follow the esophagus through the diaphragm to its junction with the stomach.

Stomach: a C-shaped organ important in food digestion and temporary storage

5. Examine the superficial structures of the abdominopelvic cavity. Beginning with the stomach, trace the rest of the digestive tract.

Small intestine: connected to the stomach and ending just before a large saclike cecum
Cecum: the initial portion of the large intestine

Large intestine: a large muscular tube coiled within the abdomen

Follow the course of the large intestine to the rectum, which is partially covered by the urinary bladder.

Rectum: passageway between the large intestine and the anus
Anus: the opening of the digestive tract to the exterior

Now lift the small intestine with the forceps to view the mesentery.

Mesentery: an apronlike serous membrane; suspends the digestive organs in the abdominal cavity. Notice that it is heavily invested with blood vessels and, more likely than not, riddled with large fat deposits.

Locate the remaining abdominal structures.

Pancreas: a diffuse gland; rests in the mesentery between the first portion of the small intestine and the stomach
Spleen: a dark red organ curving around the left lateral side of the stomach; considered part of the circulatory system and often called the red blood cell graveyard
Liver: large and brownish red; the most superior organ in the abdominal cavity, directly beneath the diaphragm

6. To locate the deeper structures of the abdominopelvic cavity, cut through the superior margin

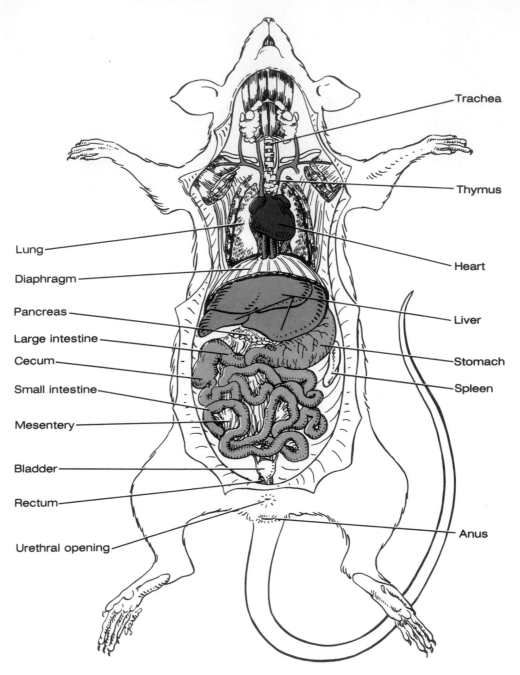

Trachea

Thymus

Lung

Heart

Diaphragm

Pancreas

Large intestine

Liver

Cecum

Stomach

Small intestine

Spleen

Mesentery

Bladder

Rectum

Anus

Urethral opening

Figure 2.3

Rat dissection: superficial organs of the thoracic and abdominal cavities.

of the stomach and the distal end of the large intestine and lay them aside. (Refer to Figure 2.4 as you work.)

Examine the posterior wall of the abdominal cavity to locate the two kidneys.

Kidneys: bean-shaped organs; retroperitoneal (behind the peritoneum)

Carefully strip away part of the peritoneum and attempt to follow the course of one of the ureters.

Ureter: tube running from the indented region of a kidney to the urinary bladder

Urinary bladder: the sac that serves as a reservoir for urine

Adrenal glands: large glands that sit astride the superior margin of each kidney; considered part of the endocrine system

7. In the midline of the body cavity lying between the kidneys are the two principal abdominal blood vessels. Identify each.

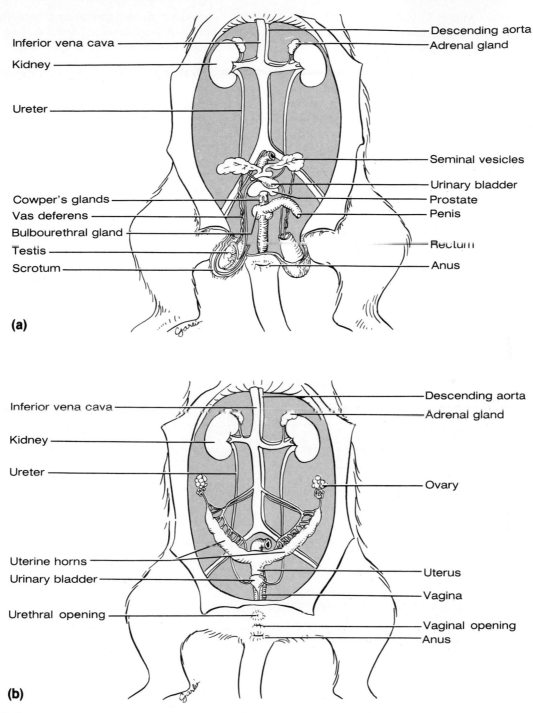

Figure 2.4

Rat dissection: deeper organs of the abdominal cavity and the reproductive structures: (a) male; (b) female.

Inferior vena cava: the large vein that returns blood to the heart from the lower regions of the body

Descending aorta: deep to the inferior vena cava; the largest artery of the body; carries blood away from the heart down the midline of the body

8. Only a cursory examination of reproductive organs will be done. First determine if the animal is a male or female. Observe the ventral body surface beneath the tail. If a saclike scrotum and a single body opening are visible, the animal is a male. If three body openings are present, it is a female. (See Figure 2.4.)

MALE ANIMAL Make a shallow incision into the **scrotum.** Loosen and lift out the oval **testis.** Exert a gentle pull on the testis to identify the slender **vas deferens**, or sperm duct, which carries sperm from the testis superiorly into the abdominal cavity and joins with the urethra. The urethra runs through the penis of the male and carries both urine and sperm out of the body. Identify the **penis,** extending from the bladder to the ventral body wall. (Figure 2.4(a) indicates other glands of the male reproductive system, but they need not be identified at this time.)

FEMALE ANIMAL Inspect the pelvic cavity to identify the Y-shaped **uterus** lying against the dorsal body wall and beneath the bladder. Follow one of the uterine horns superiorly to identify an **ovary**, a small oval structure at the end of the uterine horn. The inferior undivided part of the uterus is continuous with the vagina, which leads to the body exterior. Identify the **vaginal orifice.**

9. When you have finished your observations, store or dispose of the rat according to your instructor's directions. Wash the dissecting pan and tools with laboratory detergent, dry them, and return them to the proper storage area.

EXAMINING THE HUMAN TORSO MODEL

Examine a human torso model to identify the following organs:

Dorsal cavity: brain, spinal cord
Thoracic cavity: heart, lungs, bronchi, trachea, esophagus, diaphragm, descending aorta, inferior vena cava
Abdominopelvic cavity: liver, stomach, pancreas, spleen, small intestine, large intestine, rectum, kidneys, ureters, bladder, adrenal gland

As you observe these structures, locate the nine abdominopelvic areas studied earlier and determine which organs would be found in each area.

umbilical region _____

epigastric region _____

hypogastric region _____

right iliac region _____

left iliac region _____

right lumbar region _____

left lumbar region _____

right hypochondriac region _____

left hypochondriac region _____

Would you say that the shape and location of the human organs are similar or dissimilar to those of

the rat? _____

Assign each of the organs just identified to one of the organ system categories below.

Digestive: _____

Urinary: _____

Circulatory: _____

Reproductive: _____

Respiratory: _____

UNIT 2

THE MICROSCOPE AND ITS USES

The Microscope

EXERCISE 3

OBJECTIVES

1. To identify the parts of the microscope and list the function of each.

2. To describe and demonstrate the proper techniques for care of the microscope.

3. To define *total magnification* and *resolution*.

4. To demonstrate the proper focusing technique.

5. To define *parfocal, field,* and *depth of field.*

6. To estimate the size of objects in a field.

MATERIALS

Compound microscope
Millimeter ruler
Prepared slides of the letter *e* or newsprint
Immersion oil
Lens paper
Prepared slide of grid ruled in millimeters' (grid slide)
Prepared slide of crossed colored threads
Clean microscope slide and cover slip
Toothpicks (flat-tipped)
Physiologic saline in a dropper bottle
Methylene blue stain (dilute) in a dropper bottle
Forceps

With the invention of the microscope, biologists gained a valuable tool to observe and study structures (like cells) that are too small to be seen by the unaided eye. As a result, many of the theories basic to the understanding of biologic sciences have been established. Microscopes range in magnification from the 3× hand lens to the 1,000,000× electron microscope. This exercise will familiarize you with the workhorse of microscopes—the compound microscope—and provide you with the necessary instructions for its proper use.

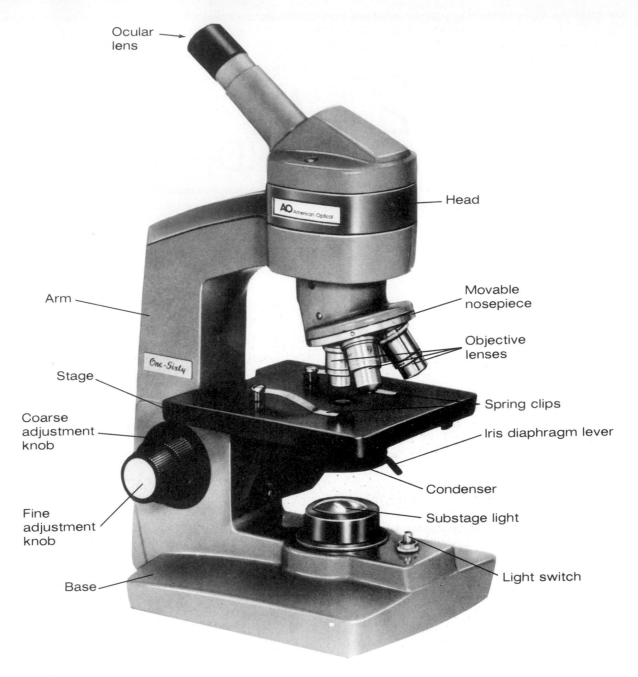

Ocular lens

Head

Arm

Movable nosepiece

Objective lenses

Stage

Spring clips

Coarse adjustment knob

Iris diaphragm lever

Condenser

Fine adjustment knob

Substage light

Base

Light switch

Figure 3.1

Compound microscope and its parts. (Courtesy of AO Scientific Instruments—Division of Warner-Lambert Technologies, Inc.)

CARE AND STRUCTURE
OF THE
COMPOUND MICROSCOPE

The compound microscope is a precision instrument and should always be handled with care. At all times you must observe the following rules for its transport, cleaning, use, and storage:

- When transporting the microscope, hold it in an upright position with one hand on its arm and the other supporting its base. Avoid jarring the instrument when setting it down.
- Use only special grit-free lens paper to clean the lenses. Clean all lenses before and after use.
- Always begin the focusing process with the lowest-power objective lens in position, changing to the higher-power lenses if necessary.
- *Never* use the coarse adjustment knob with the high-dry or oil immersion lenses.

- A cover slip must always be used with temporary (wetmount) preparations.
- Before putting the microscope in the storage cabinet, remove the slide from the stage, rotate the lowest-power objective lens into position, and replace the dust cover.
- Never remove any parts from the microscope; inform your instructor of any mechanical problems that arise.

 1. Obtain a microscope and bring it to the laboratory bench. (Use the proper carrying technique!) Compare your microscope with the illustration in Figure 3.1 and identify the following microscope parts:

Base: supports the microscope. (Note: Some microscopes are provided with an inclination joint, which allows the instrument to be tilted backward for viewing dry preparations.)

Substage light (or *mirror*): located in the base. In microscopes with a substage light source, the light passes directly upward through the microscope. If a mirror is used, light must be reflected from a separate free-standing lamp.

Stage: the platform the slide rests on while being viewed. The stage always has a hole in it to permit light to pass through both it and the specimen. Some microscopes have a stage equipped with *spring clips;* others have a clamp-type *mechanical stage.* Both hold the slide in position for viewing; in addition, the mechanical stage permits precise movement of the specimen.

Condenser: concentrates the light on the specimen. The condenser may be equipped with a height-adjustment knob that raises and lowers the condenser to vary the delivery of light. Generally, the best position for the condenser is close to the inferior surface of the stage.

Iris diaphragm lever: arm attached to the condenser that regulates the amount of light passing through the condenser. The iris diaphragm permits the best possible contrast when viewing the specimen.

Coarse adjustment knob: used to focus the specimen.

Fine adjustment knob: used for precise focusing once coarse focusing has been completed.

Head or **body tube:** supports the objective lens system (which is mounted on a movable nosepiece), and the ocular lens.

Arm: vertical portion of the microscope connecting the base and head.

Ocular (or *eyepiece*): lens contained at the superior end of the head or body tube. Observations are made through the ocular.

Nosepiece: generally carries three objective lenses.

Objective lenses: adjustable lens system that permits the use of a low-power lens, a high-dry

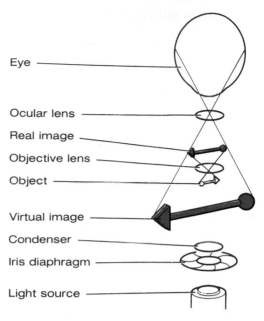

Figure 3.2

Optical system of the compound microscope. Note the real and virtual images.

lens, or an oil immersion lens. The objective lenses have different magnifying and resolving powers.

2. Examine the objectives carefully, noting their relative lengths and the numbers inscribed on their sides. On most microscopes, the low-power (l.p.) objective is the shortest and generally has a magnification of $10 \times$ (it increases the apparent size of the object by ten times or ten diameters). The high-dry (h.p.) objective is of intermediate length and has a magnification range from $40 \times$ to $50 \times$, depending on the microscope. The oil immersion objective is usually the longest of the objectives and has a magnifying power of $95 \times$ to $100 \times$. (Note: some microscopes lack the oil immersion lens but have a very low magnification lens called the **scanning lens,** which is a very short objective with a magnification of $4 \times$ to $5 \times$.)

3. Rotate the l.p. objective into position, and turn the coarse adjustment knob about 180 degrees. Note how far the stage (or objective) travels during this adjustment. Move the fine adjustment knob 180 degrees, noting again the distance that the stage (or the objective) moves.

Magnification and Resolution

The microscope is an instrument of magnification. In the compound microscope, magnification is achieved through the interplay of two lenses—the ocular lens and the objective lens. The objective lens magnifies the specimen to produce a **real image** that is projected to the ocular. This real image is magnified by the ocular lens to produce the **virtual image** seen by your eye (Figure 3.2).

The **total magnification** of any specimen being viewed is equal to the power of the ocular lens multiplied by the power of the objective lens used. For example, if the ocular lens magnifies $10 \times$ and the objective lens being used magnifies $45 \times$, the total magnification is $450 \times$.

Determine the total magnification you may achieve with each of the objectives on your microscope and record the figures on the chart below. (Note: If your microscope has a scanning lens instead of the oil immersion lens, cross out "oil immersion" and substitute "scan" on the chart.)

The compound light microscope has certain limitations. Although the level of magnification is almost limitless, the **resolution** (or resolving power), the ability to discriminate two close objects as separate, is not. The human eye can resolve objects about 100 μm or 10^6 Å apart, but the compound microscope has a resolution of only 0.2 μm under ideal conditions.

Resolving power (RP) is determined by the amount and physical properties of the visible light that enters the microscope. In general, the greater the amount of light delivered to the objective lens, the greater the resolution. The size of the objective lens aperture decreases with increasing magnification, allowing less light to enter the objective; thus you will probably find it necessary to increase the light intensity at the higher magnifications.

VIEWING OBJECTS THROUGH THE MICROSCOPE

1. Obtain a millimeter ruler and a prepared slide of the letter *e* or newsprint. Secure the slide on the stage so that the letter *e* is centered over the hole, and switch on the light source. (If the light source is not built into the base, use the curved surface of the mirror to reflect the light up into the microscope.) The condenser should be in its highest position.

2. With the l.p. objective in position over the stage, use the coarse adjustment knob to bring the objective and stage as close together as possible.

3. Looking through the ocular, adjust the light for comfort. Now use the coarse adjustment knob to focus slowly away from the *e* until it is as clearly focused as possible. Complete the focusing with the fine adjustment knob.

4. Sketch the letter in the circle just as it appears in the **field** (the area you see through the microscope).

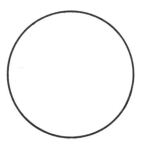

What is the total magnification? _____ $\times$

How far is the bottom of the objective from the specimen? In other words, what is the **working distance?** _____mm

(Use a millimeter ruler to make this measurement and record it in the chart below.)

How has the apparent orientation of the *e* changed top to bottom, right to left, and so on? _____

_____

5. Move the slide slowly away from you on the stage as you look through the ocular. In what direction does the image move? _____

Move the slide to the left. In what direction does the image move? _____

	Low power		High power		Oil immersion	
Magnification of the objective lenses	X		X		X	
Total magnification	X		X		X	
Detail observed						
Field size (diameter)	mm	μm	mm	μm	mm	μm
Working distance		mm		mm		mm

At first this change in orientation will confuse you, but with practice you will learn to move the slide in the desired direction with no problem.

6. Without touching the focusing knobs, increase the magnification by rotating the high-dry objective into position over the stage. Using the fine adjustment only, sharpen the focus.* What new details

become clear? _____

_____

What is the total magnification now? _____ ×

Measure the distance between the objective and the slide (the working distance) and record it on the chart.

Why should the coarse focusing knob *not* be used when focusing with the higher-powered objective

lenses? _____

_____

Is the image larger or smaller? _____

Approximately how much of the letter is visible

now? _____

Is the field larger or smaller? _____

Why is it necessary to center your object (or the portion of the slide you wish to view) before

changing to a higher power? _____

_____

Move the iris diaphragm lever while observing the

field. What happens? _____

_____

Is it more desirable to increase *or* decrease the light

when changing to a higher magnification? _____

Why? _____

_____

*Today most good laboratory microscopes are parfocal; that is, the slide should be in focus (or nearly so) at the higher magnifications once you have properly focused in l.p. If you are unable to swing the objective into position without raising the objective, your microscope is not parfocal. Consult your instructor.

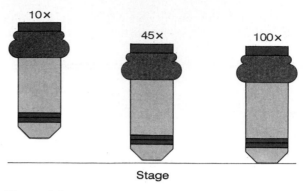

Stage

Figure 3.3

Relative working distance of the 10×, 45×, and 100× objectives.

7. Without touching the focusing knob, rotate the high-dry lens out of position so that the area of the slide over the opening in the stage is unobstructed. Place a drop of immersion oil over the *e* on the slide and rotate the immersion lens into position. Adjust the fine focus and the light for the best possible resolution. Is the field again decreased in size? _____

What is the total magnification with the immersion

lens? _____ × Is the working distance less or greater than it was when the high-dry lens was

focused? _____

Compare your observations on the relative working distances of the objective lenses with the illustration in Figure 3.3. Explain why it is desirable to begin

the focusing process in l.p. _____

_____

_____

8. Rotate the immersion lens slightly to the side and remove the slide. Clean the oil immersion lens carefully with lens paper and then clean the slide in the same manner.

DETERMINING THE SIZE OF THE MICROSCOPE FIELD

By this time you should know that the size of the microscope field decreases with increasing magnification. For future microscope work, it will be useful to determine the diameter of each of the microscope fields. This information will allow you to make a fairly accurate estimate of the size of the objects you view in any field. For example, if you have calculated the field diameter to be

4 mm and the object extends across half this diameter, you can estimate the size of the object to be approximately 2 mm.

Microscopic specimens are usually measured in micrometers and millimeters, both units of the metric system. You can get an idea of the relationship and meaning of these units from Table 3.1.

1. Return the letter *e* slide and obtain a grid slide, a slide prepared with graph paper ruled in millimeters. Each of the squares in the grid is 1 mm on each side. Focus in low power.

2. Move the slide so that one grid line touches the edge of the field on one side, and count the number of squares you can see across the diameter of the l.p. field. If you can see only part of a square, as in the accompanying diagram, estimate the part of a millimeter that the partial square represents.

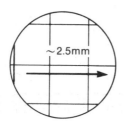

~2.5mm

For future reference, record this figure in the appropriate space marked "field size" on the summary chart on p. 18. Complete the chart by computing the approximate diameter of the high-dry and immersion fields. Say the diameter of the l.p. field (total magnification of 50×) is 2 mm. You would compute the diameter of a high-dry field with a total magnification of 100× as follows:

$$2 \text{ mm} \times 50 = X \text{ (diameter of h.p. field)} \times 100$$
$$100 \text{ mm} = 100X$$
$$1 \text{ mm} = X \text{ (diameter of the h.p. field)}$$

The formula is:
Diameter of the l.p. field (mm) × Total magnification of the l.p. field = Diameter of field X × Total magnification of field X

3. Estimate the size of the following microscopic objects. Base your calculations on the field sizes you have determined for your microscope.

Object seen in low-power field:

approximate size _____ mm.

Object seen in high-dry field:

approximate size _____ mm,

or _____ μm.

Object seen in oil immersion field:

approximate size _____ μm.

TABLE 3.1   Comparison of Metric Units of Length

Metric unit	Abbreviation	Equivalent
Meter	m	(about 39.3 in.)
Centimeter	cm	10^{-2} m
Millimeter	mm	10^{-3} m
Micrometer (or micron)	μm (μ)	10^{-6} m
Nanometer (or millimicrometer, or millimicron)	nm (mμ)	10^{-9} m
Ångstrom	Å	10^{-10} m

4.   If an object viewed with the oil immersion lens looked like the field depicted here, could you

determine its approximate size from this view? ____

How could you determine it? _____

_____

_____

_____

PERCEIVING DEPTH

Any specimen mounted on a slide has depth as well as length and width; it is rare indeed to view a tissue slide with just one layer of cells. Normally you can see two or three cell thicknesses. Therefore, it is important to learn how to determine relative depth with your microscope.*

 1.   Return the grid slide and obtain a slide of colored crossed threads. Focusing in l.p., locate a point where the threads cross.

2.   Focus down with the coarse adjustment until the threads are out of focus, then slowly focus upward again, noting which thread comes into clear focus first. This one is the lower or inferior thread. (Note: you will see both threads, so you must be very careful in determining which one comes into focus first.)

Record your observations: _____ thread

over _____

Choose another intersection and repeat the procedure: _____ thread over _____

*In microscope work the **depth of field** (the depth of the specimen clearly in focus) is greater at lower magnifications.

Repeat for the third intersection: _____

thread over _____

Which of the threads is uppermost? _____

Lowest? _____

PREPARING AND OBSERVING A WET MOUNT

1.   Obtain the following: a clean microscope slide and cover slip, a flat-tipped toothpick, a dropper bottle of physiologic saline, a dropper bottle of methylene blue stain, forceps.

2.   Place a drop of physiologic saline in the center of the slide. Using the flat end of the toothpick, gently scrape the inner lining of your cheek. Agitate the end of the toothpick containing the cheek scrapings in the drop of saline (Figure 3.4(a)). Add a small drop of the methylene blue stain to the preparation. (These epithelial cells are nearly transparent and thus difficult to see without the stain, which colors the nuclei

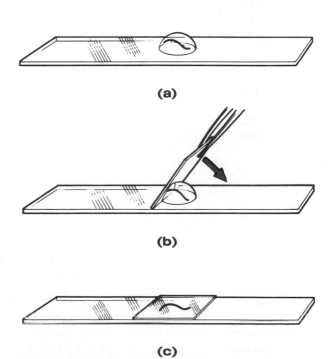

(a)

(b)

(c)

Figure 3.4

Procedure for the preparation of a wet mount: (a) The object is placed in a drop of water on a clean slide. (b) A cover slip is held at a 45° angle with forceps and (c) lowered carefully over the water and the object.

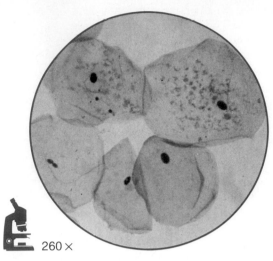

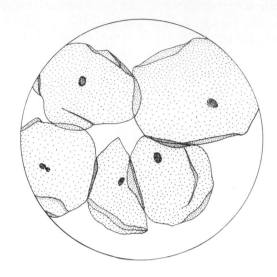

260×

Figure 3.5

Epithelial cells of the cheek cavity (surface view).

of the cells and makes them look much darker than the cytoplasm.) Stir again.

3.   Hold the cover slip with the forceps so that its inferior edge touches one side of the fluid drop (Figure 3.4(b)), then *carefully* lower the cover slip onto the preparation (Figure 3.4(c)). *Do not just drop the cover slip,* or you will trap large air bubbles under it, which will obscure the cells. *A cover slip should always be used with a wet mount* to prevent soiling the lens if you should misfocus.

4.   Place the slide on the stage and locate the cells in l.p. You will probably want to dim the light with the iris diaphragm to provide more contrast for viewing the lightly stained cells. (Furthermore, a wet mount will dry out quickly in bright light, since a bright light source is hot.)

5.   Cheek epithelial cells are very thin, six-sided cells. In the cheek, they provide a smooth, tilelike lining.

6.   Make a sketch of the epithelial cells that you observed and compare to Figure 3.5.

Approximately how large are the cheek epithe-

lial cells? _____ mm

7.   Make sure all materials are properly disposed of or returned to the proper laboratory station: clean the microscope lenses and slide, and put the dust cover on the microscope before you return it to the storage cabinet.

UNIT 3 THE CELL

The Cell— Anatomy and Division

EXERCISE
4

OBJECTIVES

1. To define *cell, organelle,* and *inclusion.*

2. To identify on a cell model or diagram the following cellular regions and to list the major function of each: nucleus, cytoplasm, and cell membrane.

3. To identify and list the major functions of the various organelles studied.

4. To compare and contrast specialized cells with the concept of the "generalized cell."

5. To define *interphase, mitosis,* and *cytokinesis.*

6. To list the stages of mitosis and describe the events of each stage.

7. To identify the mitotic phases on projected slides or appropriate diagrams.

8. To explain the importance of mitotic cell division and its product.

MATERIALS

Three-dimensional model of the "composite" animal cell or laboratory chart of cell anatomy Prepared slides of simple squamous epithelium ($AgNO_3$ stain), teased smooth muscle, human blood cell smear, and sperm

Compound microscope

Clean slide and cover slip

Physiologic saline

Flat-tipped toothpick

Janus green stain and medicine dropper

Filter paper or some other type of porous paper Forceps

Prepared slides of whitefish blastulae

Three-dimensional models of mitotic stages

Related films: *Cell Biology: Life Functions* (color, sound, 16 mm, 19 minutes, Coronet Films); *Mitosis*(color, sound, 16 mm, 24 minutes, Encyclopaedia Britannica Films); *Cell Biology: Structure and Composition* (color, sound, 16 mm, 13 minutes, Coronet Films)

In the late 1830s two German scientists, Matthias Schleiden and Theodor Schwann, introduced the "cell doctrine" or "Cell Theory," which described the *cell* as the basic structural and functional unit of all living things. The cell is still considered a basic unit, but we have come to realize that it is a very complex entity.

Perhaps the most striking property of this microscopic unit of life is its organization. A chemical analysis of cell composition reveals that it is composed primarily of the elements carbon, hydrogen, nitrogen, and oxygen, with trace amounts of several other elements. Within the cell, this inert matter takes on the unique characteristics of life and performs all the activities normally associated with living material. **Life,** then, relates to the way living matter is organized—a consideration beyond chemical composition.

The cells of the human body are highly diverse, and their differences in size, shape, and internal composition reflect their specific roles in the body. Yet cells do have many common anatomic features, and there are some functions that all must perform to maintain life. For example, all cells have the ability to metabolize (use nutrients and dispose of wastes), to grow and reproduce, to move (mobility), and to respond to a stimulus (irritability). Most of these functions are considered in detail in later exercises. This exercise focuses on structural similarities that typify the "composite" or "generalized" cell and considers only the function of cell reproduction (cell division). Cell transport (the means by which substances cross cell membranes, which is absolutely essential to cell metabolism), is dealt with separately in Exercise 5.

ANATOMY OF THE COMPOSITE CELL

In general, all cells have three major regions, or parts, that can readily be identified with a light microscope: the **nucleus,** the **cell membrane,** and the **cytoplasm.** The nucleus is usually seen as a round or oval structure near the center of the cell. It is surrounded by cytoplasm, which in turn is enclosed by the cell membrane. Since the advent of the electron microscope, even smaller cell structures—organelles—have been identified. Figure 4.1 represents the fine structure of the composite cell as revealed by the electron microscope.

Nucleus

The nucleus can quite accurately be described as the control center of the cell and is necessary for cell reproduction. A cell that has lost or ejected its nucleus (for whatever reason) is literally programmed to die because the nucleus is the site of the "genes," or genetic material—DNA.

When the cell is not dividing, the genetic material is loosely dispersed throughout the nucleus in a form called **chromatin,** which is granular or threadlike. When the cell is in the process of dividing to form daughter cells, the chromatin coils and condenses to form dense, darkly staining rodlike bodies called **chromosomes**—much in the way a stretched spring becomes shorter and thicker when relaxed. (Cell division is discussed later in this exercise.) Notice the appearance of the nucleus carefully—it is somewhat nondescript. When the nucleus appears dark and the chromatin becomes clumped, this is an indication that the cell is dying and undergoing degeneration.

The nucleus also contains one or more small round bodies, called **nucleoli,** composed primarily of proteins and ribonucleic acid (RNA). The nucleoli are believed to be storage sites for RNA and/or assembly sites for ribosomal particles (particularly abundant in the cytoplasm), which are the actual "factories" synthesizing protein.

The nucleus is bound by a double-layered porous membrane, the **nuclear membrane,** (or nuclear envelope), which is similar in composition to other cell membranes.

* **Label** the nuclear membrane, chromatin threads, and a nucleolus in Figure 4.1.

Cell Membrane

The cell membrane, or **plasma membrane,** separates cell contents from the surrounding environment. It is made up of protein and lipid (fat) and appears to have a bimolecular lipid core that protein molecules float in (see inset of Figure 4.1). Besides providing a protective barrier for the cell, the cell membrane plays an active role in determining which substances may enter or leave the cell and in what quantity. In some cells the cell membrane is thrown into minute fingerlike projections or folds called **microvilli,** which greatly increase the surface area of the cell available for absorption or passage of materials.

* **Label** microvilli on Figure 4.1.

Cytoplasm and Organelles

The cytoplasm consists of the cell contents outside the nucleus. It is the major site of most activities carried out by the cell. Suspended in the cytoplasmic material are many small structures called **organelles** (literally, small organs). The organelles are the metabolic machinery of the cell, and they are highly organized to carry out specific functions for the cell as a whole. The organelles include the ribosomes, endoplasmic reticulum, Golgi apparatus, lysosomes, mitochondria, centrioles, and cytoskeletal elements.

* Each organelle type is described briefly next. Read through this material and then correctly **label** the organelles on Figure 4.1.

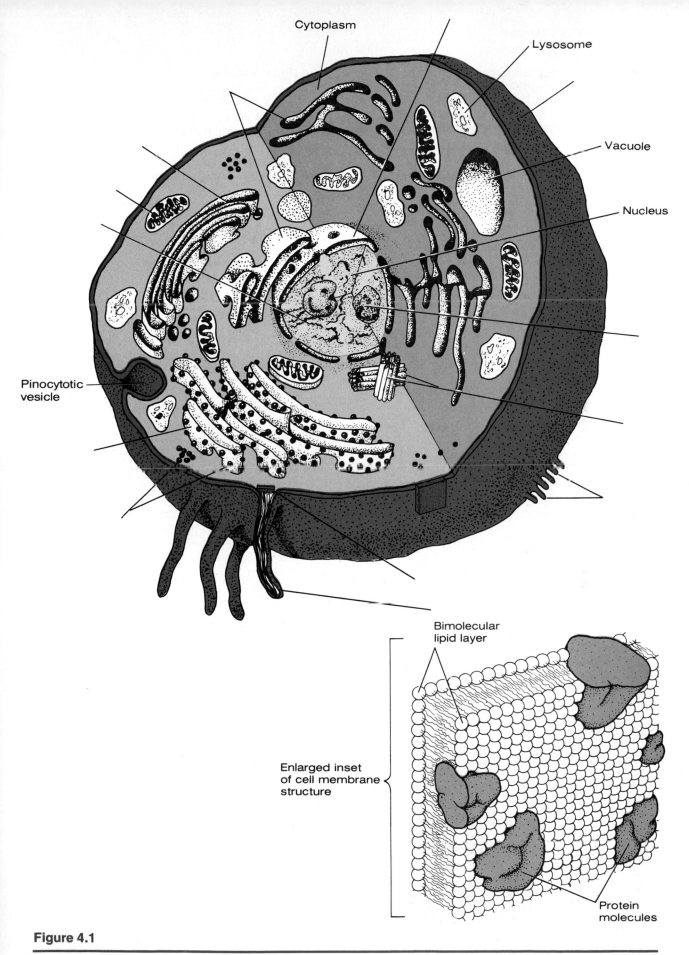

Cytoplasm

Lysosome

Vacuole

Nucleus

Pinocytotic
vesicle

Bimolecular
lipid layer

Enlarged inset
of cell membrane
structure

Protein
molecules

Figure 4.1

Anatomy of the composite animal cell. The inset shows the structural details of the cell membrane.

The **ribosomes** are tiny spherical bodies composed RNA and protein. They are the actual sites of protein synthesis. They are seen floating free in the cytoplasm or attached to a membranous structure. When they are attached, the whole ribosome-membrane complex is called the granular or rough endoplasmic reticulum.

The **endoplasmic reticulum** (ER) is a highly folded membranous system of tubules that extends throughout the cytoplasm. The ER has been observed to be continuous with the Golgi apparatus, nuclear membrane, and cell membrane. Thus it is assumed that the ER provides a system of channels for the transport of cellular substances (primarily proteins) from one part of the cell to another or to the cell exterior. The ER exists in two forms; a particular cell may have both or only one, depending on its specific functions. The granular or **rough ER,** as noted earlier, is studded with ribosomes; thus it is believed to store proteins and to deliver them to other areas of the cell. The amount of rough ER is closely correlated with the amount of protein a cell manufactures and is especially abundant in cells that produce protein products for export—for example, pancreas cells, which produce digestive enzymes destined for the small intestine. The agranular or **smooth ER** has no protein synthesis-related function but is present in conspicuous amounts in cells that produce steroid-based hormones—for example, the interstitial cells of the testes, which produce testosterone. The enzymes that catalyze the chemical reactions leading to hormone production are thought to be contained within the tubules of the smooth ER.

The **Golgi apparatus** is a stack of flattened sacs (often accompanied by bulbous ends and small vesicles) and is generally found close to the nucleus. It is now known to have a role in packaging proteins for export (the proteins are delivered to it by the rough ER) and perhaps in attaching carbohydrate groups to some of them. As the proteins accumulate in the Golgi apparatus, the sacs swell, and little vesicles filled with protein pinch off and travel to the cell membrane. On reaching the cell membrane, the vesicles fuse with it, ejecting their contents to the cell exterior.

The **lysosomes,** which appear in various sizes, are membrane-bound sacs containing an array of powerful digestive enzymes. They are believed to arise from the activities of the Golgi apparatus. The enzymes in the lysosomes are capable of digesting worn-out cell structures and foreign substances that enter the cell through phagocytosis or pinocytosis (see Exercise 5). The lysosomes also bring about some of the changes that occur during menstruation, when the uterine lining is sloughed off. Since they have the capacity of total cell destruction, the lysosomes are often referred to as the "suicide sacs" of the cell.

The **mitochondria** are generally rod-shaped bodies with a double-membrane wall; the inner membrane is thrown into folds, or cristae. Oxidative enzymes on or within the mitochondria catalyze the reactions of the Kreb's cycle and the electron transport chain (collectively called oxidative respiration), in which foods are broken down to produce energy. The released energy is captured in the bonds of ATP (adenosine triphosphate) molecules, which then diffuse out of the mitochondria to provide a ready energy supply to power the cell. Every living cell requires a constant supply of ATP for its many activities. Since the mitochondria provide the bulk of this ATP, they are referred to as the powerhouses of the cell.

The paired **centrioles** lie close to the nucleus in all animal cells capable of reproducing themselves. They are rod-shaped bodies that lie at right angles to each other; internally they consist of a system of fine tubules. During cell division, the centrioles form the mitotic spindle, a structure important for cell division.

The **cytoskeletal elements** are extremely important in cellular support and movement of substances within the cell. **Microtubules** are basically slender tubules formed of proteins called *tubulins.* Since tubulins can aggregate spontaneously to form microtubules and then disaggregate just as quickly, the microtubules have been difficult to study. Microtubules are the basis of the spindle formed by the centrioles during cell division; they act in the transport of substances down the length of elongated cells (such as neurons), and form part of the internal cytoskeleton, providing rigidity to the soft cellular substance. **Microfilaments,** elements that are ribbon or cordlike rather than hollow, are formed of contractile proteins. Because of their ability to shorten and then relax to assume a more elongated form, these are important in cell mobility and are very conspicuous in cells that are highly specialized to contract (such as muscle cells). Since the cytoskeletal structures are so labile and minute, they are rarely seen, even in electron micrographs, and are not depicted in Figure 4.1. (The exception is the microtubules of the spindle, which are very obvious during cell division (see p. 28), and the microfilaments of skeletal muscle cells (see p. 46).)

In addition to these cell structures, some cells have projections called **flagella** or **cilia** which propel the cells or allow cells to sweep substances along a tract.

The cell cytoplasm contains various other substances and structures, including stored foods (glycogen granules, lipid droplets, and other nutrients), pigment granules, crystals of various types, water vacuoles, and ingested foreign materials. But these are not part of the active metabolic machinery of the cell and are therefore called **inclusions.**

Once you have located and labeled all of these structures in Figure 4.1, examine the cell model (or cell chart) to reinforce your identifications.

OBSERVING DIFFERENCES AND SIMILARITIES IN CELL STRUCTURE

1. Obtain prepared slides of simple squamous epithelium, sperm, smooth muscle cells (teased), and human blood.

2. Observe each slide under the microscope, carefully noting similarities and differences in the cells. (The oil immersion lens will be needed to observe blood and sperm.) Distinguish the limits of the individual cells, and note the shape and position of the nucleus in each case. When you look at the human blood smear, direct your attention to the red blood cells only. Diagram your observations in the circles provided.

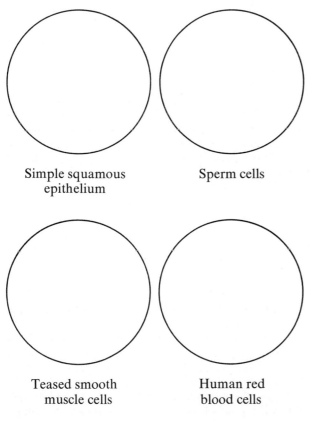

Simple squamous epithelium   Sperm cells

Teased smooth muscle cells   Human red blood cells

3. How do these four cell types differ in shape and size? _____

_____

_____

_____

How might cell shape affect cell function? _____

_____

_____

_____

_____

_____

Which cells have visible projections? _____

_____

How do these projections relate to the function of this cell? _____

_____

_____

Do any of these cells lack a cell membrane? _____

A nucleus? _____

In the cells with a nucleus, can you discern nucleoli? _____

Were you able to observe any of the organelles in these cells? _____ Why or why not?

_____

_____

_____

Many of the organelles are impossible to observe with the light microscope unless they are selectively stained. Selective or differential stains, one of which you will be using here, react to the chemical differences in the organelles in specified ways, thereby allowing you to differentiate among them.

1. Prepare a wet mount of cheek epithelial cells in physiologic saline, as in Exercise 3 (p. 21) but do not add the methylene blue stain.

2. Observe the cells under l.p. to identify the cell membrane, nucleus, and cytoplasm. Keep the light dim to increase contrast.

3. Place a drop of Janus green stain to one side of the cover slip. Place a piece of filter paper or

some other porous paper on the other side of the cover slip to remove the excess saline and to facilitate the movement of the Janus green under the cover slip.

4. Observe the cells again under the oil immersion lens. Janus green stains selectively for areas of rapid oxidation. What organelles are

reacting with the Janus green? _____

CELL DIVISION: MITOSIS AND CYTOKINESIS

Cell division in all cells other than bacteria consists of a series of events collectively called mitosis and cytokinesis. **Mitosis** is nuclear division; **cytokinesis** is the division of the cytoplasm, which begins after mitosis is nearly complete. Although mitosis is usually accompanied by cytokinesis, in some instances cytoplasmic division does not occur, leading to the formation of binucleate (or multinucleate) cells. This is relatively common in the human liver.

The process of **mitosis** results in the formation of two daughter nuclei that are genetically identical to the mother nucleus. This distinguishes mitosis from **meiosis,** a specialized type of nuclear division that occurs only in the reproductive organs (testes or ovaries). Meiosis, which yields four daughter nuclei that differ genetically and in composition from the mother nucleus, is used only for the production of eggs and sperm (gametes) for sexual reproduction. The function of cell division, including mitosis and cytokinesis in the body, is to increase the number of cells for growth and repair while maintaining their genetic heritage.

In cells about to divide, an important event precedes cell division. The genetic material (the DNA molecules composing part of the chromatin strands) is duplicated exactly during the portion of the cell cycle called **interphase.** Interphase is *not* part of mitosis; it represents the time when a cell is not actively involved in cell division. Although some people refer to interphase as the cell's resting period, this is an inaccurate description because the cell is quite active in its daily activities and is resting only from cell division. The stages of mitosis diagramed in Figure 4.2 include the following events:

Prophase: At the onset of cell division, the chromatin threads coil and shorten to form densely staining, short, barlike **chromosomes.** By the middle of prophase the chromosomes appear as double-stranded structures (each strand is a **chromatid**) connected by a small median body called a **centromere.** The centrioles separate and migrate toward opposite poles of the cell, spinning out the mitotic **spindle** between

them as they move. The spindle acts as a scaffolding for the attachment and movement of the chromosomes during later mitotic stages. The nuclear membrane and the nucleolus break down and disappear, and the chromosomes randomly attach to the spindle fibers by their centromeres.

Metaphase: A brief stage, during which the chromosomes migrate to the central plane or equator of the spindle and align along that plane in a straight line from the superior to the inferior region of the spindle (lateral view). Viewed from the poles of the cell (end view), the chromosomes appear to be arranged in a "rosette," or circle, around the widest dimension of the spindle.

Anaphase: During anaphase, the centromeres break, and the chromatids (now called chromosomes again) separate from one another and then progress slowly toward opposite ends of the cell. The chromosomes appear to be pulled by their centromere attachment, with their "arms" dangling behind them. Anaphase is complete when poleward movement ceases.

Telophase: During telophase, the events of prophase are essentially reversed. The chromosomes at the poles begin to uncoil and resume the chromatin form, the spindle disappears, a nuclear membrane forms around each chromatin mass, and nucleoli appear in each of the daughter nuclei.

Mitosis is essentially the same in all animal cells, but depending on the type of tissue, it takes from 5 minutes to several hours to complete. In most cells, centriole replication is deferred until late interphase of the next cell cycle when DNA replication begins preliminary to the onset of mitosis.

Cytokinesis, or the division of the cytoplasmic mass, begins during telophase. In animal cells, a cleavage furrow begins to form approximately over the equator of the spindle, and eventually splits or pinches the original cytoplasmic mass into two portions. Thus at the end of cell division two daughter cells exist, each smaller in cytoplasmic mass than the mother cell but genetically identical to it. The daughter cells grow and carry out the normal spectrum of metabolic processes until it is their turn to divide.

Cell division is extremely important during the body's growth period. Most cells (excluding nerve cells) undergo mitosis until puberty, when normal body size is achieved and overall body growth ceases. After this time in life, only certain cells routinely carry out cell division—for example, cells subjected to abrasion (epithelium of the skin and lining of the gut). Other cell populations—such as liver cells—stop dividing but retain this ability should some of them be removed or damaged. Skeletal muscle and nervous tissue completely lose this ability to divide and thus are severely handicapped by injury.

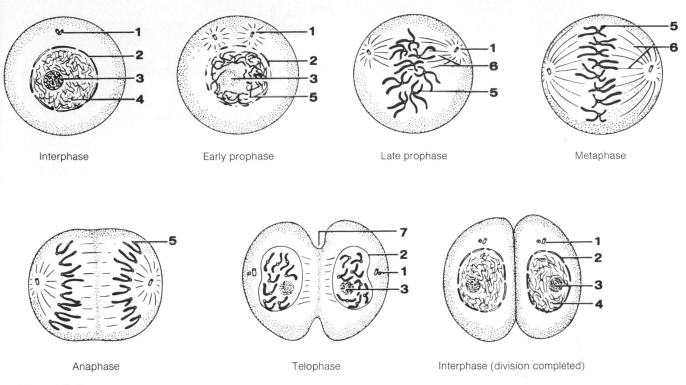

Figure 4.2

The stages of mitosis, diagrammatic views. (1 = centriole; 2 = nuclear membrane; 3 = nucleolus; 4 = chromatin; 5 = chromosomes; 6 = spindle; 7 = cleavage furrow.)

Throughout life, the body retains its ability to repair cuts and wounds and to replace some of its aged cells.

 Obtain a prepared slide of whitefish blastulae to study the stages of mitosis. The cells of each blastula (a stage of embryonic development consisting of a hollow ball of cells) are at approximately the same mitotic stage, so it may be necessary to observe more than one blastula to view all the mitotic stages. The exceptionally high rate of mitosis observed in this tissue is typical of embryos, but if occurring in specialized tissues, it can be an indication of cancerous cells, which also have an extraordinarily high mitotic rate. Examine the slide carefully, identifying the four mitotic stages and the process of cytokinesis. Compare your observations with Figure 4.2, and verify your identifications with your instructor. Then sketch your observations of each stage in the circles provided here.

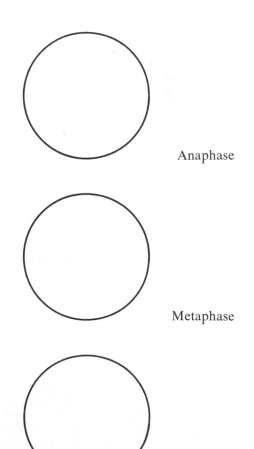

Anaphase

Metaphase

Telophase

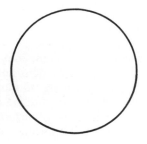

Prophase

The Cell— Transport Mechanisms and Cell Permeability

OBJECTIVES

1. To define *differential,* or *selective, permeability; diffusion (dialysis* and *osmosis); brownian motion, isotonic, hypotonic,* and *hypertonic; active transport, pinocytosis, phagocytosis,* and *permease system.*

2. To explain the processes that account for the movement of substances across the cell membrane, and to note the driving force for each.

3. To determine which way substances will move passively through a selectively permeable membrane (given appropriate information on concentration differences).

MATERIALS

For Passive Transport Experiments:
    Clean slides and cover slips
    Forceps, glass stirring rods, 15 ml graduated cylinders
    Compound microscopes
      Brownian motion: milk in dropper bottles, hotplate
      Diffusion:
        Demonstration: potassium permanganate crystals in distilled water; set up 24 hours before in a 500 cc graduated cylinder

    Four dialysis sacs
    Large beakers (500 ml)
    40% glucose solution
    Fine twine
    10% NaCl solution, boiled starch solution
    Laboratory balance
    Benedict's solution in dropper bottle
    Test tubes in racks, test tube holder
    Wax marker
    Silver nitrate ($AgNO_3$) in dropper bottle
    Lugol's iodine solution in dropper bottle
    Lancets, alcohol swabs, filter paper
    1.5% sodium chloride (NaCl) solution in dropper bottle, physiologic (mammalian) saline in dropper bottle
    Yeast suspension
    Congo red dye in dropper bottle
  Filtration:
    Ring stand, ring, clamp, filter paper, funnel, solution of uncooked starch, powdered charcoal, and copper sulfate ($CuSO_4$)
For Active Transport:
    Culture of starved ameba (*Amoeba proteus*)
    Depression slide
    Tetrahymena pyriformis culture
    Compound microscope

Because of its molecular composition, the cell membrane is selective about what passes through it. It allows nutrients to enter the cell but keeps undesirable substances out. By the same token, valuable cell proteins and other substances are kept within the cell, and excreta or wastes pass to the exterior. This property is known as **selective permeability.** Transport through the cell membrane occurs in two basic ways. In one, the cell must provide energy (ATP) to power the transport process (active transport); in the other, the transport process is driven by concentration or pressure differences (passive transport).

PASSIVE TRANSPORT

All molecules vibrate randomly (because of their inherent kinetic energy) at all temperatures above absolute zero (about –460 F). In general, the smaller the particle, the greater the kinetic energy it possesses and the faster its molecular motion. This random movement may be detected indirectly by observing a suspension. The larger particles can be seen moving randomly as they are hit and deflected by the smaller, more rapidly moving particles. The zigzag movement of the larger particles is known as **brownian motion.**

1.   Make a wet mount of milk; that is, place a small drop of milk on a slide and cover carefully with a cover slip. Allow the slide to stand on the microscope stage for about 10 minutes before observing.

2.   Observe with high power and then with the oil immersion lens. As the minute solvent (water) molecules collide with the fat globules of the milk, you can see the larger fat globules ricochet in an erratic manner (brownian motion).

3.   Place the preparation on a warm hot plate for a few seconds. Observe again. How has the *rate* of

brownian motion changed? _____

_____

_____

What would you conclude about the effect of increased temperature on the kinetic energy of

molecules? _____

_____

_____

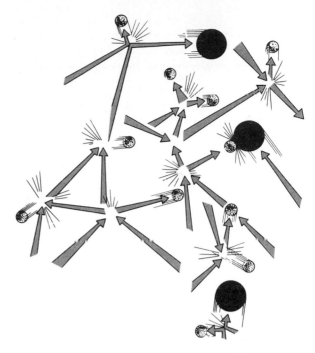

Figure 5.1

Random movement and numerous collisions, causing molecules to become evenly distributed. The small spheres represent water molecules; the large spheres represent glucose molecules.

Diffusion

If a **concentration gradient** (difference in concentration) exists, molecules eventually become evenly distributed through random molecular motion (Figure 5.1). **Diffusion** is the movement of molecules from a region of their high concentration to a region of their low concentration; the driving force is the kinetic energy of the molecules themselves.

The diffusion of particles into and out of cells is modified by the cell membrane, which constitutes a physical barrier. Molecules diffuse passively through the cell membrane if they are small enough to pass through its pores or if they can dissolve in the lipid portion of the membrane (as in the case of CO_2 and O_2). The diffusion of solutes (particles dissolved in water) through a semipermeable membrane is called **dialysis;** the diffusion of water through a semipermeable membrane is called **osmosis.** Both dialysis and osmosis, examples of diffusion phenomena, involve the movement of a substance from an area of its high concentration to one of its low concentration.

There are many examples of diffusion in nonliving systems; for example, if a bottle of ether was uncorked in the front of the laboratory, very shortly thereafter you would be nodding, as the ether molecules became distributed throughout the room. Also, if crystals of potassium permanganate dye are placed in a container of water, the dye molecules will diffuse outward from the crystals (area of

highest dye concentration). A demonstration of this second example has been set up. Observe the demonstration and answer the following questions

pertaining to it. What do you observe? _____

_____

Why did the dye molecules move? _____

_____

DIFFUSION THROUGH NONLIVING MEMBRANES   A diffusion experiment, providing information on the passage of water and solutes through semipermeable membranes, which may be applied to the study of transport mechanisms in living membrane-bound cells, is outlined below.

1.   Obtain four dialysis sacs* and four large beakers (500 ml). Number the beakers 1 to 4, and half fill all of them with distilled water except for beaker 2, which you should half fill with 40% glucose solution.

_____

*Dialysis sacs are selectively permeable membranes with pores of a particular size. The selectivity of living membranes depends on more than just pore size, but using the dialysis sacs will allow you to examine selectivity due to this factor.

2. Prepare the dialysis sacs one at a time by half filling each with 20 ml of the specified liquid, pressing out the air, and tying the neck with fine twine. Before proceeding to the next sac, quickly and carefully blot each sac dry and weigh it with a laboratory balance. Record the weight, and then drop each sac into the corresponding beaker.

- Sac 1: 40% glucose solution. Weight: _____ g

- Sac 2: 40% glucose solution. Weight: _____ g

- Sac 3: 10% NaCl solution. Weight: _____ g

- Sac 4: boiled starch solution. Weight: _____ g

Allow sacs to remain undisturbed in the beakers for 1 hour. (Use this time to continue with the rest of the exercise.)

3. After an hour, quickly blot sac 1 dry and weigh: _____ g

Has there been any change in weight? _____

Conclusions? _____

_____

_____

Place 5 ml of Benedict's solution in each of two test tubes. Put 4 or 5 drops of the beaker fluid into one test tube and 4 or 5 drops of the sac fluid into the other. Mark the tubes for identification and then place them in a beaker containing boiling water. Boil 2 minutes. Cool slowly. If a green, yellow, or rusty red precipitate forms, the test is positive, meaning that glucose is present. If negative, the solution remains the original blue color.

Was glucose still present in the sac? _____

In the beaker? _____

Conclusions? _____

_____

_____

4. Blot and weigh sac 2: _____ g

Was there an *increase* or *decrease* in weight? _____

With 40% glucose in the sac and 40% glucose in the beaker, would you expect to see any net movements of water (osmosis) or of glucose molecules (dialysis)?

_____

Why or why not? _____

_____

_____

Return the glucose solution in beaker 2 to the *stock* supply bottle before continuing.

5. Blot and weigh sac 3: _____ g

Was there any change in weight? _____

Conclusions? _____

_____

_____

Take a 5 ml sample of beaker 3 solution and put it in a test tube. Add a drop of silver nitrate. The appearance of a white precipitate or cloudiness indicates the presence of AgCl, which is formed by the reaction of $AgNO_3$ with NaCl (sodium chloride).

Results? _____

_____

Conclusions? _____

_____

6. Blot and weigh sac 4: _____ g

Was there any change in weight? _____

Conclusions? _____

_____

Take a 5 ml sample of beaker 4 solution and add a couple of drops of Lugol's iodine solution. The appearance of a black color is a positive test for the presence of starch. Did any starch diffuse from the sac into the beaker? _____ Explain: _____

_____

_____

7. In which of the test situations did net osmosis occur? _____

In which of the test situations did net dialysis occur?

_____

What conclusions can you make about the relative size of glucose, starch, NaCl, and water molecules?

_____

_____

_____

_____

With what cell structure can the dialysis sac be

compared? _____

DIFFUSION THROUGH LIVING MEMBRANES

To examine permeability properties of cell membranes, conduct the following two experiments.

Experiment 1: 1. Obtain a clean slide and cover slip, lancets, an alcohol swab, physiologic saline, mammalian Ringer's solution, test tubes (3), test tube rack, glass stirring rod, and 15 ml graduated cylinder.

2. (a) Label 3 test tubes A, B, and C, and prepare them as follows:
   A: add 2 ml 1.5% sodium chloride solution
   B: add 2 ml distilled water
   C: add 2 ml physiologic saline
   (b) Clean a fingertip with an alcohol swab, puncture it with a lancet, and add five drops of blood to each test tube. Stir each test tube with the glass rod, rinsing between each sample.
   (c) Hold each test tube in front of this printed page. _Record_ the clarity of print seen through the fluid in each tube.

Test tube A _____

Test tube B _____

Test tube C _____

3. (a) Place a very small drop of physiologic saline on a slide; clean your fingertip once again with an alcohol swab, puncture it with a lancet, and touch a small drop of blood to the saline on the slide. Tilt the slide to mix, cover with a cover slip, and immediately examine the preparation under the h.p. lens. Notice the smooth disklike shape of the red blood cells.
   (b) Now add a drop of 1.5% saline solution to the slide so that it touches one edge of the cover slip, and then carefully observe the red blood cells. What begins to happen to the normally smooth disk shape

of the red blood cells? _____

_____

_____

This crinkling-up process, called **crenation,** is due to the fact that the 1.5% saline solution is slightly hypertonic to the cell sap of the red blood cell. A **hypertonic** solution contains more solutes (thus less water) than is present in the cell. Under these circumstances, water tends to leave the cells by osmosis.

(c) Add a drop of distilled water to the edge of the cover slip. Place a piece of filter paper at the opposite edge of the cover slip; it will absorb the saline solution and draw the distilled water across the cells. Watch the red blood cells as they float across the field. Describe the change in their

appearance. _____

_____

_____

Distilled water contains _no_ solutes (therefore, it is 100% water). Distilled water and _very_ dilute solutions (that is, those containing less than 0.9% solutes) have relatively more water than is found inside a living cell. Such solutions are said to be **hypotonic** to the cell. (**Isotonic** solutions have the same solute-solvent ratio as do the cells and therefore cause no visible changes in the cells.) In a hypotonic solution, the red blood cells first "plump up," but then they suddenly start to disappear. The red blood cells burst as the water floods into them, leaving "ghosts" in their wake. This phenomenon is called **hemolysis.**

How do your observations of test tube B correlate with what you have just observed under

the microscope? _____

_____

_____

Experiment 2: Make a wet mount of yeast suspension; observe the cells under high power to determine their normal color and structure. Prepare two test tubes by placing 1 ml of the yeast suspension in each. Boil one of the tubes in a water bath for 15 seconds. Remove and add eight drops of Congo red dye to both the boiled and unboiled preparations. Prepare a wet mount from each tube.

Was the dye accepted by the unboiled cells?

_____ By the boiled cells? _____

What are your conclusions about the selectivity of living and nonliving (boiled) cell membranes?

Filtration

Filtration is a physical process by which water and solutes pass through a membrane from an area of higher hydrostatic (fluid) pressure into an area of lower hydrostatic pressure. Like diffusion, it is a passive process. For example, fluids and solutes filter out of the capillaries in the kidneys into the tubules because the blood pressure in the capillaries is greater than the fluid pressure in the tubules. Filtration is not a selective process. The amount of filtrate (fluids and solutes) formed depends almost entirely on the difference in pressure on the two sides of the membrane and on the size of the membrane pores.

 1.   Obtain the following equipment: a ring stand, ring, and ring clamp; a piece of filter paper; a beaker; and a solution of uncooked starch, powdered charcoal, and copper sulfate. Attach the ring to the ring stand with the clamp.

2.   Fold the filter paper in half twice, open it onto a cone, and place it in a funnel. Place the funnel in the ring of the ring stand and place a beaker under the funnel. Shake the starch solution, and fill the funnel with it to just below the top of the filter paper. How many drops of filtrate form in the first 10 seconds?

_____ drops

When the funnel is half empty, again count the number of drops formed in 10 seconds and record.

_____ drops

3.   After all the fluid has passed through the filter, check the filtrate and paper to see which materials were retained by the paper. (Note: If the filtrate is blue, the copper sulfate passed. Check both the paper and filtrate for black particles to see if the charcoal passed. Finally, add Lugol's iodine to a 2 cc filtrate sample in a test tube. If the sample turns blue/black with the addition of the iodine, starch is present in the filtrate.)

Passed: _____

Retained: _____

What does the filter paper represent? _____

During which counting interval was the filtration

rate greatest? _____

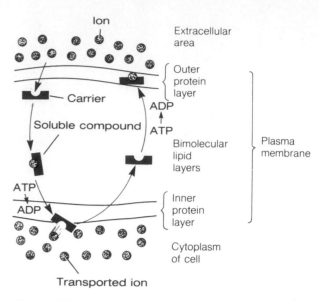

Figure 5.2

Active transport through the cell membrane. Protein carriers use ATP to energize the process.

Explain: _____

_____

_____

_____

What characteristic of the three solutes determined whether or not they passed through the filter paper?

_____

_____

_____

_____

ACTIVE TRANSPORT

Whenever a cell expends energy (ATP) to move substances across its boundaries, the process is referred to as an active transport process. The substances moved by active means are generally unable to pass by diffusion. They may be too large to pass through the pores; they may not be lipid-soluble; or they may have to move against rather than with a concentration gradient.

In one type of active transport, substances move across the membrane by combining with a protein carrier molecule; the process resembles an enzyme-substrate interaction (Figure 5.2).

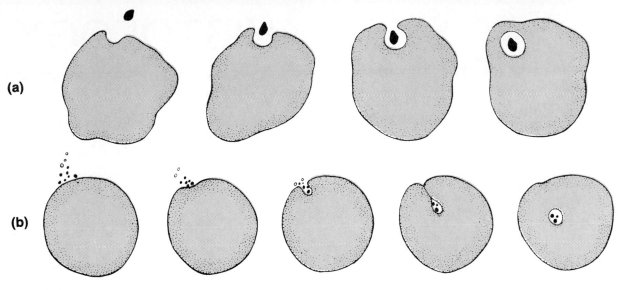

(a)

(b)

Figure 5.3

Phagocytosis and pinocytosis. (a) In phagocytosis, extensions of protoplasm (pseudopodia) flow around the external particle and enclose it within a vacuole. (b) In pinocytosis, dissolved proteins gather on the surface of the cell membrane, causing the membrane to invaginate and to incorporate the protein.

ATP is required, and in many cases the substances move against concentration or electro-chemical gradients or both. Some of the substances that are moved into the cells by the permease system are amino acids and some sugars; neither is lipid-soluble, and both are too large to pass through the pores but necessary for cell life. On the other hand, sodium ions (Na^+) are moved out of cells by active transport. There is more Na^+ outside the cell than there is inside, so the Na^+ tends to remain in the cell unless actively transported out.

Phagocytosis and pinocytosis also require ATP.

In **pinocytosis** (cell drinking), the cell membrane seems to sink beneath the material to form a small vesicle, which then pinches off into the cell interior (see Figure 5.3). Pinocytosis is most common for taking in liquids containing protein or fat.

In **phagocytosis** (cell eating), parts of the plasma membrane flow around a relatively large or solid material (for example, bacteria, cell debris) and engulf it, enclosing it within a sac. This is a rather uncommon phenomenon in the human body except for certain phagocytic or scavenger cells, such as some white blood cells and macrophages (Figure 5.3).

1.   Obtain a drop of starved *Amoeba proteus culture* and place it in the well of a depression slide.

2.   Locate an ameba under low power and then add a drop of *Tetrahymena pyriformis* culture (an ameba "meal") to the well. Keep the light as dim as possible, otherwise the ameba will "ball up" and begin to disintegrate.

3.   Observe as the ameba phagocytizes the *Tetrahymena* by forming pseudopods that engulf it. In unicellular organisms like the ameba, phagocytosis is an important food-getting mechanism, but in higher organisms, it is more important as a protective device.

UNIT 4

HISTOLOGY: BASIC TISSUES OF THE BODY

Classification of Tissues

EXERCISE
6

OBJECTIVES

1. To name the four major types of tissues in the human body and the major subcategories of each.

2. To identify the tissue subcategories through microscopic inspection or inspection of an appropriate diagram or projected slide.

3. To state the location of the various tissue types in the body.

4. To state the general functions and structural characteristics of each of the four major tissue types.

MATERIALS

Compound microscope

Prepared slides of simple squamous, simple cuboidal, simple columnar, stratified squamous (nonkeratinizing), pseudostratified ciliated, and transitional epithelium

Prepared slides of adipose, areolar, and dense fibrous connective tissue (tendon); hyaline cartilage; and bone (cross section)

Prepared slides of skeletal, cardiac, and smooth muscle (longitudinal sections)

Prepared slide of nervous tissue (spinal cord smear)

Related film: *Nature of Life: Cells, Tissues, and Organs* (color, sound, 16 mm, 15 minutes, Coronet Films

Exercise 4 describes cells as the building blocks of life and the all-inclusive functional units of unicellular organisms like amebae. But in higher organisms cells do not ordinarily operate as isolated, independent entities. In humans and other multicellular organisms, cells depend on one another and cooperate to produce homeostasis in the body.

The most complex animal starts out as a single cell, the fertilized egg, which divides almost endlessly. The thousands of cells that result become specialized for a particular function; some become supportive bone, others the transparent lens of the eye, still others skin cells, and so on. Thus a division of labor exists, with certain groups of cells highly specialized to perform functions that benefit the organism as a whole.

Cell specialization brings about great sophistication of achievement but carries with it certain hazards. When a small specific group of cells is indispensable, any inability to function on its part can paralyze or destroy the entire body. For example, the action of the heart depends on a highly differentiated group of cells in the heart muscle. If they cease functioning, the heart no longer operates efficiently and the whole body suffers or dies from lack of oxygen. The jack-of-all-trades ameba faces no such danger.

Groups of cells that are similar in structure and function are called **tissues**. The four primary tissue types—epithelium, connective tissue, nervous tissue, and muscle—have distinctive structures, patterns, and functions. The four primary tissues are further divided into subcategories.

To perform specific body functions, the tissues are organized into such **organs** as the heart, kidneys, and lungs. Most organs contain several representatives of the primary tissues, and the arrangement of these tissues determines the organ's structure and function. Thus **histology**, the study of tissues, complements a study of gross anatomy and provides the structural basis for a study of organ physiology.

The main objective of this exercise is to familiarize you with the major similarities and dissimilarities of the primary tissues and to teach you to recognize the common tissues you will encounter in units to follow. Because epithelium and some types of connective tissue will not be considered again, they are emphasized more than muscle, nervous tissue, and bone (a connective tissue), which are covered in more depth in later exercises.

EPITHELIAL TISSUE

Epithelial tissue, or epithelium, covers surfaces, and since glands almost invariably develop from epithelial membranes, they too are logically classed as epithelium. Epithelium covers the external body surface (as the epidermis), lines its cavities and tubules, and composes the various endocrine (hormone-producing) and exocrine glands of the body.

Epithelial functions include protection, absorption, filtration, and secretion. For example, the epithelium covering the body protects against bacterial and chemical damage; that lining the respiratory tract is ciliated to sweep dust and other foreign particles away from the lungs. Epithelium specialized to absorb substances lines the stomach and small intestine. In the kidney tubules, the epithelium both absorbs and filters. Secretion is a specialty of the glands.

Epithelium generally exhibits the following characteristics:

- Cells fit closely together to form membranes, or sheets of cells.
- The membranes always have one free surface.
- The cells are attached to an adhesive **basement membrane,** a structureless material secreted by the cells.
- Epithelial tissues have no blood supply of their own (are avascular) but depend on capillaries in the underlying connective tissue for a supply of food and oxygen.
- If well nourished, epithelial cells can easily regenerate themselves.

The covering and lining epithelia are classified according to two criteria—cell shape and arrangement (Figure 6.1) **Squamous** (flattened), **cuboidal** (cubelike), and **columnar** (column-shaped) epithelial cells are the general types based on shape. On the basis of arrangement, there are **simple** epithelia, consisting of one layer of cells attached to the basement membrane, and **stratified** epithelia, consisting of more than one layer of cells. The terms denoting shape and arrangement of the epithelial cells are combined to describe the epithelium fully. *Stratified epithelia are named according to the cells at the surface of the epithelial membrane,* not those resting on the basement membrane.

There are, in addition, two less easily categorized types of epithelia. **Pseudostratified** epithelium is actually a simple epithelium (one layer of cells), but because its cells extend varied distances from the basement membrane, it gives the false appearance of being stratified. **Transitional** epithelium is a rather peculiar stratified squamous epithelium formed of rounded, or "plump," cells with the ability to slide over one another to allow the organ to be stretched. Transitional epithelium is found in organs subjected to periodic distention, such as the bladder. The superficial cells are flattened (like true squamous cells) when the organ is distended and rounded when the organ is empty.

The most common types of epithelia and their most common locations in the body are diagrammed in Figure 6.2. A more complete summary of epithelia is provided in Table 6.1.

 Obtain slides of simple squamous, simple cuboidal, simple columnar, stratified squamous (nonkeratinizing), pseudostratified ciliated, and transitional epithelia. Examine each carefully, and compare your observations with the photomicrographs in Figure 6.3 (pp. 40–41). Note any modifications for specific functions, such as cilia (motile cell projections that help to move substances along the cell surface), microvilli, which increase the surface area for absorption, and/or goblet cells, which secrete lubricating mucus.

Epithelial cells forming glands are highly specialized to remove materials from the blood and to manufacture them into new materials, which they

(*Text continues on p. 42.*)

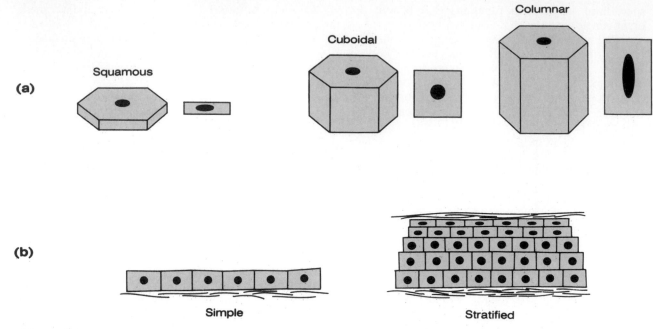

Figure 6.1

Naming epithelia (a) on the basis of cell shape and (b) on the basis of arrangement.

TABLE 6.1   Summary of Epithelia Types and Locations

Epithelia type	Common body locations
Lining epithelia	
Simple squamous	Bowman's capsule (kidney), alveoli of lungs, serous membranes
Simple cuboidal	Collecting tubules (kidney), ovary
Simple columnar	Mucous membranes of the stomach, large and small intestines
Pseudostratified columnar	Trachea mucosa (ciliated)
Stratified squamous	Vagina and esophagus mucous membrane linings (nonkeratinized)
Stratified cuboidal	Sweat glands
Stratified columnar	Male urethral mucosa
Transitional	Mucous membrane linings of the bladder and ureters
Covering epithelia	
Stratified squamous	Epidermis of the skin (keratinized)
Glandular epithelia	
Endocrine glands	Thyroid, parathyroids, anterior pituitary, adrenal cortex
Exocrine glands	Salivary, sweat, and sebaceous glands; liver, pancreas

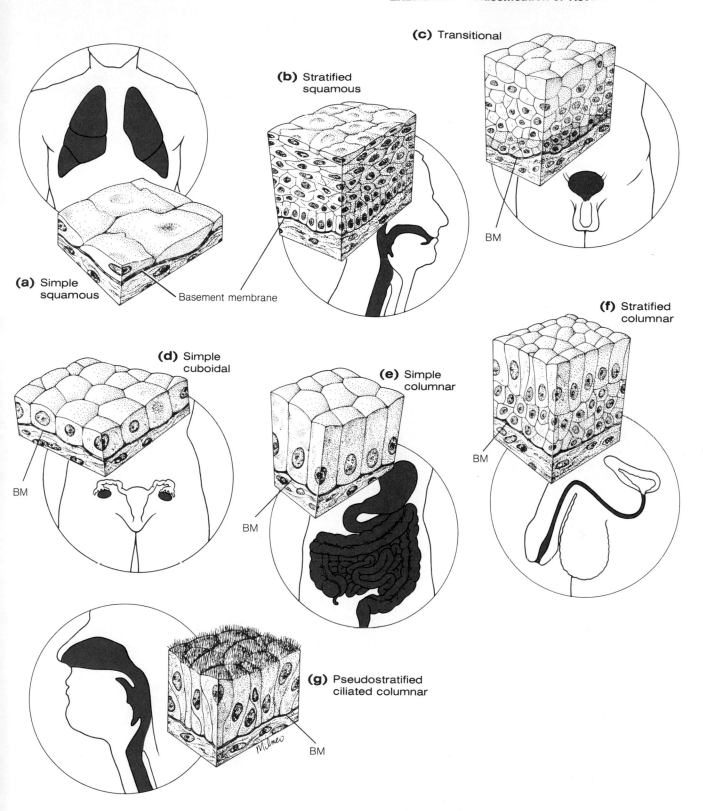

(c) Transitional

(b) Stratified squamous

(a) Simple squamous

Basement membrane

BM

(d) Simple cuboidal

(e) Simple columnar

(f) Stratified columnar

BM

BM

BM

(g) Pseudostratified ciliated columnar

BM

Figure 6.2

Types of epithelia (BM = basement membrane). Types a, d, e, and g are simple epithelia; types b, c, and f are stratified epithelia.

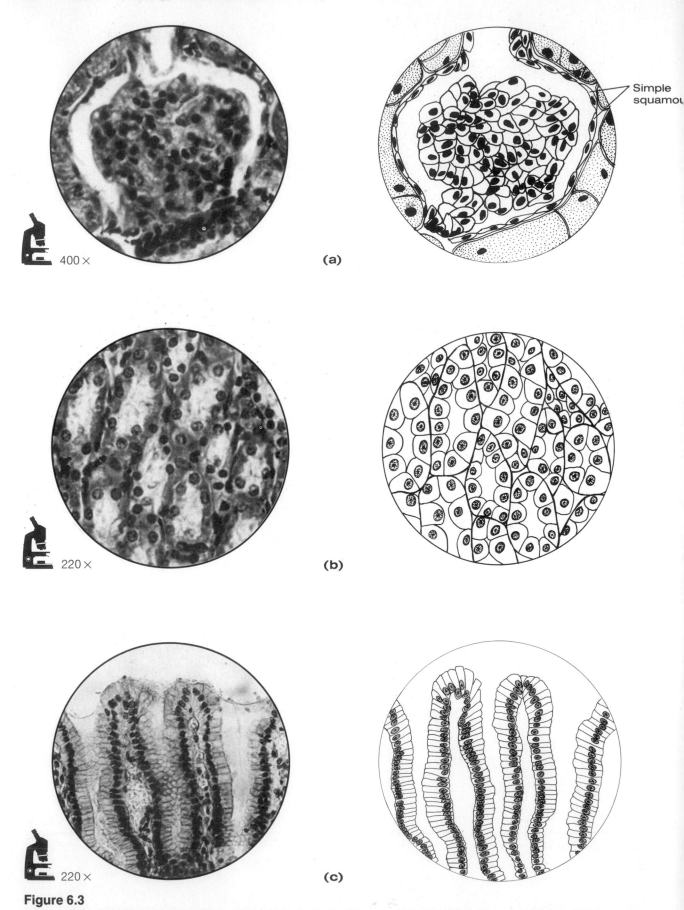

Simple
squamou

(a)

(b)

(c)

Figure 6.3

Types of typical epithelia. (a) Simple squamous. (b) Simple cuboidal. (c) Simple columnar.

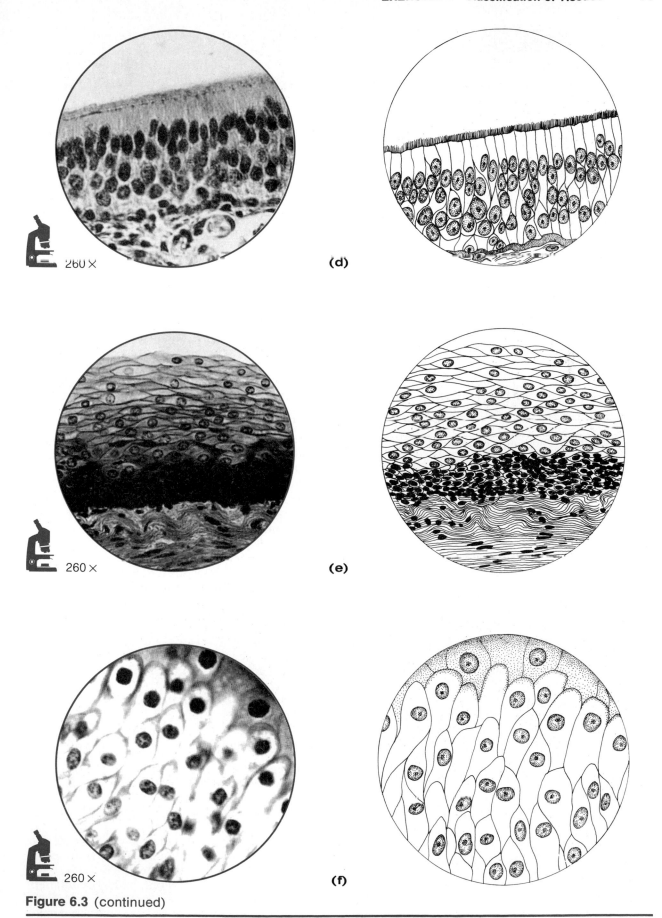

Figure 6.3 (continued)

Types of typical epithelia (d) Pseudo-stratified ciliated. (e) Stratified squamous. (f) Transitional.

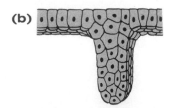
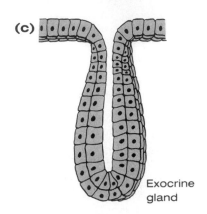
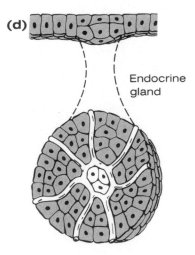

Figure 6.4

Formation of endocrine and exocrine glands from epithelial sheets. (a) Epithelial cells grow and push into the underlying tissue. (b) A cord of epithelial cells forms. (c) In an exocrine gland, a lumen (cavity) forms. The inner cells form the duct, the outer cells produce the secretion. (d) In the formation of an endocrine gland, the connecting cells forming the duct atrophy, leaving the secretory cells with no connection to the epithelial surface. However, they do become heavily invested with blood and lymphatic vessels that receive the secretions.

then secrete. There are two types of glands, as shown in Figure 6.4. The **endocrine glands** lose their surface connection (duct) as they develop; thus they are referred to as ductless glands. Their secretions (all hormones) are extruded directly into the blood or the lymphatic vessels that weave through the glands. The **exocrine glands** retain their ducts, and their secretions empty through these ducts to an epithelial surface. The exocrine glands—including the sweat and oil glands, liver, and pancreas—are both external and internal; they will be discussed in conjunction with the organ systems to which their products are functionally related.

CONNECTIVE TISSUE

Connective tissue is found in all parts of the body as discrete structures or as part of various body organs. It is the most abundant and widely distributed of the tissue types.

The connective tissues perform a variety of functions, but they primarily protect, support, and bind together other tissues of the body. For example, bones are composed of connective tissue (**osseous** tissue), and they protect and support other body tissues and organs. The ligaments and tendons (**dense fibrous** connective tissue) bind the bones together or bind skeletal muscles to bones.

Areolar connective tissue is a soft packaging material that cushions and protects body organs. Fat (**adipose**) tissue provides insulation for the body tissues and a source of stored food. Blood-forming (**hemopoietic**) tissue replenishes the body's supply of red blood cells. In addition, connective tissue also serves a vital function in the repair of all body tissues since many wounds are repaired by connective tissue in the form of scar tissue.

The characteristics of connective tissue include the following:

- With a few exceptions (cartilages, tendons, and ligaments), connective tissues are well vascularized.
- Connective tissues are composed of many types of cells.
- There is a great deal of nonliving material (matrix) between the cells of connective tissue.

The nonliving material between the cells—the **matrix**—deserves a bit more explanation. It is produced by the cells and then extruded. It may be liquid, semisolid, gellike, or very hard. The matrix is primarily responsible for the strength associated with connective tissue, but there is variation. At one extreme, hemopoietic and adipose tissues are composed mostly of cells. At the opposite extreme, bone and cartilage have very few cells and large amounts of matrix.

Fibers of various types and amounts are deposited in and form a part of the matrix material. They

TABLE 6.2 Structural Characteristics, Location, and Function of Connective Tissues*

Tissue type	Matrix type	Cell types	Location	Function
Connective Tissue Proper				
Areolar (loose) connective tissue	Gellike, with all three types of fibers	Varied—fibroblasts (fiber-forming cells), macrophages, mast cells, plasma cells	Widely distributed; forms basement membranes of epithelia; packages organs	Cushions organs; its macrophages phagocytize bacteria
Adipose connective tissue	Gellike but very sparse; a few elastic and collagenic fibers	Closely packed adipose (fat) cells, often called "signet ring" cells because nucleus is pushed to the side by a large fat-filled vacuole	Under skin (sub-cutaneous tissue), around kidneys and eyeballs; also found in specific "fat depots" (buttocks, breasts, etc.)	Provides reserve of food; insulates against heat loss; supports and pro-tects the organs it encloses
Dense fibrous connective tissue				
Regular	Primarily parallel collagenic fibers, some elastic fibers	Fibroblasts	Tendons and ligaments	Withstands great stress when force is applied in one direction
Irregular	Primarily irregularly arranged, collagenic and elastic fibers	Fibroblasts	Dermis of the skin; submucosa of digestive tract; cap-sules of joints; fascia of muscles	Protects; able to withstand stretching
Cartilage				
Hyaline	Amorphous but firm; fibers form dense imperceptible network	Chondrocytes lie in matrix cavities (lacunae); usually arranged in group-ings of two to four cells; avascular	Embryonic skeleton; covers ends of long bones; costal car-tilages of ribs and cartilages of nose, trachea, and larynx	Supports and rein-forces
Fibrocartilage	Less firm than in hyaline cartilage; regular arrangement of collagenic fibers	Same as above	Intervertebral disks; disks of knee joint	Strength; absorbs shock
Elastic	Most flexible carti-lage owing to many elastic fibers	Same as above	Cartilages support-ing external ear and epiglottis	Maintains shape of a structure while allowing flexibility
Bone, or osseous tissue	Hard, calcified; con-tains many collagen fibers	Osteocytes in lacunae, arranged concentrically around a central canal carrying nutrients; very well vascularized	Skeleton	Supports and pro-tects body organs; provides lever system for muscles; stores materials; provides site for hemopoiesis
Blood (considered separately in Exercise 29)				

*Blood-forming tissue (i.e., red marrow and lymphoid tissue) that are histologically complex are not summarized here.

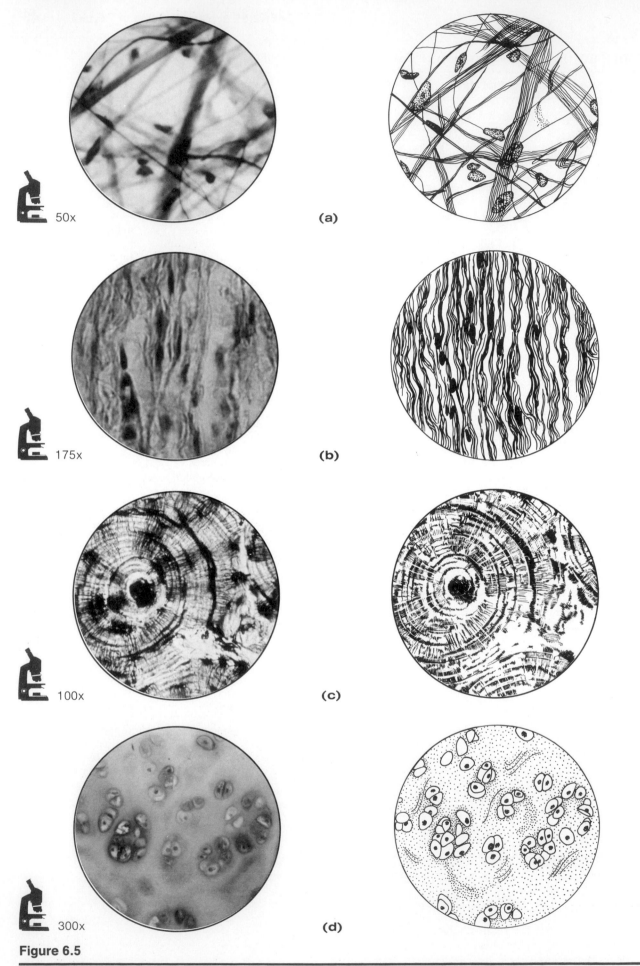

(a)

(b)

(c)

(d)

50x

175x

100x

300x

Figure 6.5

Types of connective tissue. (a) Areolar connective tissue. (b) Dense fibrous connective tissue. (c) Bone or osseous connective tissue. (d) Hyaline cartilage.

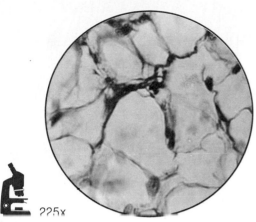

225x

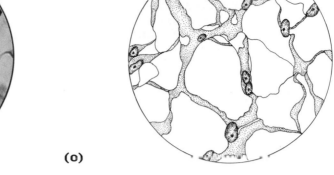

(o)

Figure 6.5 (continued)

Types of connective tissue. (e) Adipose connective tissue.

include **collagenic** (white) fibers, **elastic** (yellow) fibers, and **reticular** (fine collagenic) fibers.

There are four main types of connective tissue, all of which typically have large amounts of matrix—these are connective tissue proper, cartilage, bone, and blood. Table 6.2 lists the general characteristics, location, and function of the connective tissues found in the body.

 Obtain prepared slides of adipose, areolar, and dense fibrous connective tissue (tendon): hyaline cartilage; and osseous connective tissue (bone). Compare your observations with the views in Figure 6.5.

While examining the areolar connective tissue, notice how much "empty space" there appears to be and distinguish between the collagen fibers and the coiled elastic fibers. Also, try to locate a **mast cell,** which has large darkly staining granules in its cytoplasm. This cell type releases histamine that makes capillaries quite permeable during inflammatory reactions and allergies and thus is partially responsible for that "runny nose" of allergies.

In adipose tissue, locate a cell in which the nucleus can be seen pushed to one side by the large fat-filled vacuole (signet ring cell) and notice how little matrix there is in fat or adipose tissue.

Distinguish between the living cells and the matrix in the dense fibrous, hyaline cartilage, and bone preparations.

MUSCLE TISSUE

Muscle tissue is highly specialized to contract (shorten) in order to produce movement of some body parts. As you might expect, muscle cells tend to be quite elongated, providing a long axis for contraction. The three basic types of muscle tissue are described briefly here; skeletal muscle is treated more completely in a later exercise.

Skeletal muscle, the "meat," or flesh, of the body, is attached to the skeleton. It is under voluntary control (consciously controlled), and its contraction moves the limbs and other external body parts. The cells of skeletal muscles are long, cylindrical, and multinucleate (more than one nucleus); they have obvious striations (stripes).

Cardiac muscle is found only in the heart. As it contracts, the heart acts as a pump, propelling the blood through the blood vessels. Structurally, cardiac muscle is quite similar to skeletal muscle (striations are also visible), but cardiac cells are branching cells that interdigitate (fit together) at tight junctions called **intercalated disks**. These structures allow the cardiac muscle to act as a unit. Cardiac muscle is under involuntary control, which means that we cannot voluntarily or consciously control the operation of the heart.

Smooth muscle, or visceral muscle, is found in the walls of hollow organs and blood vessels. Its contraction generally constricts or dilates the lumen (cavity) of the structure, thus propelling substances along predetermined pathways. Smooth muscle cells are quite different in appearance from those of skeletal or cardiac muscle. No striations are visible, and the uninucleate cells are spindle-shaped.

 Obtain and examine prepared slides of skeletal, cardiac, and smooth muscle. Note their similarities and dissimilarities in both your observations and in the illustrations in Figure 6.6.

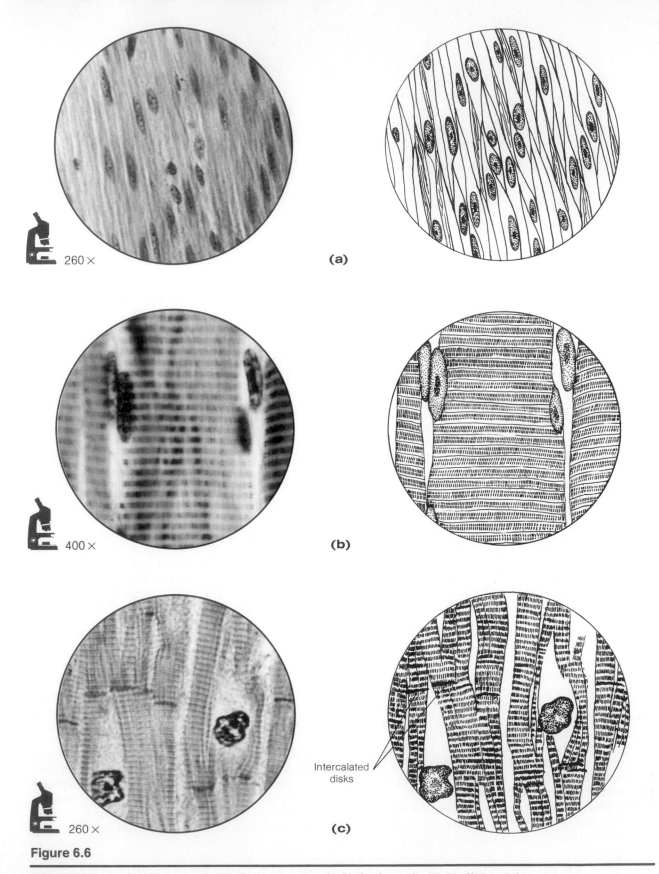

260×   (a)

400×   (b)

Intercalated
disks

260×   (c)

Figure 6.6

Muscle tissues. (a) Isolated, teased smooth muscle cells. (b) Skeletal muscle. (c) Cardiac muscle.

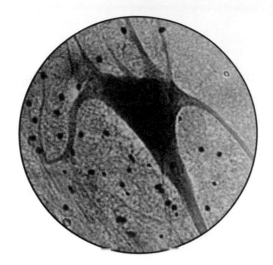

110×

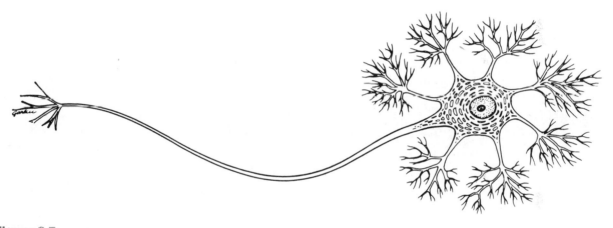

Figure 6.7

The photomicrograph shows neurons from an ox spinal cord smear. A diagrammatic view of a neuron is shown at the bottom.

NERVOUS TISSUE

Nervous tissue is composed of cells called **neurons,** which are highly specialized to receive stimuli (irritability) and to conduct waves of excitation, or impulses, to all parts of the body (conductivity). The structure of neurons is markedly different from that of all other body cells. Their cytoplasm is drawn out into long extensions as much as 3 feet (about 1 m), which allows a single neuron to conduct an impulse over relatively long distances. More detail about the anatomy of the different classes of neurons and about the special supporting cells that also are a part of nervous tissue appears in Exercise 17.

 Obtain a prepared slide of a spinal cord smear. Locate a neuron and compare it to Figure 6.7. Keep the light dim—this will help you see the cellular extensions.

UNIT 5

THE INTEGUMENTARY SYSTEM AND BODY MEMBRANES

The Integumentary System

EXERCISE 7

OBJECTIVES

1. To recount several important functions of the skin, or integumentary system.

2. To recognize and name from observation of an appropriate model, diagram, projected slide, or microscopic specimen the following skin structures: epidermis (and note relative positioning of the stratum germinativum and stratum corneum), dermis (papillary and reticular layers), hair follicles and hair, sebaceous glands, and sudoriferous glands.

3. To name the four layers of the epidermis and describe the characteristics of each.

4. To compare the properties of the epidermis to those of the dermis.

5. To describe the distribution and function of the skin derivatives—sebaceous glands, sudoriferous glands, and hair

6. To differentiate between eccrine and apocrine sudoriferous glands.

7. To enumerate the factors determining skin color.

8. To describe the function of melanin.

MATERIALS

Skin model (three-dimensional, if available)
Compound microscope
Prepared slide of human skin with hair follicles
Related film: *Human Skin* (color, 8 mm film loop, Encyclopaedia Brittanica Films)

The skin, or integument, is often considered an organ system because of its extent and complexity; it is much more than an external body covering. Architecturally, the skin is a marvel; it is tough yet pliable. This characteristic enables it to withstand constant insult from outside agents.

The skin has many functions, most (but not all) concerned with protection. It insulates and cushions the underlying body tissues and protects the entire body from mechanical damage (bumps and cuts), chemical damage (acids, alkalis, and the like), thermal damage (heat), and bacterial invasion (by virtue of its acid mantle and continuous surface). The uppermost layer of the skin (the cornified layer) is hardened, preventing water loss from the body surface. The skin's abundant capillary network (under the control of the nervous system) plays an important role in regulating heat loss from the body surface.

The skin has other functions as well. For example, it acts as a miniexcretory system; urea, salts, and water are lost through the skin pores. The skin is also the site of vitamin D synthesis for the body. Finally, the cutaneous sense organs are located in the dermis.

BASIC STRUCTURE
OF THE SKIN

Structurally, the skin consists of two kinds of tissue. The outer epidermis is made up of stratified squamous epithelium, and the underlying dermis is made up of dense irregular connective tissue. These layers are firmly cemented together. Friction, however, such as the rubbing of a poorly fitting shoe, may cause them to separate, resulting in a blister. Immediately under the dermis is the subcutaneous tissue (primarily adipose tissue, which is not considered part of the skin). The main skin areas and structures are described below.

As you read, locate the following structures on Figure 7.1 and on a skin model.

Epidermis

The avascular epidermis consists of several layers of cells.

The uppermost layer is the **stratum corneum,** sometimes called the *horny layer* because it consists of flattened keratinized cells. Keratin is a protein with waterproofing properties; thus this layer provides a natural "raincoat" for the body and prevents water loss from the deeper tissues. Keratinized cells are dead cells. They are constantly rubbing and flaking off and being replaced by the division of deeper cells.

Immediately below the stratum corneum is the **stratum lucidum,** the cells of which appear clear because of an accumulation of eleidin (a keratin precursor molecule).

The **stratum granulosum,** below the stratum lucidum, is the area in which the cells begin to die owing to their accumulation of eleidin and their increasing distance from the dermal blood supply.

The **stratum germinativum** is immediately adjacent to the dermis and thus contains the only epidermal cells that receive adequate nourishment (via diffusion of nutrients from the dermis). These cells are constantly undergoing cell division; millions of new cells are produced daily. As the daughter cells are pushed upward, away from the source of nutrition, they gradually die, and their soft protoplasm becomes increasingly keratinized.

Melanin, a brown pigment, is produced by special cells (melanocytes) found in the basal layers of the stratum germinativum. The skin tans because of an increase in melanin production when the skin is exposed to sunlight. The melanin provides a protective pigment "umbrella" over basal cell nuclei, thus shielding their genetic material (DNA) from the damaging effects of ultraviolet radiation. A concentration of melanin in one spot is commonly called a *freckle.*

Skin color is the result of two factors—the amount of melanin production and the degree of oxygenation of the blood. People with high levels of melanin production have brown-toned skin; in light-skinned people, who have less melanin, the dermal blood supply flushes through the rather transparent cell layers above, thus giving the skin a rosy glow. When the blood is poorly oxygenated or does not circulate abundantly in the dermis, the skin takes on a bluish, or **cyanotic,** cast.

Dermis

The connective tissue making up the dermis consists of two principal regions—the papillary and reticular areas—and like the epidermis, it varies in thickness. For example, the skin is particularly thick on the palms of the hands and soles of the feet and is quite thin on the eyelids.

The **papillary layer** is the upper dermal region. It is very uneven and has conelike projections from its superior surface, the **dermal papillae,** which attach it to the epidermis above. These projections are reflected in fingerprints, which are unique patterns of ridges that remain unchanged throughout life. Abundant capillary networks in the papillary layer furnish nutrients for the epidermal layers and allow heat to radiate to the skin surface. The pain and touch receptors are also found here.

The **reticular layer** is the deepest skin layer. It contains many arteries and veins, sweat and sebaceous glands, and pressure receptors.

Both the papillary and reticular layers are heavily invested with collagenic and elastic fibers. The latter give skin its exceptional elasticity in youth. In old age, the number of elastic fibers decreases and the subcutaneous layers lose fat, which leads to

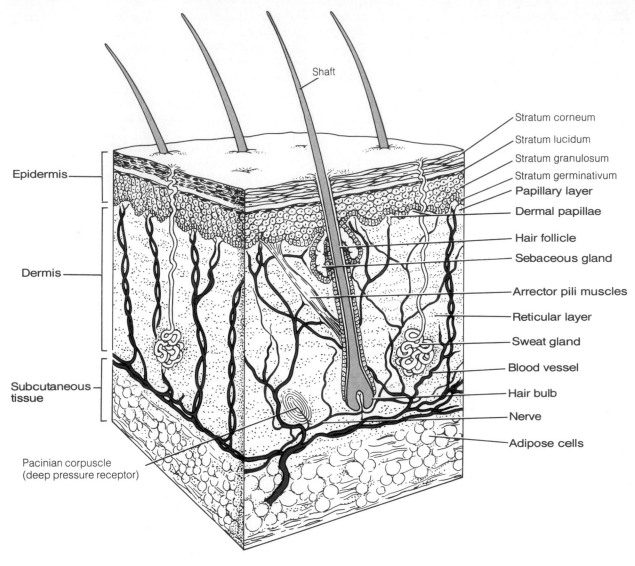

Figure 7.1

Three-dimensional view of the skin.

wrinkling and inelasticity of the skin. Fibroblasts, adipose cells, various types of macrophages (which are important in the body's defense), and other cell types are found throughout the dermis.

The dermis has an abundant blood supply, which allows it to play a role in the regulation of body temperature. When body temperature is high, the arterioles dilate, and the capillary network of the dermis becomes engorged with the heated blood. Thus body heat is allowed to radiate from the skin surface. If the environment is cool and body heat must be conserved, the arterioles constrict so that blood bypasses the capillary networks. Any restriction of the normal blood supply to the skin results in cell death and, if severe enough, skin ulcers. *Bedsores* (*decubitus ulcers*) occur in bedridden patients who are not turned regularly enough. The weight of the body exerts pressure on the skin, especially over bony projections, which leads to restriction of the blood supply and death of tissue.

The dermis is also richly provided with lymphatic vessels and a nerve supply. Many of the nerve endings bear highly specialized receptor organs that, when stimulated by environmental changes, transmit messages to the central nervous system for interpretation. (These receptors are discussed in depth in Exercise 21.)

APPENDAGES OF THE SKIN

The appendages of the skin—hair, nails, and cutaneous glands—are all derivatives of the epidermis, but they reside in the dermis. They originate from the stratum germinativum and grow downward into the deeper skin regions.

Cutaneous Glands

The cutaneous glands fall primarily into two categories: the sebaceous glands and the sudoriferous glands. The **sebaceous glands** are found nearly all over the skin, with the exception of the palms of the hands and the soles of the feet. Their ducts usually empty into a hair follicle, but some open directly onto the skin surface.

The product of the sebaceous glands, called **sebum,** is a mixture of oily substances and fragmented cells. The sebum is a lubricant that keeps the skin soft and moist (a natural skin cream) and keeps the hair from becoming brittle. The sebaceous glands become particularly active during puberty thus, the skin tends to become oilier during this period of life. *Blackheads* are accumulations of dried sebum and bacteria; *acne* is due to active infection of the sebaceous glands.

Epithelial openings, called pores, are the outlets for the **sudoriferous glands** (sweat glands). These exocrine glands are widely distributed in the skin. Sudoriferous glands are subcategorized on the basis of the composition of their secretions. The **eccrine glands**, which are distributed all over the body, produce clear perspiration, consisting primarily of water, salts (NaCl), and urea. The **apocrine glands,** found predominantly in the axillary and genital areas, secrete a milky protein-based substance (also containing water, salts, and urea) that is an ideal nutrient medium for microorganisms generally found on the skin.

The sweat glands, under the control of the nervous system, are an important part of the body's heat-regulating apparatus. They secrete perspiration when the external temperature or body temperature is high. When this water-based substance evaporates, it carries excess body heat with it. Thus perspiration acts as an emergency system of heat liberation when the capillary cooling system is not sufficient or is unable to maintain homeostasis.

Hair

Hairs are found over the entire body surface, with the exception of the palms of the hands, the soles of the feet, and the lips. A hair, enclosed in a hair **follicle,** is also an epithelial structure. The portion of the hair enclosed within the follicle is called the **root**; the portion projecting from the scalp surface is called the **shaft**. The hair is formed by mitosis of the well-nourished germinal epithelial cells at the basal end of the follicle (the hair **bulb**). As the daughter cells are pushed farther away from the growing region, they die and become keratinized; thus the bulk of the hair shaft, like the bulk of the epidermis, is dead material.

A hair consists of a central region (medulla) surrounded by a protective cortex. Abrasion of the cortex results in "split ends," in which the medulla is no longer bound together by the cortex. Hair color is a manifestation of the amount of pigment (generally melanin) within the hair cortex.

If you look carefully at the structure of the hair follicle (see Figure 7.1), you will see that it generally is slanted. Small bands of smooth muscle cells—**arrector pili**—connect each side of the hair follicle to the papillary layer of the dermis. When these muscles contract (during cold or fright), the hair is pulled upright, dimpling the skin surface with "goose bumps." This phenomenon is especially dramatic in a scared cat, whose fur actually stands on end to increase its apparent size. The activity of the arrector pili muscles also exerts pressure on the sebaceous glands surrounding the follicle, causing a small amount of sebum to be released.

Nails

Nails, the hornlike derivatives of the epidermis, consist of a root (adhering to an epithelial nail bed) and a body. The germinal cells in the nail root divide to produce its growth; these cell divisions occur below the lunula, the crescent- or half-moon–shaped region at the proximal end of the nail. Like the hair shaft, the nail is composed of nonliving material. The nails are nearly colorless but appear pink because of the blood supply in the nail bed. When someone is cyanotic due to a lack of oxygen in the blood, the nails take on a blue cast.

 Obtain a prepared slide of human skin and study it carefully under the microscope, observing as many of the structures diagrammed in Figure 7.1 as possible. Diagram and label your observations in the space provided here.

How is this stratified squamous epithelium different from that observed in Exercise 6? _____ _____ _____

How do these differences relate to the functions of these two similar epithelia? _____ _____ _____

Classification of Membranes

OBJECTIVES

1. To name and recognize by microscopic examination epidermal, mucous, and serous membranes.

2. To list the general functions of each membrane type and note its location in the body.

3. To compare the structure and function of the major membrane types.

MATERIALS

Compound microscope
Prepared slides of trachea (cross section) and small intestine (cross section)
Prepared slide of serous membrane
Longitudinally cut fresh beef joint (if available)

The body membranes, which cover surfaces, line body cavities, and form protective (and often lubricating) sheets around organs, fall into two major categories. These are the **epithelial membranes** (consisting of the epidermis of the skin and the mucous and serous membranes) and the **synovial membranes** (composed entirely of connective tissue). The epidermis is introduced in Exercise 7 in conjunction with a broader topic—the skin or integumentary system. The other membranes are considered briefly in this exercise.

EPITHELIAL MEMBRANES

Mucous Membranes

The mucous membranes (mucosae) are composed of epithelial cells resting on a layer of loose (areolar) connective tissue called the *lamina propria*. They line all body cavities that open to the body exterior—the respiratory, digestive, and urinary tracts. Although mucous membranes often secrete mucus, this is not a requirement. The mucous membranes of both the digestive and respiratory tracts secrete mucus, but that of the urinary tract does not.

 Examine a slide made from a cross section of the trachea and of the small intestine. Draw each in the appropriate circle, and fully identify each epithelial type.

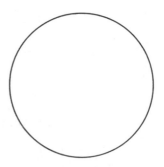

Mucosa of trachea

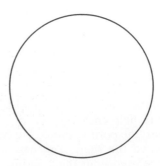

Mucosa of small intestine

Goblet cells are columnar epithelial cells with a large mucus-containing vacuole (goblet) in the cytoplasm. Which of these mucous membranes contains goblet cells? _____

_____

How do the roles of these mucous membranes differ?

_____

_____

_____

_____

How are they the same? _____

_____

_____

Would you say that mucous membranes have a high

regenerative capacity? _____ Why or

why not? _____

_____

_____

Serous Membranes

The serous membranes (serosae) are also epithelial membranes. They are generally composed of two layers—a layer of simple squamous epithelium on a basement membrane. The serous membranes generally occur in twos. The parietal layer lines a body cavity, and the visceral layer covers the outside of the organs in that cavity. In contrast to the mucous membranes, which line open body cavities, the serous membranes line body cavities that are closed to the exterior (with the exception of the female peritoneal cavity, which the fallopian tubes enter, and the dorsal body cavity). The serosae secrete a thin fluid (serous fluid), which lubricates the organs and body walls and thus reduces friction as the organs slide across one another and against the body cavity walls. A serous membrane also lines the interior of blood vessels (endothelium) and the heart (endocardium). In capillaries, the entire wall is composed of serosa that serves as a selectively permeable membrane between the blood and the tissue fluid of the body.

 Examine a prepared slide of a serous membrane and diagram it in the circle provided here.

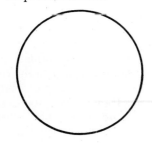

Figure 8.1

A typical synovial joint.

What kind of cells that you have seen many times

before are these shaped like? _____

_____

What are the specific names of the serous membranes of the heart and heart cavity (respec-

tively)? _____

_____

The lungs and thoracic cavity (respectively)? _____

_____

The abdominal viscera and visceral cavity (re-

spectively)? _____

_____

SYNOVIAL MEMBRANES

Synovial membranes, unlike the mucous and serous membranes, are composed of connective tissue; they contain no epithelial cells. These membranes line the cavities surrounding the joints, providing a smooth surface and secreting a lubricating fluid. They also line smaller sacs of connective tissue (bursae) and tendon sheaths, both of which cushion structures moving against each other, as during muscle activity. Figure 8.1 illustrates the positioning of a synovial membrane in the joint cavity.

If a freshly sawed beef joint is available, examine the interior surface of the joint capsule to observe the smooth texture of the synovial membrane.

UNIT 6

Bone Classification, Structure, and Relationships: An Overview

EXERCISE

9

OBJECTIVES

1. To list at least three functions of the skeletal system.

2. To identify the four main kinds of bones.

3. To identify surface bone markings and function.

4. To identify the major anatomic areas on a sagitally cut long bone (or diagram of one).

5. To identify the major regions and structures of a haversian system in a histologic specimen of compact bone (or diagram of one).

6. To explain the role of the inorganic salts and organic matrix in providing flexibility and hardness to bone.

MATERIALS

Numbered disarticulated bones that demonstrate classic examples of the four bone classifications (long, short, flat, and irregular)

Long bone sawed longitudinally (beef bone from a slaughterhouse, if possible, or prepared laboratory specimen)

Compound microscope

Prepared slide of ground bone (cross section)

Long bone soaked in 10% nitric acid until flexible

Long bone baked at 250 F for more than 2 hours

The skeleton is constructed of two of the most supportive tissues found in the human body—cartilage and bone. In embryos, the skeleton is predominantly composed of hyaline cartilage, but in the adult, most of the cartilage is replaced by more rigid bone. Cartilage persists only in such isolated areas as the bridge of the nose, the larynx, the joints, and parts of the ribs.

Besides supporting and protecting the body as an internal framework, the skeleton also provides a system of levers with which the skeletal muscles work to move the body. In addition,

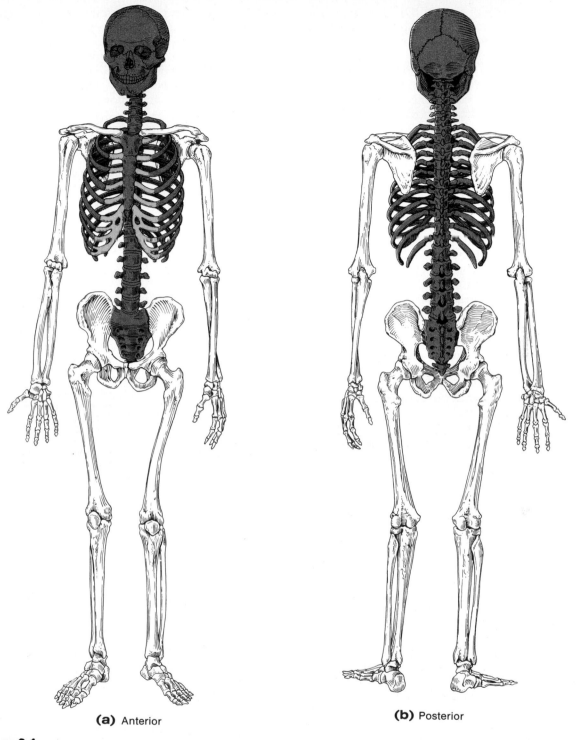

(a) Anterior

(b) Posterior

Figure 9.1

The human skeleton. The shaded area is the axial skeleton; the unshaded area is the appendicular skeleton.

the bones store such substances as lipids and calcium. Finally, the red marrow cavities of bones provide a site for hemopoiesis (blood cell formation).

The skeleton is made up of bones that are connected at joints, or articulations. The skeleton is subdivided into two divisions: the **axial** skeleton (those bones that lie around the body's center of gravity) and the **appendicular skeleton** (those of the limbs, or appendages) (Figure 9.1).

Before beginning your study of the skeleton, imagine for a moment that your bones have turned to putty. What if you were running when this metamorphosis took place? Now imagine your bones forming a continuous metal framework within your body, somewhat like a network of plumbing pipes. What problems could you envision with this arrangement? These images should help you understand how well the skeletal system provides support and protection, as well as facilitates movement.

BONE MARKINGS

Even a casual observation of the bones will reveal that bone surfaces are not featureless smooth areas but are scarred with an array of bumps, holes, and ridges. These bone markings reveal where muscles, tendons, and ligaments were attached and where blood vessels and nerves passed. Bone markings fall into two categories: projections, or processes which grow out from the bone, and depressions, or cavities, which are indentations in the bone.

The projections of bone are of several types:

Condyle: rounded convex projection that articulates with another bone
Crest: narrow ridge of bone serving as a site for muscle attachment
Epicondyle: raised area on a condyle
Head: extension carried on a narrow neck that takes part in forming a joint
Line: similar to a crest but less pronounced
Ramus: armlike bar, extending from the bone body, that takes part in joint formation
Spine: sharp, slender site of muscle attachment
Trochanter: very large, usually irregularly shaped site of muscle attachment
Tubercle: small rounded site of muscle attachment
Tuberosity: large rounded projection or roughened area that serves as a site of muscle attachment

There are also several types of depressions:

Fissure: narrow slitlike opening that often serves as a nerve passageway
Foramen: opening through a bone that serves as a passageway for blood vessels and nerves
Fossa: shallow depression that often forms a socket for another bone in a joint
Meatus: canallike passageway that often serves as a conduit for a nerve
Sinus: cavity within bone, often filled with air and lined with a mucous membrane
Sulcus: groove or furrow

CLASSIFICATION OF BONES

The 206 bones of the adult skeleton are composed of two basic kinds of osseous tissue that differ in their texture. **Compact** bone looks smooth and homogeneous; **spongy** (or cancellous) bone is composed of small spicules of bone and lots of open space. Bones may be classified further on the basis

of their relative gross anatomy into four groups: long, short, flat, and irregular bones.

The **long bones** are generally longer than wide, consisting of a shaft with heads at either end. Long bones are composed predominantly of compact bone. The **short bones** are generally cube-shaped, and they contain more spongy bone than compact bone.

As the name implies, the **flat bones** are generally thin and flattened, with two thin layers of compact bone sandwiching a layer of spongy bone between them. Bones that do not fall into one of the preceding categories are classified as **irregular bones**.

Some anatomists also recognize two other subcategories of bones. **Sesamoid bones** are small bones formed in tendons. The kneecaps are sesamoid bones, but the other members of this group are found in different locations depending on the individual. **Wormian bones** are tiny bones between major cranial bones.

 Examine the isolated (disarticulated) bones (numbered) on display. See if you can find specific examples of the bone markings described above. Then classify each of the bones into one of the four anatomic groups by recording its number in the chart that follows.

Long	Short	Flat	Irregular

GROSS ANATOMY OF THE TYPICAL LONG BONE

 1. Obtain a fresh beef long bone that has been sawed along its longitudinal axis, or use a cleaned dry bone. With the help of Figure 9.2, identify the shaft, or **diaphysis**. Note its smooth surface, which is composed of compact bone. If you are using a fresh specimen, pull away the **periosteum**, or fibrous membrane covering, to view the bone surface. The periosteum is the source of the blood vessels and nerves that invade the bone.

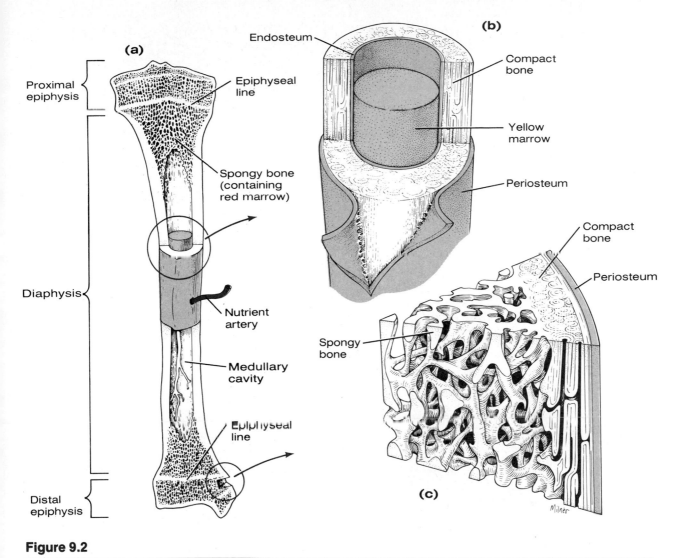

Figure 9.2

Tibia (long bone). (a) Anterior view with longitudinal section cut away at both ends; (b) cross section of shaft (diaphysis); (c) pie-shaped three-dimensional view of spongy bone.

2. Now inspect the **epiphysis**, the end of the long bone. Note that it is composed of a thin layer of compact bone filled with spongy bone.

3. Identify the **articular cartilage** covering the epiphyseal surface in place of the periosteum. Since it is composed of glassy hyaline cartilage, it provides a smooth surface to prevent friction at joint surfaces.

4. If the animal was still young and growing, you will be able to see a thin area of hyaline cartilage, the **epiphyseal disk**, which provides for longitudinal growth of the bone during youth. Once the long bone has stopped growing, these areas are replaced with bone and may appear as thin, barely discernible remnants—the **epiphyseal lines.**

5. In an adult animal, the interior or cavity of the shaft is essentially a storage region for adipose tissue, or **yellow marrow.** In the infant, this area is involved in forming blood cells, and so **red marrow** is found in the marrow cavities. In adult bone, the red marrow is confined to the interior of the epiphyses, where it occupies the spaces between the spicules of spongy bone.

6. If you are examining a fresh bone, look carefully to see if you can distinguish the delicate **endosteum** lining the shaft. In a living bone, osteoclasts (bone-destroying cells) are found on the inner surface of the endosteum, against the compact bone of the diaphysis. As the bone grows in diameter on its external surface, it is constantly being broken down on its inner surface. Thus the thickness of the compact bone layer composing the shaft remains relatively constant.

MICROSCOPIC STRUCTURE OF COMPACT BONE

As you have seen, spongy bone has a spiky, open-work appearance, resulting from the arrangement

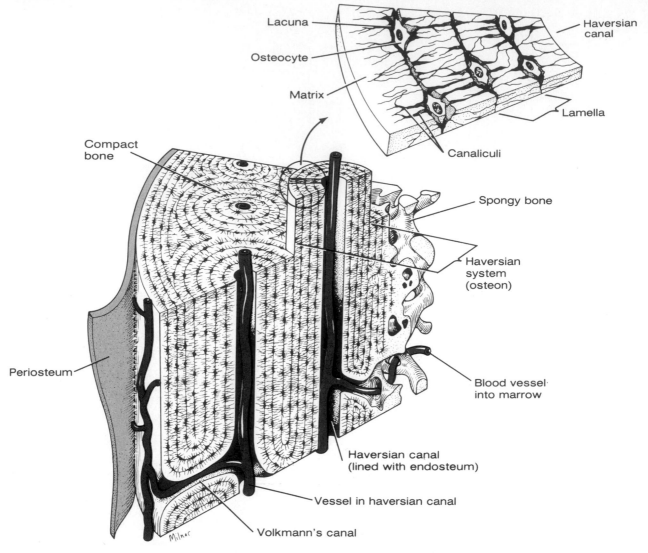

Figure 9.3

Histologic structure of a long bone (three-dimensional).

of the spicules of bony material, or trabeculae, that compose it, while compact bone appears to be dense and homogeneous. Microscopic examination of compact bone, however, reveals that it is riddled with passageways carrying blood vessels, nerves, and lymphatic vessels that provide the living bone cells with needed substances and a way to eliminate wastes.

 1. Obtain a prepared slide of ground bone and examine it under low power. Using Figure 9.3 as a guide, focus on a **haversian canal.** The haversian canal runs parallel to the long axis of the bone and carries blood vessels, nerves, and lymph vessels through the bony matrix. Identify the **osteocytes** (mature bone cells) in **lacunae**

(chambers), which are arranged in concentric circles (concentric **lamellae**) around the haversian canal. A haversian canal and all the concentric lamellae surrounding it are referred to as a **haversian system,** or **osteon.** Also identify **canaliculi,** tiny canals radiating outward from a haversian canal to the lacunae of the first lamella and then from lamella to lamella. The canaliculi form a dense transportation network through the hard bone matrix, connecting all the living cells of the haversian system to the nutrient supply. You may need a higher-power magnification to see the fine canaliculi.

2. Note the **Volkmann's canals** in Figure 9.3. These canals run into the compact bone and marrow cavity from the periosteum at right angles to the shaft.

With the haversian canals, the Volkmann's canals complete the communication pathway between the bone interior and its external surface.

CHEMICAL COMPOSITION OF BONE

Bone is one of the hardest materials in the body. Although bone is relatively light, it has a remarkable ability to resist tension and shear forces that continually act on it. An engineer would tell you that a cylinder (like bone) is one of the strongest structures for its mass. Thus nature has given us an extremely strong, exceptionally simple (almost crude), and flexible supporting system without sacrificing mobility.

The hardness of bone is due to the inorganic calcium phosphate salts deposited in its ground substance. Its flexibility comes from the organic constituents of the matrix, particularly the collagenic fibers.

 Obtain a bone sample that has been soaked in nitric acid and one that has been baked. Heating removes the organic part of bone, while acid dissolves out the minerals. Do the treated bones retain the structure of untreated specimens? _____

Gently apply pressure to each bone sample.

What happens to the heated bone? _____

_____

The bone treated with acid? _____

_____

What does the nitric acid appear to remove from the bone? _____

_____

What does baking appear to do to the bone?

_____

_____

_____

In rickets, the bones are not properly calcified. Which of the demonstration specimens would more closely resemble the bones of a child with rickets? _____

_____

_____

The Axial Skeleton

OBJECTIVES

1. To name the three bone groups composing the axial skeleton (skull, thorax, and vertebral column).

2. To identify the bones composing the axial skeleton, either by examining the isolated bones or by pointing them out on an articulated skeleton, or skull, and to name the important bone markings on each.

3. To distinguish by examination the different types of vertebrae.

4. To discuss the importance of the intervertebral fibrous disks and spinal curvatures.

5. To differentiate among lordosis, kyphosis, and scoliosis.

MATERIALS

Intact skull and Beauchene skull
X-rays of individuals with scoliosis, lordosis, and kyphosis (if available)
Articulated skeleton
Isolated cervical, thoracic, and lumbar vertebrae, sacrum, and coccyx
Disarticulated bones of the thorax

The axial skeleton (the shaded portion of Figure 9.1) can be divided into three parts: the skull, the vertebral column, and the thorax.

THE SKULL

The skull is composed of two sets of bones: the cranium, or cranial vault, encloses and protects the fragile brain tissue; the facial bones present the eyes in an anterior position and form the base for the facial muscles, which make it possible for us to present our feelings to the world. All but one of the bones of the skull are joined by interlocking joints called *sutures;* the mandible, or jawbone, is attached to the rest of the skull by a freely movable joint.

The bones of the skull, shown in Figures 10.1 through 10.4, are described below. As you read through this material, identify each bone on an intact (and/or Beauchene) skull. Note that important bone markings are listed beneath the bones on which they appear.

The Cranium

The cranium is composed of eight large flat bones. *With the exception of two paired bones (the parietals and the temporals), all are single bones.* Sometimes the six ossicles of the middle ear are also considered part of the cranium.

FRONTAL See Figures 10.1, 10.2, 10.4. Anterior portion of cranium; forms the forehead and superior part of the orbit.

Supraorbital foramen: opening above each orbit allowing blood vessels and nerves to pass.
Glabella: smooth area between the eyes.

PARIETAL See Figures 10.1 and 10.2. Posterolateral to the frontal bone, forming sides of cranium.

Sagittal suture: midline articulation point of the two parietal bones.
Coronal suture: point of articulation of parietals with frontal bone.

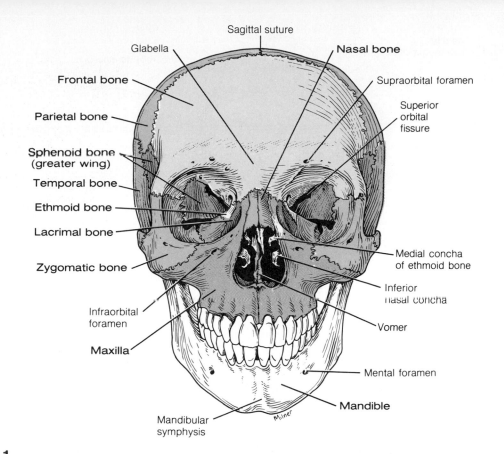

Figure 10.1

Human skull, anterior view.

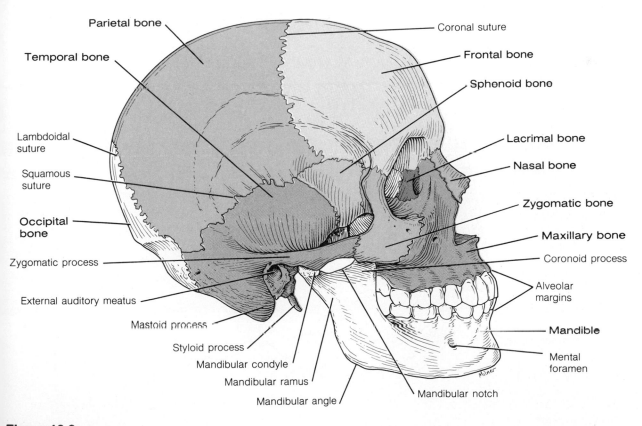

Figure 10.2

Human skull, lateral view.

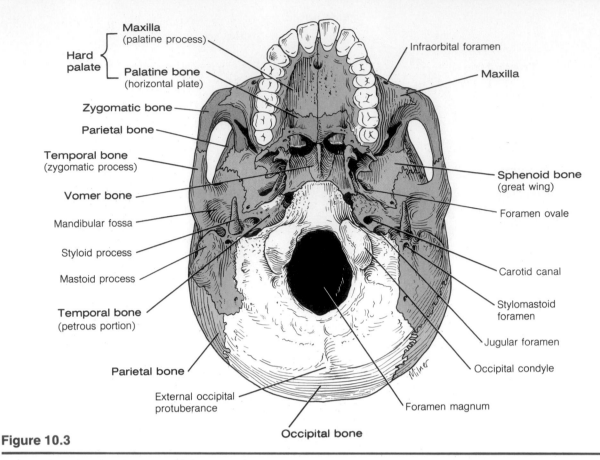

Figure 10.3

Human skull, inferior view; mandible removed.

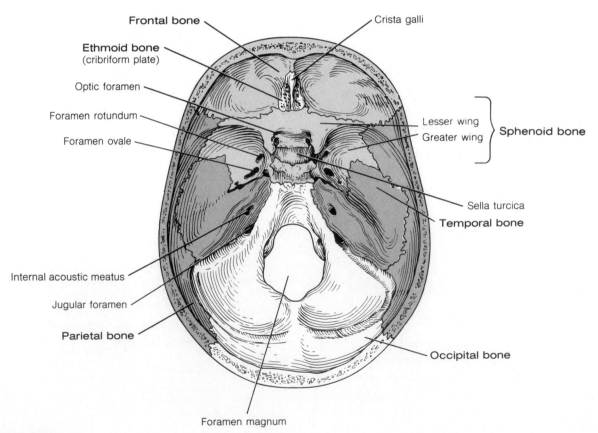

Figure 10.4

Human skull, superior view; top of the cranium removed.

TEMPORAL See Figures 10.1 through 10.4. Inferior to parietal on lateral skull. The temporals can be divided into two parts: the **squamous portion** adjoins the parietals; the **petrous portion** forms the lateral inferior aspect of the skull.

Squamous suture: point of articulation of temporal with parietal bone.
External auditory meatus: canal leading to eardrum and middle ear.
Styloid process: needlelike projection inferior to external auditory meatus; attachment point for muscles and ligaments.
Zygomatic process: bridgelike projection joining the zygomatic bone (cheekbone) anteriorly.
Mastoid process: rough projection inferior and posterior to external auditory meatus; attachment site for muscles. The mastoid process is full of air cavities and is so close to the middle ear, a trouble spot for infections, that it often becomes infected too, a condition referred to as mastoiditis. Should this troublesome condition persist, the mastoid process may be surgically removed. Because this area is separated from the brain by only a very thin layer of bone, an ear infection that has spread to the mastoid process can inflame the brain coverings or the meninges. This latter condition is known as *meningitis.*
Mandibular fossa: rounded depression anterior to external auditory meatus; forms the socket for the mandibular condyle, the point where the mandible (lower jaw) joins the cranium.
Jugular foramen: opening medial to styloid process through which blood vessels and cranial nerves IX, X, and XI pass.
Carotid canal: opening, medial to the styloid process, through which the carotid artery passes to reach the brain.
Stylomastoid foramen: tiny opening between the mastoid and styloid processes through which the seventh cranial nerve leaves the brain.
Internal acoustic meatus: opening on posterior aspect of temporal bone allowing passage of two cranial nerves (see Figure 10.4).

OCCIPITAL See Figures 10.2, 10.3, 10.4. Most posterior bone of cranium—forms floor and back wall.

Lambdoidal suture: site of articulation of occipital bone and parietal bones.
Foramen magnum: large opening in base of occipital, which allows the spinal cord to join with the brain.
Occipital condyles: rounded projections lateral to the foramen magnum, which articulate with the first cervical vertebra (atlas).
External occipital protuberance: midline prominence posterior to the foramen magnum.

SPHENOID See Figures 10.1 through 10.4. Bat-shaped bone forming the anterior plateau of the cranial cavity across the width of the skull.

Greater wings: portions of the sphenoid seen exteriorly anterior to the temporal and forming a portion of the orbits of the eyes.
Superior orbital fissures: jagged openings in orbits providing passage for cranial nerves III, IV, V, and VI to enter the orbit where they serve the eye.

The sphenoid bone can be seen in its entire width if the top of the cranium (calverium) is removed (see Figure 10.4).

Sella turcica (Turk's saddle): small depression in sphenoid midline in which the pituitary gland rests in the living person.
Lesser wings: bat-shaped portion of sphenoid anterior to sella turcica.
Foramen rotundum: opening lateral to sella turcica providing passage for a branch of the fifth cranial nerve.
Foramen ovale: opening posterior to the sella turcica providing passage for a branch of the fifth cranial nerve.
Optic foramina: openings in the base of the lesser wings through which the optic nerves enter the orbits to serve the eyes.

ETHMOID See Figures 10.1, 10.2, and 10.4. Irregularly shaped bone anterior to the sphenoid. Forms the roof of the nasal cavity, upper nasal septum, and part of the medial orbit walls.

Crista galli (cock's comb): vertical projection providing a point of attachment for the dura mater (outermost membrane covering of the brain).
Cribriform plates: bony plates lateral to the crista galli through which olfactory fibers pass to the brain from the nasal mucosa.
Superior and medial conchae (turbinates): thin plates of bone extending laterally from the perpendicular plate of the ethmoid into the nasal cavity. The shelflike conchae provide for a more efficient air flow through the nasal cavity and greatly increase the surface area of the mucosa that covers them, thus increasing the ability of the mucosa to warm and humidify the incoming air.

Facial Bones

There are 14 bones composing the face, and 12 of these are paired. *Only the mandible and vomer are single bones.* An additional bone, although not a facial bone, is considered here because of its location. Refer to Figures 10.1 through 10.4 to find the structures described below.

MANDIBLE See Figures 10.1 and 10.2. The lower jaw bone, which articulates with the temporal bones, providing the only freely movable joints of the skull.

Body: horizontal portion; forms the chin.
Rami: extend superiorly from the body on either side.

Mandibular condyle: point of articulation of the mandible with the mandibular fossa of the temporal bone.

Coronoid process: jutting anterior portion of the ramus; site of muscle attachment.

Angle: point at which ramus meets the body.

Mental foramina: prominent openings on the body that transmit blood vessels and nerves.

Alveoli: sockets on superior margin of mandible in which the teeth lie.

Mandibular symphysis: anterior median depression indicating point of mandibular fusion.

MAXILLAE See Figures 10.1, 10.2, and 10.3. Two bones fused in a median suture, which form the upper jawbone and part of the orbits. All facial bones, with the exception of the mandible, join the maxillae. Thus they represent the main, or keystone, bones of the face.

Alveoli: sockets on the inferior margin in which teeth lie.

Palatine processes: form the anterior parts of the hard palate.

Infraorbital foramen: opening under the orbit carrying nerves to the nasal region.

PALATINE See Figure 10.3. Paired bones posterior to the palatine processes; form posterior hard palate.

ZYGOMATIC See Figures 10.1, 10.2, and 10.3. Lateral to the maxilla; forms the portion of the face commonly called the cheekbone, and forms part of the lateral orbit.

LACRIMAL See Figures 10.1 and 10.2. Fingernail-sized bones forming a part of the medial orbit walls between the maxilla and the ethmoid. Each lacrimal bone bears a groove or furrow, which serves as a passageway for tears (*lacrima* means "tear").

NASAL See Figures 10.1 and 10.2. Small rectangular bones forming the upper part of the bridge of the nose.

VOMER See Figure 10.1. Irregularly shaped bone in median plane of nasal cavity; forms the posterior and inferior nasal septum.

INFERIOR CONCHAE (turbinates) See Figure 10.1. Thin curved bones protruding from the lateral walls of the nasal cavity; serve the same purpose as the turbinate portions of the ethmoid bone (described earlier).

HYOID BONE See Figure 10.5. Not really considered a skull bone. Located in the throat above the larynx; serves as a point of attachment for many tongue and mouth muscles. Does not articulate with any other bone, and is thus

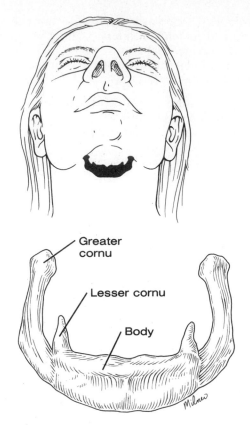

Figure 10.5

Hyoid bone.

unique. Horseshoe-shaped with a body and two pairs of horns, or **cornua.**

Palpate the following areas on yourself:

• Mastoid process of the temporal bone
• Temporomandibular joints (Open and close your jaws to locate these.)
• Hyoid bone (Place a thumb and finger high behind the lateral edge of the mandible and squeeze medially.)

Four facial bones (maxillary, sphenoid, ethmoid, and frontal) contain sinuses (air cavities lined with mucosa), which lead into the nasal passages (see Figure 10.6). These paranasal sinuses lighten the facial bones and may act as resonance chambers for speech. The maxillary sinus is the largest of the sinuses found in the skull.

THE VERTEBRAL COLUMN

The vertebral column, extending from the skull to the pelvis, forms the major axial support of the body. Additionally, it surrounds and protects the delicate spinal cord while allowing the spinal

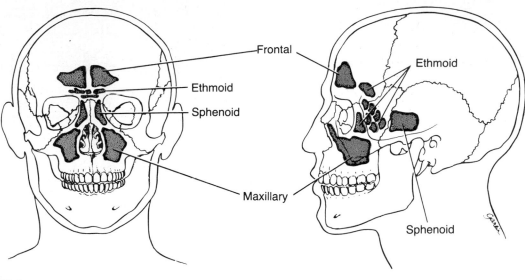

Figure 10.6

Paranasal sinuses.

nerves to issue from the cord via openings between adjacent vertebrae. The term *vertebral column* might suggest to some people a rather rigid supporting rod, but this is far from the truth. The vertebral column consists of 24 single bones (**vertebrae**) and two composite, or fused, bones (the sacrum and coccyx) that are connected in such a way as to provide a flexible curved structure (Figure 10.7).

The vertebrae are separated by pads of fibrocartilage, **intervertebral disks,** that cushion the vertebrae and absorb shocks. Each disk is composed of two major regions, a central gelatinous *nucleus pulpulsus* that behaves like a fluid, and an outer ring of encircling collagen fibers called the *anulus fibrosis,* which stabilizes the disk and contains the pulpulsus. As a person ages, the water content of the disks decreases (as it does in other tissues throughout the body), and the disks become thinner and less compressible. This situation, along with other degenerative changes such as weakening of the ligaments and tendons of the vertebral column, predisposes older people to *ruptured disks.* A ruptured disk is a situation in which the nucleus pulpulsus herniates through the anulus portion and compresses adjacent nerves.

The presence of the disks and the S-shaped or springlike construction of the vertebral column prevent shock to the head in walking and running and allow for flexibility in the movement of the body trunk. The thoracic and sacral curvatures of the spine are referred to as primary curvatures, since they are present at birth. Later the secondary curvatures are formed. The cervical curvature appears when the baby begins to raise its head, and the lumbar curvature develops when the baby begins to walk.

1. Note the normal curvature of the vertebral column in your laboratory specimen, and then examine Figure 10.8, which depicts three abnormal spinal curvatures—scoliosis, kyphosis, and lordosis. These abnormalities may result from disease or poor posture. Also examine x-ray sheets showing these same conditions in a living patient, if they are available.

2. Using an articulated spinal column (or an articulated skeleton), examine the freedom of movement between two lumbar vertebrae separated by an intervertebral disk.

When the fibrous disk is properly positioned, are the spinal cord or peripheral nerves impaired in

any way? _____

Remove the disk and put the two vertebrae back together. What happens to the nerve?

_____

Can you explain why a ruptured or herniated disk, a relatively common disorder of the vertebral column,

would be very painful? _____

_____

_____

_____

_____

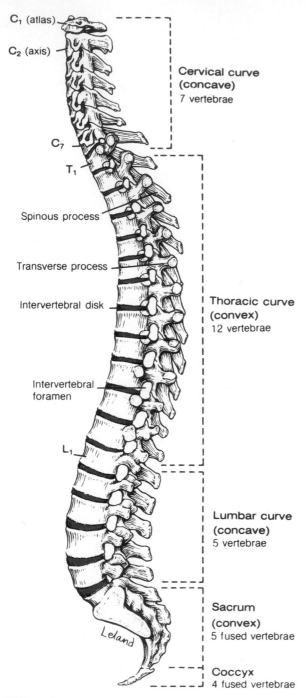

Figure 10.7

The vertebral column. Thin disks between the thoracic vertebrae allow great flexibility in the thoracic column; thick disks between the lumbar vertebrae reduce flexibility.

Structure of a Typical Vertebra

Although they differ in size and specific features, all vertebrae have some structures in common (Figure 10.9).

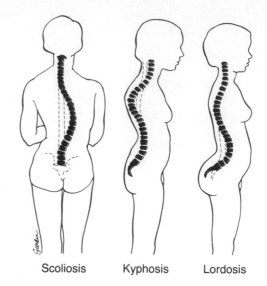

Scoliosis   Kyphosis   Lordosis

Figure 10.8

Abnormal spinal curvatures.

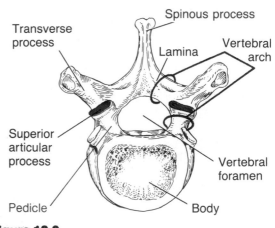

Figure 10.9

A typical vertebra, superior view (inferior articulating surfaces not shown).

Body (or centrum): rounded central portion of the vertebra, which faces anteriorly in the human vertebral column.

Vertebral arch: this is composed of pedicles, laminae, and a spinous process. It represents the junction of all posterior extensions from the vertebral body.

Vertebral foramen: conduit for the spinal cord.

Transverse processes: two lateral projections from the vertebral body.

Spinous process: single medial and posterior projection.

Superior and inferior articular processes: paired projections lateral to the vertebral foramen that

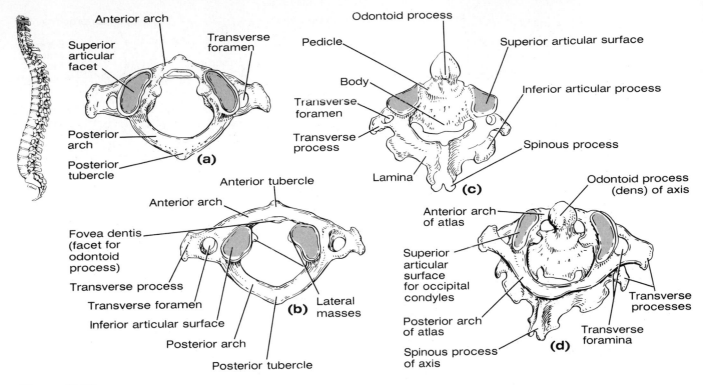

Figure 10.10

Cervical vertebrae C_1 and C_2. (a) Superior view of the atlas (C_1); (b) inferior view of the atlas (C_1); (c) superior view of the axis (C_2); (d) superior lateral view of the articulated atlas and axis.

allow articulation with adjacent vertebrae. The superior articular processes face toward the spinous process, whereas the inferior articular processes face away from the spinous process.

Intervertebral foramina: notches on the inferior surfaces of the pedicles that create openings for spinal nerves to leave the spinal cord between adjacent vertebrae.

Figures 10.10 and 10.11 show how specific vertebrae differ; refer to them as you read the following sections.

Cervical Vertebrae

The seven cervical vertebrae (referred to as C_1 to C_7) form the neck portion of the vertebral column. The first two cervical vertebrae (atlas and axis) are highly modified to perform special functions (see Figure 10.10). The **atlas** (C_1 in a widespread system of reference) lacks a body, and its lateral processes contain large concave depressions on their superior surfaces that receive the occipital condyles of the skull. This joint enables you to nod yes. The **axis** (C_2) acts as a pivot for the rotation of the atlas (and skull) above. It bears a large vertical process, the **odontoid process,** or **dens,** which serves as the pivot point. The articulation between C_1 and C_2 allows you to rotate your head from side to side.

The "typical" cervical vertebrae (C_3 through C_7) are distinguished from the thoracic and lumbar vertebrae by several features (see Figure 10.11(a)). They are the smallest, lightest vertebrae; the vertebral foramen is triangular; and the spinous process is short and usually bifurcated, or divided into two branches. The spinous process of C_7 is not branched, however, and is substantially longer than that of the other cervical vertebrae. Transverse processes of the cervical vertebrae are wide, and they contain foramina through which the vertebral arteries pass superiorly on their way to the occipital region of the brain. Any time you see these foramina in a vertebra, you should know immediately that it is a cervical vertebra. There are no articular facets on the transverse processes.

Thoracic Vertebrae

The 12 thoracic vertebrae (referred to as T_1 through T_{12}) may be recognized by the following structural characteristics. As you can see in Figure 10.11(b), they have a larger body than the cervical vertebrae. The body is somewhat heart-shaped, with two articulating surfaces, or costal facets, on each side (one superior, the other inferior) close to the origin of the vertebral arch. The vertebral foramen is oval or round, and the spinous process is long, with a sharp downward hook. The closer the thoracic vertebra is to the

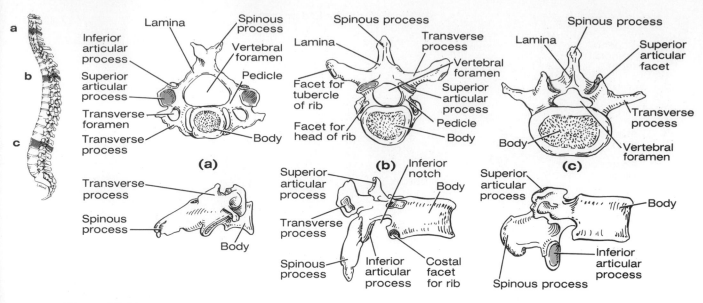

Figure 10.11

Comparison of typical cervical, thoracic, and lumbar vertebrae (superior view above, lateral view below): (a) cervical vertebra; (b) thoracic vertebra; (c) lumbar vertebra.

lumbar region, the less sharp and shorter is the spinous process. Articular facets on the transverse processes articulate with the tubercles of the ribs.

Lumbar Vertebrae

The five lumbar vertebrae (L_1 through L_5) have massive blocklike bodies and short, thick, hatchet-shaped spinous processes extending backward horizontally (see Figure 10.11(c)). The superior articular facets are concave medially; the inferior ones are convex laterally. Since most of the stress of the vertebral column occurs in the lumbar region, these are the sturdiest of the vertebrae.

The spinal cord ends at the superior edge of L_2, but the outer covering of the cord, filled with cerebrospinal fluid, extends an appreciable distance beyond. Thus a lumbar puncture (for examination of the cerebrospinal fluid) or the administration of "saddle block" anesthesia for childbirth is normally done between L_3 and L_4 or L_4 and L_5, where there is no chance of injuring the delicate spinal cord.

The Sacrum

The sacrum (Figure 10.12) is a composite bone formed from the fusion of five vertebrae. Superiorly it articulates with L_5, and inferiorly it connects with the coccyx. The **median sacral crest** is a remnant of the spinous processes of the fused vertebrae. The winglike **alae,** formed from the fusion of the transverse processes, articulate laterally with the hip bones. The sacrum is slightly concave anteriorly and forms the posterior border of the pelvis. Four ridges cross the anterior part of the sa-

crum, and **sacral foramina** are located at either end of these ridges. These foramina allow for the passage of blood vessels and nerves. The vertebral canal continues inside the sacrum as the **sacral canal.** The **sacral promontory** (anterior border of the body of S_1) is an important anatomic landmark for obstetricians.

The Coccyx

The coccyx (see Figure 10.12) is formed from the fusion of three to five small irregularly shaped vertebrae. It is literally the human tailbone, a vestige of the tail that other vertebrates have. The coccyx is attached to the sacrum by ligaments.

Obtain examples of each type of vertebra and examine them carefully, comparing them to Figures 10.10, 10.11, and 10.12.

THE THORAX

The thorax is composed of the sternum, ribs, and thoracic vertebrae (Figure 10.13). It is referred to as the thoracic cage because of its appearance and because it forms a protective cone-shaped enclosure around the organs of the thoracic cavity (heart and lungs, for example).

The Sternum

The sternum (breastbone), a typical flat bone, is a result of the fusion of the **manubrium, body**

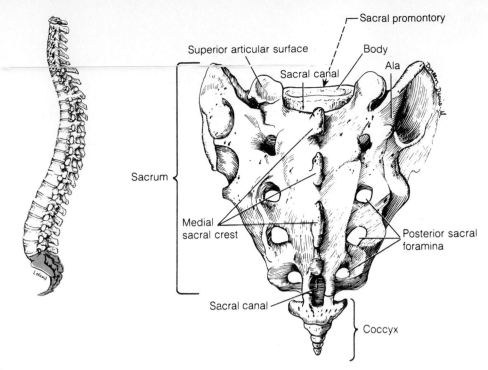

Figure 10.12

Sacrum and coccyx, posterior view.

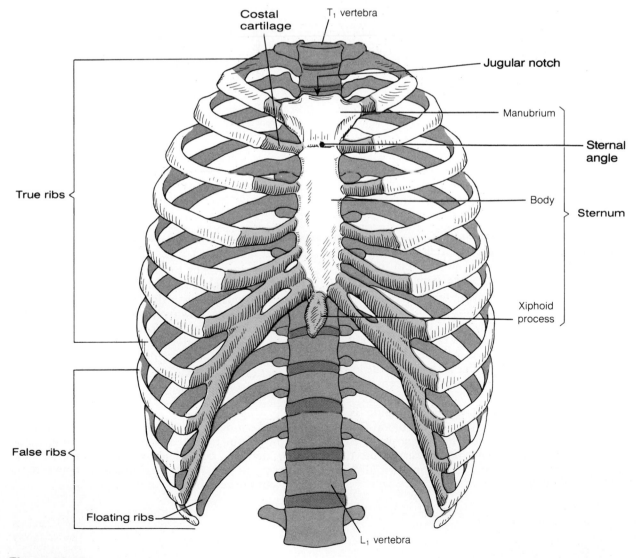

Figure 10.13

Bony thorax, anterior view.

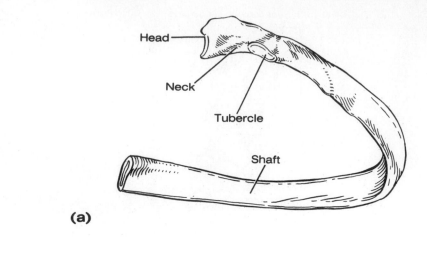

(a)

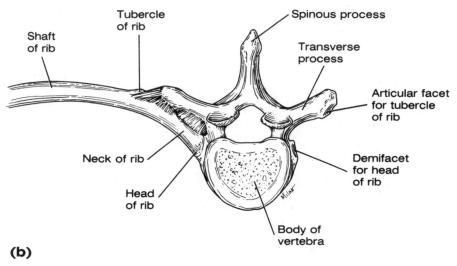

(b)

Figure 10.14

(a) Typical true rib from right side of body (costal cartilage not shown); (b) superior view of the articulation between a rib and a thoracic vertebra.

(gladiolus), and **xiphoid process.** It is attached to the first seven pairs of ribs. The sternum has two important bony landmarks—the **jugular notch** and the **sternal angle**. The jugular notch (concave upper border of the manubrium) can be palpated easily; generally it is at the level of the third thoracic vertebra. The sternal angle is a result of the manubrium and body meeting at a slight angle to each other, so that a transverse ridge is formed at the level of the second ribs, providing a handy reference point for counting ribs.

Because of its accessibility, the sternum is a favored site for obtaining samples of blood-forming (hemopoietic) tissue for the diagnosis of suspected blood diseases. A needle is inserted into the marrow of the sternum and the sample withdrawn (sternal puncture).

The Ribs

The 12 pairs of ribs form the walls of the thoracic cage (see Figures 10.13 and 10.14). All of the ribs articulate posteriorly with the vertebral column and then curve downward and toward the anterior body surface. The *"true ribs,"* the first seven pairs, attach directly to the sternum by costal cartilages. *"False ribs,"* the next five pairs, have indirect cartilage attachments to the sternum or no sternal attachment at all, as in the case of the last two pairs (which are also called floating ribs).

First take a deep breath to expand your chest. Notice how your ribs seem to move outward and how your sternum rises. Then examine an articulated skeleton to observe the relationship between the ribs and the vertebrae.

The Appendicular Skeleton

OBJECTIVES

1. To identify on an articulated skeleton the bones of the shoulder and pelvic girdles and their attached limbs.

2. To arrange unmarked, disarticulated bones in their proper relative position to form the entire skeleton.

3. To differentiate between a male and a female pelvis.

4. To discuss the common features of the human appendicular girdles (pectoral and pelvic), and to note how their structure relates to their specialized functions.

5. To identify specific bone markings in the appendicular skeleton.

MATERIALS

Articulated skeletons
Disarticulated skeletons (complete)
Articulated pelves (male and female for comparative study)

The appendicular skeleton (the unshaded portion of Figure 9.1) is composed of the 126 bones of the appendages and the pectoral and pelvic girdles, which attach the limbs to the axial skeleton.

 Carefully examine each of the bones described and identify the characteristic bone markings of each. The markings aid in determining whether a bone is the right or left member of its pair. *This is a very important instruction because, before completing this laboratory exercise, you will be constructing your own skeleton.*

BONES OF THE SHOULDER GIRDLE AND UPPER EXTREMITY

The Shoulder Girdle

The shoulder girdle or pectoral girdle (Figure 11.1) consists of four bones—the two anterior clavicles and the two posterior scapulae—and functions to attach the arms to the axial skeleton. In addition, the bones of the shoulder girdle serve as attachment points for many trunk and neck muscles.

The **clavicle,** or collarbone, is a slender doubly curved bone, rounded on the medial end, which attaches to the sternal manubrium, and flattened on the lateral end, where it articulates with the scapula to form a part of the shoulder joint. The clavicle serves as an anterior brace, or strut, to hold the arm away from the top of the thorax and helps prevent dislocation of the shoulder.

The **scapulae,** or shoulder blades, are generally triangular and are commonly called the "wings" of humans. Each scapula has a flattened body and two important processes—the **acromion process** (the enlarged end of the spine of the scapula) and the beaklike **coracoid process.** The scapular notch at the base of the coracoid process allows nerves to pass. The acromion process connects with the clavicle; the coracoid process points anteriorly over the tip of the shoulder joint and serves as a point of attachment for some of the muscles of the upper limb. The scapula has no direct attachment to the axial skeleton but is loosely held in place by trunk muscles. The scapula has three angles: superior, inferior, and lateral. The scapula also has three named borders: the superior; the medial, or vertebral; and the lateral, or axillary. Several shallow depressions (fossae) appear on both sides of the scapula and are named according to location. The

71

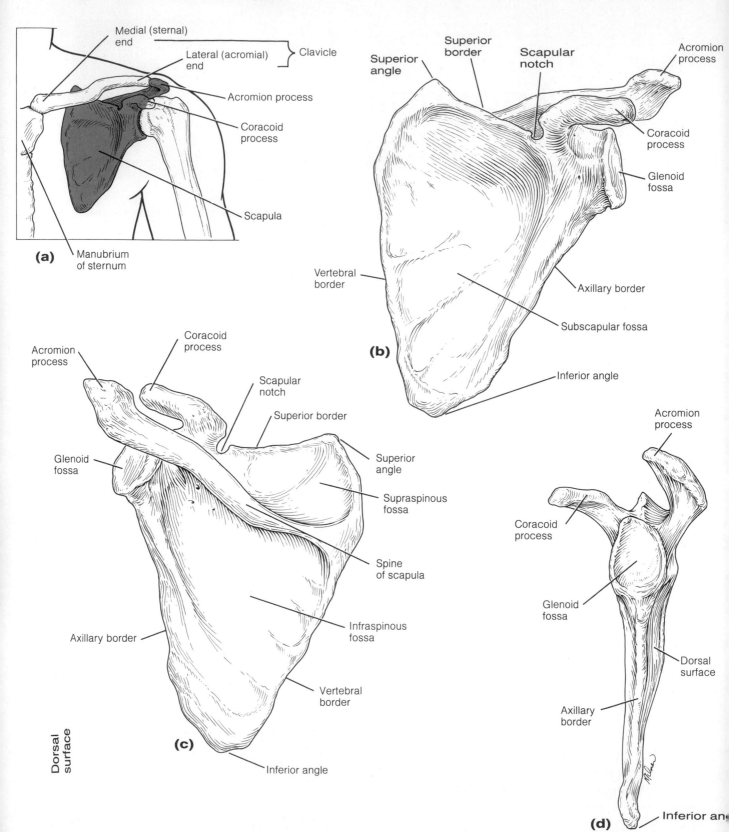

Figure 11.1

Bones of the shoulder girdle. (a) Left shoulder girdle articulated to show the relationship of the girdle to the bones of the thorax and upper arm; (b) left scapula, anterior view; (c) left scapula, posterior view; (d) left scapula, lateral view.

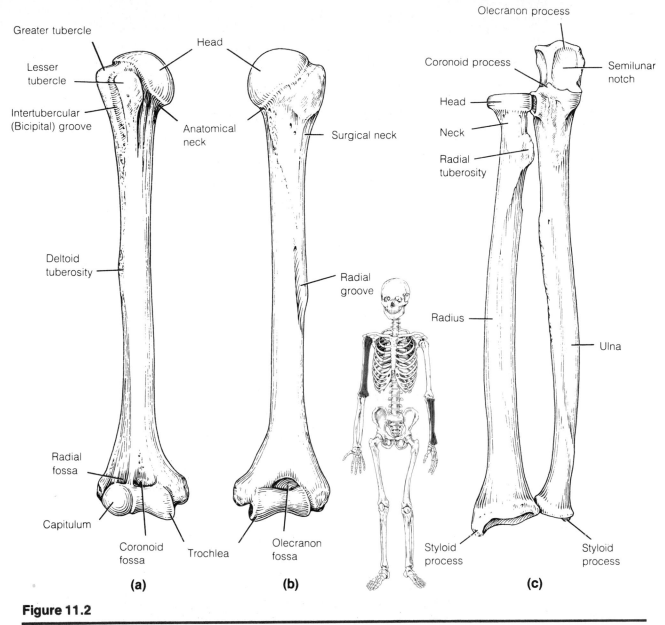

Figure 11.2

Bones of the right upper arm and forearm: (a) humerus, anterior view; (b) humerus, posterior view; (c) radius and ulna, anterior view.

glenoid fossa, a shallow socket that receives the head of the arm bone, is located in the lateral angle.

The shoulder girdle is exceptionally light and allows the upper limb a degree of mobility not observed anywhere else in the body. This is due to the following factors:

- There is but a single point on each shoulder girdle for attachment to the axial skeleton—the sternoclavicular joints.
- The relative looseness of the scapular attachment allows it to slide back and forth against the thorax with muscular activity.
- The glenoid fossa is shallow, and reinforcement of the shoulder joint is relatively poor.

However, this exceptional flexibility exacts a price: the shoulder girdle is very susceptible to dislocation.

The Upper Arm

The upper arm (Figure 11.2) consists of a single bone—the **humerus,** a typical long bone. At its proximal end is the rounded head, which fits into the shallow glenoid fossa of the scapula. The head is separated from the shaft by the anatomic neck and the more constricted surgical neck, which is a common site of fracture. Opposite the head are two prominences, the **greater** and **lesser tubercles,** separated by a groove (the **intertubercular** or **bicipital groove**) that guides the tendon of the biceps muscle

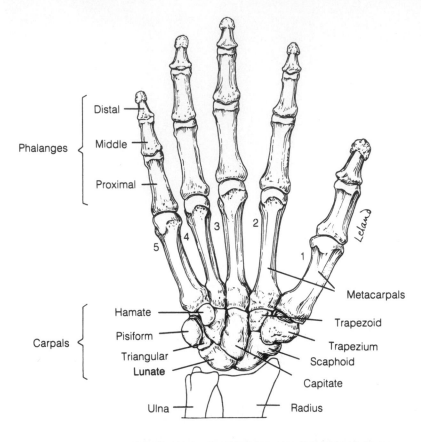

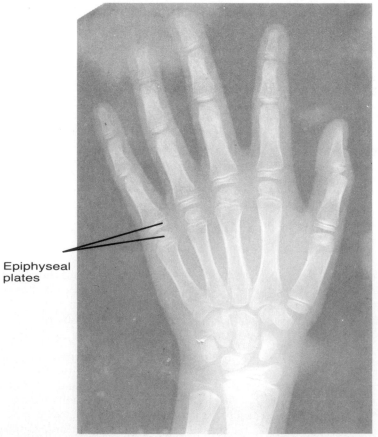

Figure 11.3

Bones of the right hand and wrist, anterior view.

to its point of attachment (the rim of the glenoid fossa). In the midpoint of the shaft is a roughened area, the **deltoid tuberosity,** where the large fleshy shoulder muscle, the deltoid, attaches. Just inferior to the deltoid tuberosity is the radial groove, which indicates the pathway of the radial nerve.

At the distal end of the humerus are two condyles—the medial **trochlea** (looking rather like a spool), which articulates with the ulna and the lateral **capitulum,** which articulates with the radius of the forearm. Above the trochlea on the anterior surface is a depression, the **coronoid fossa;** on the posterior surface is the **olecranon fossa.** These two depressions allow the corresponding processes of the ulna to move freely when the elbow is flexed and extended. A small radial fossa, lateral to the coronoid fossa, receives the head of the radius when the elbow is flexed.

The Forearm

Two bones, the radius and the ulna, compose the skeleton of the forearm, or antibrachium (see Figure 11.2). In the anatomic position, the **radius** is in the lateral position in the forearm. Proximally, the disk-shaped head of the radius articulates with the capitulum of the humerus. Just below the head, on the medial aspect of the shaft, is a prominence called the **radial tuberosity,** the point of attachment for the tendon of the biceps muscle of the arm.

The **ulna** is the medial bone of the forearm. Its proximal end bears the anterior **coronoid process** and the posterior **olecranon process,** which are separated by the **semilunar notch;** together they grip the trochlea of the humerus in a plierslike joint. The smaller distal end of the ulna bears a small medial **styloid process,** which serves as a point of attachment for the ligaments of the wrist.

The Wrist

The wrist (Figure 11.3) is referred to anatomically as the **carpus,** and the eight bones composing it are the **carpals.** The carpals are arranged in two irregular rows of four bones each. In the proximal row (lateral to medial) are the scaphoid, lunate, triangular, and pisiform bones; the scaphoid and lunate articulate with the distal end of the radius. In the distal row are the trapezium, trapezoid, capitate, and hamate. The carpals are bound closely together by ligaments, which restrict movements between them.

The Hand

The hand, or manus (see Figure 11.3), consists of two groups of bones: the **metacarpals** (bones of the palm) and the **phalanges** (bones of the fingers). The metacarpals are numbered 1 to 5 from the thumb side of the hand toward the little finger. When the fist is clenched, the heads of the metacarpals become prominent as the knuckles. Each hand contains 14 phalanges. There are 3 phalanges in each finger except the thumb, which has only proximal and distal phalanges.

BONES OF THE PELVIC GIRDLE AND LOWER EXTREMITY

The Pelvic Girdle

The pelvic girdle, or pelvis (Figure 11.4), is formed by two **os coxae** (hipbones), together with the sacrum and coccyx. In contrast to the bones of the shoulder girdle, those of the pelvic girdle are heavy and massive, and they are attached securely to the axial skeleton. The sockets for the heads of the femurs (thighbones) are deep and heavily reinforced by ligaments to ensure a stable, strong limb attachment. The ability to bear weight is more important here than extreme mobility and flexibility; the combined weight of the upper body rests on the pelvis (specifically, where it meets the L_5 vertebra).

Each os coxa is a result of the fusion of three bones—the ilium, ischium, and pubis—which are distinguishable in the young child. The **ilium,** which connects posteriorly with the sacrum (at the **sacroiliac joint**), is a large flaring bone forming the major portion of the os coxa; when you put your hands on your hips, you are palpating the ilia. The upper margin of the ilium, the **iliac crest,** is roughened; it terminates in the **anterior superior spine** and the **posterior superior spine.** Two inferior spines are located below these. The shallow **iliac fossa** marks its internal surface, and a shallow ridge, the **iliopectineal line,** outlines the pelvic inlet.

The **ischium** is the "sit-down" bone, forming the most inferior and posterior portion of the os coxa. The most outstanding marking on the ischium is the **ischial tuberosity,** which receives the weight of the body when sitting. The **ischial spine,** superior to the ischial tuberosity, is an important anatomic landmark of the pelvic cavity. Two other anatomic features are the **lesser** and **greater sciatic notches,** which allow passage of nerves and blood vessels to and from the leg.

The **pubis** is the most anterior portion of the os coxa. The fusion of the **rami** of the pubic bone anteriorly and the ischium posteriorly forms a bar of bone enclosing the **obturator foramen,** through which large blood vessels and nerves run from the pelvic cavity into the leg. The pubic bones of each hipbone fuse anteriorly at the ridgelike **pubic crest** to form a cartilaginous joint called the **pubic symphysis.**

The ilium, ischium, and pubis fuse at the deep hemispherical socket called the **acetabulum** (literally, "vinegar cup"), which receives the head of the thighbone.

1. Before continuing with the bones of the lower limbs, take the time to examine an articulated pelvis. Note how each os coxa articulates with the sacrum posteriorly and how the two os coxae join at the pubic symphysis. The

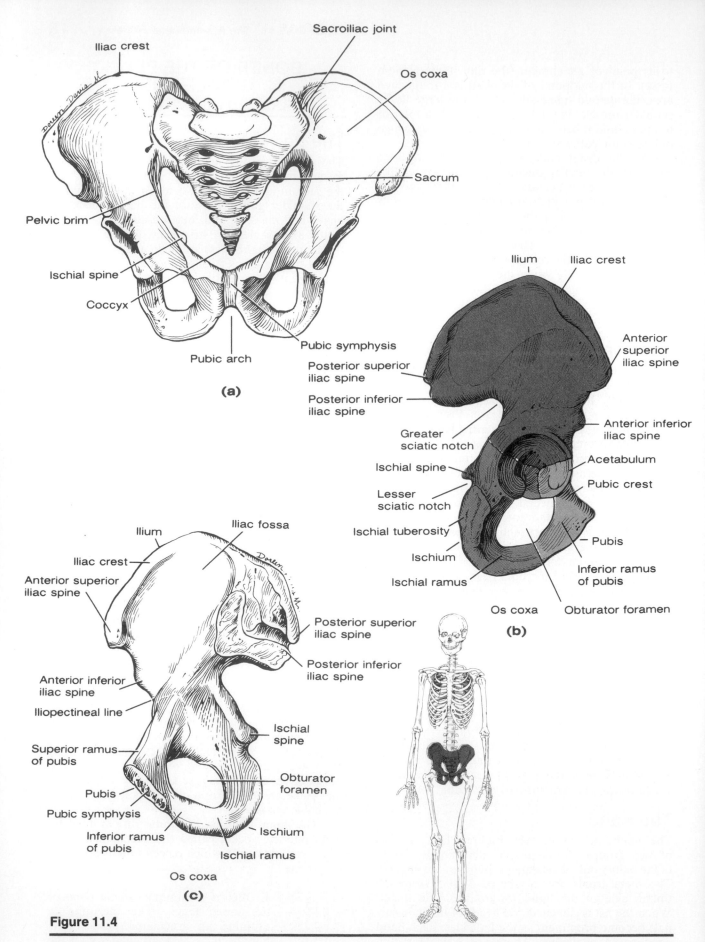

Figure 11.4

Bones of the pelvic girdle: (a) articulated pelvis, showing the two os coxae and the sacrum; (b) right os coxa, lateral view, showing the point of fusion of the ilium, ischium, and the pubic bones; (c) right os coxa, medial view.

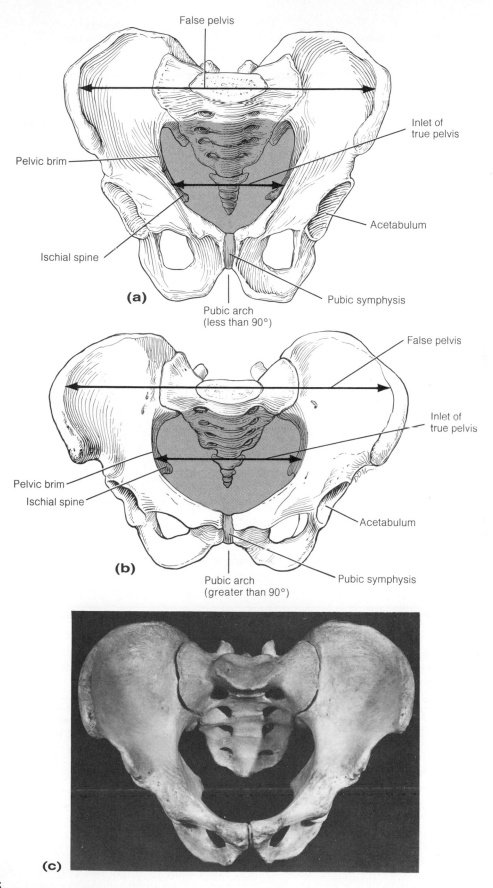

False pelvis

Inlet of
true pelvis

Pelvic brim

Acetabulum

Ischial spine

Pubic symphysis

(a)

Pubic arch
(less than 90°)

False pelvis

Inlet of
true pelvis

Pelvic brim

Ischial spine

Acetabulum

Pubic symphysis

(b)

Pubic arch
(greater than 90°)

(c)

Figure 11.5

Comparison of male and female pelves, anterior views: (a) male; (b) female; (c) photograph of male pelvis

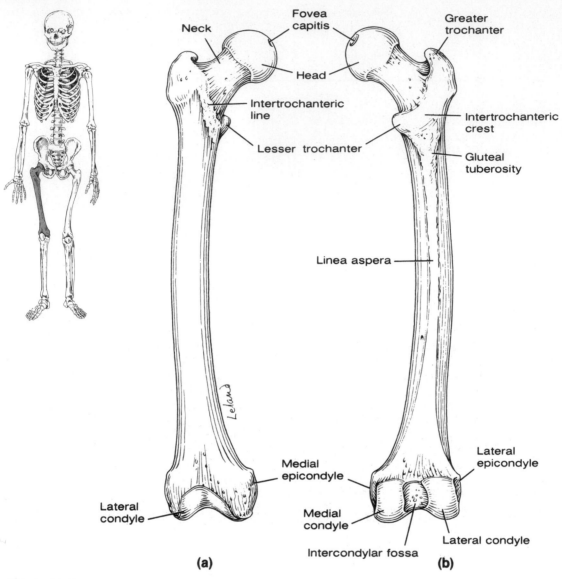

Figure 11.6

Bones of the right thigh: (a) femur, anterior view; (b) posterior view.

sacroiliac joint, because of the pressure it must bear, is often a site of lower back problems.

2.   The pelvis is divided into two regions. The **true pelvis,** surrounded by bone, is inferior to the flaring portion of the ilia (the pelvic brim) and contains the pelvic organs. The **false pelvis** lies superior to the true pelvis and encompasses only the area medial to the flaring portions of the ilia. Identify these areas. The true pelvis of the woman is of great importance, because it must be large enough to allow for the passage of an infant's head during childbirth. Thus, the dimensions of the cavity, the outlet (inferior opening), and inlet (superior opening) of the true pelvis are critical and are carefully measured by the obstetrician.

3.   Of course, individual pelvic structure varies, but there are differences between the male and female pelves that are fairly constant (see Figure 11.5). Examine male and female pelves for the following differences:

- The female inlet is larger and more circular.
- The female pelvis as a whole is shallower, and the bones are lighter and thinner.
- The female sacrum is broader and less curved, and the pubic arch is more rounded.
- The female acetabula face more anteriorly and are farther apart, and the ilia flare more laterally.
- The female ischial spines are shorter, thus enlarging the pelvic outlet.

The Upper Leg

The **femur,** or thighbone (Figure 11.6), is the sole bone of the thigh or upper leg and is the heaviest, strongest bone in the body. Its proximal end bears a ball-like head, a neck, and **greater** and **lesser**

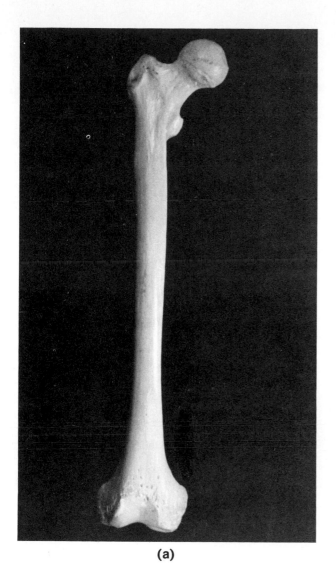

(a)

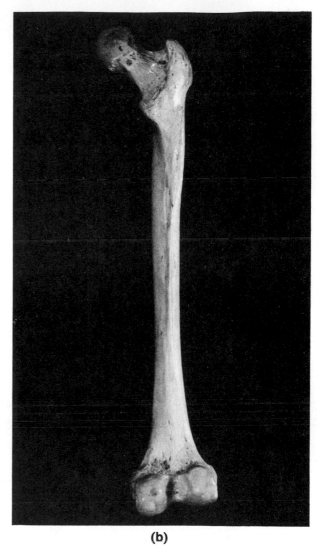

(b)

Figure 11.6 (continued)

Photos of bones of the right thigh: (a) femur, anterior view (b) posterior view.

trochanters (separated posteriorly by the **intertro-chanteric crest** and anteriorly by the **intertro-chanteric line**). The head of the femur articulates with the acetabulum of the hipbone in a deep secure socket that is heavily reinforced with ligaments. However, the neck of the femur is a common fracture site, especially in the elderly.

The femur inclines medially as it runs downward to the lower leg bones; this brings the knees in the line with the body's center of gravity or maximum weight. The medial course of the femur is even more noticeable in females because of the wider female pelvis.

Distally the femur terminates in the **lateral** and **medial epicondyles.** The more inferior lateral and medial condyles, which are separated by the **intercondylar fossa,** articulate with the tibia below.

The trochanters and trochanteric crest, as well as the **gluteal tuberosity** and the **linea aspera** located on the shaft, serve as sites for muscle attachment.

The Lower Leg

Two bones, the tibia and the fibula, form the lower leg (see Figure 11.7). The **tibia,** or shinbone. is the larger and more medial of the two lower leg bones. At the proximal end, the **medial** and **lateral condyles** (separated by the **intercondylar eminence**) receive the distal end of the femur to form the knee joint. The **tibial tuberosity,** a roughened protrusion on the anterior tibial surface (just below the condyles), serves as the site for attachment of the patellar (kneecap) ligament. Small facets on its superior and inferior lateral surface articulate with the fibula. Distally, a process called the **medial malleolus** forms the inner bulge of the ankle, and the smaller distal end articulates with the talus bone

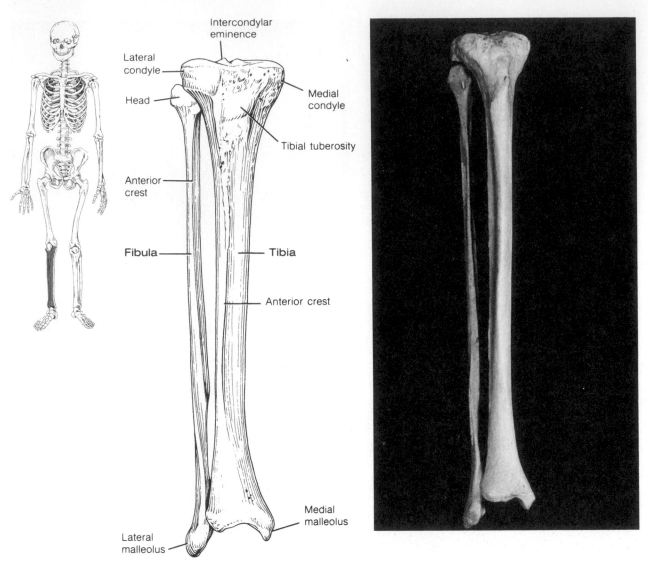

Figure 11.7

Bones of the right leg: tibia and fibula, anterior view.

of the foot. The anterior surface of the tibia is a sharpened ridge (anterior crest) that is relatively unprotected by muscles; thus it is easily felt beneath the skin.

The **fibula,** which lies parallel to the tibia, takes no part in forming the knee joint. Its proximal head articulates with the lateral condyle of the tibia. The fibula is thin and sticklike with a sharp anterior crest. It terminates distally in the **lateral malleolus,** which forms the outer part of the ankle.

The Ankle

The ankle, or **tarsus** (Figure 11.8), is composed of seven tarsal bones—the **calcaneus, talus,** navicular, cuboid, and lateral, medial, and intermediate cuneiforms. Body weight is concentrated on the two largest tarsals, the calcaneus (heel bone), and the talus, which lies between the tibia and the calcaneus.

The Foot

The bones of the foot include the five **metatarsals,** which form the instep, and the 14 **phalanges,** which form the toes (see Figure 11.8). Like the fingers of the hand, each toe has three phalanges except the large toe, which has two.

The bones in the foot are arranged to produce three strong arches—two longitudinal arches (medial and lateral) and one transverse arch (Figure 11.9). Ligaments, binding the foot bones together, and tendons of the foot muscles hold the bones firmly in the arched position but still allow a certain degree of give. Weakened arches are referred to as fallen arches or flat feet.

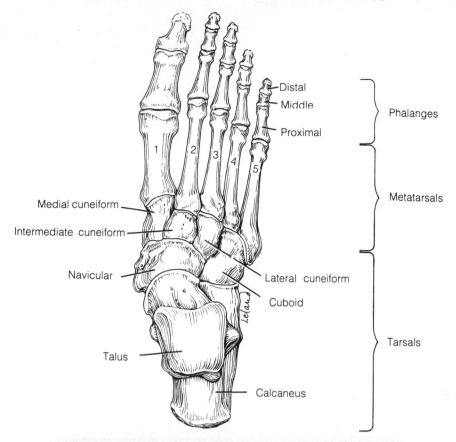

Phalanges
- Distal
- Middle
- Proximal

1 2 3 4 5

Metatarsals

Medial cuneiform

Intermediate cuneiform

Navicular

Lateral cuneiform

Cuboid

Talus

Tarsals

Calcaneus

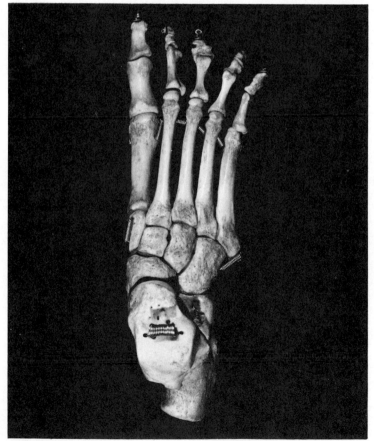

Figure 11.8

Bones of the right ankle and foot, superior view.

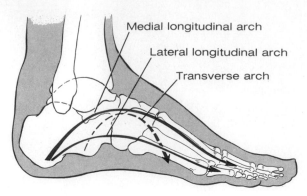

Medial longitudinal arch

Lateral longitudinal arch

Transverse arch

Figure 11.9

Arches of the foot.

1.   When you have finished examining the disarticulated bones of the appendicular skeleton, note their relationships in the articulated skeleton.

2.   Arrange the disarticulated bones on the laboratory bench in their proper relative positions to form an entire skeleton. Careful observations of the bone markings should help you distinguish between right and left members of bone pairs. Ask the instructor to check your arrangement to ensure that it is correct.

The Fetal Skeleton

OBJECTIVES

1. To define *fontanel* and discuss its function and fate in the fetus.

2. To demonstrate important differences between the fetal and adult skeletons.

MATERIALS

Isolated fetal skull
Fetal skeleton
Adult skeleton

The human fetus about to be born has 275 bones, many more than the 206 bones found in the adult skeleton. This is because many of the bones described as single bones in the adult skeleton (for example, the os coxae, sternum, and sacrum) have not yet fully ossified and fused in the fetus.

1. Obtain a fetal skull and study it carefully. Make observations as needed to answer the following questions. Does it have the same bones as the adult skull? How does the size of the fetal face relate to the cranium? How does this compare to what is seen in the adult? Why do the centers of some of the fetal skull bones protrude outward?

2. Indentations between the bones of the fetal skull, called **fontanels,** are fibrous cartilage membranes. These areas will become bony (ossify) as the fetus ages, completing the process by the age of 20 to 22 months. The fontanels allow the fetal skull to be compressed slightly during birth and also allow for brain growth during late fetal life. Locate the following fontanels on the fetal skull with the aid of Figure 12.1: anterior or frontal fontanel, posterolateral or mastoid fontanel, anterolateral or sphenoidal fontanel, and posterior or occipital fontanel.

3. Obtain a fetal skeleton and examine it carefully, noting differences between it and an adult skeleton. Pay particular attention to the vertebrae, sternum, frontal bone of the cranium, patellae (kneecaps), os coxae, carpals and tarsals, and rib cage.

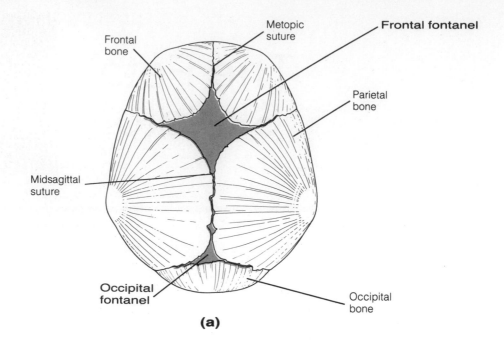

(a)

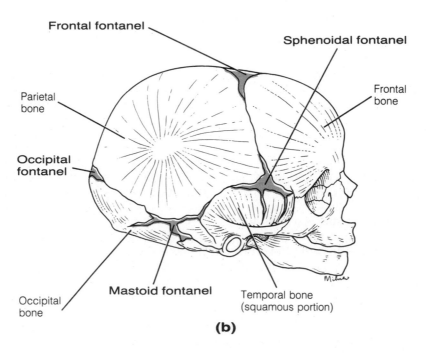

(b)

Figure 12.1

The fetal skull: (a) superior view; (b) lateral view.

Articulations and Body Movements

OBJECTIVES

1. To name the three major categories of joints, and to compare the structure and mobility of the categories.

2. To identify the types of movement seen in diarthrotic joints.

3. To identify some of the causes of joint problems.

4. To define *origin* and *insertion*.

5. To demonstrate or identify the various body movements.

MATERIALS

Articulated skeleton, skull
Diarthrotic beef joint (fresh)
Anatomic chart of joint types (if available)
X-rays of normal and arthritic joints (if available)

With one exception (the hyoid bone), every bone in the body is connected to or forms a joint with at least one other bone. Joints, or articulations, perform two functions for the body: holding the skeletal bones together and allowing the rigid skeletal system some flexibility so that gross body movements can occur.

TYPES OF JOINTS

All joints consist of bony regions separated by cartilage or fibrous connective tissue. The joints are classified as synarthroses, amphiarthroses, and diarthroses, according to the degree of movement they allow.

Synarthroses (Immovable Joints)

The synarthroses (Figure 13.1) are fibrous joints that allow essentially no movement.

The two major types of fibrous joints are sutures and syndesmoses. In **sutures** the irregular edges of the bones interlock and are united by fibrous connective tissue, as in most joints of the skull. In **syndesmoses** the articulating bones are connected by dense fibrous tissue, but the bones do not interlock. The joint at the distal end of the tibia and fibula is an example of a syndesmosis.

Examine a human skull again. Note that adjacent bone surfaces do not actually touch, but are separated by fibrous connective tissue. Also examine a skeleton and anatomic chart of joint types for examples of syndesmoses.

Amphiarthroses (Slightly Movable Joints)

The amphiarthroses (see Figure 13.1) are cartilaginous joints that allow a slight degree of movement. In **symphyses** (*symphyses* means "a growth together") the bones are connected by a broad, flat disk of fibrocartilage. The intervertebral joints and the pubic symphysis of the pelvis are symphyses. In **synchondroses** the bony portions are united by hyaline cartilage. Synchondroses are usually temporary joints, eventually replaced by bone. The best examples of synchondroses are the epiphyseal disks seen in the long bones of growing children and the articulations of the ribs and sternum.

Identify the amphiarthroses on a human skeleton and on an anatomic chart of joint types.

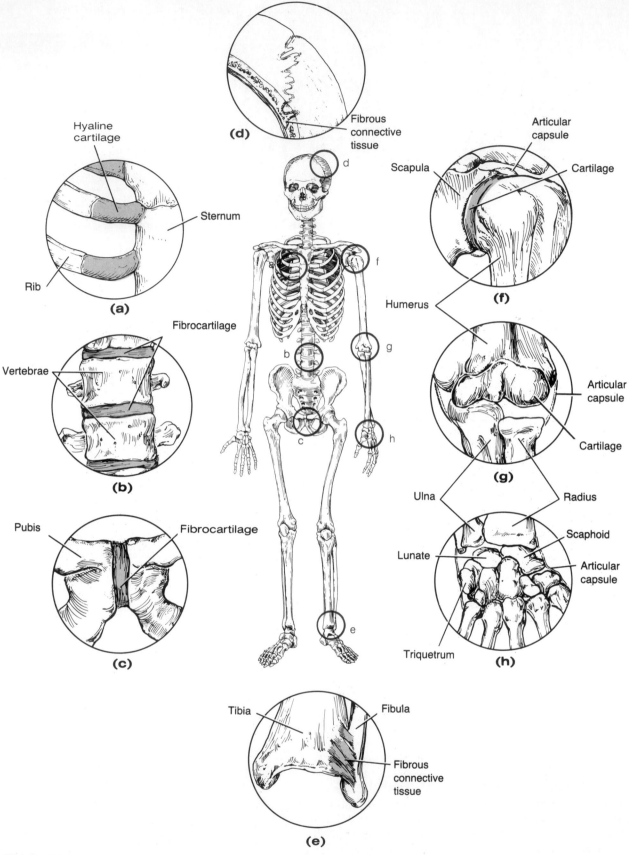

Figure 13.1

Types of joints. Joints to the left of the skeleton are amphiarthrotic (slightly moveable); joints above and below the skeleton are synarthrotic (immovable); joints to the right of the skeleton are diarthrotic (freely moveable). (a) Synchondrosis (costal cartilages of the ribs); (b) symphysis (intervertebral disks of fibrocartilage connecting the vertebrae; (c) symphysis (cartilaginous pubic symphysis connecting the pubic bones anteriorly); (d) suture (fibrous connective tissue connecting the interlocking skull bones); (e) syndesmosis (fibrous connective tissue connecting the distal ends of the tibia and fibula); (f) synovial joint (multiaxial shoulder joint); (g) synovial joint (uniaxial elbow joint); (h) synovial joints (biaxial intercarpal joints of the hand).

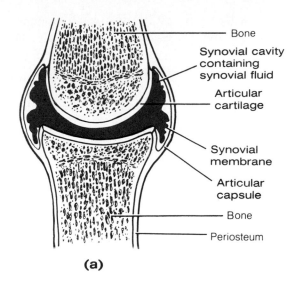

(a)

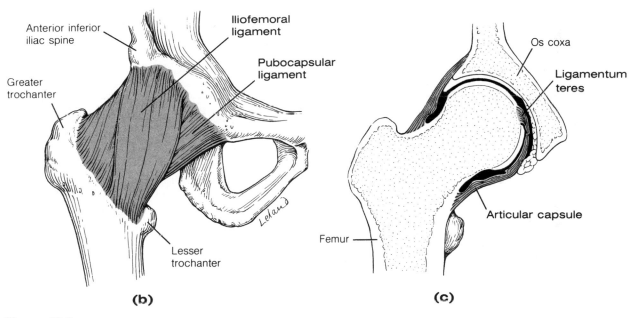

(b)   **(c)**

Figure 13.2

Diarthrotic joints: (a) major components of diarthrotic joints; (b) ligamentous reinforcements of right hip joint, anterior view; (c) right hip joint, frontal section view.

Diarthroses (Freely Movable Joints)

Diarthroses (see Figure 13.1), also called *synovial joints, are characterized* by the greatest degree of freedom or flexibility. This flexibility varies; some diarthroses can move in only one plane, and others can move in several directions (multiaxial movement). Most of the joints in the body are diarthroses.

All diarthrotic joints are characterized by the following structural characteristics (Figure 13.2):

- The joint surfaces are enclosed by a sleeve or capsule of fibrous connective tissue.

- The interior of this capsule is lined with a smooth connective tissue membrane, called *synovial membrane,* which produces a lubricating fluid to reduce friction.

- Articulating surfaces of the bones forming the joint are covered with hyaline (articular) cartilage.

- The fibrous capsule may or may not be reinforced with ligaments and may or may not contain bursae (fluid-filled sacs that reduce friction where tendons cross bone).

- Fibrocartilage pads may or may not be present within the capsule.

Examine a fresh beef joint to identify the structural features of diarthrotic joints.

Because there are so many diarthroses, they have been divided into the following subcategories:

- Gliding: articulating surfaces are flat or slightly curved, allowing sliding movements in one or two planes. Examples are the intercarpal and intertarsal joints and the sternoclavicular joint.
- Hinge: the rounded process of one bone fits into the concave surface of another to allow movement in one plane (uniaxial), usually flexion and extension. Examples are the knee and elbow.
- Pivot: the rounded or conical surface of one bone articulates with a shallow depression or foramen in another bone to allow uniaxial rotation, as in the joint between the atlas and axis (C_1 and C_2).
- Ellipsoidal or condyloid: the oval condyle of one bone fits into an ellipsoidal depression in another bone, allowing biaxial (two-way) movement. The wrist joint and the metacarpal-phalangeal joints (knuckles) are examples.
- Saddle: articulating surfaces are saddle-shaped; the articulating surface of one bone is convex, and the reciprocal surface is concave. Saddle joints, which are biaxial, include the joint between the thumb metacarpal and the trapezium of the wrist.
- Ball and socket: the ball-shaped head of one bone fits into a cuplike depression of another. These are multiaxial joints, allowing movement in all directions and pivotal rotation. Examples are the shoulder and hip joints.

Examine the articulated skeleton, anatomic charts, and yourself to identify the subcategories of diarthrotic joints. Make sure you understand the terms *uniaxial*, *biaxial,* and *multiaxial.*

JOINT DISORDERS

Most of us don't think about our joints until something goes wrong with them. Joint pains and malfunctions may be caused by a variety of things. For example, a hard blow to the knee can cause bursitis, or "water on the knee," due to damage to or inflammation of the bursa or synovial membrane. Slippage of a fibrocartilage pad or the tearing of a ligament may result in a painful condition that persists over a long period, since these poorly vascularized structures heal so slowly.

Sprains and dislocations are other types of joint problems. In a sprain, the ligaments or tendons reinforcing a joint are damaged by excessive stretching or are torn away from the bony attachment. Since both ligaments and tendons are cords of dense connective tissue with a poor blood supply, sprains heal slowly and are quite painful. Dislocations occur when bones are forced out of their normal position in the joint cavity. They are normally accompanied by torn or stressed ligaments and considerable inflammation. The process of returning the bone to its proper position, called reduction, should be done only by a physician. Attempts by the untrained person to "snap the bone back into its socket" are often more harmful than helpful.

Advancing years also take their toll on joints. Weight-bearing joints in particular eventually begin to degenerate. Adhesions (fibrous bands) may form between the surfaces where bones join, and extraneous bone tissue (spurs) may grow along the joint edges. Such degenerative changes lead to the complaint so often heard from the elderly: "My joints are getting so stiff. . . ."

If possible compare an x-ray of an arthritic joint to one of a normal joint.

BODY MOVEMENTS

Every muscle of the body is attached to bone (or other connective tissue structures) at two points—the **origin** (the stationary, immovable, or less movable attachment) and the **insertion** (the movable attachment). Body movement occurs when muscles contract across diarthrotic or synovial joints. When the muscle contracts and its fibers shorten, the insertion moves toward the origin. The type of movement depends on the construction of the joint (uniaxial, biaxial, or multiaxial) and on the placement of the muscle relative to the joint. The most common types of body movements are described below and illustrated in Figure 13.3.

Attempt to demonstrate each movement as you read through the following material:

Flexion: a movement, generally in the sagittal plane, that decreases the angle of the joint and lessens the distance between the two bones. Flexion is typical of hinge joints (bending the knee or elbow), but is also common at ball-and-socket joints (bending forward at the hip).

Extension: a movement that increases the angle of a joint and the distance between two bones or parts of the body (straightening the knee or elbow). Extension is the opposite of flexion. If extension is greater than 180 degrees (bending the trunk backward), it is termed *hyperextension.*

Abduction: movement of a limb away from the midline or median plane of the body, generally

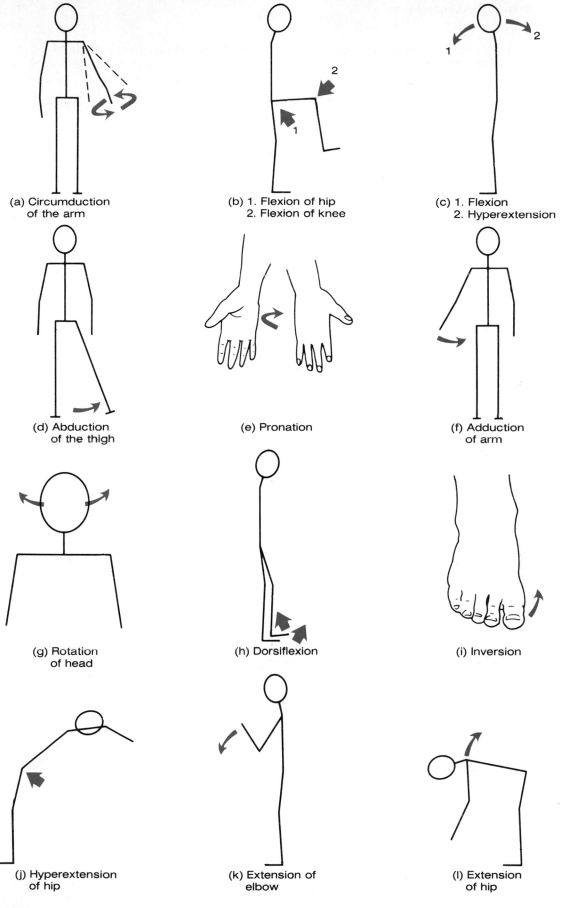

(a) Circumduction of the arm

(b) 1. Flexion of hip
2. Flexion of knee

(c) 1. Flexion
2. Hyperextension

(d) Abduction of the thigh

(e) Pronation

(f) Adduction of arm

(g) Rotation of head

(h) Dorsiflexion

(i) Inversion

(j) Hyperextension of hip

(k) Extension of elbow

(l) Extension of hip

Figure 13.3

Movements occurring at joints of the body.

on the frontal plane, or the fanning movement of fingers or toes when they are spread apart.

Adduction: movement of a limb toward the midline of the body. Adduction is the opposite of abduction.

Rotation: movement of a bone around its longitudinal axis without lateral or medial displacement. Rotation, a common movement of ball-and-socket joints, also describes the movement of the atlas around the odontoid process of the axis.

Circumduction: a combination of flexion, extension, abduction, and adduction commonly observed in ball-and-socket joints like the shoulder. The proximal end of the limb remains stationary, and the distal end moves in a circle. The limb as a whole outlines a cone.

Pronation: movement of the palm of the hand from an anterior or upward-facing position to a posterior or downward-facing position. This action moves the distal end of the radius across the ulna.

Supination: movement of the palm from a posterior position to an anterior position (the anatomic position). Supination is the opposite of pronation. During supination, the radius and ulna are parallel.

The last four terms refer to movements of the foot:

Inversion: a movement that results in the medial turning of the sole of the foot.

Eversion: a movement that results in the lateral turning of the sole of the foot; the opposite of inversion.

Dorsiflexion: a movement of the ankle joint in a dorsal direction (standing on one's heels).

Plantarflexion: a movement of the ankle joint in which the foot is flexed downward (standing on one's toes or pointing the toes).

UNIT 7 THE MUSCULAR SYSTEM

Microscopic Anatomy, Organization, and Classification of Skeletal Muscle

EXERCISE
14

OBJECTIVES

1. To visually identify the three types of muscle tissue found in the body (by microscopic inspection or in a diagram or photomicrograph).

2. To describe similarities and differences in the three types of muscle tissue and to note where they are found.

3. To describe the structure of skeletal muscle from gross to microscopic levels.

4. To define and explain the role of the following:
actin	perimysium
myosin	aponeurosis
fiber	tendon
myofibril	endomysium
myofilament	epimysium

5. To describe the structure of a myoneural junction and explain its role in muscle function.

6. To define: *agonist* (prime mover), *antagonist, synergist,* and *fixator.*

7. To cite criteria used in naming muscles.

MATERIALS

Three-dimensional model of skeletal muscle cells (if available)

Forceps

Dissecting needles

Microscope slides and cover slips

Frog Ringer's solution in dropper bottles

Pithed frog, thigh muscle exposed

Compound microscope

Histologic slides of skeletal muscle (longitudinal and cross section), skeletal muscle showing myoneural junctions, smooth muscle and cardiac muscle (all longitudinal section)

Three-dimensional models of smooth and cardiac muscle (if available)

Three-dimensional model of skeletal muscle showing myoneural junction (if available)

Related film: *Muscles* (color, sound, 16 mm, 30 minutes, CRM Educational Films)

The bulk of the body's muscle is called skeletal muscle because it is attached to the skeleton (or underlying connective tissue). Skeletal muscle influences body contours and shape, allows you to smile and frown, provides a means of locomotion, and enables you to manipulate the environment. The balance of the body's muscle—smooth and cardiac muscle—is the major component of the walls of hollow organs and the heart, where it is involved with the transport of materials within the body.

Each of the three muscle types has a structure and function uniquely suited to its task in the body. But since the term *muscular system* applies specifically to skeletal muscle, the primary objective of this exercise is to investigate the structure and function of skeletal muscle. The structure of the other two muscle types is considered less rigorously as a means of comparison.

Skeletal muscle is also known as voluntary muscle (because it is under our conscious control) and as striated muscle (because it appears to be striped). As you might guess from both of these alternative names, skeletal muscle has some very special characteristics. Thus an investigation of skeletal muscle should begin at the cellular level.

THE CELLS OF SKELETAL MUSCLE

Skeletal muscle is composed of relatively large, long cylindrical cells ranging from 10 to 100 μm in diameter and up to 6 cm in length. However, the cells of large, hard-working muscles like the antigravity muscles of the hip are extremely coarse, ranging up to 25 cm in length, and can be seen with the naked eye.

Skeletal muscle cells (Figure 14.1(a)) are multinucleate: multiple oval nuclei can be seen just beneath the cell membrane (called the sarcolemma in these cells). The nuclei are pushed peripherally by the longitudinally arranged **myofibrils,** which nearly fill the sarcoplasm (Figure 14.1(b)). Alternating light (I) and dark (A) bands along the length of the perfectly aligned myofibrils give the muscle fiber as a whole its striped appearance.

Electron microscope studies have revealed that the myofibrils are made up of even smaller thread-like structures called **myofilaments** (Figure 14.1 (b) and (d)). The myofilaments are composed of two varieties of contractile proteins—**actin** and **myosin** —which slide past each other during muscle activity to bring about shortening or contraction of the muscle cells. It is the highly specific arrangement of the myofilaments within the myofibrils that is responsible for the banding pattern in skeletal muscle. The actual contractile units of muscle, called **sarcomeres,** extend from the middle of one I band (its Z line) to the middle of the next along the length of the myofibrils (see Figure 14.1 (d)).

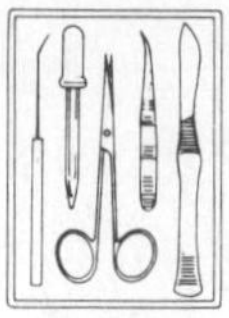

1. Look at the three-dimensional model of skeletal muscle cells, noting the relative shape and size of the cells. Identify the nuclei, myofibrils, and light and dark bands.

2. Obtain forceps, two dissecting needles, slide and cover slip, and a dropper bottle of Ringer's solution. With forceps, remove a very small piece of muscle from the thigh of a pithed frog. Place the tissue on a clean microscope slide and add a drop of the Ringer's solution. Pull the muscle fibers apart with the dissecting needles (tease them) until you have a fluffy looking mass of tissue. Cover the teased tissue with a cover slip and observe under the high-power lens of a microscope. Look for the banding pattern. Observe again under high power. Regulate the light carefully to obtain the highest possible contrast.

3. Compare your observations with what can be seen with professionally prepared muscle tissue. Obtain a slide of skeletal muscle (longitudinal section), and view it under high power. From your observations, draw a small section of a muscle fiber in the space provided here. *Label* the nuclei, cell membrane, and A and I bands.

What structural details become apparent with the

prepared slide? _____

_____

Figure 14.1

Structure of skeletal muscle cells: (a) muscle fibers, longitudinal and transverse views, and photomicrograph, (b) portion of a skeletal muscle cell (one myofibril has been extended and disrupted to indicate its myofilament composition); (c) one sarcomere of the myofibril; (d) banding pattern in the sarcomere. (From H.E. Huxley, "The Contraction of Muscle." Copyright   November 1958 by Scientific American, Inc. All rights reserved.)

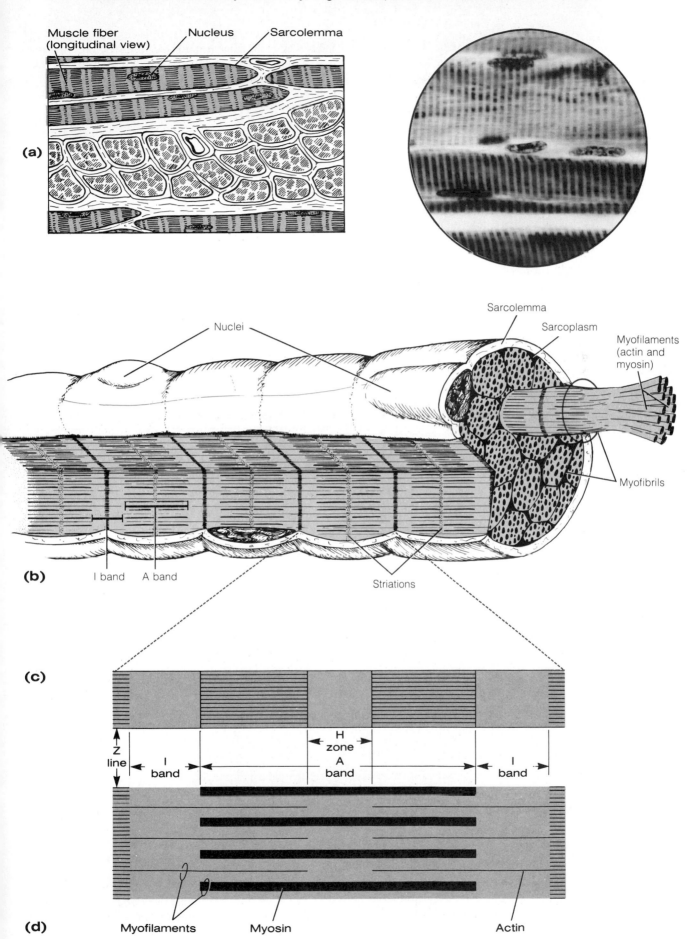

(a)

Muscle fiber (longitudinal view) Nucleus Sarcolemma

(b)

Nuclei Sarcolemma Sarcoplasm Myofilaments (actin and myosin)

Myofibrils

I band A band Striations

(c)

Z line I band H zone A band I band

(d)

Myofilaments Myosin Actin

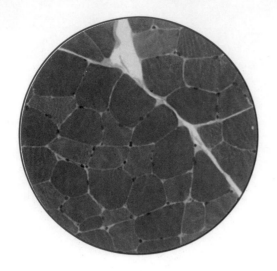

we refer to as skeletal muscles (Figure 14.2). Each muscle fiber is enclosed in a delicate, connective tissue sheath called an **endomysium.** Several sheathed muscle fibers are wrapped by a collagenic membrane called a **perimysium** to form a bundle of fibers called a **fascicle,** or **fasciculus.** A large number of fascicles are bound together by a substantially coarser "overcoat" of connective tissue wrappings called an **epimysium,** or **deep fascia,** which sheathes the entire muscle. These epimysia blend into the strong cordlike **tendons** or sheetlike **aponeuroses,** which attach muscles to each other or indirectly to bones.

The tendons perform several functions, two of the most important being to provide durability and to conserve space. Since tendons are tough collagenic connective tissue, they can span rough bony prominences, which would destroy the more delicate muscle tissues. And because of their relatively small size, more tendons than fleshy muscles can pass over a joint.

In addition to supporting and binding the muscle fibers and providing strength to the muscle as a whole, the connective tissue wrappings provide a route for the entry and exit of nerves and blood vessels that serve the muscle fibers. The larger, more powerful muscles have more connective tissue than the muscles involved in fine or delicate movements have. As we age, the amount of muscle fiber decreases and the amount of connective tissue increases; thus the skeletal muscles gradually become more sinewy, or "stringier."

 Obtain a slide showing a cross section of skeletal muscle tissue. Identify and sketch the muscle fibers, endomysium, perimysium, epimysium, and nuclei (if visible).

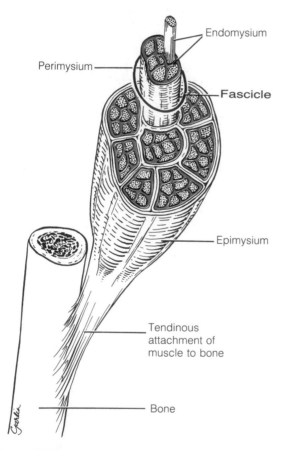

Endomysium

Perimysium

Fascicle

Epimysium

Tendinous attachment of muscle to bone

Bone

Figure 14.2

Connective tissue coverings of skeletal muscle.

ORGANIZATION OF SKELETAL MUSCLE CELLS INTO MUSCLES

Muscle fibers are soft and surprisingly fragile. Thus thousands of muscle fibers are bundled together with connective tissue to form the organs

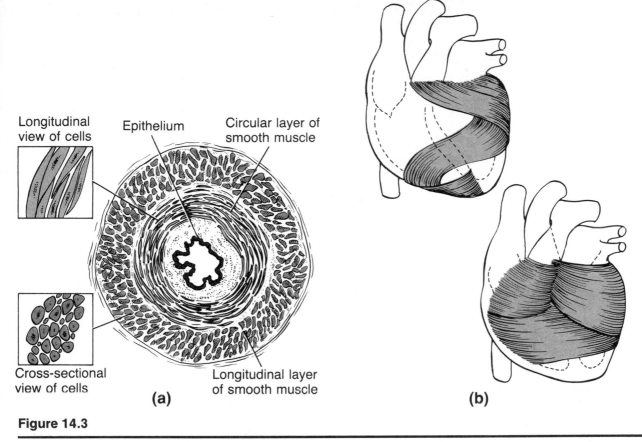

Figure 14.3

Arrangement of smooth and cardiac muscle fibers. (a) Diagrammatic representation of a cross section of a sperm duct, vas deferens; smooth muscle cells are shown in both longitudinal and cross-sectional views. (b) Longitudinal view of the heart chambers, showing the spiral arrangement of the cardiac muscle fibers.

Many people think of muscles in terms of the fusiform shape, that is, an enlarged "belly" of longitudinally arranged muscle fibers that converges at either end in a tendon. However, muscles vary considerably in the arrangement of their fibers, according to their location in the body. In some, fibers are arranged in a fan shape; others have parallel, circular, or featherlike arrangements.

SMOOTH MUSCLE

Smooth muscle is considered involuntary, which means that it cannot consciously be controlled. Smooth muscle is found in the walls of the blood vessels and internal organs (stomach, small and large intestine, bladder, and so on), where it functions to propel substances along a predetermined tract.

Smooth muscle cells are arranged in sheets or layers. Most often, there are two layers oriented at right angles, one running circularly and the other longitudinally (Figure 14.3 (a)). When the two layers alternately contract, they change the size and shape of the organ, allowing it to carry out a particular body function.

Although smooth muscle is a major structural component of the hollow organs of the body, its function is difficult to study and easily neglected in the laboratory. The specific physiologic characteristics of smooth muscle are dealt with in Exercise 39.

1. Observe a three-dimensional model of smooth muscle. Note the fusiform shape of the cells and the single central nucleus.

2. Obtain a slide of smooth muscle tissue in a longitudinal section and observe the muscle layers under high power. Note how the tapering muscle cells dovetail.

How many nuclei are present in each cell? _____

Are there any obvious striations? _____

Draw several smooth muscle cells and label the nucleus, sarcoplasm, and sarcolemma.

CARDIAC MUSCLE

Cardiac muscle is found in only one place—the heart. The heart acts as a vascular pump, propelling blood to all tissues of the body; cardiac muscle is thus very important to life. Cardiac muscle is involuntary, thus ensuring a constant blood supply.

The cardiac cells, only sparingly invested in connective tissue, are arranged in spiral or figure-8 shaped bundles (Figure 14.3 (b)). When the heart contracts, its internal chambers become smaller (or are temporarily obliterated), forcing the blood into the large arteries leaving the heart.

 1. Observe the three-dimensional model of cardiac muscle, noting its branching cells and the areas where the cells interdigitate, the **intercalated disks.** These two structural features provide a continuity to cardiac muscle not seen in other muscle tissues and allow close coordination of heart activity.

2. Note the similarities and differences between cardiac muscle and skeletal muscle.

3. Obtain and observe a longitudinal section of cardiac muscle under high power, and draw a small section of the tissue in the space provided in the right column. Label the nucleus, striations, intercalated disks, and sarcolemma.

THE MYONEURAL JUNCTION

Cardiac muscle has a special "pacemaker" that initiates the activity of heart muscle, and smooth muscle contraction may be elicited by "pacemaker cells" sensitive to local chemical stimuli. But all voluntary muscle cells are always stimulated by nerve impulses via motor neurons. The junction between a nerve fiber (axon) and a muscle fiber is called a **myoneural** or neuromuscular **junction** (Figure 14.4).

Each motor axon breaks up into many branches before it terminates, and each of these branches stimulates a muscle cell. Thus a single neuron may stimulate many muscle fibers. Together, a neuron and all the muscle cells it stimulates make up the functional structure called the **motor unit.**

Each motor axon terminates in structures called **end plates,** each of which has numerous projections called **sole feet.** The neuron and muscle fiber membranes, close as they are, do not actually touch. They are separated by a small gap of 300 to 500 Å. called the **synaptic cleft** or **synaptic gutter.**

Within the sole foot are many mitochondria and numerous small vesicles containing a neurotransmitter called acetylcholine. When a nerve impulse reaches the end plate, a few of these vesicles liberate their contents into the synaptic cleft. The acetylcholine rapidly diffuses across the junction and combines with the receptors on the sarcolemma. If sufficient acetylcholine has been released, a transient change in the permeability of the sarcolemma briefly allows more sodium ions to diffuse into the muscle fiber. The sudden influx of sodium ions results in the depolarization of the sarcolemma and subsequent contraction of the muscle fiber.

After repolarization of its membrane, the muscle fiber can be stimulated to contract again. Even while the sarcolemma is being depolarized, the acetylcholine is being inactivated by an enzyme on the sarcolemma close to the receptor sites. Therefore, a single neuron impulse produces only one muscle fiber contraction, and sustained depolariza-

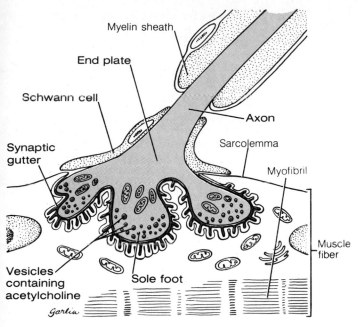

Myelin sheath

End plate

Schwann cell

Axon

Synaptic gutter

Sarcolemma

Myofibril

Muscle fiber

Vesicles containing acetylcholine

Sole foot

Garlia

Figure 14.4

The myoneural junction.

tion of the sarcolemma in the absence of additional nerve impulses is prevented.

1. If possible, examine a three-dimensional model of skeletal muscle cells that illustrates the myoneural junction. Identify the structures described above.

2. Obtain a slide of skeletal muscle stained to show the end plates of a motor unit. Examine the slide under high power to identify the axonal fibers extending leashlike to the muscle cells. Follow one of the axonal fibers to its terminus to identify the oval-shaped motor end plates. Sketch a small section in the space provided, labeling the motor axon, its terminal branches, motor end plates, and muscle fibers.

CLASSIFICATION OF SKELETAL MUSCLES

Naming Skeletal Muscles

Remembering the names of the skeletal muscles is a monumental task, but certain clues help. Muscles are named on the basis of the following criteria:

- Direction of muscle fibers: some muscles are named in reference to some imaginary line, usually the midline of the body or the longitudinal axis of a limb bone. The fibers of a muscle that includes the term *rectus* (straight) in its name run parallel to that imaginary line. For example, the rectus abdominus is the straight muscle of the abdomen. Likewise, the terms *transverse* and *oblique* indicate that the muscle fibers run at right angles and obliquely (respectively) to the imaginary line.

- Relative size of the muscle: terms such as *maximus* (largest), *minimus* (smallest), *longus* (long), and *brevis* (short) are often used in the names of muscles—as in gluteus maximus and gluteus minimus.

- Location of the muscle: some muscles are named according to the bone they are associated with. For example, the temporalis muscle overlies the temporal bone.

- Number of origins: when the term *biceps, triceps,* or *quadriceps* forms part of a muscle name, you can generally assume that the muscle has two, three, or four origins (respectively). For example, the biceps muscle of the arm has two heads or origins.

- Location of the muscle's origin and insertion: for example, the sternocleidomastoid muscle has its origin on the sternum (*sterno*) and clavicle *(cleido)* and inserts on the mastoid process of the temporal bone.

- Shape of the muscle: for example, the deltoid muscle is roughly triangular (*deltoid* means "triangle"), and the trapezius muscle resembles a trapezoid.

- Action of the muscle: for example, all the adductor muscles of the anterior thigh bring about its adduction, and all the extensor muscles of the wrist extend the wrist.

Types of Muscles

Most often, body movements are not a result of the contraction of a single muscle but are the coordinated action of several muscles acting together. Muscles that are primarily responsible for producing a particular movement are called **prime movers,** or **agonists.**

Muscles that oppose or reverse a movement are called **antagonists.** When a prime mover is active, the fibers of the antagonist are stretched and in the relaxed state. The antagonist can also regulate the prime mover by providing some resistance, to prevent overshoot or to stop its action.

It should be noted that antagonists can be prime movers in their own right; for example, the biceps of the arm (a prime mover of elbow flexion) is antagonized by the triceps (a prime mover of elbow extension).

Synergists contribute substantially to the action of agonists by reducing undesirable or unnecessary movement. When a muscle crosses two or more joints, its contraction would cause movement in all the joints spanned unless synergists were there to stabilize them. For example, the finger-flexor muscles cross both the wrist and the phalangeal joints. You can make a fist without bending your wrist only because synergist muscles stabilize the wrist joint and allow the prime mover to exert its force at the finger joints.

Fixators or fixation muscles are specialized synergists. They immobilize the origin of a prime mover so that all the tension is exerted at the insertion. You may recall from Exercise 11 that the scapula is held to the axial skeleton medially only by muscles and is quite freely movable. Many of the muscles that move the arm originate on the scapula, and if they are to be effective, movement of the scapula must be minimized. The fixator muscles in that region, which run from the axial skeleton to the scapula, stabilize the scapula. Additionally, muscles that help maintain posture are fixators.

Gross Anatomy of the Muscular System

OBJECTIVES

1. To name and locate the major muscles of the human body (on a torso model, laboratory chart, or diagram) and state the action of each.

2. To explain how muscle actions are related to their location.

3. To name muscle origins and insertions as required by the instructor.

4. To identify antagonists of the major prime movers.

5. To name and locate muscles on a dissection animal.

6. To recognize similarities and differences between human and pig musculature.

MATERIALS

Disposable gloves or protective skin cream
Preserved and injected pig (one for every two to four students)
Dissecting trays and instruments
Name tag and large plastic bag
Paper towels
Embalming fluid (formalin)
Human torso model or large anatomic chart showing human musculature

IDENTIFICATION OF HUMAN MUSCLES

Muscles of the Head and Neck

The muscles of the head serve many specific functions. For instance, the muscles of facial expression differ from most skeletal muscles because they insert into the skin (or other muscles) rather than into bone. As a result, they move the facial skin, allowing a wide range of emotions to be shown on the face. Other muscles of the head are the muscles of mastication, which manipulate the mandible during chewing, and the six muscles of the orbit, which aim the eye. (The orbital muscles are studied in conjunction with the anatomy of the eye in Exercise 24.) Neck muscles are primarily concerned with the movement of the head and shoulder girdle. The head and neck muscles are discussed in Tables 15.1 and 15.2 and shown in Figures 15.1 and 15.2.

Carefully read the description of each muscle and attempt to visualize what happens when the muscle contracts. Once you have read through the tables and have identified the head and neck muscles in Figures 15.1 and 15.2, use a torso model or an anatomic chart to again identify as many of these muscles as possible. Then carry out the following palpations on yourself:

• To demonstrate how the temporalis works, clench your teeth. The masseter can also be palpated at this time at the angle of the jaw.

Muscles of the Trunk

The trunk musculature includes muscles that move the vertebral column; anterior thorax muscles that act to move ribs, head, and arms; and muscles of the abdominal wall that play a role in the movement of the vertebral column but more importantly form the "natural girdle," or the major portion, of the abdominal body wall.

 The trunk muscles are described in Tables 15.3 and 15.4 and shown in Figures 15.3 and 15.4. As before, identify the muscles in the figure as you read the tabular descriptions and then identify them on the torso or laboratory chart. When you have completed this study, work with a partner to demonstrate the operation of the following muscles. One of you can demonstrate the movement (the following steps are addressed to this partner); the other can supply the necessary resistance and palpate the muscle being tested.

1. Start by fully abducting the arm and extending the elbow. Now try to adduct the arm against resistance. You are exercising the latissimus dorsi.

2. To observe the deltoid, attempt to abduct your shoulder against resistance. Now attempt to elevate your shoulder against resistance; you are contracting the upper portion of the trapezius.

3. The pectoralis major comes into play when you press your hands together at chest level with your elbows widely abducted.

Arm Muscles

The arm muscles fall into three groups: those that move the upper arm, those causing movement at the elbow, and those effecting movements of the wrist and hand.

The muscles that cross the shoulder joint to insert on the humerus and cause movement of the upper arm are primarily trunk muscles (subscapularis, supraspinatus and infraspinatus, deltoid, and so on) that originate on the axial skeleton or shoulder girdle. These muscles are included with the trunk muscles.

The second group of arm muscles, which cross the elbow joint to bring about movement of the forearm, consists of the muscles forming the musculature of the humerus. These muscles arise primarily from the humerus and insert in forearm bones. They are responsible for flexion, extension, pronation, and supination. The origins, insertions, and actions of these muscles are summarized in Table 15.5 and the muscles are shown in Figure 15.5.

The third group composes the musculature of the forearm. For the most part, these muscles insert on the digits, producing movements at the wrist and fingers. In general, muscles acting on the wrist and hand can more easily be identified if their insertion tendons are located first. These muscles are described in Table 15.6 and illustrated in Figure 15.6.

First study the tables and figures, then see if you can identify these muscles on a torso model or anatomic chart.

Complete this portion of the exercise with palpation demonstrations as outlined next.

- To observe the biceps brachii, attempt to flex your forearm (hand supinated) against resistance. The tendons can also be felt.
- If you acutely flex your elbow and then try to extend it against resistance, you can demonstrate the action of your triceps brachii.

Muscles of the Lower Extremity

Muscles that act on the lower extremity cause movement at the hip, knee, and foot joints. Since the human pelvic girdle is composed of heavy fused bones that allow little movement, no special group of muscles is necessary to stabilize it. This is unlike the shoulder girdle, where many muscles (mainly trunk muscles) are necessary to stabilize the scapulae.

Muscles acting on the thigh (femur) cause various movements at the multiaxial hip joint (flexion, extension, rotation, abduction, and adduction). These include the iliopsoas (described in Table 15.7), the adductor group, and other muscles summarized in Table 15.8 and illustrated in Figures 15.7 and 15.8.

Muscles acting on the leg form the major musculature of the thigh. (Anatomically the term *leg* refers only to that portion between the knee and the ankle.) The thigh muscles cross the knee to allow its flexion and extension. They include the hamstrings and the quadriceps and are described in Table 15.9 and illustrated in Figures 15.7 and 15.8. Since these muscles also have attachments on the pelvic girdle, they can cause movement at the hip joint.

The muscles originating on the leg and acting on the foot and toes are described in Table 15.10 and shown in Figures 15.9 and 15.10. Identify the muscles as instructed previously. Complete this exercise by performing the following palpation demonstrations.

- You can demonstrate the quadriceps femoris by trying to extend the knee against resistance. Note how the patellar tendon reacts. The biceps femoris comes into play when you flex your knee against resistance.
- Now stand on your toes. Have your partner palpate the lateral and medial heads of the gastrocnemius and follow it to its insertion in the Achilles tendon.

(Text continues on p. 120.)

TABLE 15.1 Major Muscles of Human Head (see Figure 15.1)

Muscle	Comments	Origin	Insertion	Action
Facial expression				
Epicranius: frontalis and occipitalis	Bipartite muscle consisting of frontalis and occipitalis, which covers dome of skull	Cranial aponeurosis (frontalis); occipital bone (occipitalis)	Skin of eyebrows and root of nose (frontalis) and cranial aponeurosis (occipitalis)	With aponeurosis fixed, frontalis raises eyebrows; occipitalis fixes aponeurosis and pulls scalp posteriorly
Orbicularis oculi	Sphincter muscle of eyelids	Frontal and maxillary bones and ligaments around orbit	Encircles orbit and inserts in tissue of eyelid	Various parts can be activated individually; closes eyes, produces blinking, squinting, and draws eyebrows downward
Orbicularis oris	Multilayered sphincter muscle of lips with fibers that run in many different directions	Arises indirectly from maxilla and mandible; fibers blended with fibers of other oral muscles	Encircles mouth; inserts into muscle and skin at angles of mouth	Closes mouth; purses and protrudes lips (kissing muscle)
Corrugator supercilii	Small muscle; activity associated with that of orbicularis oculi	Arch of frontal bone above nasal bone	Skin of eyebrow	Draws eyebrows medially; wrinkles skin of forehead vertically
Levator labii superioris	Thin muscle between orbicularis oris and inferior eye margin	By three heads from the zygomatic, infraorbital margin of maxilla, root of nose	Skin and muscle of upper lip and border of nostril	Raises and furrows upper lip; flares nostril (as in disgust)
Zygomaticus— major and minor	Extends diagonally from corner of mouth to cheekbone	Zygomatic bone	Skin and muscle at corner of mouth	Raises lateral corners of mouth upward (smiling muscle)
Depressor labii inferioris	Small muscle from lower lip to jawbone	Body of mandible lateral to its midline	Skin and muscle of lower lip	Draws lower lip downward
Depressor anguli oris	Small muscle lateral to depressor labii inferioris	Body of mandible below incisors	Skin and muscle at angle of mouth below insertion of zygomaticus	Zygomaticus antagonist; draws corners of mouth downward and laterally
Mentalis	One of muscle pair forming V-shaped muscle mass on chin	Incisor fossa of mandible	Skin of chin	Protrudes lower lip; wrinkles chin
Buccinator	Horizontal cheek muscle; principal muscle of cheek; deep to the masseter	Molar region of maxilla and mandible	Orbicularis oris	Draws corner of mouth laterally; compresses cheek (as in whistling); holds food between teeth during chewing
Mastication				
Masseter	Extends across jawbone; can be palpated on forcible closure of jaws	Zygomatic process and arch	Angle and ramus of mandible	Closes jaw and elevates mandible
Temporalis	Fan-shaped muscle over temporal bone	Temporal fossa	Coronoid process of mandible	Closes jaw and elevates mandible
Buccinator	(See muscles of facial expression)			

TABLE 15.1 continued   Major Muscles of Human Head (see Figure 15.1)

Muscle	Comments	Origin	Insertion	Action
Pterygoid—medial	Runs between internal (medial) surface of mandible (thus largely concealed by that bone and bones of face)	Sphenoid, palatine, and maxillary bones	Medial surface of mandibular ramus and angle	Synergist of temporalis and masseter; closes and elevates mandible; in conjunction with lateral pterygoid, aids in grinding movements of jaw
Pterygoid—lateral	Superior to medial pterygoid	Greater wing of sphenoid bone	Mandibular condyle	Protracts jaw (moves it forward); in conjunction with medial pterygoid, aids in grinding movements of jaw

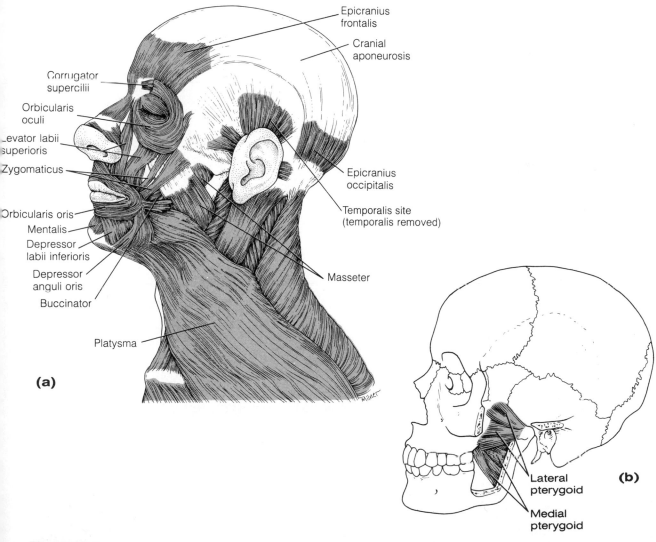

Figure 15.1

Muscles of the human head. (a) Superficial muscles of the human face and neck; (b) superficial muscles removed and zygomatic arch and mandible cut away to reveal the deep pterygoid muscles.

TABLE 15.2 Anterior Muscles of the Human Neck (see Figure 15.2)

Muscle	Comments	Origin	Insertion	Action
Superficial				
Platysma	Unpaired muscle: thin, sheetlike superficial neck muscle, not strictly a head muscle but plays role in facial expression (see Fig. 15.1)	Fascia of chest (over pectoral muscles and deltoid)	Lower margin of mandible, skin, and muscle at corner of mouth	Depresses mandible; pulls lower lip back and down; i.e., produces downward sag of the mouth
Sternocleidomastoid	Two-headed muscle located deep to platysma on anterolateral surface of neck; fleshy parts on either side form anterior triangle of neck	Manubrium of sternum and medial portion of clavicle	Mastoid process of temporal bone	Simultaneous contraction of both muscles of pair causes flexion of neck forward, generally against resistance (as when lying on the back); acting independently, rotate head toward shoulder on opposite side
Scalene—anterior, medial, and posterior	Located more on lateral than anterior neck; deep to platysma	Transverse processes of C_1 to C_5	Posterolaterally on first two ribs	Flex and slightly rotate neck; elevate first two ribs (aid in inspiration)
Deep				
Digastric	Consists of two bellies united by an intermediate tendon; assumes a V-shaped configuration under chin; attached by a connective tissue loop to hyoid bone	Lower margin of mandible (anterior belly) and mastoid process (posterior belly)	Hyoid bone	Acting in concert, elevate hyoid bone and may depress mandible
Mylohyoid	Just deep to digastric; forms floor of mouth	Medial surface of mandible	Hyoid bone	Elevate hyoid bone and base of tongue
Sternohyoid	Runs most medially along neck; straplike	Posterior surface of manubrium	Lower margin of body of hyoid bone	Acting with sternothyroid and omohyoid (all below hyoid bone), depresses larynx and hyoid bone if mandible is fixed; may also flex skull
Sternothyroid	Lateral to sternohyoid; straplike	Manubrium and medial end of clavicle	Thyroid cartilage of larynx	(see sternohyoid, above)
Omohyoid	Straplike with two bellies; lateral to sternohyoid	Superior surface of scapula	Hyoid bone	(see sternohyoid, above)

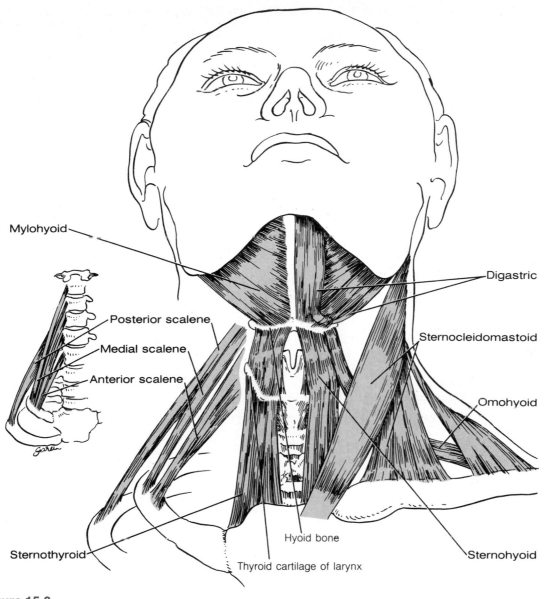

Figure 15.2

Muscles of the human neck. Superficial muscles are shown on the right; deeper muscles are shown on the left.

TABLE 15.3 Anterior Muscles of Human Thorax, Shoulder, and Abdominal Wall (see Figure 15.3)

Muscle	Comments	Origin	Insertion	Action
Pectoralis major	Large fan-shaped muscle covering upper portion of chest	Clavicle, sternum, and cartilage of first six ribs and aponeurosis of external oblique muscle	Fibers converge to insert by short tendon into greater tubercle of humerus	Adducts, medially rotates arm; with arm fixed, pulls chest upward (thus also acts in forced inspiration)
Serratus anterior	Deep and superficial portions; beneath and inferior to pectoral muscles on lateral rib cage	Lateral aspect of first to eighth (or ninth) ribs	Vertebral border of anterior surface of scapula	Moves scapula forward and down toward chest wall
Deltoid	Fleshy triangular muscle forming shoulder muscle mass	Lateral third of clavicle; acromion process and spine of scapula	Deltoid tubercle of humerus	Acting as a whole, abducts arm; when only specific fibers are active, can aid in flexion, extension, and rotation of humerus
Pectoralis minor	Flat, thin muscle directly beneath and obscured by pectoralis major	Upper border of third, fourth, and fifth ribs, near their costal cartilages	Corocoid process of scapula	With ribs fixed, draws scapula forward and downward; with scapula fixed, draws rib cage upward
Intercostals—external	11 pairs lie between ribs; fibers run obliquely toward sternum	Inferior border of rib above (not shown in figure)	Superior border of rib below	Pulls ribs toward one another to elevate rib cage; aids in inspiration
Intercostals—internal	11 pairs lie between ribs; fibers run at right angles to those of external intercostals	Inferior border of rib above (not shown in figure)	Superior border of rib below	Draws ribs together to depress rib cage; aids in expiration; antagonistic to external intercostals
Rectus abdominis	Medial superficial muscle, extends from pubis to rib cage; ensheathed by fascia of oblique muscles; segmented	Pubic crest	Xiphoid process and costal cartilages of fifth through seventh ribs	Flexes vertebral column; increases abdominal pressure; depresses ribs
External oblique	Most superficial lateral muscles; fibers run downward and medially; ensheathed by an aponeurosis	Anterior surface of last eight ribs	Linea alba* and iliac crest	(see Rectus abdominis, above)
Internal oblique	Fibers run at right angles to those of external oblique, which it underlies	Lumbodorsal fascia, iliac crest, and inguinal ligament	Linea alba, pubic crest, and costal cartilages of last three ribs	(see Rectus abdominis, above)
Transversus abdominus	Innermost muscle of abdominal wall; fibers run horizontally	Inguinal ligament, iliac crest, and cartilages of last five or six ribs	Linea alba and pubic crest	Compresses abdominal contents

*The linea alba ("white line") is a narrow, tendinous sheath that runs along the middle of the abdomen from the sternum to the pubic symphysis. It is formed by the fusion of the aponeurosis of the external oblique and transversus muscles.

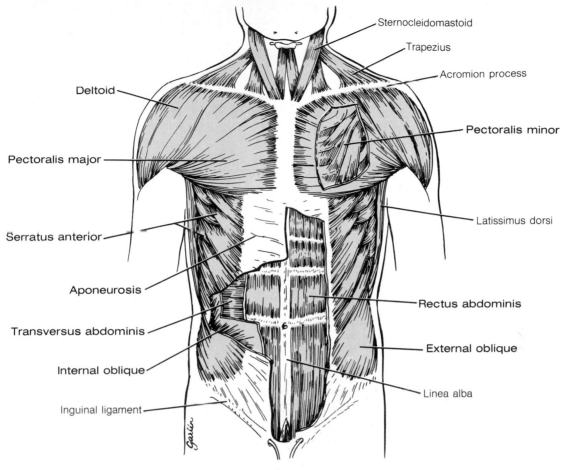

Figure 15.3

Muscles of the human trunk and shoulder (anterior). Portions of the superficial muscles are cut away to reveal the deeper muscles.

TABLE 15.4 Posterior Muscles of Human Trunk (see Figure 15.4)

Muscle	Comments	Origin	Insertion	Action
Trapezius	Most superficial muscle of posterior neck and trunk; very broad origin and insertion	Occipital bone ligamentum nuchae and spines of C_7 and all thoracic vertebrae	Acromion and spinous process of scapula; lateral third of clavicle	Extends head; adducts scapula and stabilizes it; upper fibers elevate scapula; lower fibers depress it
Latissimus dorsi	Broad flat muscle of lower back (lumbar region); extensive superficial origins	Indirect attachment to spinous processes of lower six thoracic vertebrae, lumbar vertebrae, sacrum, and iliac crest	Floor of intertubercular groove of humerus	Extends and medially rotates arm; adducts arm; depresses scapula; brings arm down in power stroke, as in striking a blow
Infraspinatus	Partially covered by deltoid and trapezius	Infraspinous fossa of scapula	Greater tubercle of humerus	Lateral rotation of humerus
Teres minor	Small muscle inferior to infraspinatus	Lateral margin of scapula	Greater tuberosity of humerus	Lateral rotation of the humerus
Teres major	Located inferiorly to teres minor	Posterior surface at inferior angle of scapula	Crest of lesser tubercle of humerus	Extends, medially rotates, and adducts humerus
Supraspinatus	Obscured by trapezius and deltoid	Supraspinous fossa of scapula	Greater tubercle of humerus	Initiates abduction of humerus
Levator scapulae	Located at back and side of neck, deep to trapezius	Transverse processes of C_1 through C_4	Superior vertebral border of scapula	Acting individually with fixed scapula, flexes neck to the same side; raises and adducts scapula
Rhomboids—major and minor	Beneath trapezius and posterior to levator scapulae; run from vertebral column to scapula	Spinous processes of C_7 and T_1 through T_4	Vertebral border of scapula	Act together to pull scapula medially (retraction) and to elevate it
Capitis—splenius, semispinalis, and longissimus	Deep muscles of posterior neck; splenius overlies other two capitis muscles and is found deep to levator scapulae	Ligamentum nuchae and spinous processes of C_7 through T_3 (splenius); transverse processes of C_7 through T_{12} (semispinalis); and transverse processes of T_1 through T_3 and articular processes of C_4 through C_7 (longissimus)	Mastoid process of temporal bone and occipital bone (splenius); occipital bone (semispinalis); and mastoid process (longissimus)	Acting together, extend or hyperextend head; acting independently (right versus left), abduct and rotate head
Erector spinae	A long composite muscle composed of iliocostalis (most lateral), longissimus, and spinalis (most medial)	Sacrum, iliac crest, lumbar vertebrae, and lower two thoracic vertebrae	Ribs and transverse processes of vertebrae approximately six segments above origin	All act to extend and abduct vertebral column; fibers of the longissimus also extend head

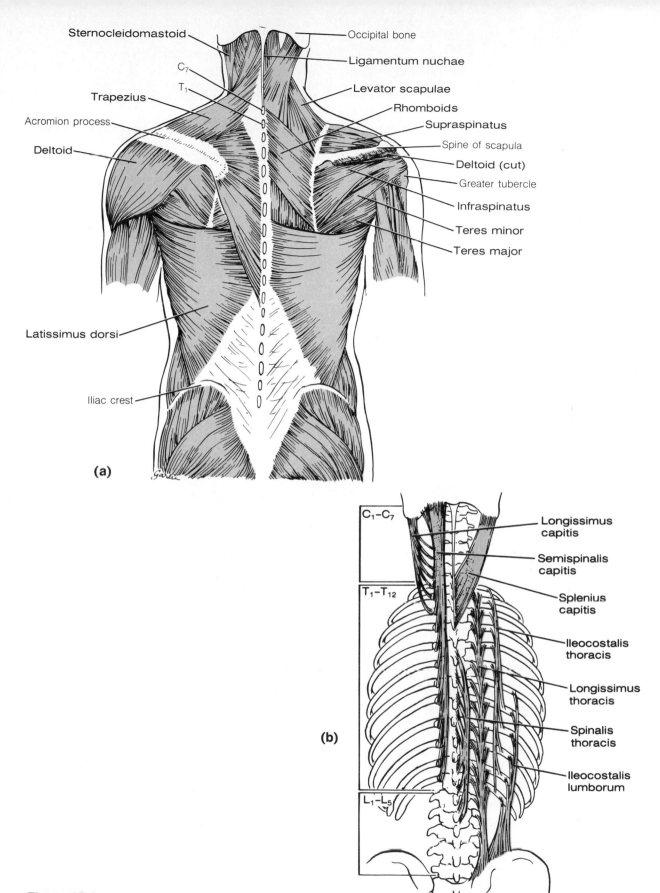

Sternocleidomastoid
Occipital bone
Ligamentum nuchae
C₇
T₁
Levator scapulae
Trapezius
Rhomboids
Acromion process
Supraspinatus
Deltoid
Spine of scapula
Deltoid (cut)
Greater tubercle
Infraspinatus
Teres minor
Teres major
Latissimus dorsi
Iliac crest

(a)

C₁–C₇
Longissimus capitis
Semispinalis capitis
Splenius capitis
T₁–T₁₂
Ileocostalis thoracis
Longissimus thoracis
(b)
Spinalis thoracis
Ileocostalis lumborum
L₁–L₅

Figure 15.4

Posterior human trunk musculature. (a) Muscles of the neck, trunk, and shoulder; superficial muscles are shown on the left, deeper muscles are shown on the right. (b) Deep muscles of the trunk.

TABLE 15.5 Muscles of Human Humerus that Act on the Forearm (see Figure 15.5)

Muscle	Comments	Origin	Insertion	Action
Triceps brachii	Sole, large fleshy muscle of posterior humerus; three-headed origin	Long head: inferior margin of glenoid fossa; lateral head: posterior humerus; medial head: distal radial groove	Olecranon process of ulna	Powerful forearm extensor; antagonist of brachialis and biceps brachii
Biceps brachii	Most familiar muscle of anterior humerus because it bulges when forearm is flexed	Short head: coracoid process; tendon of long head runs in bicipital groove over top of shoulder joint	Radial tuberosity	Flexion (powerful) of elbow and supination of hand; flexion of the shoulder; "it turns the corkscrew and pulls the cork"
Brachioradialis	Superficial muscle of lateral forearm	Lateral ridge at distal end of humerus	Base of styloid process of radius	Forearm flexor
Brachialis	Immediately deep to biceps brachii	Distal portion of anterior humerus	Coronoid process of ulna	A major flexor of forearm

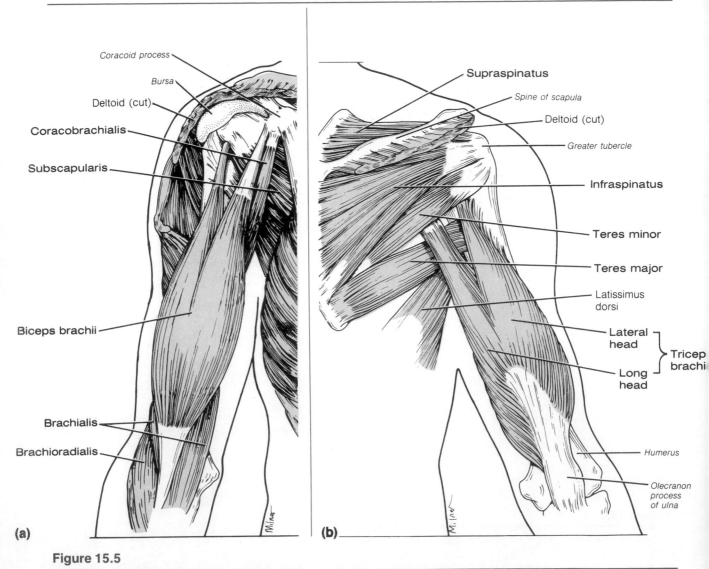

Figure 15.5

Muscles of the right human upper arm: (a) anterior view; (b) posterior view.

TABLE 15.6   Muscles of Human Forearm that Act on Hand and Fingers (see Figure 15.6)

Muscle	Comments	Origin	Insertion	Action
Anteromedial				
Flexor carpi radialis	Superficial; runs diagonally across forearm	Medial epicondyle of humerus	Base of second metacarpal	Powerful flexor of wrist; adducts hand
Flexor carpi ulnaris	Superficial; medial to flexor carpi radialis	Medial epicondyle of humerus and olecranon process of ulna	Base of fifth metacarpal	Powerful flexor of wrist
Palmaris longus	Small fleshy muscle with a long tendon; between flexors carpi radialis and ulnaris	Medial epicondyle of humerus	Palmar aponeurosis	Flexes wrist
Flexor digitorum superficialis	Deeper muscle; overlain by muscles named above; visible at distal end of forearm	Medial epicondyle of humerus, medial surface of ulna, and anterior border of radius	Middle phalanges of second through fifth fingers	Flexes wrist and middle phalanges of second through fifth fingers
Flexor pollicis longus	Deep muscle of anterior forearm; distal to and paralleling lower margin of flexor digitorum superficialis	Anterior surface of radius, medial epicondyle of humerus, and interosseous membrane	Distal phalanx of thumb	Flexes thumb (*pollix* is Latin for "thumb"); weak flexor of wrist
Flexor digitorum profundus	Deep muscle; overlain entirely by flexor digitorum superficialis	Anteromedial surface of ulna and interosseous membrane	Distal phalanges of second through fifth fingers	Flexes distal phalanges (and wrist)
Pronator teres	Seen in a superficial view between proximal margins of brachioradialis and flexor carpi ulnaris	Medial epicondyle of humerus and coronoid process of ulna	Midshaft of radius	Flexes forearm and acts synergistically with pronator quadratus to pronate hand
Pronator quadratus	Deep muscle of distal forearm	Distal portion of ulna	Lateral surface of radius, distal end	Pronates hand
Posterolateral				
Extensor carpi radialis longus	Superficial; parallels brachioradialis on lateral forearm	Lateral superior condylar ridge of humerus	Base of second metacarpal	Extends and abducts wrist (extends forearm)
Extensor carpi radialis brevis	Posterior to extensor carpi radialis longus	Lateral epicondyle of humerus	Base of third metacarpal	Extends and abducts wrist (extends forearm)
Extensor carpi ulnaris	Superficial; medial posterior forearm	Lateral epicondyle of humerus	Base of fifth metacarpal	Extends and abducts wrist
Extensor digitorum	Superficial; between extensor carpi ulnaris and extensor carpi radialis brevis	Lateral epicondyle of humerus	By four tendons into distal phalanges of second through fifth fingers	Extends fingers and wrist
Extensor pollicis longus	Deep muscle overlain by extensor carpi ulnaris	Dorsal shaft of ulna and interosseous membrane	Distal phalanx of thumb	Extends thumb (assisted by extensor, pollicis brevis)
Abductor pollicis	Deep muscle; lateral and parallel to extensor pollicis longus	Posterior surface of radius and ulna	First metacarpal	Abducts thumb
Supinator	Deep muscle at posterior aspect of elbow	Lateral epicondyle of humerus	Proximal end of radius	Supinates hand; antagonistic to pronator muscles

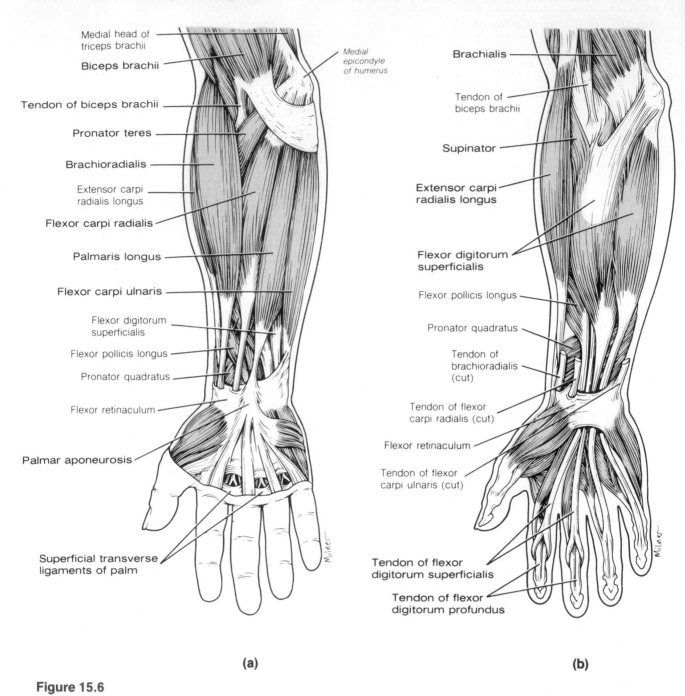

Medial head of
triceps brachii

Biceps brachii

Tendon of biceps brachii

Pronator teres

Brachioradialis

Extensor carpi
radialis longus

Flexor carpi radialis

Palmaris longus

Flexor carpi ulnaris

Flexor digitorum
superficialis

Flexor pollicis longus

Pronator quadratus

Flexor retinaculum

Palmar aponeurosis

Superficial transverse
ligaments of palm

Medial
epicondyle
of humerus

Brachialis

Tendon of
biceps brachii

Supinator

Extensor carpi
radialis longus

Flexor digitorum
superficialis

Flexor pollicis longus

Pronator quadratus

Tendon of
brachioradialis
(cut)

Tendon of flexor
carpi radialis (cut)

Flexor retinaculum

Tendon of flexor
carpi ulnaris (cut)

Tendon of flexor
digitorum superficialis

Tendon of flexor
digitorum profundus

(a)

(b)

Figure 15.6

Muscles of the right human forearm: (a) superficial anterior muscles; (b) second-layer anterior muscles; (c) superficial posterior muscles; (d) deep posterior muscles.

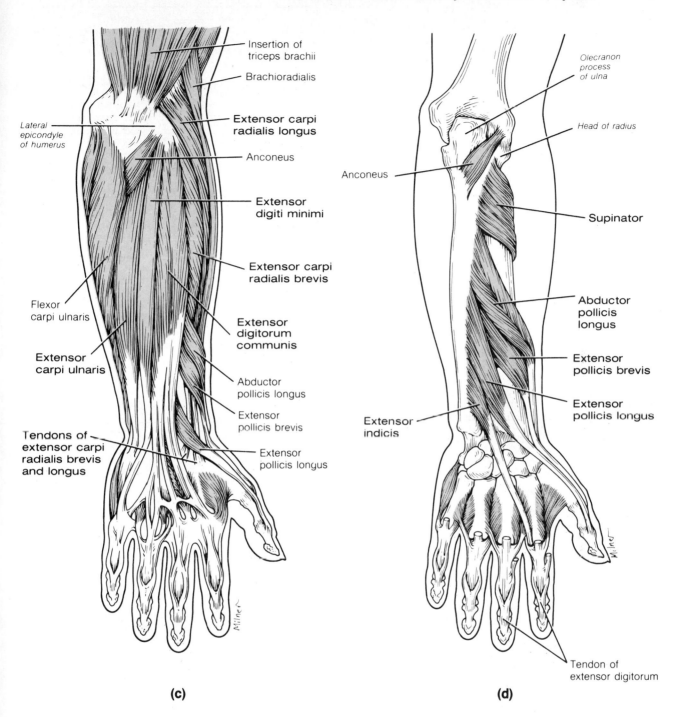

Insertion of triceps brachii

Brachioradialis

Lateral epicondyle of humerus

Extensor carpi radialis longus

Anconeus

Extensor digiti minimi

Extensor carpi radialis brevis

Flexor carpi ulnaris

Extensor digitorum communis

Extensor carpi ulnaris

Abductor pollicis longus

Extensor pollicis brevis

Tendons of extensor carpi radialis brevis and longus

Extensor pollicis longus

(c)

Olecranon process of ulna

Head of radius

Anconeus

Supinator

Abductor pollicis longus

Extensor pollicis brevis

Extensor pollicis longus

Extensor indicis

Tendon of extensor digitorum

(d)

TABLE 15.7 Posterior Muscles of Human Abdominal Wall and Pelvic Region (see Figure 15.7)

Muscle	Comments	Origin	Insertion	Action
Quadratus lumborum	Forms greater portion of posterior abdominal wall	Iliac crest and iliolumbar fascia	Inferior border of twelfth rib	Flexes vertebral column laterally when acting separately; when pair acts jointly, lumbar vertebral column is extended and ribs are fixed
Iliopsoas—iliacus and psoas major	Two closely related muscles; fibers pass under inguinal ligament to insert into femur via a common tendon	Iliac fossa (iliacus); transverse processes, bodies, and disks of lumbar vertebrae (psoas major)	Lesser trochanter of femur	Flex trunk on thigh; major flexor of hip (or thigh on pelvis when pelvis is fixed)

TABLE 15.8 Muscles Acting on Human Thigh (see Figures 15.7 and 15.8)

Muscle	Comments	Origin	Insertion	Action
Iliopsoas	(See Table 15.7)			Major flexor of hip
Gluteus maximus	Largest and most superficial of gluteal muscles (which form buttock mass)	Ilium, sacrum, and coccyx	Gluteal tuberosity of femur and iliotibial tract*	Complex, powerful hip extensor (most effective when hip is flexed, as in climbing stairs—but not as in walking); antagonist of iliopsoas; laterally rotates thigh
Gluteus medius	Covered by gluteus maximus	Upper lateral surface of ilium	Greater trochanter of femur	Abducts and medially rotates thigh
Gluteus minimus	Smallest and deepest gluteal muscle	Inferior surface of ilium	Greater trochanter of femur	Abducts and medially rotates thigh
Tensor fasciae latae	Enclosed between fascia layers of thigh	Anterior aspect of iliac crest	Iliotibial band of fascia lata	Flexes, abducts, and medially rotates thigh
Piriformis	Located on posterior aspect of hip joint	Anterolateral surface of sacrum	Superior border of greater trochanter of femur	Primarily rotates thigh laterally
Adductors—magnus, longus, and brevis	Large muscle mass forming medial aspect of thigh; arise from front of pelvis and insert at various levels on femur	Ischial and pubic rami (magnus); pubis near pubic symphysis (longus); and inferior ramus of pubis (brevis)	Linea aspera and medial epicondyle of femur (magnus); linea aspera above magnus (longus); linea aspera above longus (brevis)	Adduct and medially rotate thigh; flex hip
Pectineus	Overlies adductor brevis on proximal thigh	Pectineal line of pubis	Inferior to lesser trochanter of femur	Adducts, flexes, and laterally rotates thigh
Gracilis	Straplike superficial muscle of medial thigh	Inferior ramus of pubis	Medial surface of head of tibia	Adducts thigh; flexes and medially rotates leg

*The iliotibial tract, a thickened lateral portion of the fascia lata, ensheaths all the muscles of the thigh. It extends as a tendinous band from the iliac crest to the knee.

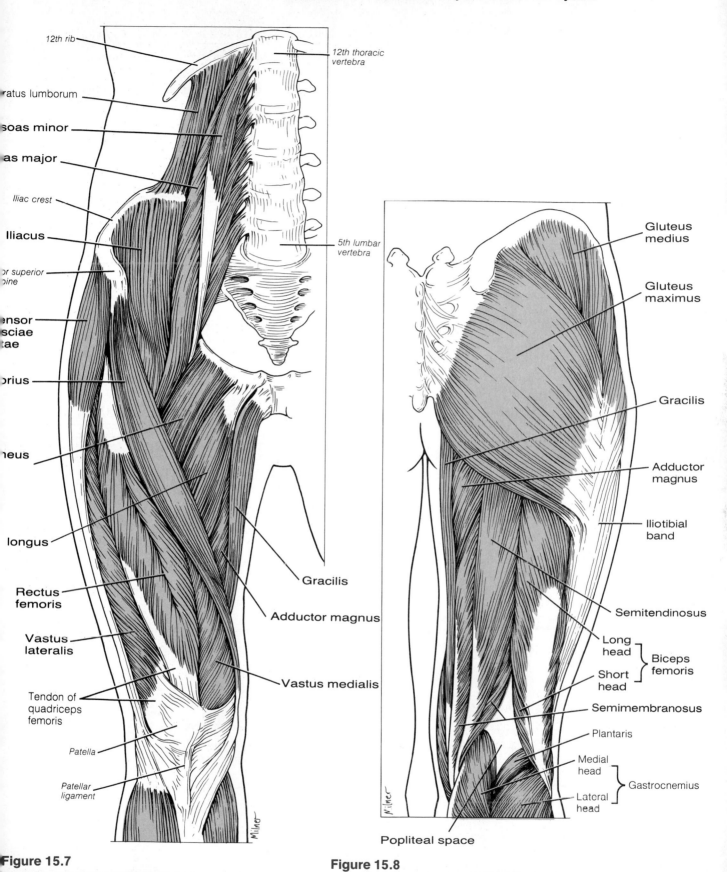

Figure 15.7

Pelvic and thigh muscles of the human, right anterior view.

Figure 15.8

Superficial muscles of the human hip and thigh, right posterior view.

TABLE 15.9 Muscles Acting on Human Leg (Knee Joint) (see Figures 15.7 and 15.8)

Muscle	Comments	Origin	Insertion	Action
Sartorius	Straplike superficial muscle running obliquely across anterior surface of thigh to knee	Anterior superior iliac spine	By an aponeurosis from medial aspect of proximal tibia	Flexes and laterally rotates thigh; flexes knee; known as "tailor's muscle" because it helps bring about cross-legged position in which tailors are often depicted
Gracilis	(see Table 15.8)			Adduction, flexion, and medial rotation of thigh
Hamstrings*				
Biceps femoris	Most lateral muscle of group; arises from two heads	Ischial tuberosity (long head); linea aspera and distal femur (short head)	Tendon passes laterally to insert into head of fibula and tibia	Extends and adducts hip; laterally rotates thigh; flexes knee
Semitendinosus	Medial to biceps femoris	Ischial tuberosity	Medial aspect of upper tibial shaft	Extends thigh; flexes knee
Semimembranosus	Deep to semitendinosus	Ischial tuberosity	Medial condyle of tibia	Extends thigh; flexes knee
Quadriceps†				
Rectus femoris	Superficial muscle of thigh; runs straight down thigh; only muscle of group to cross hip joint; arises from two heads	Anterior inferior iliac spine and superior margin of acetabulum	Tibial tuberosity	Extends knee and flexes thigh at hip
Vastus lateralis	Forms lateral aspect of thigh	Greater trochanter and linea aspera	Tibial tuberosity	Extends knee
Vastus medialis	Forms medial aspect of thigh	Linea aspera	Tibial tuberosity	Extends knee
Vastus intermedius	Obscured by rectus femoris; lies between vastus lateralis and vastus medialis on anterior thigh	Anterior and lateral surface of femur	Tibial tuberosity	Extends knee

*The hamstrings are the fleshy muscles of the posterior thigh. The name comes from the butchers' practice of using the tendons of these muscles to hang hams for smoking. As a group, they are powerful extensors of the hip; they counteract the powerful quadriceps by stabilizing the knee joint when standing.

†The quadriceps form the flesh of the anterior thigh and have a common insertion in the tibial tuberosity via the patellar tendon. They are powerful leg extensors enabling humans to kick a football, for example.

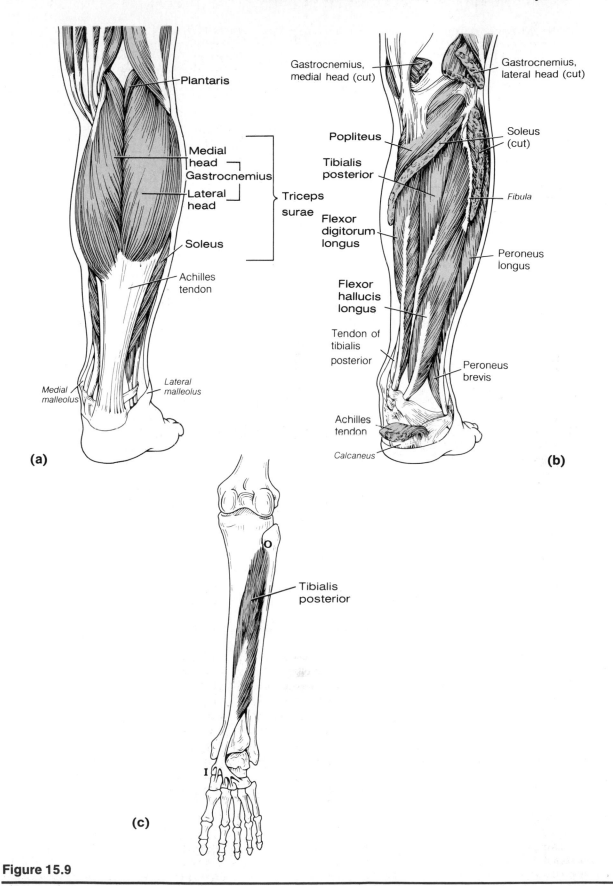

Figure 15.9

Calf muscles of the human leg (right posterior): (a) superficial view. (b) Deep muscles, triceps surae removed. (c) Diagrammatic view of the location of the deep tibialis posterior muscle (all other calf muscles removed). (0 = origin; I = insertion.)

TABLE 15.10 Muscles Acting on Human Foot and Ankle (see Figures 15.9 and 15.10)

Muscle	Comments	Origin	Insertion	Action
Superficial posterior				
Triceps surae	Muscle pair that shapes posterior calf		Via common tendon (Achilles) into heel	Plantarflex foot
Gastrocnemius	Superficial muscle of pair; two prominent bellies	By two heads from medial and lateral condyles of femur	Calcaneus via Achilles tendon	Crosses knee joint; thus also flexes knee
Soleus	Deep to gastrocnemius	Proximal portion of tibia and fibula	Calcaneus via Achilles tendon	
Deep posterior				
Tibialis posterior	Deep to soleus	Superior portion of tibia and fibula and interosseous membrane	Tendon passes obliquely to medial side of ankle and under arch of foot; inserts into several tarsals and meta-tarsals, 2–4	Plantarflexes and inverts foot
Flexor digitorum longus	Runs medial to and partially overlies tibialis posterior	Posterior surface of tibia	Distal phalanges of second through fifth toes	Flexes toes; plantarflexes and inverts foot
Flexor hallucis longus	Lies lateral to inferior aspect of tibialis posterior	Distal portion of fibula shaft	Tendon runs under foot to insert on distal phalanx of great toe	Flexes great toe; plantarflexes and inverts foot
Lateral aspect				
Peroneus longus	Superficial lateral muscle; overlies fibula	Head and upper portion of fibula	By long tendon under foot to first metatarsal	Plantarflexes and everts foot, as part of peronei group
Peroneus brevis	Smaller muscle; deep to peroneus longus	Distal portion of fibula shaft	By tendon running behind lateral malleolus to insert on proximal end of fifth metatarsal	Plantarflexes and everts foot, as part of peronei group
Peroneus tertius	Small muscle; deep to peroneus brevis	Distal anterior surface of fibula	Tendon passes anteriorly to lateral malleolus; inserts on dorsum of fifth metatarsal	Dorsiflexes and everts foot
Anterior aspect				
Tibialis anterior	Superficial muscle of anterior leg; parallels sharp anterior margin of tibia	Lateral condyle and upper 2/3 of tibia; interosseous membrane	By tendon into inferior surface of first cuneiform and metatarsal	Dorsiflexes and inverts foot
Extensor digitorum longus	Anterolateral surface of leg; posterior to tibialis anterior	Lateral condyle of tibia; proximal 3/4 of fibula; interosseous membrane	Tendon divides into four parts; insert into medial and distal phalanges, toes 2–5	Dorsiflexes and everts foot; extends toes
Extensor hallucis longus	Deep to extensor digitorum longus and tibialis anterior	Anterior medial shaft of fibula and interosseous membrane	Tendon inserts on distal phalanx of great toe	Dorsiflexes foot and extends great toe

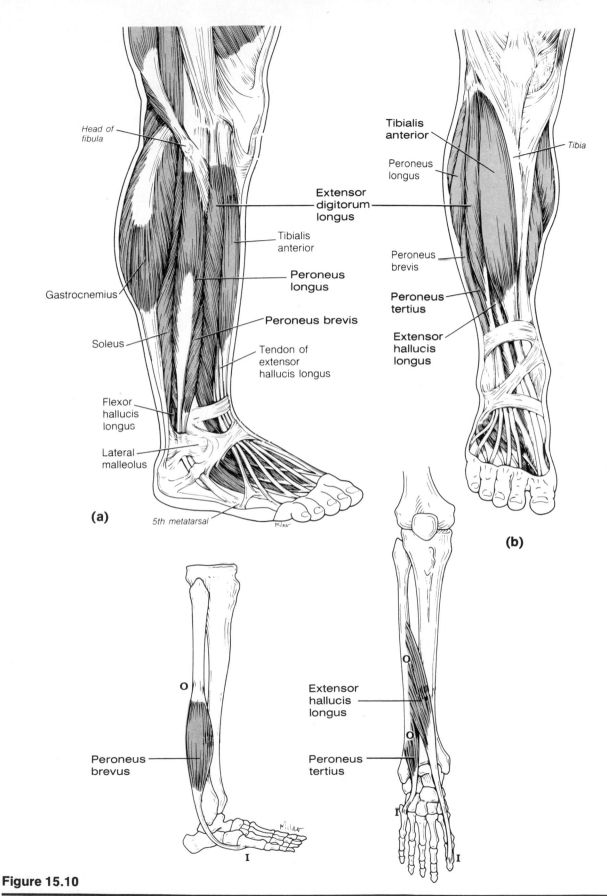

Head of fibula

Gastrocnemius

Soleus

Flexor hallucis longus

Lateral malleolus

Extensor digitorum longus

Tibialis anterior

Peroneus longus

Peroneus brevis

Tendon of extensor hallucis longus

(a)

5th metatarsal

Tibialis anterior

Peroneus longus

Extensor digitorum longus

Peroneus brevis

Peroneus tertius

Extensor hallucis longus

Tibia

(b)

O

Peroneus brevus

I

Extensor hallucis longus

Peroneus tertius

O

O

I

I

Figure 15.10

Muscles of the human right leg. (a) Muscles of the lateral compartment; (b) Muscles of the anterior compartment. Small diagrams show isolated muscles from each compartment. (0 = origin; I = insertion.)

DISSECTION AND IDENTIFICATION OF FETAL PIG MUSCLES

The skeletal muscles of all mammals are named in a similar fashion. However, some muscles that are separate in lower animals are fused in humans, and some muscles present in lower animals are lacking in humans. This exercise involves dissection of the pig musculature in conjunction with the study of human muscles, to enhance your knowledge of the human muscular system. Since the aim is to become familiar with the muscles of the human body, you should pay particular attention to the similarities between pig and human muscles. However, pertinent differences will be pointed out as they are encountered.

Preparing the Fetal Pig for Dissection

The preserved laboratory animals purchased for dissection have been embalmed with a solution (usually containing formalin) to prevent deterioration of the tissues. The animals generally are delivered in plastic bags containing a small amount of embalming fluid. <u>Do not dispose of this fluid when removing the pig.</u> It is important to keep the pig's tissues moist, as it will be used for dissection exercises until the end of the course. The embalming fluid may cause your eyes to smart and may dry your skin, but these small irritants are more desirable than working with a dissection specimen that has become hard and odoriferous due to bacterial action. You can alleviate the skin irritation by using disposable gloves or a skin protective cream before dissection.

1.   Obtain a fetal pig, dissection tray, dissection instruments, and a name tag. Mark the name tag with the names of the members of your group, and set it aside. Attach the name tag to the plastic bag at the end of the laboratory session so that you may identify your animal in subsequent laboratories.

2.   Place the pig dorsal side down on the dissecting tray. To secure the animal to the dissecting tray, make a loop knot with twine around one upper limb, carry the twine under the inferior surface of the tray, and secure the opposing limb. Repeat for the lower extremities.

3.   Make a short, shallow midventral incision at the base of the throat to penetrate the skin. From this point on, use scissors. Continue the cut the length of the ventral body surface to the umbilicus. Cut laterally around the umbilicus on each side, and continue with two incisions (flanking the median line to the pelvic region (Figure 15.11)).

4.   Continue each incision from the pelvic region down the medial surface of each leg to the hoof, and cut the skin completely around the ankles.

5.   Return to the neck. Make an incision completely around the neck. Then cut down each foreleg to the wrist. Completely cut the skin around the wrists.

6.   Now loosen the skin from the loose connective tissue (superficial fascia) that binds it to the underlying structures. With one hand, grasp the skin on one side of the midline ventral incision. Using your fingers, the closed scissor tips, or a blunt probe, break through the "cottony" connective tissue fibers to release the skin from the muscle beneath. Work toward the dorsal surface and then upward toward the neck region. As you pull the skin from the body, you should see small, white cordlike structures that extend from it to the muscles at fairly regular intervals. These are the cutaneous nerves, which serve the skin. You will also see (particularly on the ventral surface) that a thin layer of muscle fibers adheres to the skin. These are the cutaneous muscles, which enable the pig to move or twitch the skin to get rid of irritants such as insects. Do not remove the skin from the head, since the muscles of the pig are not sufficiently similar to human head muscles to merit study.

7.   Complete the skinning process by releasing the skin from the forelimbs, lower torso, and hindlimbs in the same manner.

8.   Inspect your skinned pig. Note that it is difficult to see any cleavage lines between the muscles because of the overlying connective tissue. Note the neck incision where the pig's blood vessels were injected with latex. If time allows, begin on the side opposite the incision, using forceps or fingers, and carefully remove as much gelatinous material, fat, and fascia from the surface of the muscles as possible. Since you will dissect only one side of the animal's musculature, you need to perform this clearing process only on one side. If the muscle dissection exercises are to be done at a later laboratory session, carefully rewrap the pig's skin around its body. Then dampen several paper towels in embalming fluid, and wrap these around the animal. Return the pig to the plastic bag, adding more embalming fluid if necessary, seal the bag, and attach your name tag. Use the same technique for preparing your pig for storage each time the pig is used. Place the pig in the storage container designated by your instructor.

9.   Before leaving the laboratory, dispose of any tissue remnants in the "organic debris" container, wash the dissecting tray and instruments with soapy water, rinse and dry, and wash down the laboratory bench.

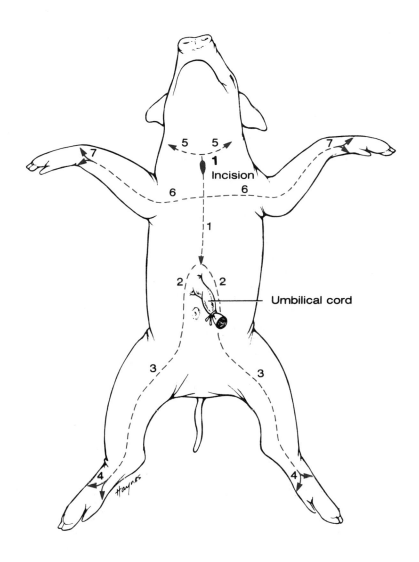

Figure 15.11

Incisions to be made in skinning a pig (numbers indicate sequence).

Dissection of Pig Trunk and Neck Muscles

The proper dissection of muscles involves carefully separating one muscle from another so that they can be identified and transecting the superficial muscles to study the deeper layers.

In general, when you want to transect a muscle, you should completely free it from all adherent connective tissue along its length and then cut through about halfway between its origin and insertion attachments.

Before you begin dissection, observe your skinned pig. If you look carefully, you will see changes in the direction of muscle fibers, which will help you locate the muscle borders. As a rule, all the fibers of one muscle are held together by a connective tissue sheath (epimysium) and run in the same general direction. Pulling in slightly different directions on two adjacent muscles will usually enable you to expose the normal cleavage line between them. Once you have identified the cleavage lines, use a blunt probe to break the connective tissue connections between them and separate the muscles from one another. If the muscles separate as clean, distinct bundles, your procedure is probably correct. If you observe a ragged or "chewed-up" appearance, you are probably tearing a muscle apart rather than separating it from the adjacent muscles. Only those muscles most easily identified and separated out will be identified in this exercise because of time considerations.

ANTERIOR NECK MUSCLES

1. Using Figure 15.12 as a guide, examine the anterior neck surface of the pig and identify the following superficial neck muscles. (The platysma belongs in this group but was probably removed during the skinning process.) The **sternomastoid** (also called the **sternocephalic** in the pig) is a narrow muscle extending medially from the mastoid process to the sternum along the side of the neck. The large external jugular vein, which drains the head, should be obvious crossing the anterior aspect of this muscle. The **mylohyoid** muscle parallels the bottom aspect of the chin, and the **digastric** muscle flanks the mylohyoid laterally. You can also identify the fleshy **masseter** muscle, which flanks the digastric muscle laterally, although it is not one of the neck muscles. The **sternohyoid** is a narrow straplike muscle running medially along the ventral aspect of the neck. Note the large masses of thymus tissue lying beneath these muscles.

2. The deeper muscles of the anterior neck of the pig are small and straplike and hardly worth the effort of the dissection; however, you can see one of these deeper muscles with a minimum of effort. Transect the sternomastoid and the sternohyoid approximately at mid-belly. Reflect the cut ends back to identify the **sternothyroid,** which runs downward on the anterior surface of the throat just deep and lateral to the sternohyoid muscle. Notice that after it originates from the sternum, this muscle separates into two parts that insert into the lateral and ventral surfaces of the thyroid cartilage of the larynx.

CHEST MUSCLES In the pig, the chest, or pectoral, muscles adduct the arm just as they do in humans. However, humans have only two pectoral muscles, and the pig has three—the superficial pectoral, the posterior deep pectoral, and the anterior deep pectoral. Because of their relatively great degree of fusion, the pig's pectoral muscles give the appearance of a single muscle. The pectoral muscles are rather difficult to dissect and identify, as they do not separate from one another easily. Refer to Figure 15.12 as you work.

The **superficial pectoral** is a thin flat band of muscle that can be seen arising from the sternum just anterior to the posterior deep pectoral, discussed below.

The diagonally oriented fibers of the **posterior deep pectoral** can be seen as inferior to the superficial pectoral. The muscle originates on the ventral sternum and costal cartilages of the fourth to ninth ribs and runs obliquely to insert on the proximal humerus. Slip the handle of your scalpel between the superficial pectoral and the posterior deep pectoral muscles to separate them.

The **anterior deep pectoral** is a thin straplike muscle deep to the superficial pectoral muscle. It extends superiorly from the posterior deep pectoral to run over the shoulder to the dorsal scapular region.

Identify and trace out the origins and insertions of the muscles of the abdominal wall, all of which have similar functions as the same-named muscles in humans.

MUSCLES OF THE ABDOMINAL WALL Refer to Figure 15.12 in the following muscle dissections. Avoid the umbilical area and the male external genitalia if the pig is a male.

1. The **rectus abdominus** is a superficial longitudinal band of muscle running immediately lateral to the midline of the body on the abdominal surface. Note that the four transverse tendinous intersections of the rectus abdominus are absent or difficult to identify in the pig. Identify the **linea alba,** which separates the rectus abdominus muscles. Note the relationship of the rectus abdominis to the other abdominal muscles and their fascia.

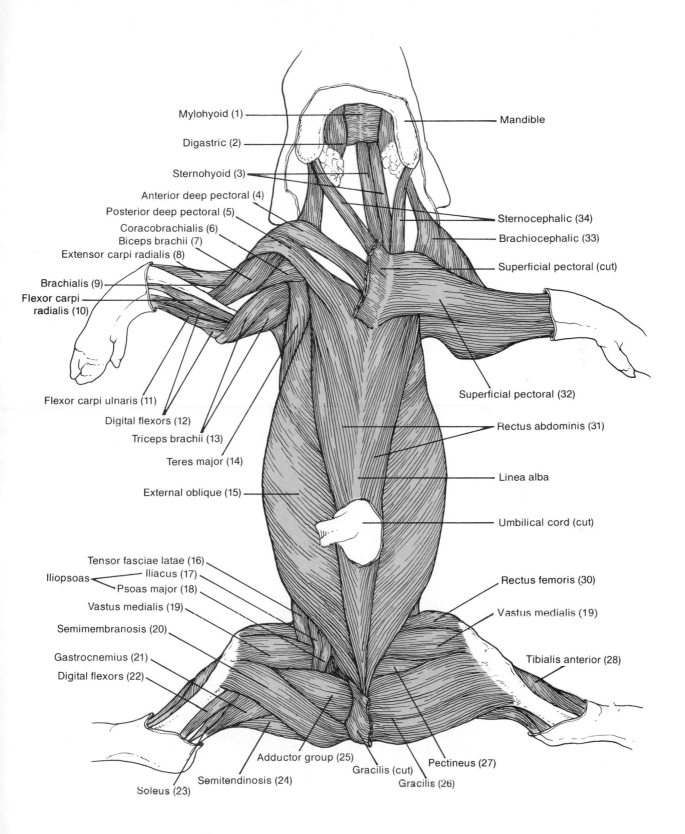

Figure 15.12

(a) Muscles of the fetal pig's trunk and neck (anterior). Superficial muscles are shown on the right side; deeper muscles are shown on the left (Sartorius muscles of the thigh have been removed.)

Figure 15.12 (continued)

(b) Photo of muscles of the anterior trunk and neck. (Photo by Jack Scanlon.)

2. The **external oblique** is a sheetlike muscle immediately lateral (and running beneath) the rectus abdominus. (See also Figure 15.13.) Transect the external oblique to reveal the anterior attachment of the rectus abdominus and the **internal oblique.** Reflect the external oblique, and observe this deeper muscle. Note the fiber direction of this muscle. How does it compare to that of the external oblique?

_____

3. Transect the internal oblique muscle to reveal the fibers of the **transversus abdominus,** with its fibers running transversely across the abdomen. The transversus is very thin and may be difficult to separate from the internal oblique.

SUPERFICIAL MUSCLES OF THE POSTERIOR TRUNK AND NECK Refer to Figure 15.13 as you dissect the following superficial muscles of the posterior surface.

1. Turn your pig to its lateral surface and begin your dissection. Humans have one large single trapezius muscle, but the pig has three separate muscles: the **clavotrapezius,** the **acromiotrapezius,** and the **spinotrapezius.** The most superior muscle of the group is the clavotrapezius, which is homologous to that part of the human trapezius that inserts into the clavicle. Slip a probe under this muscle and follow it to its apparent origin. Where does it appear

to originate? _____

Is this similar to its origin in man? _____

The fibers of the clavotrapezius form the superior portion of a much longer muscle called the **brachiocephalic** muscle; the ventral portion is called the **clavobrachialis** muscle. In humans, the portion of the trapezius muscle represented by the clavotrapezius inserts into the clavicle; however, since the clavicle is lacking in the pig, the clavotrapezius is continuous with the clavobrachialis, which inserts into the distal end of the humerus, where it appears to blend with the fibers of the anterior deep pectoral muscle. The two fused portions of this muscle act in concert to move the forelimb of the pig anteriorly. Release the clavotrapezius muscle from the adjoining muscles.

2. The **acromiotrapezius** is a short fan-shaped muscle posterior to the clavotrapezius. It originates from the cervical vertebral spines and inserts on the scapular spine by a broad and easily identified aponeurosis. The **spinotrapezius,** the most posterior of the trapezius muscles in the pig, runs from the thoracic vertebrae to the scapula to insert by a thin tendon into the scapula. Pull on the acromiotrape-

zius and spinotrapezius muscles. Do they appear to have the same function(s) in the pig as in humans?

3. The **latissimus dorsi,** a broad fan-shaped muscle covering a sizable portion of the lateral surface of the posterior trunk, is partially covered by the spinotrapezius. Its fibers run downward and anteriorly around the lateral thorax to insert into the humerus (as in humans).

4. The **deltoid** in the pig is a thin muscle band extending from the posterior scapula to the proximal end of the humerus, which adducts the humerus. Contrast the deltoid muscle of the pig to the thick fleshy muscles seen in humans.

DEEP MUSCLES OF THE POSTERIOR TRUNK AND NECK

1. In preparation, transect the latissimus dorsi, the acromiotrapezius, the spinotrapezius, and the deltoid muscle and reflect them back. Refer to Figure 15.13 as you work. The **supraspinatus** muscle, a portion of which lies deep to the deltoid, can be seen superficially anterior to the caudal border of the deltoid and posterior to the anterior deep pectoral. It originates on the scapular spine and clothes the scapula before inserting on the humerus by dual insertion. Reflect the cut ends of the deltoid muscle to reveal the thicker **infraspinatus** muscle, which lies deep to it. It originates on the lateral-posterior aspect of the scapula, occupying the region beneath the scapular spine, and inserts on the proximal humerus. It abducts and rotates the forelimb laterally. The **serratus ventralis** muscle arises deep to the pectoral muscles and covers the lateral surface of the ribcage. It is easily identified by its fingerlike muscular origins, which arise on the fourth to eighth ribs. It is best seen by raising the forelimb up and away from the body wall.

2. Reflect the cut ends of the spinotrapezius and acromiotrapezius muscles to expose the **rhomboid muscles,** which originate from several separate muscle slips on the cervical and thoracic vertebrae and extend to the scapula. Note that these muscles insert into the medial border of the scapula rather than into the scapular spine, the insertion site of the trapezius muscles. The cephalic portion of this group, which extends downward from the occipital bone to attach to the scapula, is the **rhomboid capitus,** which has no counterpart in the human body. The other muscles in this group are homologous to the rhomboid muscles of humans.

3. The **splenius muscle,** located deep to the rhomboid muscles, is a large flat muscle occupying most

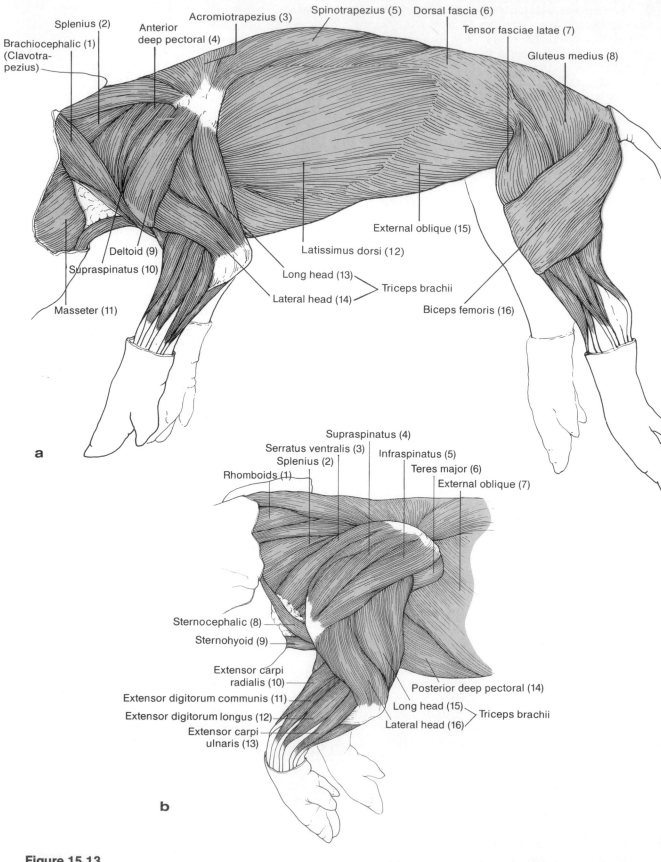

Splenius (2)
Brachiocephalic (1) (Clavotra-pezius)
Anterior deep pectoral (4)
Acromiotrapezius (3)
Spinotrapezius (5)
Dorsal fascia (6)
Tensor fasciae latae (7)
Gluteus medius (8)

Deltoid (9)
Supraspinatus (10)
Masseter (11)

External oblique (15)
Latissimus dorsi (12)
Long head (13)
Lateral head (14)
Triceps brachii
Biceps femoris (16)

a

Supraspinatus (4)
Serratus ventralis (3)
Splenius (2)
Infraspinatus (5)
Teres major (6)
Rhomboids (1)
External oblique (7)

Sternocephalic (8)
Sternohyoid (9)

Extensor carpi radialis (10)
Extensor digitorum communis (11)
Extensor digitorum longus (12)
Extensor carpi ulnaris (13)

Posterior deep pectoral (14)
Long head (15)
Lateral head (16)
Triceps brachii

b

Figure 15.13

(a) Diagram of the superficial musculature of fetal pig (lateral aspect). (b) Diagram of the muscles of the left shoulder, with superficial muscles removed.

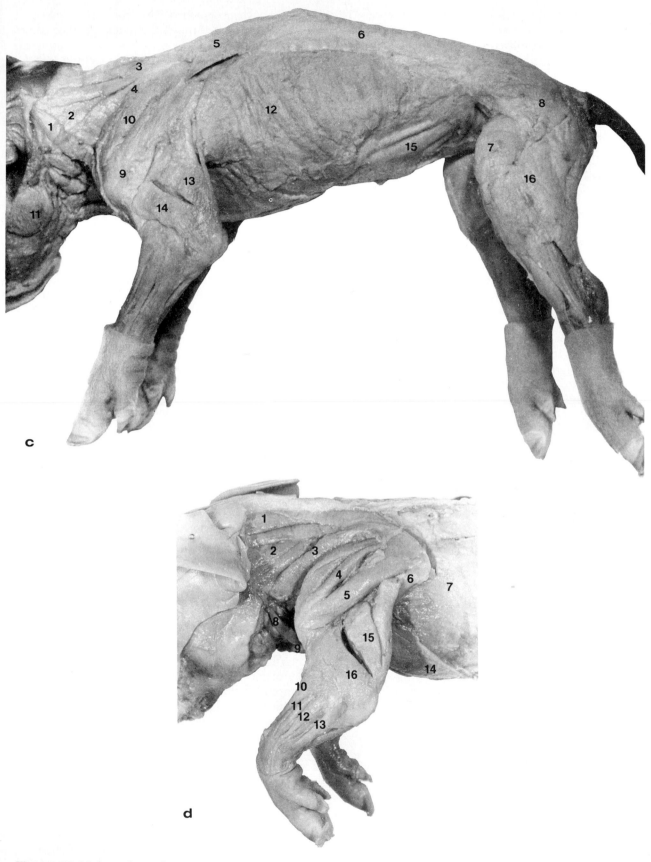

Figure 15.13 (continued)

Photos of the superficial musculature of fetal pig: (c) lateral aspect; (d) left shoulder, with superficial muscles removed.

of the side of the neck close to the vertebrae. It originates on the ligamentum nuchae and inserts into the occipital bones. It functions to raise the head.

Dissection of Pig Forelimb Muscles

Refer to Figures 15.12 and 15.13 as you study these muscles.

UPPER FORELIMB MUSCLES

 1. The triceps muscle of the pig (**triceps brachii**) can be easily identified if the pig is placed on its side. It is a large fleshy muscle covering the posterior aspect and much of the side of the humerus. As in humans, this muscle arises from three heads, which originate from the humerus and scapula and insert jointly into the olecranon process of the ulna. Remove the fascia from the upper region of the lateral arm surface to identify the lateral and long heads of the triceps. The long head is medial to the lateral head on the posterior surface of the arm. The medial head can be exposed by transection and reflection of the lateral head. How does the function of this muscle in the pig compare to its function in humans?

2. The **brachialis** can be located anterior to the lateral head of the triceps muscle. Identify its origin on the humerus, and trace its course as it crosses the elbow laterally and inserts on the ulna. It flexes the foreleg of the pig.

3. In the pig, the **biceps brachii** muscle is a small, slender spindle-shaped muscle that is covered almost completely by the clavobrachialis portion of the brachiocephalic muscle. It originates from the scapula in the area anterior to the glenoid fossa and inserts into the proximal radius and ulna. It flexes the forelimb of the pig. Follow this muscle to its origin. Does the biceps have two heads in the pig?

4. The **coracobrachialis** of the pig is rather insignificant and can be seen as a very small muscle crossing the ventral aspect of the shoulder joint. It runs beneath the biceps brachii to insert on the humerus and has the same function as its homologue in humans.

LOWER FORELIMB MUSCLES
Identification of the forelimb muscles is difficult, because of the tough fascia sheath that encases them.

1. Remove as much of this connective tissue as possible, and cut through the ligaments that secure the tendons at the wrist (transverse carpal ligaments) so that you will be able to follow the various muscles to their insertions. Only selected forelimb muscles will be identified in the pig.

2. Begin your identification of the forelimb muscles by examining the lateral surface of the forelimb. The muscles of this region are very much alike in appearance and are difficult to identify unless a definite order is followed. Thus you will begin with the most posterior muscles and proceed to the anterior aspect. Remember to carefully check the tendons of insertion to verify your muscle identification in each case. These muscles are illustrated in Figure 15.13.

3. Follow the **extensor carpi ulnaris** muscle from the lateral epicondyle of the humerus to the ulnar side of the fifth metacarpal. Often this muscle has a shiny insertion tendon that helps in its identification.

4. The **extensor digitorum lateralis,** or **longus,** muscle is similar in appearance to the extensor carpi ulnaris and lies just anterior to it. The extensor digitorum lateralis arises from the distal humerus and inserts by a divided tendon into the digits. This muscle has no human counterpart.

5. You can see the **extensor digitorum communis** for its entire length along the lateral surface of the forelimb just anterior to the extensor digitorum lateralis. Trace it to its four tendons, which insert on the second to the fifth digits. The extensor digitorum communis and the two extensor muscles described above act together to extend the digits.

6. The **extensor carpi radialis** is anterior to the extensor digitorum communis and is sometimes partially covered by it. It originates from the distal end of the humerus and inserts into the distal end of the radius. It rotates the foot of the pig.

7. The ribbonlike muscle seen on the lateral surface of the humerus and passing down the foreleg to insert on the styloid process of the radius is the **brachioradialis.** It acts to rotate the forelimb. (If your removal of the fascia was not very careful, this muscle may have been removed in the clearing process.)

8. Turn the pig so that you can observe the ventral forelimb muscles, mostly flexors and pronators, easily. For the following series of muscle identifications refer to Figure 15.12. Your examination of these muscles will be cursory, so you need not attempt to trace out the origins and insertions of these muscles. In the pig, as in humans, most of these muscles arise from the medial epicondyle of the humerus. Identify the following muscles or muscle

groups by beginning at the posterior surface of the medial forearm and working anteriorly. The **digital flexor muscles** are two thin muscles seen flanking the flexor carpi ulnaris, anteriorly and posteriorly. The **flexor carpi radialis** is located anteriorly to the **flexor carpi ulnaris** and acts with it to flex the wrist.

Dissection of Pig Hindlimb Muscles

With the pig on its side, remove the fat and fascia from all the thigh surfaces, but do not cut through or remove the **fascia lata,** which is a tough white aponeurosis covering the anterolateral surface of the thigh from the hip to the leg.

MUSCLES OF THE POSTERO-LATERAL HINDLIMB

Referring to Figures 15.12, 15.13, and 15.14, identify the following muscles of the hip, thigh, and posterior shank.

1. The **tensor fascia lata** is wide and thin at its superior end, where it originates on the iliac crest and narrows to a long tendon that inserts into the fascia lata. Its action is to tighten the fascia lata. Transect this muscle. The **gluteus medius** is a thick muscle lying beneath and posterior to the tensor fascia lata. It originates from the lumbodorsal and gluteal fascias and inserts into the greater trochanter of the femur. It acts to abduct the thigh.

2. The **gluteus maximus** muscle in the pig is very thin and is found overlying the anterior portion of the gluteus medius muscle with which it acts synergistically. No attempt will be made to distinguish the limits of the individual gluteal muscles.

3. The **hamstring muscles,** a group in the lower extremity, consist of the biceps femoris, the semitendinosus, and the semimembranosus muscles. The **biceps femoris** is a large triangular muscle of the thigh that covers most of its posterolateral surface. Trace it from its origin on the ischium to its insertion on the distal femur and proximal tibia. It abducts and extends the thigh and flexes the shank. Transect this muscle and reflect its cut ends to identify the following muscles lying deep to it. The **semitendinosus** is a thick narrow muscle band lying deep to the posterior border of the biceps femoris. Contrary to what its name implies ("half-tendon"), this muscle is primarily muscular and fleshy except at its insertion. Trace this muscle from its origin on the pelvic girdle to its insertion on the proximal tibia. It acts to bend the knee. The **semimembranosus,** a muscle lying medial to the semitendinosus, is the

most posteriorly placed muscle of the thigh. The distal end of this broad muscle overlies and is closely associated with the adductor magnus muscle. Trace it from its origin on the pelvic girdle to its insertion on the proximal tibia. Before continuing with the identification of the muscles of the shank, identify the large **vastus lateralis** muscle, which is anterior to the semimembranosus muscle and part of the quadriceps muscle group to be dealt with later.

How does the semimembranosus compare with its

human homologue? _____

_____

4. Remove the heavy fascia covering the lateral surface of the shank and proceed from posterior to anterior aspect to identify the following muscles on the posterolateral shank (leg). Reflect the lower portion of the biceps femoris muscle to see the origin of the following muscles. The **triceps surae** is a large composite muscle of the calf, which is homologous to the same-named muscle in humans. The **gastrocnemius** portion of the triceps surae is seen as the largest muscle on the shank. As in humans, it has two heads and inserts via the Achilles tendon into the calcaneus. Run a probe beneath this muscle and then transect it to reveal the **soleus,** which is deep to the gastrocnemius.

5. The **peroneus muscles** are long thin muscles that arise from the proximal tibia and distal femur and run to tarsal insertions. The **peroneus longus** extends along the lateral surface of the tibia and the **peroneus tertius** along the tibia's anterior surface. They act to flex the ankle.

6. The **extensor digitorum longus** is a thin elongated muscle anterior to the peroneus muscles. It inserts by divided tendon into the digits, and its action is the same as in humans. The **tibialis anterior** is the most anterior muscle of the shank and is in direct contact with the tibia along its whole length. Its proximal end encases the extensor digitorum longus muscle. Trace it from its origin on the proximal tibia to the second metatarsal, where it inserts by a strong tendon. It acts to flex the ankle.

MUSCLES OF THE ANTEROMEDIAL HINDLIMB

1. Turn the pig on its dorsal surface to identify the following muscles (refer to Figure 15.12). The **sartorius** is seen superficially on the anterior medial thigh as a thin strap of muscle tissue running from the pelvis to the proximal end of the tibia. The femoral artery can be identified running along its posterior edge. This muscle acts to adduct the thigh and flex the hip. Transect this muscle.

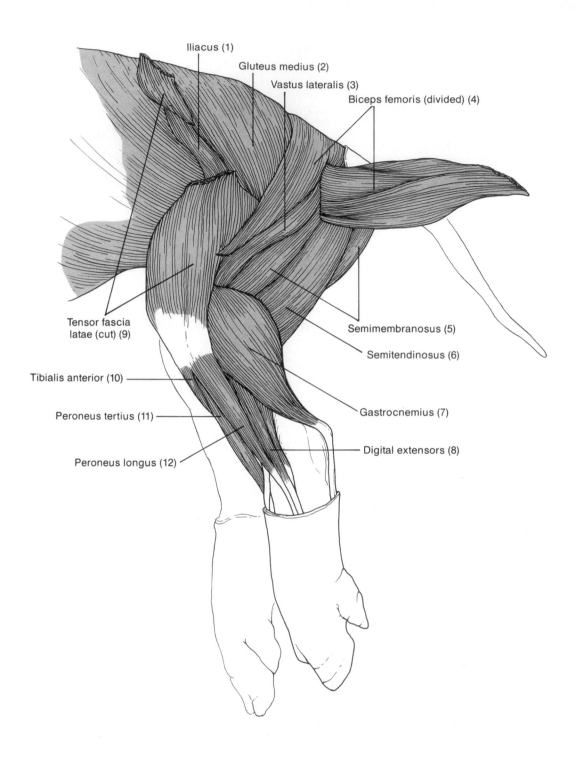

Iliacus (1)

Gluteus medius (2)

Vastus lateralis (3)

Biceps femoris (divided) (4)

Tensor fascia latae (cut) (9)

Tibialis anterior (10)

Peroneus tertius (11)

Peroneus longus (12)

Semimembranosus (5)

Semitendinosus (6)

Gastrocnemius (7)

Digital extensors (8)

Figure 15.14

(a) Diagram of the muscles of the left posterolateral thigh and leg of the fetal pig.

Figure 15.14 (continued)

(b) Photo of the muscles of the left posterolateral thigh and leg.

2.   The **quadriceps** muscles insert in common into the patella and act to extend the shank. The **vastus medialis** is seen just superior to the sartorius muscle on the medial thigh surface. It originates from the head of the femur. The **rectus femoris,** a thick bandlike muscle running along the anterior surface of the femur, is partially overlain by the vastus lateralis (identified earlier), which clothes the antero-lateral aspect of the thigh. The rectus femoris takes its origin from the anterior ilium.

What is the origin of the rectus femoris in humans?

_____

_____

3.   The **gracilis,** a broad muscle covering the posterior portion of the medial aspect of the thigh, originates on the pubic symphysis and inserts on the medial proximal tibial surface. Its action is to adduct the thigh in the pig. Free and transect this muscle to view the muscles lying deep to it.

How does this compare with the human gracilis?

_____

_____

4.   The **adductor magnus** lies deep to the gracilis. Separate and distinguish this muscle from the semi-membranosus muscle, which lies inferior to it and with which it is closely associated.

5.   Examine the superior margin of the adductor magnus to locate the long slender **pectineus** muscle. It is normally covered by the sartorius muscle distally. It originates on the pubic bone and inserts on the medial aspect of the femur. This muscle is similar in action to its human homologue and acts to adduct the thigh.

6.   Just superior to the pectineus muscle, you can see a small portion of the iliopsoas, a composite muscle whose action is similar to that of its counter-part in humans. The dissection of the remaining musculature of the anterior shank in the pig is not included here; however, you might want to examine the general location of the digital flexor muscles in Figure 15.12.

Muscle Physiology

OBJECTIVES

1. To observe muscle contraction on the microscopic level and describe the role of ATP and various ions in muscle contraction.

2. To define and explain the physiologic basis of the following:

   action potential
   subthreshold or subliminal stimulus
   threshold or minimal stimulus
   maximal stimulus
   treppe or staircase phenomenon
   wave summation
   multiple motor unit summation
   tetanus
   muscle fatigue
   absolute refractory period
   depolarization
   repolarization

3. To trace the events that result from the electrical stimulation of a muscle.

4. To explain why the "all or none" law is reflected in the activity of a single muscle cell but not in an intact skeletal muscle.

5. To recognize that the graded response of skeletal muscle is a function of the number of muscle fibers stimulated and the frequency of the stimulus.

6. To name and describe the phases of a muscle twitch.

7. To distinguish between a muscle twitch and a sustained (tetanic) contraction and to describe their importance in normal muscle activity.

8. To demonstrate how the kymograph or polygraph can be used to obtain pertinent and representative recordings of various physiologic events of skeletal muscle activity.

9. To explain the significance of muscle tracings obtained during experimentation.

MATERIALS

ATP muscle kits (glycerinated rabbit psoas muscle*; ATP and salt solutions obtainable from Carolina Biological Supply, Item #20-3525)
Petri dishes
Microscope slides
Cover glasses
Millimeter scale
Compound microscope
Dissecting microscope
Small beaker (50 ml)
Frog Ringer's solution
Scissors
Metal needle probes
Pointed glass probes
Medicine dropper
Cotton thread
Forceps
Glass or porcelain plate
Live bullfrog
Apparatus A or B
A: kymograph, kymograph paper (smoking stand, burner, and glazing fluid if using a smoke-writing apparatus), skeletal muscle lever, signal magnet, laboratory stand, clamp, electronic stimulators
B: physiograph (polygraph), polygraph paper and ink, myograph, pin and clip electrodes, stimulator output extension cable, transducer cable, straight pins, frog board, laboratory stand, clamp

*Note to Instructor: At the beginning of the lab, the muscle bundle should be removed from the test tube and cut into ~2-cm lengths. Both the cut muscle segments and the entubed glycerol should be put into a petri dish. One muscle *segment* is sufficient for each two to four students making observations.

MUSCLE ACTIVITY

The contraction of skeletal and cardiac muscle can be considered in terms of three events—the electrical excitation of the muscle cell, the excitation-contraction coupling, and the shortening of the muscle cell due to the sliding of the myofilaments within it.

At rest, all cells maintain a membrane potential in which the cell interior is approximately -80 to -90 millivolts (mv) compared with the cell exterior. This electrical difference is a result of the differences in the concentrations of sodium (Na^+) and potassium (K^+) ions on the two sides of the membrane. Intracellular potassium concentration is much greater than its extracellular concentration, and intracellular sodium concentration is considerably less than its extracellular concentration. A cell's potential for doing electrical work—its **resting potential**—depends on this unequal concentration of potassium and sodium ions on either side of the cell membrane.

Action Potential

When a muscle cell is stimulated, the sarcolemma becomes temporarily permeable to sodium, which rushes into the cell. This sudden influx of sodium ions reverses the membrane potential. That is, the cell interior becomes positively charged at that point, an event called **depolarization.** When depolarization reaches a certain level, a depolarization wave, or **action potential,** travels along the sarcolemma, conducting the electrical impulse from one end of the cell to the other.

As the influx of sodium ions occurs, the sarcolemma becomes impermeable to sodium and permeable to potassium ions. Potassium ions leak out of the cell, restoring the resting potential but not the original ionic conditions, an event called **repolarization.** The repolarization wave follows the depolarization wave across the cell membrane.

Until repolarization of the membrane has been completed, the muscle cell cannot be stimulated to contract again. In this condition it is said to be in the **absolute refractory period.** Repolarization is followed by activation of the sodium-potassium pump, which actively transports potassium ions into the cell and sodium ions out of the cell to reestablish the ionic concentrations of the resting state.

Contraction

The propagation of the action potential along the sarcolemma causes the release of calcium (Ca^{++}) ions from storage depots (tubules of the sarcoplasmic reticulum) within the muscle cell. When the calcium ions reach the myofilaments, they act as an ionic trigger that initiates contraction, and the actin and myosin filaments slide past each other. Once the action potential has ceased, the calcium ions are almost immediately reabsorbed into the tubules of the sarcoplasmic reticulum. Instantly the muscle cell relaxes.

The events of the contraction process can most simply be summarized as follows: muscle cell contraction is initiated by depolarization of the membrane owing to the influx of Na^+ into the cell. This electrical event is coupled with the sliding of the myofilaments—contraction—by the release of Ca^{++}. Keep in mind this sequence of events as you conduct the experiments.

OBSERVATION OF MUSCLE FIBER CONTRACTION

 In this simple observational experiment, you will have the opportunity to review your understanding of muscle cell anatomy and to watch the fiber contracting (and not contracting) in response to the presence of certain chemicals (ATP, and potassium and magnesium ions).

1.   Obtain the following materials from the supply area: 2 glass teasing needles, 3 glass microscope slides and cover glasses, millimeter scale or ruler, dropper vials containing the following solutions: a. 0.25% ATP in triply distilled water; b. 0.25% ATP plus 0.05M KCl plus 0.001M MgCl$_2$ in distilled water; and c. 0.05M KCl plus 0.001M MgCl$_2$ in distilled water; a petri dish, and a previously cut muscle bundle segment. While you are at the supply area, place the muscle bundle in the petri dish and pour a small amount of glycerol (the fluid in the supply petri dish) over your muscle segment. Also obtain both a compound and a dissecting microscope and bring them to your laboratory bench.

2.   Using the fine glass needles, tease the muscle segment to separate its fibers. The objective is to isolate *single* muscle cells or fibers for observation. Be patient and work carefully so that the fibers do not get torn during this isolation procedure.

3.   Transfer one or more of the fibers (or the thinnest strands you have obtained) on to a clean microscope slide with a glass needle, and cover it with a cover slip. Examine the fiber under low and then high power magnifications to observe the striations and the smoothness of the fibers when they are in the relaxed state.

4.   Transfer three or four fibers in a small amount of glycerol to a second clean microscope slide with a glass needle. Using the needle as a prod, carefully position the fibers so that they are parallel to one another and as straight as possible. Place this slide under a dissecting microscope and measure the length of each of the fibers with a millimeter scale held adjacent to each fiber. Record these lengths on the chart that follows.

	Muscle fiber 1	Muscle fiber 2	Muscle fiber 3
Initial length (mm)			
Contracted length (mm)			
% of shortening			

5. Flood the fibers (situated under the dissecting microscope) with several drops of the solution containing ATP + potassium ions and magnesium ions. Watch the reaction of the fibers after adding the solution. After 30 seconds (or slightly longer), remeasure each fiber and record the observed lengths on the chart. Also, observe the fibers to see if any width changes have occurred. Calculate the degree (or percentage) of contraction by using the following simple formula, and record this information on the chart also.

initial length (mm) − contracted length (mm) =

_____ degree of contraction (mm)

then: $\dfrac{\text{degree of contraction (mm)}}{\text{initial length (mm)}} \times 100 =$

_____% contraction

6. Carefully transfer one of the contracted fibers to a clean microscope slide, cover with a cover glass, and observe with the compound microscope. Mentally compare your initial observations with the view you are observing now. What differences do you see?

(Be specific.) _____

_____

_____

What zones (or bands) have disappeared? _____

7. Repeat steps 3 to 6, using clean slides and fresh muscle cells. First use the solution of ATP in distilled water (no salts) and then use the solution containing only salts (no ATP) for the third series.

What degree of contraction was observed when ATP was applied in the absence of potassium and

magnesium ions? _____

What degree of contraction was observed when the muscle fibers were flooded with a K^+ and Mg^{++}-

containing solution that lacked ATP? _____

What conclusions can you draw about the importance of ATP, and potassium and magnesium ions to

the muscle contractile process? _____

INDUCTION OF CONTRACTION IN THE FROG GASTROCNEMIUS MUSCLE

Physiologists have learned a great deal about the way muscles function by isolating muscles from laboratory animals and then stimulating these muscles to observe their responses. Various stimuli—electrical shock, temperature changes, extremes of pH, certain chemicals—elicit muscle activity, but laboratory experiments of this type typically use electrical shock. This is because it is easier to control the onset and cessation of electrical shock, as well as the strength of the stimulus.

Preparing a Muscle for Experimentation

The preparatory work that precedes the recording of muscle activity tends to be quite time-consuming. If you work in teams of two or three, the work can be divided. While one of you is setting up the recording apparatus (kymograph or physiograph), one or two students can dissect the frog leg. Experimentation should begin as soon as the dissection has been completed.

Various types of apparatus are used to record muscle contraction. All include a way to mark time intervals, a way to indicate exactly when the stimulus was applied, and a way to measure the magnitude of the contraction response. Instructions are provided here for setting up kymograph (Figure 16.1) and physiograph (Figure 16.2) apparatus. Specific instructions for use of recording apparatus during recording will be provided by your instructor.

 1. Before beginning the frog dissection, have the following supplies ready at your laboratory bench: a small beaker containing 20 to 30 ml of frog Ringer's solution, scissors, a metal needle probe, a glass probe with a pointed tip, a medicine dropper, cotton thread, forceps, and a glass or porcelain plate.

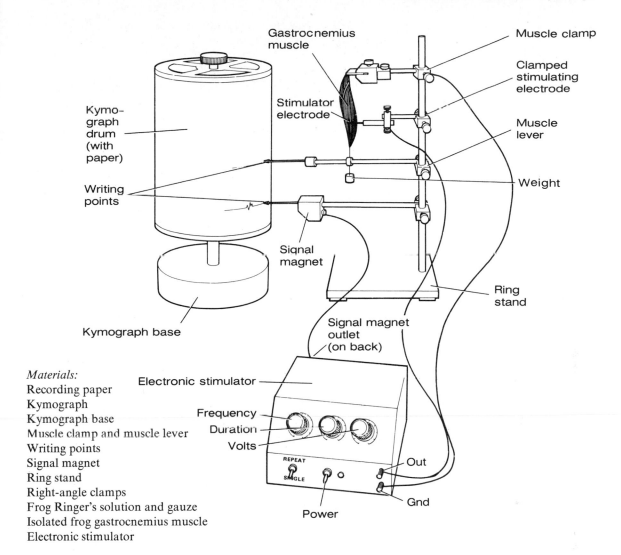

Materials:
Recording paper
Kymograph
Kymograph base
Muscle clamp and muscle lever
Writing points
Signal magnet
Ring stand
Right-angle clamps
Frog Ringer's solution and gauze
Isolated frog gastrocnemius muscle
Electronic stimulator

1. Attach muscle clamp to supporting rod and secure cut femor in clamp.

2. Position and clamp muscle lever to supporting rod beneath muscle tissue. Tie thread ligating the Achilles tendon to hook on the muscle lever so that muscle hangs vertically. Keep muscle moist with Ringer's solution from this point until completion of preparations and experimentation.

3. Hang a 10-g weight on muscle lever just below thread attachment to provide proper tension on lever. Muscle lever should rest horizontally. Writing tip should lightly touch paper surface on drum. Test its position by pulling lightly and directly upward on thread. Writing tip should remain in contact with drum surface.

4. Clamp signal magnet to supporting rod. Tip of its writing stylus should line up directly beneath that of muscle lever. *Precise alignment is important* so that a correlation can be seen between the point of stimulus and onset of muscle contractions.

5. Position stimulator to one side so that area immediately in front of recording apparatus is free for manipulations during and after experimentation. Turn all its switches to OFF and all dials to left. Plug the stimulator into the power outlet.

6. Connect two gray wires from signal magnet outlet (back of stimulator) to binding posts on signal magnet. To operate signal magnet, toggle switch (upper left on front of stimulator) must be ON (flipped up). Signal magnet will indicate up to 30 pulses per second when it is operating. If you use frequencies greater than 30 pulses per second during experimentation, turn signal magnet toggle switch to OFF (down) position.

7. Wash platinum tips of stimulator electrode in distilled water and dry. Connect electrode to white and black output terminals of stimulator and clamp electrode to supporting rod so that its tips rest on muscle tissue.

Figure 16.1

Kymograph setup for frog gastrocnemius experiments.

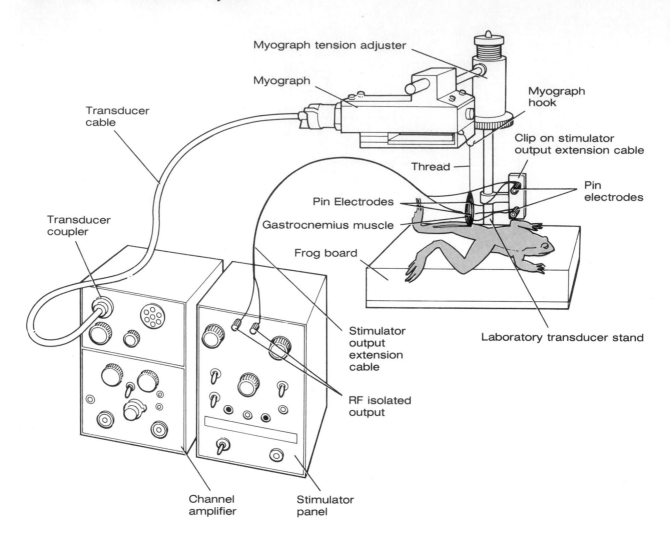

Myograph tension adjuster

Myograph

Transducer cable

Myograph hook

Clip on stimulator output extension cable

Thread

Transducer coupler

Pin electrodes

Pin Electrodes

Gastrocnemius muscle

Frog board

Stimulator output extension cable

Laboratory transducer stand

RF isolated output

Channel amplifier

Stimulator panel

Materials:
Channel amplifier and stimulator transducer cable
Stimulator panel and stimulator output extension cable
Myograph
Myograph tension adjuster
Transducer stand
Two pin electrodes
Frog board and straight pins
Prepared frog (gastrocnemius muscle freed and
    Achilles tendon ligated with thread)
Frog Ringer's solution

1. Connect myograph to transducer stand and attach frog board to stand.

2. Attach transducer cable to myograph and to input connection on amplifier channel.

3. Attach stimulator output extension cable to output on stimulator panel (red to red, black to black).

4. Using clip at opposite end of extension cable, attach cable to bottom of transducer stand adjacent to frog board.

5. Attach two pin electrodes securely to electrodes on clip.

6. Place knee of prepared frog in clip on frog board and secure by inserting a straight pin through tissues of frog. Keep frog muscle moistened with Ringer's solution.

7. Attach thread ligating the Achilles tendon of frog to myograph leaf-spring hook.

8. Adjust position of myograph on stand to produce a constant tension on thread attached to muscle (taut but not tight). Gastrocnemius muscle should hang vertically directly below myograph hook.

9. Insert free ends of pin electrodes into muscle, one at proximal end and other at distal end.

Figure 16.2

Physiograph setup for frog gastrocnemius experiments.

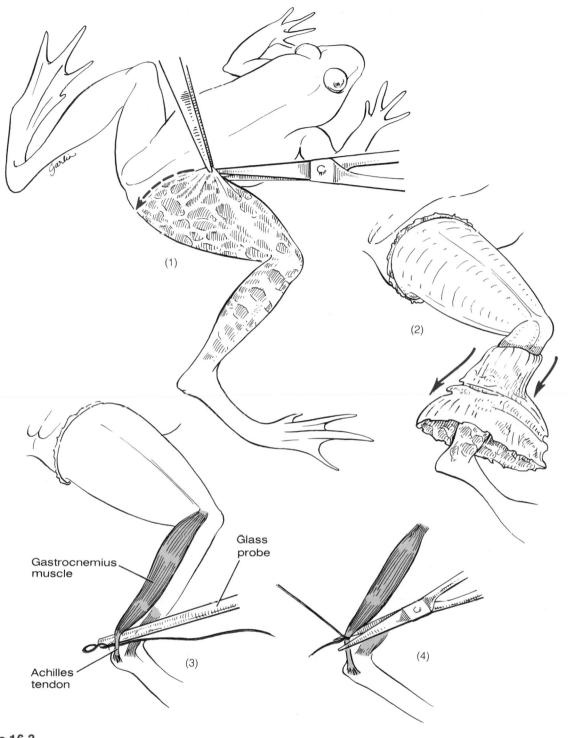

Figure 16.3

Preparation of the frog gastrocnemius muscle. Numbers indicate sequence of manipulations.

2. Obtain a frog. Hold it in your left hand, its dorsal side toward your palm. Grasp it firmly around the neck with your thumb and forefinger. Place one blade of the scissors into the frog's mouth and quickly cut off the tip of its head, well posterior to its eyes. Insert the dissecting needle into the spinal cord cavity and rotate it to destroy all spinal cord connections. When this is done properly, the frog's hindlegs will extend convulsively and then relax completely. Preparation of the frog in this manner renders it unable to feel pain and prevents reflex movements (like hopping) that would interfere with the experiments.

3. Place the frog, ventral surface down, on the glass plate. Make an incision into the skin approximately midthigh (Figure 16.3), and then continue the cut completely around the thigh. Grasp the skin with the forceps and strip it from the leg and hindfoot. The skin tends to adhere more at the joints, but a careful, persistent pulling motion—somewhat like pulling off a nylon stocking—will enable you to remove it in one piece. *From this point on, the exposed muscle tissue should be kept continuously moistened* with the Ringer's solution to prevent spontaneous twitches.

4. Identify the gastrocnemius muscle (the fleshy muscle of the posterior calf) and the Achilles tendon that secures it to the heel.

5. Slip a glass probe under the gastrocnemius muscle and run it along the entire length and under the Achilles tendon to free them from the underlying tissues.

6. Cut a piece of thread about 10 in. long and use the glass probe to slide the thread under the Achilles tendon. Knot the thread firmly around the tendon and then sever the tendon distal to the thread. If you are using a physiograph, the frog is now ready for experimentation (see Figure 16.2). If you are using a kymograph, the gastrocnemius muscle must be completely isolated, as described in step 7.

7. Cut away the fibulotibial bone just distal to the knee. Expose the femur of the thigh and cut it completely through at midthigh. Remove as much of the thigh muscle tissue as possible by carefully cutting it away with the scissors. The isolated gastrocnemius muscle can now be mounted on the muscle bar, and the stimulating electrodes of the kymograph can be attached (see Figure 16.1). About halfway through the laboratory period, dissect the second leg for use.

Recording Muscle Activity

The **"all or none" law** of muscle physiology states that a muscle cell will contract maximally when stimulated adequately. Skeletal muscles, however,

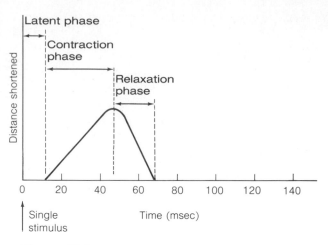

Figure 16.4

Tracing of a muscle twitch.

consisting of thousands of muscle cells, react to stimuli with graded responses. Thus muscle contractions can be slight or vigorous, depending on the requirements of the task. The graded responses (different degrees of shortening) of a skeletal muscle depend on the number of muscle cells being stimulated. In the intact organism, the number of motor units firing at any one time determines how many muscle cells will be stimulated; in this laboratory, the frequency and strength of an electrical current determines the response.

A single contraction of skeletal muscle is called a **muscle twitch.** A tracing of a muscle twitch (Figure 16.4) shows three distinct phases: latent, contraction, and relaxation. The **latent phase** is the interval from the stimulus application until the muscle begins to shorten. Although no activity is indicated on the tracing during this phase, important electrical and chemical changes are occurring within the muscle. During the **contraction period,** the muscle fibers shorten and the tracing shows an increasingly higher needle deflection, and the tracing peaks. During the **relaxation period,** represented by a downward curve, the muscle fibers relax and lengthen. On a slowly moving recording surface, the single muscle twitch appears as a spike (rather than a bell-shaped curve, as in Figure 16.4), but on a rapidly moving recording surface, the three distinct phases just described become recognizable.

DETERMINING THE MINIMAL, OR THRESHOLD, STIMULUS

1. Assuming that you have already set up the recording apparatus, set the time marker to deliver one pulse per second and the paper speed at a slow rate, approximately 0.1 cm per second.

2. Set the duration control on the stimulator at 7 msec, multiplier × one and the voltage control at zero v, multiplier × one. Turn the knob of the stimulator fully clockwise.

3. Administer single stimuli to the muscle at 1- to 2-sec intervals, beginning with 0.1 v and increasing by 0.1 v between successive stimuli until a contraction is obtained (shown by a spike on the paper).

At what voltage did contraction occur?

_____v

The voltage at which the first perceptible contractile response is obtained is called the **threshold,** or **minimal, stimulus.** All stimuli applied prior to this point are termed **subthreshold,** or **subliminal, stimuli,** because at those voltages no response was elicited.

4. Stop the recording and mark the record to indicate the threshold stimulus and voltage and time. *Do not remove the record from the recording surface;* continue with the next experiment. *Remember:* keep the muscle preparation moistened with Ringer's solution at all times.

THE TREPPE, OR STAIRCASE, PHENOMENON As a muscle is stimulated to contract, a curious phenomenon is observed in the tracing pattern of the first few twitches. Even though the stimulus intensity is unchanged, the height of the individual spikes increases in a stepwise manner—producing a sort of staircase pattern. This phenomenon is not well understood, but the following explanation has been offered: in the muscle cell's resting state, there is less Ca^{++} in the sarcoplasmic reticulum, and the enzyme systems in the muscle cell are less efficient than after it has contracted a few times. As the muscle tissue begins to contract, Ca^{++} moves into the cells from the extracellular fluid. The heat generated by muscle activity increases the efficiency of the enzyme systems, and the muscle becomes more efficient and contracts more vigorously. This is the physiologic basis of the warm-up period prior to competition in sports events.

1. Set up the apparatus as in the previous experiment, but set the voltage control at 10 v.
2. Deliver single stimuli at 1-sec intervals until the strength of contraction does not increase further.
3. Stop the recording apparatus and mark the record *treppe.* Note also the number of contractions (and seconds) required to reach the constant contraction magnitude. Record the voltage you used and the time when you completed this experiment. Continue on to the next experiment.

GRADED MUSCLE RESPONSE TO INCREASED STIMULUS INTENSITY

1. Follow the previous set-up instructions, but set the voltage control at the threshold voltage (as determined in the first experiment).
2. Deliver single stimuli at 1- or 2-sec intervals. Initially increase the voltage between shocks by 0.5 v; then increase the voltage by 1 to 2 v between shocks as the experiment continues, until contraction height increases no further. Stop the recording apparatus.

What voltage produced the highest spike (thus the maximal strength of contraction)?

_____v

This voltage, called the **maximal stimulus** (for *your* muscle specimen), is the weakest stimulus at which all muscle cells are being stimulated. As the voltage was increased to this point, more and more muscle cells (motor units) were activated, resulting in a stronger and stronger contraction. Past this point, an increase in the intensity of the stimulus will not produce any greater contractile response. **Multiple motor unit summation** is the process by which the increased contractile strength reflects the relative number of muscle cells stimulated.

3. Mark the record *multiple motor unit summation* and note the maximal stimulus voltage and the time you completed the experiment. Continue on to the next experiment.

TIMING THE MUSCLE TWITCH

1. Follow the previous set-up directions, but set the voltage for the maximal stimulus (as determined in the preceding experiment) and set the paper advance or recording speed at maximum (about 20 cm per second on the physiograph; 2 cm per second with the kymograph).
2. Deliver single stimuli at 2- to 3-sec intervals to obtain several "twitch" curves. Stop the recording.
3. Determine the duration of the latent, contraction, and relaxation phases of the twitches and record here:

Duration of latent period: _____ sec

Duration of contraction period: _____ sec

Duration of relaxation period: _____ sec

4. Label the record to indicate the point of stimulus, the beginning of contraction, the end of contraction, and the end of relaxation.
5. Allow the muscle to rest (but keep it moistened) before continuing with the next experiment.

GRADED MUSCLE RESPONSE TO INCREASED STIMULUS FREQUENCY

Muscles subjected to frequent stimulation, without a chance to relax, exhibit two kinds of responses—wave summation and tetanus—depending on the level of frequency.

Wave Summation: If a muscle is stimulated with a rapid series of stimuli of the same intensity before it has had a chance to relax completely, the response to the second and subsequent stimuli will be greater than to the first stimulus. This phenomenon, called wave summation, occurs because the muscle is already in a partially contracted state when the stimuli are delivered.

1. With the recorder running at slow speed and the stimulus intensity set 20 v higher than the threshold stimulus, stimulate the muscle at a rate of 15 to 25 stimuli per second.
2. Shut off the recorder and label the record as *wave summation.* Note also the time, the voltage intensity, and the frequency.

Tetanus: Stimulation of a muscle at an even higher frequency will produce a "fusion" of the summated twitches; in effect, a single sustained contraction is achieved. Tetanus is a feature of normal skeletal muscle functioning; the single muscle twitch is primarily a laboratory phenomenon.

1. To demonstrate tetanus, maintain the conditions you used for wave summation except for the frequency of stimulation. Set the stimulator to deliver 60 stimuli per second.
2. As soon as you obtain a single smooth, sustained contraction (with no evidence of relaxation), discontinue stimulation and shut off the recorder.
3. Label the tracing with the conditions of experimentation, the time, and the area of tetanus.

MUSCLE FATIGUE

Muscle fatigue, the loss of the ability to contract, is believed to be a result of the oxygen debt that occurs in the tissue after prolonged activity (through the accumulation of such waste products as lactic acid as well as the depletion of ATP). True muscle fatigue rarely occurs in the body, because it is most often preceded by a subjective feeling of fatigue. Furthermore, fatigue of the myoneural junctions typically precedes fatigue of the muscle.

1. To demonstrate muscle fatigue, set up an experiment like the tetanus experiment, but allow the stimulus to be delivered until the muscle completely relaxes and the contraction curve returns to the base line.
2. Measure the time interval between the beginning of complete tetanus and the beginning of fatigue (when the tracing begins its downward curve). Mark the record appropriately.

3. Determine the time required for complete fatigue to occur (the time interval from the beginning of fatigue until the return of the curve to the base line). Mark the record appropriately.
4. Allow the muscle to rest (keeping it moistened with Ringer's solution) for 10 min, and then repeat the experiment.

What was the effect of the rest period on the fatigued muscle? _____

_____

What is the physiologic basis for this reaction?

_____

_____

5. Discard the muscle preparation or frog in the appropriate organic debris container.

THE EFFECT OF LOAD ON SKELETAL MUSCLE

When the fibers of a skeletal muscle are slightly stretched by a weight or tension, the muscle responds by contracting more forcibly and thus is capable of doing more work. If the load is increased beyond the optimum, the latent period becomes longer, contraction height decreases, and relaxation (fatigue) occurs more quickly. With excessive stretching, the muscle is unable to develop any tension and no contraction occurs. Since the filaments no longer overlap at all with this degree of stretching, the sliding force cannot be generated.

If your equipment allows you to add more weights to the muscle specimen or to increase the tension on the muscle, perform the following experiment to determine the effect of loading on skeletal muscle.

1. Mount a fresh muscle preparation to the muscle bar, but do not weight the muscle or apply additional tension to it.
2. Determine the threshold stimulus for the muscle and then set the stimulator to deliver 0.1 v (100 mv) more.
3. Stimulate the unweighted muscle with single shocks at 1- to 2-sec intervals to achieve three or four muscle twitches.
4. Stop the recording apparatus and add 10 g of weight or tension to the muscle. Restart the recording and stimulate again to obtain three or four spikes.
5. Repeat the previous step five more times, increasing the weight by 10 g each time until the total load on the muscle is 60 g.
6. Discontinue recording and remove the tracing. Mark the curves on the record to indicate the load (in grams).

7. Measure the height of contraction (in millimeters) of the sequence of twitches obtained with each load, and insert this information into the chart below.
8. Compute the work done by the muscle for each twitch (load) sequence.

Weight of load (g) × distance load lifted (mm) = work done

Enter these calculations into the chart.

Load (g)	Distance load lifted (mm)	Work done
0		
10		
20		
30		
40		
50		
60		

9. Plot a line graph of work done against the weight on the accompanying grid.

Work

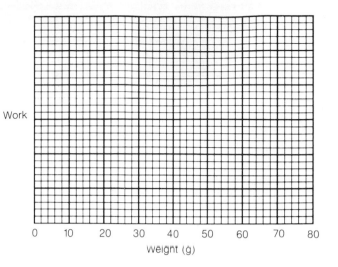

weight (g)

10. Dismantle all apparatus and prepare the equipment for storage. Dispose of the frog remains in the appropriate container. Inspect your records of the experiments and make sure each is fully labeled with the experimental conditions, the date, and the names of those who conducted the experiments. (If you used a smoked-drum kymograph recording system, "fix" the records before leaving the laboratory.) Attach a tracing (or a photocopy of the tracing) for each experiment to this page for future reference.

UNIT 8

THE NERVOUS SYSTEM

Histology of Nervous Tissue

OBJECTIVES

1. To differentiate between the function of neurons and neuroglia and to list the functional characteristics of each.

2. To list four types of neuroglia cells.

3. To identify the following anatomic characteristics of a neuron on an appropriate diagram or projected slide:

   Cell body (perikaryon)
   Nucleus and nucleolus
   Neuron fibers (axons, dendrites)
   Neurilemma
   Nodes of Ranvier
   Nissl bodies
   Neurofibrils
   Myelin sheath
   Axonal end bulbs
   Axon hillock

4. To state the functions of axons, dendrites, axonal end bulbs, neurofibrils, and myelin sheaths.

5. To explain how a nerve impulse is transmitted from one neuron to another.

6. To explain the role of Schwann cells in the formation of the myelin sheath.

7. To classify neurons according to structure (unipolar, bipolar, multipolar) and function (motor, sensory, internuncial).

8. To distinguish between a nerve and a tract and between a ganglion and a nucleus.

9. To describe the structure of a nerve, identifying the connective tissue coverings (endoneurium, perineurium, and epineurium) and citing their functions.

MATERIALS

Model of a "typical" neuron (if available)
Compound microscope
Histologic slides of an ox spinal cord smear and teased myelinated nerve fibers

Prepared slides of Purkinje cells (cerebellum), pyramidal cells (cerebrum), and a dorsal root ganglion
Prepared slide of a nerve (cross section)

The nervous system is the master integrating and coordinating system, continuously monitoring and processing sensory information both from the external environment and from within the body. Every thought, action, and sensation is a reflection of its activity. Like a computer, it processes and integrates new "inputs" with information previously fed into it ("programmed") to produce an appropriate response ("readout"). However, no man-made computer can possibly compare in complexity and scope to the human nervous system.

Despite its complexity, nervous tissue is made up of just two principal cell populations: the **neurons** and the **neuroglia** (glial cells). The *neuroglia*, literally "nerve glue," are essentially supportive cells. They include astrocytes, oligodendrocytes, microglia, and ependymal cells (Figure 17.1). These cells serve the needs of the neurons by acting as supportive and protective cells, myelinating cells, and phagocytes. In addition, they probably serve a nutritive function by acting as a selective barrier between the capillary blood supply and the neurons. Although neuroglia resemble neurons in some ways (they have fibrous cellular extensions), they are not capable of generating and transmitting nerve impulses, a capability that is highly developed in neurons.

NEURON ANATOMY

The delicate **neurons** are the structural units of nervous tissue. They are highly specialized to exhibit irritability and conductivity, both essential for the effective transmission of messages (nerve impulses) from one part of the body to another.

Although neurons differ structurally, they have many identifiable features in common (Figure 17.2). All have a **cell body** from which slender processes or fibers extend. Nerve cell bodies, which are found only in the CNS (brain or spinal cord) or in **ganglia** (collections of nerve cell bodies outside the CNS), make up the gray matter of the nervous system. The neuron processes running through the CNS form **tracts** of white matter; outside the CNS they form the peripheral **nerves**.

The nerve cell body contains a large round nucleus surrounded by cytoplasm (neuroplasm). The cytoplasm is riddled with neurofibrils and with darkly staining structures called Nissl bodies. **Neurofibrils** are believed to have a support and nutritive function; **Nissl bodies**, an elaborate type of endoplasmic reticulum, are involved in the metabolic activities of the cell.

Neuron processes that conduct impulses *toward* the nerve cell body are called **dendrites**; those that

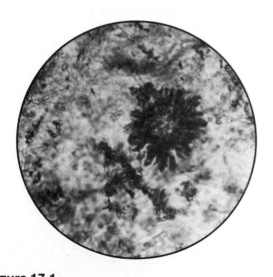

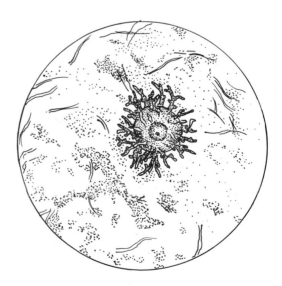

(a)

Figure 17.1

Types of neuroglia: (a) astrocyte.

(continued.)

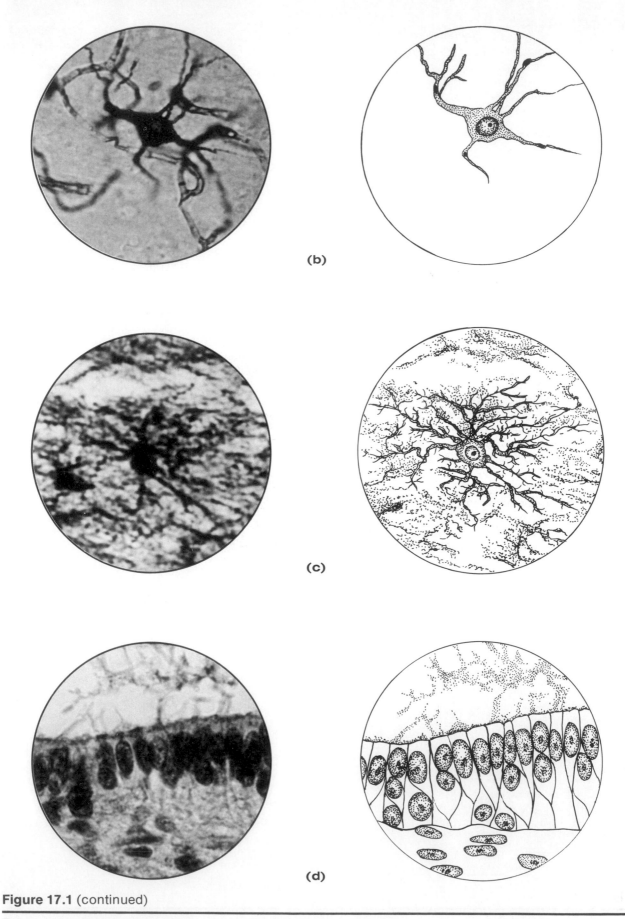

Figure 17.1 (continued)

Types of neuroglia: (b) oligodendrocytes; (c) microglia; (d) ependymal cells.

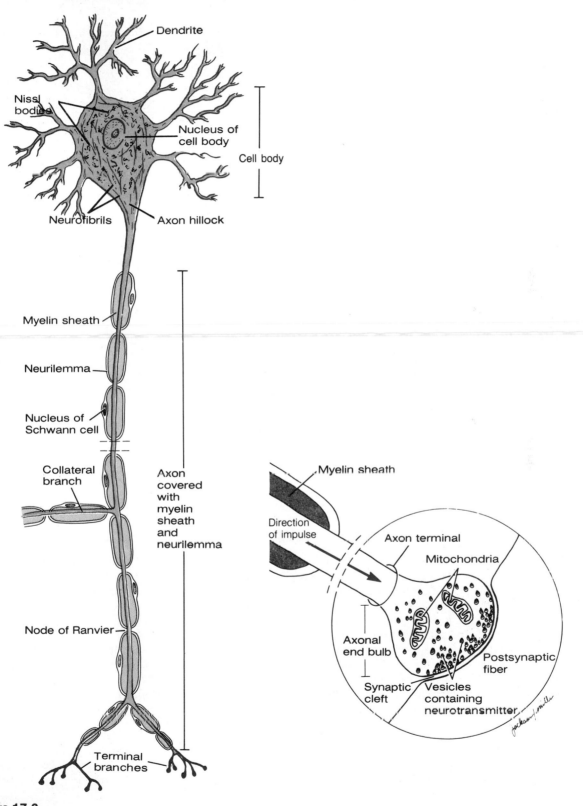

Figure 17.2

Structure of a typical motor neuron (inset shows synapse enlarged).

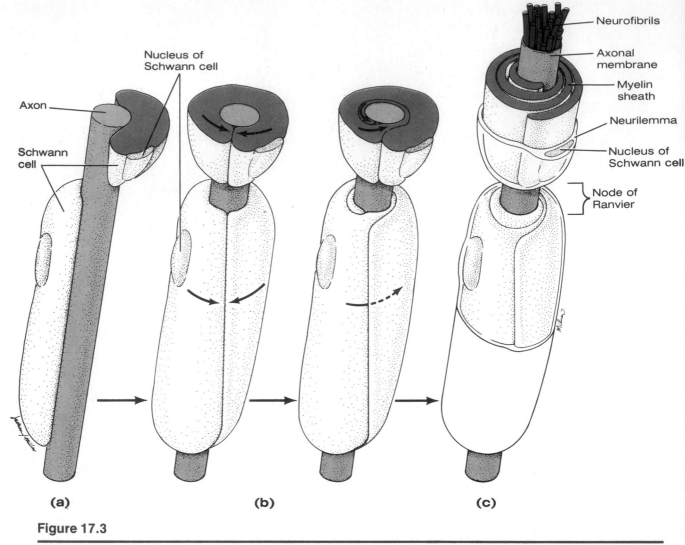

Figure 17.3

Myelination of neuron processes by individual Schwann cells: (a) a Schwann cell becomes apposed to the axon; (b) the Schwann cell coils around the axon; (c) the process is completed, and a myelinated fiber is formed.

carry impulses *away from* the nerve cell body are called **axons**. Typically, neurons have only one axon (which may branch into **collaterals**) but may have many dendrites, depending on the neuron type.

In general, a neuron is excited by other neurons when their axons release neurotransmitters close to its dendrites or nerve cell body. The impulse travels across the nerve cell body and down the axon. As Figure 17.2 shows, the axon fiber begins at a slightly enlarged cell body structure called the **axon hillock** and ends in many small structures called **axonal end bulbs,** or synaptic knobs. These end bulbs store the neurotransmitter chemical in tiny vesicles. The enlarged view of the axonal end bulb reveals that it is separated from the cell body or dendrites of the next neuron by a tiny gap called the **synaptic cleft**. Although they are close, there is no actual physical contact between neurons. When an impulse reaches the axonal end bulbs, the synaptic vesicles rupture and release the neurotransmitter into the synapse. The neurotransmitter then diffuses across the syn-

apse to bind to membrane receptors on the next neuron, initiating the action potential.*

Most long nerve fibers (both axons and dendrites) are covered with a fatty material called myelin, and such fibers are referred to as **myelinated fibers**. Axons leaving the CNS are typically heavily myelinated by special cells called **Schwann cells**, which wrap themselves tightly around the axon jelly-roll fashion (Figure 17.3). During the wrapping process, the cytoplasm is squeezed from between adjacent layers of the Schwann cell membranes, so that when the process is completed a tight core of cell membrane material (protein-lipoid material) encompasses the axon. This wrapping is the **myelin sheath**. The bulk of the Schwann cell cytoplasm ends up just beneath the outermost portion of its

*Specialized synapses in skeletal muscle are called myoneural junctions. They are discussed in Exercise 14.

cell membrane. The peripherally exposed part of the Schwann cell membrane is referred to as the **neurilemma**. Since the myelin sheath is formed by many individual Schwann cells, it has gaps or is a discontinuous sheath; the indentations in the sheath are called **nodes of Ranvier** (see Figure 17.2).

Within the CNS, myelination is accomplished by glial cells. These CNS sheaths do not exhibit the nodes of Ranvier seen in fibers myelinated by Schwann cells. Because of its chemical composition, myelin acts to insulate the fibers and greatly increases the speed of neurotransmission by neuron fibers.

1. Study the typical neuron shown in Figure 17.2, noting the structural details described above, and then identify these structures on a neuron model.

2. Obtain a prepared slide of the ox spinal cord smear, which has large, easily identifiable neurons. Study one representative neuron under oil immersion and identify the nerve cell body, the nucleus, the large prominent nucleoli, and the granular Nissl bodies. If possible, distinguish the axon from the many dendrites. Sketch the cell in the space provided here, and label the important anatomic details you have observed.

3. Obtain a prepared slide of teased myelinated nerve fibers and identify the following: nodes of Ranvier, neurilemma, axis cylinder (the axon itself), Schwann cell nuclei, and myelin sheath.

Do the nodes seem to occur at consistent intervals,

or are they irregularly distributed? _____

Explain the significance of this finding: _____

_____

_____

_____

_____

Sketch a portion of a myelinated nerve fiber in the space provided here, illustrating two or three nodes of Ranvier. Label the axis cylinder, myelin sheath, nodes, and neurilemma.

NEURON CLASSIFICATION

Neurons may be classified on the basis of structure or of function. Figure 17.4 depicts both schemes.

Basis of Structure

Neurons may be differentiated according to the number of processes attached to the nerve cell body. In **unipolar neurons**, one very long process extends from the nerve cell body. Physiologically, only the most distal portion of the process acts as a dendrite; the rest acts as an axon. All neurons that conduct impulses toward the CNS are unipolar.

Bipolar neurons have two processes—one axon and one dendrite—attached to the nerve cell body. This neuron type is quite rare, typically found only as part of the receptor apparatus of the eye, ear, and olfactory mucosa.

Many processes issue from the cell body of **multipolar neurons**, all classified as dendrites except for a single axon. Most neurons in the brain and spinal cord (CNS neurons) and those whose axons carry impulses away from the CNS fall into this last category.

 Obtain prepared slides of Purkinje cells of the cerebellar cortex, pyramidal cells of the cerebral cortex, and a dorsal root ganglion. As you observe them under the microscope, try to pick out the anatomic details depicted in Figure 17.5. Note that the motor neurons of the cerebral and cerebellar tissues (both brain tissues) are extensively branched; in contrast, the neurons of the dorsal root ganglion are more rounded. You may also be able to identify astrocytes (neuroglia) in the brain tissue slides if you examine them closely.

Which of these neuron types would be classified as

multipolar neurons? _____

Which as unipolar? _____

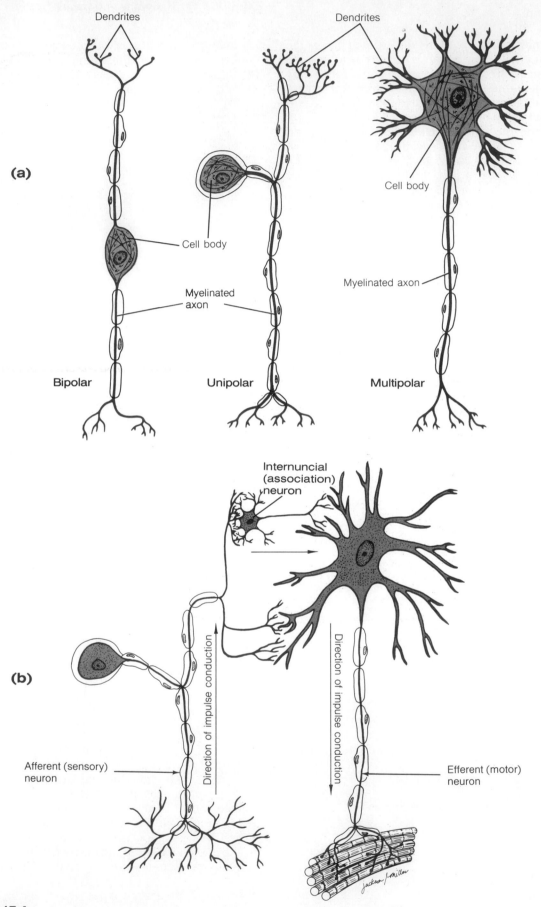

Figure 17.4

Classification of neurons: (a) on the basis of structure; (b) on the basis of function.

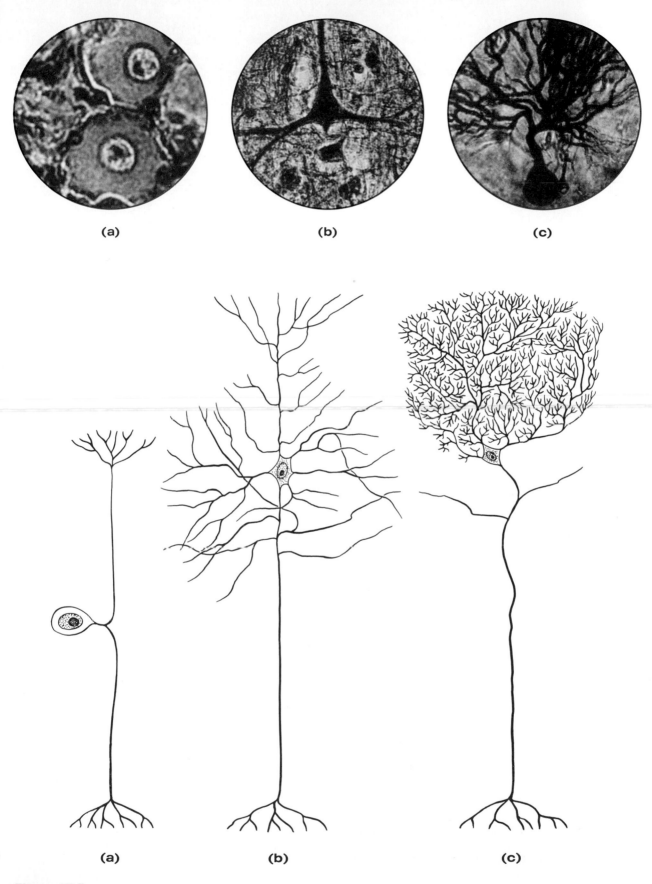

Figure 17.5

Structure of selected neurons: (a) neurons of the dorsal root ganglion; (b) pyramidal cell of the cerebral cortex; (c) Purkinje cells of the cerebellum.

Basis of Function

In general, neurons carrying impulses from the sensory receptors in the internal organs (viscera) or in the skin are termed **afferent**, or **sensory**, **neurons** (refer back to Figure 17.4). The dendritic endings of sensory neurons are generally equipped with specialized receptors that are stimulated by specific changes in their immediate environment. The structure and function of these receptors is considered separately in Exercise 21 (General Sensation). The nerve cell bodies of sensory neurons are always found in a ganglion outside the CNS.

Neurons carrying activating impulses from the CNS to the viscera and/or body muscles and glands are termed **efferent**, or **motor**, **neurons**. The cell bodies of motor neurons are always located in the CNS.

The third functional category of neurons are the **association**, or **internuncial**, **neurons**, which connect sensory and motor neurons. Their nerve cell bodies, like those of motor neurons, are always located within the CNS.

Can you correlate the structural classifications with the functional classifications? (In other words, does the number of processes have anything to do with

a neuron's function?) _____

_____

_____

_____

STRUCTURE OF A NERVE

A nerve is a bundle of neuron fibers or processes wrapped in connective tissue coverings that extends to and/or from the CNS and visceral organs or structures of the body periphery (such as skeletal muscles, glands, and skin).

Within a nerve, each fiber is surrounded by a delicate connective tissue sheath called an **endoneurium**, which insulates it from the other neuron processes adjacent to it. (The endoneurium is often mistaken for the myelin sheath; it is instead an additional sheath that surrounds the myelin sheath.) Groups of fibers are bound by a coarser connective tissue, called the **perineurium**, to form bundles of fibers called **fascicles**. Finally, all the fascicles are bound together by a tough, white fibrous connective tissue sheath called the **epineurium** to form the cordlike nerve (Figure 17.6). In addition to the connective tissue wrappings, blood vessels and lymphatic vessels serving the fibers also travel within a nerve.

Like neurons, nerves are classified according to the direction in which they transmit impulses. Nerves carrying both sensory (afferent) and motor

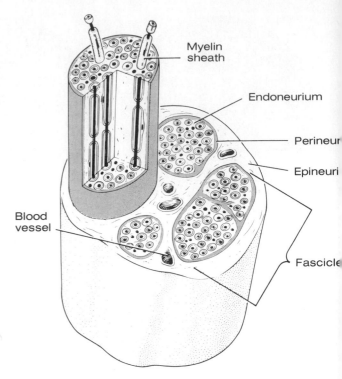

Figure 17.6

Structure of a nerve cross section, showing connective tissue wrappings.

(efferent) fibers are called **mixed nerves**; all spinal nerves are mixed nerves. Nerves that carry only sensory processes and conduct impulses only toward the CNS are referred to as **afferent**, or **sensory**, **nerves**. A few of the cranial nerves are pure sensory nerves, but the majority are mixed nerves. The ventral roots of the spinal cord, which carry only motor fibers, can be considered **efferent**, or **motor, nerves.**

Examine under the compound microscope a prepared cross section of a peripheral nerve. Sketch the nerve, identifying and labeling nerve fibers, myelin sheaths, fascicles, and endoneurium, perineurium, and epineurium sheaths.

Gross Anatomy of the Brain and Cranial Nerves

EXERCISE 18

OBJECTIVES

1. To identify or locate the following brain structures on a dissected specimen, human brain model (or slices), or appropriate diagram, and to state their functions:

   - *cerebral hemisphere structures:* lobes, important fissures, lateral ventricles, basal nuclei, corpus callosum, fornix, septum pellucidum
   - *brainstem structures:* thalamus, intermediate mass, hypothalamus, optic chiasma, pituitary gland, mammillary bodies, pineal body, choroid plexus of the third ventricle, foramen of Monro; corpora quadrigemina, cerebral aqueduct, cerebral peduncles; pons, medulla, fourth ventricle
   - *cerebellum structures:* cerebellar hemispheres, vermis, arbor vitae

2. To describe the composition of the gray and white matter.

3. To locate the well-recognized functional areas of the human cerebral hemispheres.

4. To define *gyri,* and *fissures* (sulci).

5. To identify the three meningeal layers and state their function, and to locate the falx cerebri, falx cerebelli, and tentorium cerebelli.

6. To state the function of the arachnoid villi and dural sinuses.

7. To discuss the formation, circulation, and drainage of cerebrospinal fluid, describing its function and composition and the effects of obstructed circulation or drainage.

8. To identify at least four pertinent anatomic differences between the human brain and that of the sheep (or other mammal)

9. To identify the cranial nerves by number and name on an appropriate model or diagram, stating the origin and function of each.

MATERIALS

Human brain model (dissectible)
Preserved human brain (if available)
Coronally sectioned human brain slice (if available)
Preserved sheep brain (meninges and cranial nerves intact)
Dissecting tray and instruments

When viewed alongside all nature's animals, humans are indeed unique; the key to their uniqueness is found in the brain. Only in humans has the brain region called the cerebrum become so elaborated and grown out of proportion that it overshadows other brain areas. The lower animals are primarily concerned with informational input and response for the sake of survival and preservation of the species, but human beings devote considerable time to nonsurvival ends. They are the only animals who manipulate abstract ideas and search for knowledge for its own sake, who are capable of emotional response and artistic creativity, or who can anticipate the future and guide their lives according to ethical and moral values. For all this, humans can thank their overgrown cerebrum.

We can be considered composite reflections of our brain's experience. If all past sensory input could mysteriously and suddenly be "erased," we would be unable to walk, talk, or communicate in any manner. Spontaneous movement would be apparent, as in a fetus, but no voluntary integrated function of any type would be possible. Clearly we would cease to be the same individuals.

Because of the complexity of the nervous system, its anatomic structures are usually considered in terms of two principal divisions: the **central nervous system** (CNS) and the **peripheral nervous system** (PNS). The central nervous system consists of the brain and spinal cord, which primarily interpret incoming sensory information and issue instructions based on past experience. The periph-

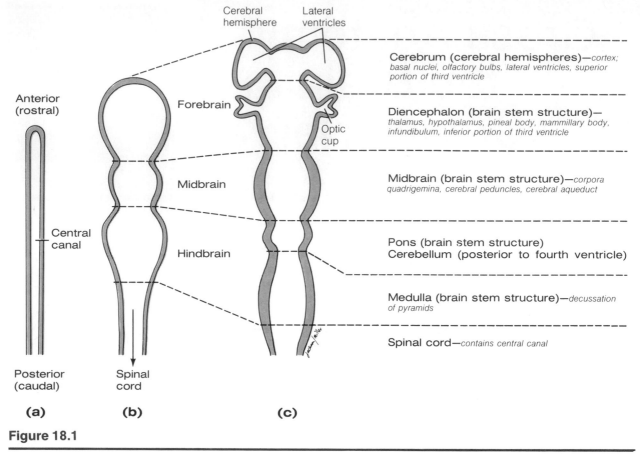

Cerebral hemisphere · Lateral ventricles

Anterior (rostral)

Forebrain

Cerebrum (cerebral hemispheres)—*cortex; basal nuclei, olfactory bulbs, lateral ventricles, superior portion of third ventricle*

Diencephalon (brain stem structure)— *thalamus, hypothalamus, pineal body, mammillary body, infundibulum, inferior portion of third ventricle*

Optic cup

Midbrain

Midbrain (brain stem structure)—*corpora quadrigemina, cerebral peduncles, cerebral aqueduct*

Central canal

Hindbrain

Pons (brain stem structure) Cerebellum (posterior to fourth ventricle)

Medulla (brain stem structure)—*decussation of pyramids*

Spinal cord—*contains central canal*

Posterior (caudal)

Spinal cord

(a) **(b)** **(c)**

Figure 18.1

Embryonic development of the brain. Adult brain structures developing from forebrain, midbrain, and hindbrain subdivisions. (a) Neural tube precedes (b) subdivisions of neural tube, which precede (c) the adult brain structures.

eral nervous system consists of the cranial and spinal nerves, ganglia, and sensory receptors. These structures serve as the communication lines carrying impulses—from the sensory receptors to the CNS and from the CNS to the appropriate glands or muscles.

In this exercise both CNS (brain) and PNS (cranial nerves) structures will be studied because of their close anatomic relationship.

THE HUMAN BRAIN

During embryonic development of all verte-brates, the CNS first makes its appearance as a simple tubelike structure, the **neural tube,** that extends down the dorsal medial plane. By the fourth week, the human brain begins to form as an expan-sion of the anterior or rostral end of the neural tube (the end toward the head). Shortly thereafter, constrictions appear, dividing the developing brain into three major regions—the forebrain, midbrain, and hindbrain (Figure 18.1). The remainder of the neural tube becomes the spinal cord.

During fetal development, two anterior outpock-etings extend from the forebrain and grow rapidly to form the cerebral hemispheres. Because of space

restrictions imposed by the skull, the cerebral hemi-spheres grow posteriorly and inferiorly and finally end up enveloping and obscuring the rest of the forebrain and midbrain structures. Somewhat later in development, the dorsal portion of the hindbrain also enlarges to produce the cerebellum. The central canal of the neural tube, which remains continuous throughout the brain and cord, becomes enlarged in four regions of the brain to form chambers called **ventricles.**

External Anatomy

Generally, in studying the major brain areas in the laboratory, the brain is considered in terms of three major regions: the cerebral hemispheres, brain stem and cerebellum. The correlation between these three anatomic regions and the structures of the forebrain, midbrain, and hindbrain is also outlined in Figure 18.1.

CEREBRAL HEMISPHERES The cerebral hemi-spheres are the most superior portion of the brain (Figure 18.2). Their entire surface is thrown into elevated ridges of tissue called **gyri** that are separated by depressed areas called **fissures,** or **sulci.** Many of these fissures and gyri are important anatomic landmarks.

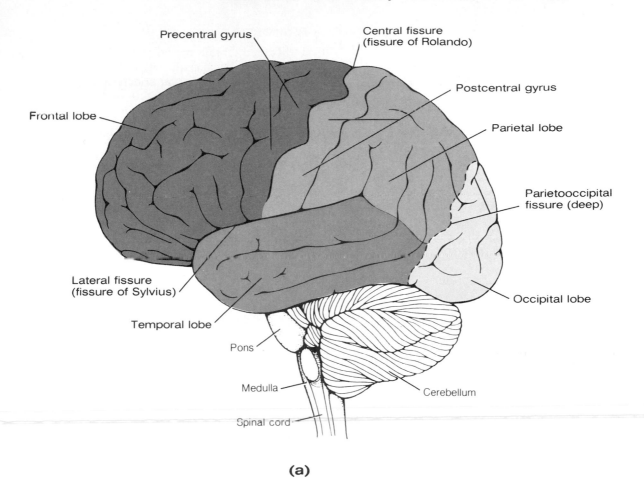

(a)

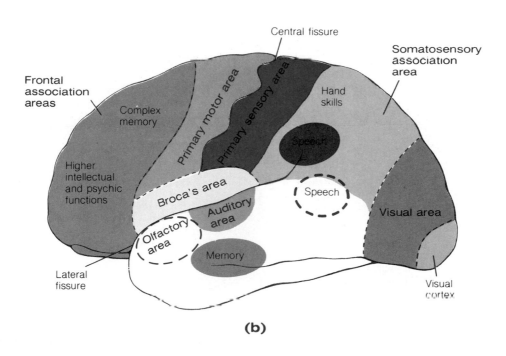

(b)

Figure 18.2

Left lateral view of the human brain: (a) major structural areas; (b) major functional areas of the left cerebral hemisphere.

The cerebral hemispheres are divided by a single deep fissure, the **longitudinal fissure.** The **central fissure** (fissure of Rolando) divides the **frontal lobe** from the **parietal lobe,** and the **lateral fissure** (fissure of Sylvius) separates the **temporal lobe** from the parietal lobe. The **parieto-occipital** fissure, which divides the **occipital lobe** from the parietal lobe, is not visible externally. Note that the cerebral lobes are named for the cranial bones that lie over them.

The functional areas of the cerebral hemispheres have also been located. The **primary** (somatic) **sensory area** is located in the **postcentral gyrus** of the parietal lobe. Impulses traveling from the body's sensory receptors (such as those for pressure, pain, and temperature) are localized in this area of the brain. ("This information is from my big toe.") Immediately posterior to the primary sensory area is the **somatosensory association area,** in which the meaning of incoming stimuli is analyzed. ("Ouch! I have a *pain* there.") Thus, the somatosensory association area allows you to become aware of pain, coldness, a light touch, and the like.

Impulses from the special sense organs are interpreted in other specific areas also noted in Figure 18.2(b). For example, the visual area is located in the posterior portion of the occipital lobe, the auditory area is located in the temporal lobe in the gyrus bordering the lateral fissure, and the olfactory area is located deep within the temporal lobe along its medial surface.

The **primary motor area,** which is responsible for conscious or voluntary movement of the skeletal muscles, is located in the **precentral gyrus** of the frontal lobe. A specialized motor speech area called **Broca's area** is found at the base of the precentral gyrus just above the lateral fissure. Damage to this area (which is located only in one cerebral hemisphere, usually the left) reduces or eliminates the ability to articulate words.

Areas involved in higher intellectual reasoning are believed to lie in the anterior portions of the frontal lobes; complex memories appear to be stored in the temporal lobe. A rather poorly defined region at the junction of the temporal, parietal, and occipital lobes is the **speech initiation area,** in which vocal response to incoming thoughts (speech and language relationships) is initiated.

The cell bodies of cerebral neurons involved in these functions are found only in the outermost gray matter of the cerebrum, the area called the **cerebral cortex.** The balance of cerebral tissue—the deeper white matter, or the **cerebral medulla**—is composed of fiber tracts carrying impulses to or from the cortex.

Using a model of the human brain (and a preserved human brain, if available), identify the areas and structures of the cerebrum described above.

BRAIN STEM

1. Turn the brain model so the ventral surface of the brain stem can be viewed. Using Figure 18.3 as a guide, start superiorly and identify the **olfactory bulbs** and **tracts, optic nerves, optic chiasma** (where the fibers of the optic nerves partially cross over), **optic tract, pituitary gland,** and **mammillary bodies.**
2. Continue inferiorly to identify the **cerebral peduncles** (fiber tracts in the **midbrain** connecting the pons below with cerebrum above), the pons, and the medulla oblongata. *Pons* means "bridge," and the **pons** consists primarily of motor and sensory fiber tracts connecting the brain with lower CNS centers. The lowest brain stem region, the **medulla,** is also composed primarily of fiber tracts. You can see the **decussation of pyramids,** a crossover point for the major motor tract (pyramidal tract) descending from the motor areas of the cerebrum to the cord, on the medulla's anterior surface. The medulla also houses many vital autonomic centers involved in the control of heart rate, respiratory rhythm, and blood pressure as well as involuntary centers involved in the initiation of vomiting, swallowing, and so on.

CEREBELLUM

1. Turn the brain model so you can see the dorsal aspect. Identify the large cauliflower-like cerebellum, which projects dorsally from under the occipital lobe of the cerebrum. Note that, like the cerebrum, the cerebellum has two hemispheres and a convoluted surface. It also has an outer cortex made up of gray matter with an inner region of white matter.
2. Remove the cerebellum to view the **corpora quadrigemina,** located on the posterior aspect of the midbrain. The two superior prominences are the **superior colliculi** (visual reflex centers), and the two smaller inferior prominences are the **inferior colliculi** (auditory reflex centers).

Internal Anatomy

The deeper structures of the brain have also been well mapped. Like the external structures, these can be studied in terms of the three major regions.

CEREBRAL HEMISPHERES

1. Take the brain model apart so you can see a median sagittal view of the internal brain structures (Figure 18.4). Observe the cerebrum closely to see the extent of the outer cortex (gray matter), which contains the nerve cell bodies of cerebral neurons. (The pyramidal cells of the cerebral motor cortex are representative of the neurons seen in the precentral gyrus.)
2. Now observe the deeper area of white matter, or the medulla, which is composed of fiber tracts.

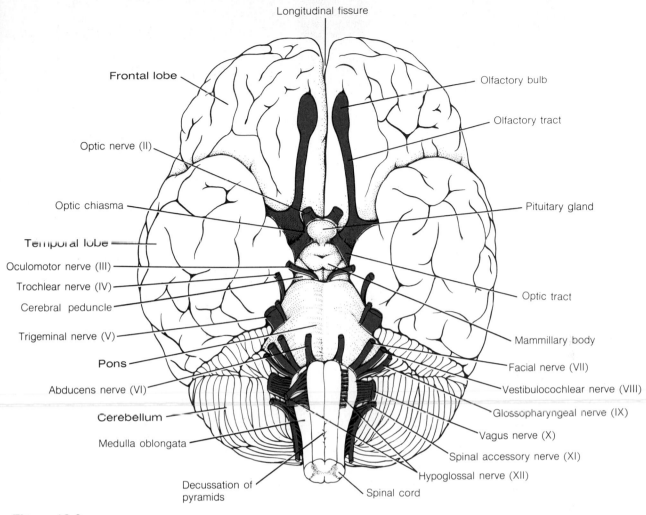

Longitudinal fissure
Frontal lobe
Olfactory bulb
Olfactory tract
Optic nerve (II)
Optic chiasma
Pituitary gland
Temporal lobe
Oculomotor nerve (III)
Trochlear nerve (IV)
Cerebral peduncle
Optic tract
Trigeminal nerve (V)
Pons
Mammillary body
Abducens nerve (VI)
Facial nerve (VII)
Vestibulocochlear nerve (VIII)
Cerebellum
Glossopharyngeal nerve (IX)
Vagus nerve (X)
Medulla oblongata
Spinal accessory nerve (XI)
Hypoglossal nerve (XII)
Decussation of pyramids
Spinal cord

Figure 18.3

Ventral surface of the human brain.

The fiber tracts found in the cerebral medulla are named association tracts if they connect two portions of the same hemisphere, projection tracts if they run between the cerebral cortex and the lower brain or spinal cord, and commissures if they run from one hemisphere to another. Observe the large **corpus callosum,** the major commissure connecting the cerebral hemispheres. The corpus callosum arches above the structures of the diencephalon. Note also the **fornix,** a band-like fiber tract concerned with olfaction as well as limbic system functions, and the membranous **septum pellucidum,** which separates the lateral ventricles of the cerebral hemispheres.

3. In addition to the gray matter of the cerebral cortex, there are several "islands" of gray matter, the **basal nuclei,** buried deep within the white matter of the cerebral hemispheres, flanking the lateral and third ventricles. You can see the basal nuclei if you have an appropriate dissectible model or a coronally sectioned human brain slice. Otherwise, Figure 18.5 will suffice.

The basal nuclei, which are important subcortical motor nuclei (and part of the extra pyramidal system), are involved in the regulation of voluntary motor activities. The most important of them are the arching, comma-shaped **caudate nucleus,** the **claustrum,** the **amygdaloid nucleus** (located at the tip of the caudate nucleus), and the **lentiform nucleus,** which is composed of the **putamen** and **globus pallidus nuclei.** A band of projection fibers coursing down from the precentral (motor) gyrus combines with sensory fibers traveling to the sensory cortex to form a broad band of fibrous material called the **internal capsule.** The internal capsule ramifies through the substance of the basal nuclei, giving them a striped appearance. This is why the caudate nucleus and the lentiform nucleus are sometimes referred to as the **corpus striatum,** or "striped body."

4. Note the relationship of the lateral ventricles and corpus callosum to the diencephalon structures; that is, hypothalamus, thalamus, and third ventricle—from the coronal viewpoint.

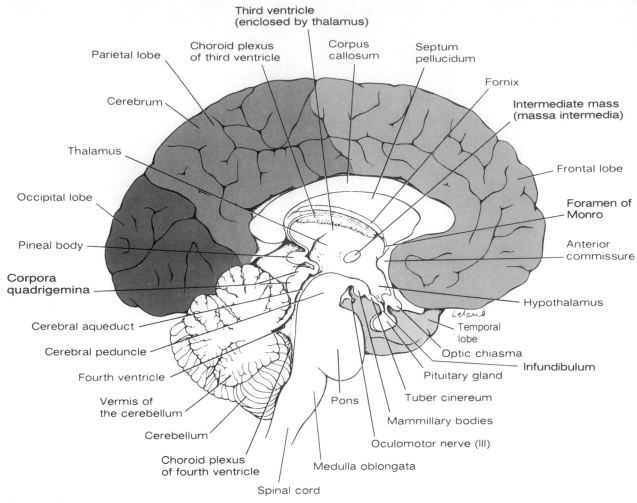

Third ventricle
(enclosed by thalamus)

Parietal lobe

Choroid plexus
of third ventricle

Corpus
callosum

Septum
pellucidum

Fornix

Cerebrum

Intermediate mass
(massa intermedia)

Thalamus

Frontal lobe

Occipital lobe

Foramen of
Monro

Pineal body

Anterior
commissure

Corpora
quadrigemina

Hypothalamus

Cerebral aqueduct

Leland

Temporal
lobe

Cerebral peduncle

Optic chiasma

Fourth ventricle

Infundibulum

Pituitary gland

Vermis of
the cerebellum

Tuber cinereum

Cerebellum

Pons

Mammillary bodies

Choroid plexus
of fourth ventricle

Medulla oblongata

Oculomotor nerve (III)

Spinal cord

Figure 18.4

Midsagittal section of the human brain.

BRAIN STEM   Figure 18.4 shows the internal structures of the brain stem.

1. The diencephalon, generally considered the most superior portion of the brain stem, is embryologically part of the forebrain, along with the cerebral hemispheres. The major structures of the diencephalon are the thalamus, hypothalamus, and epithalamus. The **thalamus** consists of two large lobes of gray matter that laterally enclose the shallow third ventricle of the brain. A slender stalk of thalamic tissue, the **massa intermedia,** or **intermediate mass,** connects the two thalamic lobes and bridges the ventricle. The thalamus is a relay station for sensory impulses passing upward to the cortical sensory areas for localization and interpretation. Locate also the **foramen of Monro,** a tiny orifice connecting the third ventricle with the lateral ventricle on the same side.

2. The **hypothalamus** makes up the floor and a part of the inferior lateral walls of the third ventricle. It is an important autonomic center involved in the regulation of body temperature, water balance, and fat and carbohydrate metabolism as well as in many other activities and drives. Locate also the pituitary gland, or **hypophysis,** which hangs from the anterior floor of the hypothalamus by a slender stalk, the **infundibulum.** (The pituitary gland is usually not present in preserved brain specimens.) In life, the pituitary rests in the fossa of the sella turcica portion of the sphenoid bone. Its function is discussed in Exercise 27.

    Anterior to the pituitary, the optic chiasma portion of the optic pathway to the brain can also be identified. The **mammillary bodies,** hypothalamic reflex centers for olfaction in lower mammals, bulge exteriorly from the floor of the hypothalamus posterior to the pituitary gland. In humans, the mammillary bodies appear to be involved in feeding reflexes such as licking the lips and swallowing.

3. The **epithalamus** forms the roof of the third ventricle and is the most dorsal portion of the diencephalon. Important structures in the epithalamus are the **pineal body,** or **epiphysis**

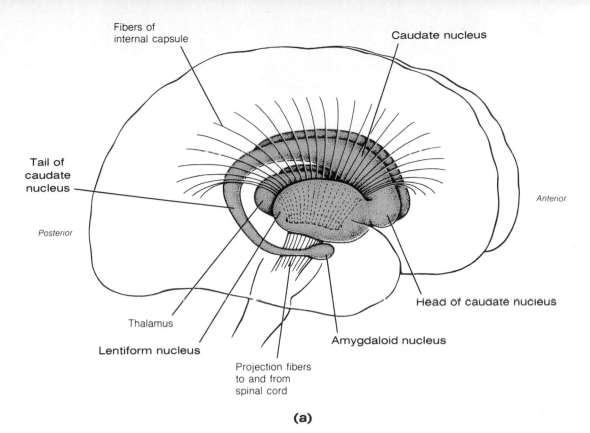

(a)

(b)

Figure 18.5

Location of the basal nuclei, (a) three-dimensional view of the basal nuclei; showing their position within the cerebrum; (b) frontal section of the human brain.

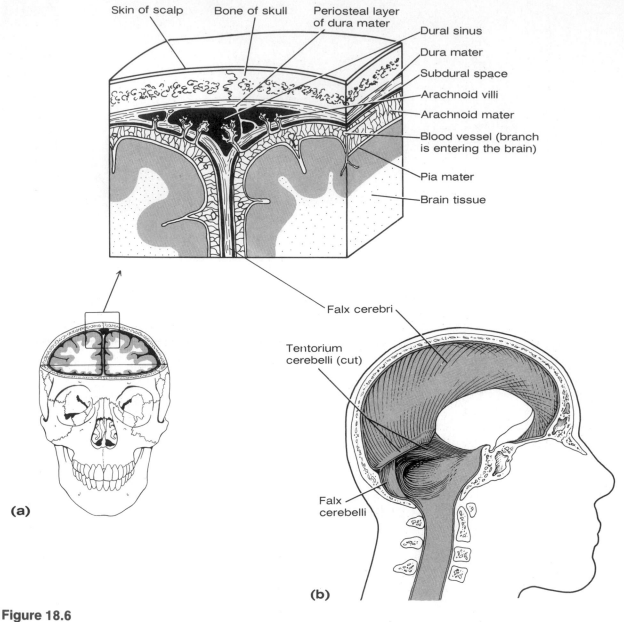

Skin of scalp   Bone of skull   Periosteal layer of dura mater

Dural sinus
Dura mater
Subdural space
Arachnoid villi
Arachnoid mater
Blood vessel (branch is entering the brain)
Pia mater
Brain tissue

Falx cerebri

Tentorium cerebelli (cut)

Falx cerebelli

(a)

(b)

Figure 18.6

Meninges of the brain: (a) relationship to the sagittal sinus; (b) dural extensions.

(a neuroendocrine structure), and the **choroid plexus** of the third ventricle. The choroid plexuses, knotlike collections of capillaries within each ventricle, secrete the cerebrospinal fluid.

4. Now trace the short midbrain from the mammillary bodies to the rounded pons below. The **cerebral aqueduct** is a slender canal traveling through the midbrain; it connects the third ventricle to the fourth ventricle in the hindbrain below. The cerebral penduncles and the rounded corpora quadrigemina make up the midbrain tissue anterior and posterior (respectively) to the cerebral aqueduct.

5. Locate the hindbrain structures. Trace the rounded pons to the medulla oblongata below, and identify the fourth ventricle posterior to

these structures. Attempt to identify the single median foramen of Magendie and the two lateral foramina of Luschka (see page 162). These three orifices, found in the roof of the fourth ventricle, serve as conduits for cerebrospinal fluid to circulate into the subarachnoid space from the fourth ventricle.

CEREBELLUM Examine the cerebellum. Note that it is composed of two lateral hemispheres connected by a midline lobe called the **vermis.** As in the cerebrum, the cerebellum has an outer cortical area of gray matter and an inner area of white matter. The treelike branching of the cerebellar white matter is referred to as the **arbor vitae,** or tree of life. The cerebellum is concerned with the unconscious coordination of skeletal muscle activity and the control of balance and equilibrium. Fibers converge on the cerebellum from the equilibrium apparatus of the inner ear, visual pathways, proprioceptors of the tendons and skeletal muscles, and from many other areas. Thus the cerebellum remains constantly aware of the position and state of tension of the various body parts.

Meninges of the Brain

The brain (and spinal cord) are covered and protected by three connective tissue membranes called **meninges** (Figure 18.6). The outermost meninx is the leathery **dura mater,** a double-layered membrane. One of its layers is attached to the inner surface of the skull, forming the periosteum; the other forms the outermost brain covering and is continuous with the dura mater of the spinal cord.

The dural layers are fused together except in three areas where the inner membrane extends inward to form a septum that secures the brain to structures inside the cranial cavity. One such extension, the **falx cerebri,** dips into the longitudinal fissure between the cerebral hemispheres to attach to the crista galli of the ethmoid bone of the skull. The cavity created at this point is the large **sagittal dural sinus,** which collects blood draining from the brain tissue. The **falx cerebelli,** separating the two cerebellar hemispheres, and the **tentorium cerebelli,** separating the cerebrum from the cerebellum below, are two other important inward folds of the inner dural membrane.

The middle meninx, the weblike **arachnoid mater,** underlies the dura mater and is partially separated from it by the **subdural space.** Threadlike projections bridge the **subarachnoid space** to attach the arachnoid mater to the innermost meninx, the **pia mater.** The delicate pia mater is extensively vascularized and clings tenaciously to the surface of the brain, following its convolutions.

In life, the subarachnoid space is filled with cerebrospinal fluid. Specialized projections of the arachnoid tissue called **arachnoid villi** protrude through the dura mater to allow the cerebrospinal fluid to drain back into the venous circulation via the dural sinuses.

Meningitis, an inflammation of the meninges, is a serious threat to the brain because of the intimate association between the brain and meninges. Meningitis is often diagnosed by taking a sample of cerebrospinal fluid from the subarachnoid space.

Cerebrospinal Fluid

The cerebrospinal fluid, much like plasma in composition, is continually formed by the choroid plexuses, small capillary knots hanging from the roof of the ventricles of the brain. The cerebrospinal fluid in and around the brain forms a watery cushion that protects the delicate brain tissue against blows to the head.

Within the brain, the cerebrospinal fluid circulates from the two lateral ventricles (in the cerebral hemispheres) into the third ventricle via the foramina of Monro, and then through the cerebral aqueduct of the midbrain into the fourth ventricle in the hindbrain (Figure 18.7). A portion of the fluid reaching the fourth ventricle continues down the central canal of the spinal cord, but the bulk of it circulates into the subarachnoid space, exiting through three openings in the walls of the fourth ventricle (the two foramina of Luschka and the foramen of Magendie). The fluid returns to the blood in the dural sinuses through the arachnoid villi.

Ordinarily, cerebrospinal fluid forms and drains at a constant rate. However, under certain conditions—for example, obstructed drainage or circulation resulting from tumors or anatomic deviations—the cerebrospinal fluid begins to accumulate and exerts increasing pressure on the brain.

CRANIAL NERVES

The cranial nerves are part of the peripheral nervous system and not part of the brain proper, but they are most appropriately identified in conjunction with the study of brain anatomy. The 12 pairs of cranial nerves primarily serve the head and neck. Only one pair (the vagus nerves) extends to the thoracic and abdominal cavities. All but the first two pairs (olfactory and optic nerves) arise from the brain stem and pass through foramina in the base of the skull to reach their destination

The cranial nerves are numbered consecutively, and in most cases their names reflect the major structures they control. The cranial nerves are described by name, number, origin, course, and function in Table 18.1. This information should be committed to memory. The last column of the table describes techniques for testing cranial nerves, which is an important part of any neurologic examination. You need not memorize these tests, but this information may help you understand cranial nerve function, especially as it pertains to some aspects of brain function.

 Observe the anterior surface of the brain model to identify the cranial nerves. Figure 18.3 may also aid you in this study.

Most cranial nerves are mixed nerves (containing both motor and sensory fibers). However, close scrutiny of Table 18.1 will reveal that three pairs of cranial nerves (optic, olfactory, and vestibulocochlear) are purely sensory in function.

You may recall that the nerve cell bodies of neurons are always located within the central

(*Text continues on p. 165.*)

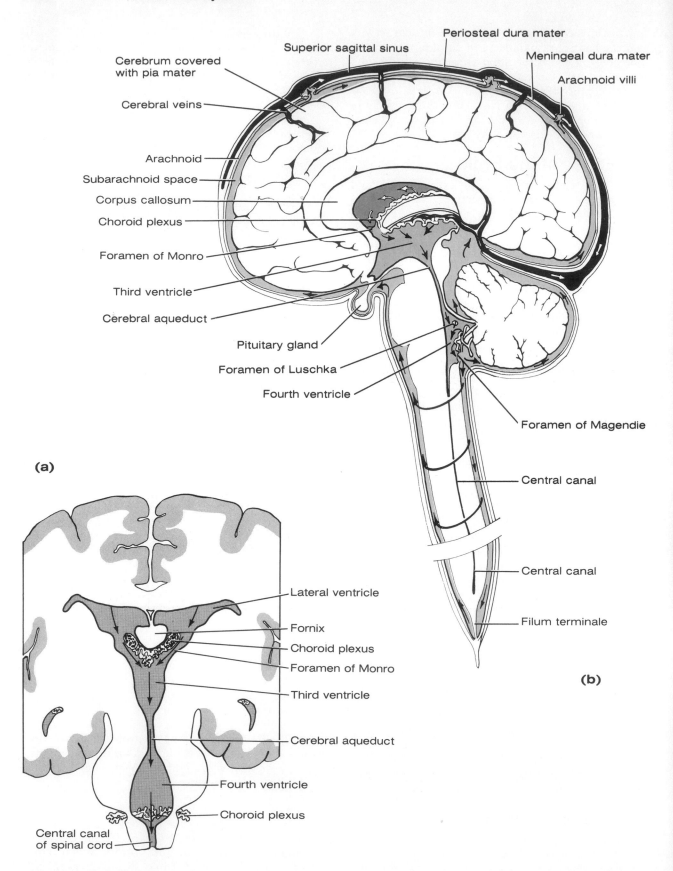

Figure 18.7

The cerebrospinal fluid surrounds the brain and spinal cord (circulatory pathway indicated by arrows): (a) frontal view of lateral ventricles (illustrates the flow through the foramina of Munro); (b) sagittal view of the brain and cord.

TABLE 18.1   The Cranial Nerves

Name and number	Origin and course	Function	Testing
I. Olfactory	Fibers arise from olfactory mucosa and run through cribriform plate of ethmoid bone to synapse with olfactory bulbs.	Purely sensory—carries impulses associated with sense of smell.	Person is asked to sniff aromatic substances, such as oil of cloves and vanilla, and to identify each.
II. Optic	Fibers arise from retina of eye and pass through optic foramen in sphenoid bone. Fibers of two optic nerves then take part in forming optic chiasma (with partial crossover of fibers) after which they continue on to thalamus as the optic tracts. Final fibers of this pathway travel from the thalamus to the optic cortex as the optic radiation.	Purely sensory—carries impulses associated with vision.	Vision and visual field are determined with eye chart and by testing the point at which the person first sees an object (finger) moving into the visual field. Fundus of eye viewed with ophthalmoscope to detect papilledema (swelling of optic disk), or point at which optic nerve leaves the eye and to observe blood vessels.
III. Oculomotor	Fibers emerge from midbrain and exit from skull via superior orbital fissure to run to eye.	Somatic motor fibers to superior, inferior, and medial rectus muscles, which direct eyeball and levator palpebrae muscles of eyelid; parasympathetic fibers to iris and smooth muscle controlling lens shape (control reflex responses to varying light intensity and focusing of eye for near vision); contains proprioceptive sensory fibers carrying impulses from extrinsic eye muscles.	Pupils are examined for size, shape, and equality. Pupillary reflex is tested with penlight (pupils should constrict when illuminated). Convergence for near vision is tested, as is subject's ability to follow objects up, down, side to side, and diagonally.
IV. Trochlear	Fibers emerge from midbrain and exit from skull via superior orbital fissure to run to eye.	Provides somatic motor fibers to superior oblique muscle (an extrinsic eye muscle); conveys proprioceptive impulses from same muscle to brain.	Tested in common with cranial nerve III.
V. Trigeminal	Fibers emerge from pons and form three divisions, which exit separately from skull: mandibular division through foramen ovale in sphenoid bone, maxillary division via foramen rotundum in sphenoid bone, and ophthalmic division through superior orbital fissure of eye socket.	Both motor and sensory for face; conducts sensory impulses from skin of face and anterior scalp, from mucosae of mouth and nose, and from surface of eyes; mandibular division also contains motor fibers that innervate muscles of mastication and muscles of floor of mouth.	Sensations of pain, touch, and temperature are tested with safety pin and hot and cold objects. Corneal reflex tested with wisp of cotton. Motor branch assessed by asking person to clench his teeth, open mouth against resistance, and move jaw side to side.
VI. Abducens	Fibers leave inferior region of pons and exit from skull via superior orbital fissure to run to eye.	Carries motor fibers to lateral rectus muscle of eye and proprioceptive fibers from same muscle to brain.	Tested in common with cranial nerve III.

(continued.)

TABLE 18.1 continued

Name and number	Origin and course	Function	Testing
VII. Facial	Fibers leave pons and travel through temporal bone via internal acoustic meatus, exiting via stylomastoid foramen to reach facial region.	Mixed—supplies somatic motor fibers to muscles of facial expression and parasympathetic motor fibers to lacrimal and salivary glands; carries sensory fibers from taste receptors of anterior portion of tongue.	Anterior two-thirds of tongue is tested for ability to taste sweet (sugar), salty, sour (vinegar), and bitter (quinine) substances. Symmetry of face is checked. Subject is asked to close eyes, smile, whistle, and so on. Tearing is assessed with ammonia fumes.
VIII. Vestibulocochlear	Fibers run from inner-ear equilibrium and hearing apparatus, housed in temporal bone, through internal acoustic meatus to enter brain stem just below pons.	Purely sensory—vestibular branch transmits impulses associated with sense of equilibrium from vestibular apparatus and semicircular canals; cochlear branch transmits impulses associated with hearing from cochlea.	Hearing is checked by air and bone conduction using tuning fork.
IX. Glossopharyngeal	Fibers emerge from medulla and leave skull via jugular foramen to run to throat region.	Mixed—somatic motor fibers serve pharyngeal muscles, and parasympathetic motor fibers serve salivary glands; sensory fibers carry impulses from pharynx, tonsils, posterior tongue (taste buds), and pressure receptors of carotid artery.	Position of the uvula is checked. Gag and swallowing reflexes are checked. Subject is asked to speak and cough. Posterior third of tongue may be tested for taste.
X. Vagus	Fibers emerge from medulla and pass through jugular foramen and descend through neck region into thorax and abdominal regions.	Fibers carry somatic motor impulses to pharynx and larynx and sensory fibers from same structures; very large portion is composed of parasympathetic motor fibers, which supply heart and smooth muscles of abdominal visceral organs; transmits sensory impulses from viscera.	As for cranial nerve IX (IX and X are tested in common, since they both innervate muscles of throat and mouth.)
XI. Spinal accessory	Fibers arise from medulla and superior aspect of spinal cord and travel through jugular foramen to reach muscles of neck and back.	Provides somatic motor fibers to sternocleidomastoid and trapezius muscles and to muscles of soft palate, pharynx, and larynx (spinal and medullary fibers respectively); proprioceptive impulses are conducted from these muscles to brain.	Sternocleidomastoid and trapezius muscles are checked for strength by asking person to rotate head and shrug shoulders against resistance.
XII. Hypoglossal	Fibers arise from medulla and exit from skull via hypoglossal canal to travel to tongue.	Carries somatic motor fibers to muscles of tongue and proprioceptive impulses from tongue to brain.	Person is asked to protrude and retract tongue. Any deviations in position are noted.

nervous system or in specialized collections of nerve cell bodies (ganglia) outside the CNS. Nerve cell bodies of the sensory cranial nerves are located in ganglia; those of the mixed cranial nerves are found both within the brain and in peripheral ganglia.

Several cranial nerve ganglia are named here. Using your textbook or an appropriate reference, name the cranial nerve the ganglion is associated with and state its location.

Cranial nerve ganglion	Cranial nerve	Site of ganglion
gasserian		
geniculate		
inferior		
superior		
spiral		
vestibular		

DISSECTION OF THE SHEEP BRAIN

The brain of any mammal is enough like the human brain to warrant comparison. Obtain a sheep brain, dissecting pan, and instruments, and bring them to your laboratory bench.

1. Place the intact sheep brain ventral surface down on the dissecting pan and observe the dura mater. Feel its consistency and note its toughness. Cut through the dura mater along the line of the longitudinal fissure (which separates the cerebral hemispheres) to enter the sagittal sinus. Gently force the cerebral hemispheres apart laterally to expose the corpus callosum deep to the longitudinal fissure.

2. Carefully remove the dura mater and examine the superior surface of the brain. Note that its surface is thrown into convolutions (fissures and gyri), just as the human brain is. Locate the arachnoid meninx, which appears on the brain surface as a delicate "cottony" material spanning fissures. In contrast, the innermost meninx, the pia mater, closely follows the cerebral contours.

Dorsal Structures

1. Refer to Figure 18.8 (b) and (c) as a guide in identifying the following structures. The cerebral hemispheres should be easy to locate. How do the size of the sheep's cerebral hemispheres and the depth of the fissures compare to those in the human brain? _____ _____ _____

2. Carefully examine the cerebellum. Note that it is not divided longitudinally, in contrast to the human cerebellum, and that its fissures are oriented differently.

    What dural falx is missing that is present in humans? _____ _____ _____

3. Locate the three pairs of cerebellar peduncles, fiber tracts that connect the cerebellum to other brain structures, by lifting the cerebellum dorsally away from the brain stem. The most posterior pair, the inferior cerebellar peduncles, connect the cerebellum to the medulla. The middle cerebellar peduncles attach the cerebellum to the pons, and the superior cerebellar peduncles run from the cerebellum to the midbrain.

4. To expose the dorsal surface of the midbrain, gently spread the cerebrum and cerebellum apart, as shown in Figure 18.9. Identify the corpora quadrigemina, which appear as four rounded prominences on the dorsal midbrain surface.

    What is the function of the corpora quadrigemina? _____ _____ _____

    Also locate the pineal body, which appears in the midline just anterior to the corpora quadrigemina.

Ventral Structures

Figure 18.9 (a) and (c) show the important features of the ventral surface of the brain.

1. Look for the clublike olfactory bulbs, anteriorly on the inferior surface of the frontal lobes of the cerebral hemispheres. Axons of olfac-

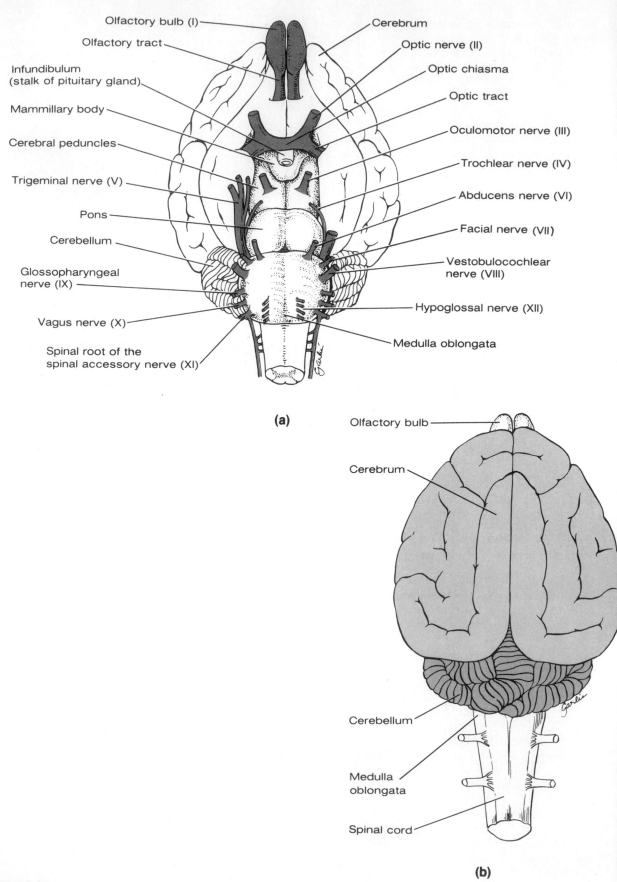

Olfactory bulb (I)
Olfactory tract
Infundibulum
(stalk of pituitary gland)
Mammillary body
Cerebral peduncles
Trigeminal nerve (V)
Pons
Cerebellum
Glossopharyngeal
nerve (IX)
Vagus nerve (X)
Spinal root of the
spinal accessory nerve (XI)

Cerebrum
Optic nerve (II)
Optic chiasma
Optic tract
Oculomotor nerve (III)
Trochlear nerve (IV)
Abducens nerve (VI)
Facial nerve (VII)
Vestobulocochlear
nerve (VIII)
Hypoglossal nerve (XII)
Medulla oblongata

(a)

Olfactory bulb
Cerebrum
Cerebellum
Medulla
oblongata
Spinal cord

(b)

Figure 18.8

Intact sheep brain: (a) ventral view; (b) dorsal view.

(continued.)

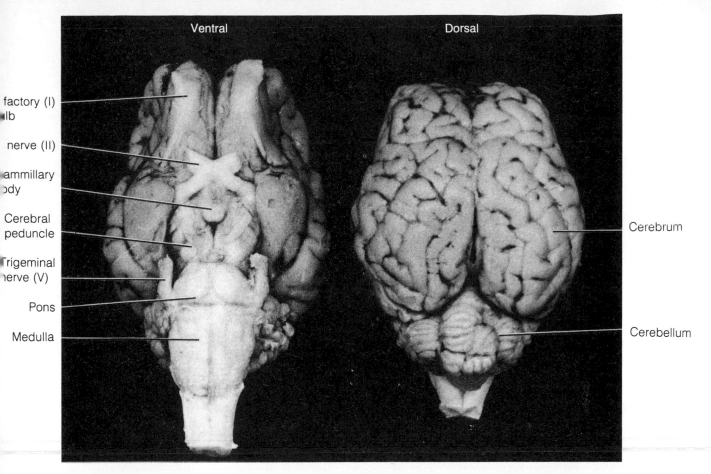

Ventral   Dorsal

Olfactory (I) bulb

Optic nerve (II)

Mammillary body

Cerebral peduncle

Trigeminal nerve (V)

Pons

Medulla

Cerebrum

Cerebellum

Figure 18.8 (continued.)

Intact sheep brain: (c) photograph showing ventral and dorsal views. (Photo courtesy of Ann Allworth.)

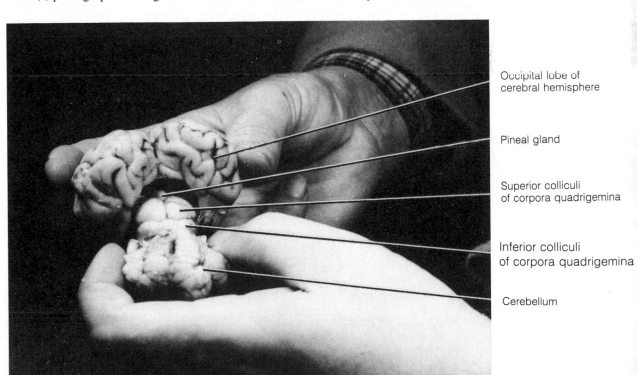

Occipital lobe of cerebral hemisphere

Pineal gland

Superior colliculi of corpora quadrigemina

Inferior colliculi of corpora quadrigemina

Cerebellum

Figure 18.9

Means of exposing the dorsal midbrain structures of the sheep brain. (Photo courtesy of Ann Allworth.)

tory neurons run from the nasal mucosa through the perforated cribriform plate of the ethmoid bone to synapse with the olfactory bulbs.

How does the size of these olfactory bulbs

compare with those of humans? _____

_____

_____

Is the sense of smell more important as a protective or a food-getting sense in sheep *or*

humans? _____

2. The optic nerve (II) carries sensory impulses from photoreceptor cells of the retina of the eye. Thus this cranial nerve is involved in the sense of vision. Identify the optic nerves, optic chiasma, and optic tracts.
3. Posterior to the optic chiasma, two structures protrude from the ventral aspect of the hypothalamus—the infundibulum (stalk of the pituitary gland) immediately posterior to the optic chiasma and the mammillary body. Notice that the mammillary body is a single rounded eminence; in humans it is a double structure.
4. Identify the cerebral peduncles on the ventral aspect of the midbrain just posterior to the mammillary body of the hypothalamus. The cerebral peduncles are fiber tracts connecting the cerebrum and medulla. Identify the large oculomotor nerves (III), which arise from the ventral midbrain surface, and the tiny trochlear nerves (IV), which can be seen at the junction of the midbrain and pons. Both these cranial nerves provide motor fibers to extrinsic muscles of the eyeball.
5. Move posteriorly from the midbrain to identify first the pons and then the medulla oblongata, both hindbrain structures composed primarily of ascending and descending fiber tracts.
6. Return to the junction of the pons and midbrain and proceed posteriorly to identify the following cranial nerves, all arising from the pons: the trigeminal nerves (V), which are involved in chewing and sensations of the head and face; the abducens nerves (VI), which abduct the eye (and thus work in conjunction with cranial nerves III and IV); and the large facial nerves (VII), which are involved in taste sensation, gland function (salivary and lacrimal glands), and facial expression.
7. Continue posteriorly to identify the purely sensory vestibulocochlear nerves (VIII), which are involved in the sensations of hearing and equilibrium; the glossopharyngeal nerves (IX), which contain motor fibers innervating throat

structures and sensory fibers transmitting taste stimuli (in conjunction with cranial nerve VII); the vagus nerves (X), often called "wanderers," which serve many organs of the head, thorax, and abdominal cavity; the spinal accessory nerves (XI), which serve muscles of the neck, larynx, and shoulder; and the hypoglossal nerves (XII), which stimulate tongue and neck muscles. Note that the spinal accessory nerves arise from both the medulla and the spinal cord.

Internal Structures

1. The internal structure of the brain can only be examined after further dissection. Place the brain ventral side down on the dissecting pan and make a cut completely through it in a superior to inferior direction. Cut through the longitudinal fissure, corpus callosum, and midline of the cerebellum. Refer to Figure 18.10 as you work.

2. The thin nervous tissue membrane immediately ventral to the corpus callosum that separates the lateral ventricles is the septum pellucidum. Pierce this membrane and probe the lateral ventricle cavity. The fiber tract ventral to the septum pellucidum and anterior to the third ventricle is the fornix.

How does the size of the fornix in this brain

compare with the human fornix? _____

_____

Why do you suppose this is so? (Hint: What is

the function of this band of fibers?) _____

_____

_____

_____

3. Identify the thalamus, which forms the walls of the third ventricle and is located posterior and ventral to the fornix. The intermediate mass spanning the ventricular cavity appears as an oval protrusion of the thalamic wall. Anterior to the intermediate mass, locate the foramen of Monro, a canal connecting the lateral ventricle on the same side with the third ventricle.

4. The hypothalamus forms the floor of the third ventricle. Identify the optic chiasma, infundibulum, and mammillary body on its exterior surface. You can see the pineal body at the superior-posterior end of the third ventricle just beneath the junction of the corpus callosum and fornix.

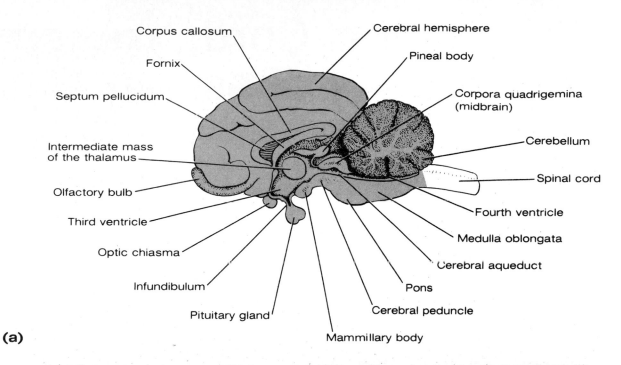

(a)

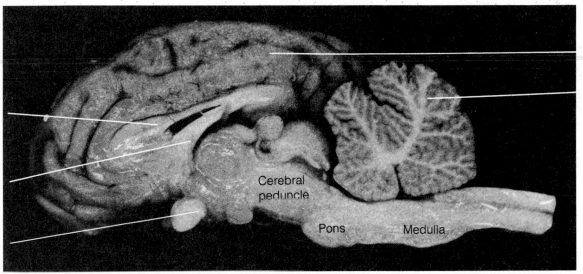

(b)

Figure 18.10

Sagittal section of the sheep brain showing internal structures: (a) diagrammatic view; (b) photograph. (Courtesy of Ann Allworth.)

5. The midbrain can be located by identifying the corpora quadrigemina that form its dorsal roof. Follow the cerebral aqueduct (the narrow canal connecting the third and fourth ventricles) through the midbrain tissue to the fourth ventricle. Identify the cerebral peduncles, which form its anterior walls.

6. The pons and medulla can be found anterior to the fourth ventricle. The medulla continues into the spinal cord without any obvious anatomic change, but the point at which the fourth ventricle narrows to a small canal is generally accepted as the beginning of the spinal cord.

7. The cerebellum can be seen posterior to the fourth ventricle. Note its internal treelike arrangement of white matter, the arbor vitae.

8. Check with your instructor to determine if cow spinal cord sections (preserved) are available for the spinal cord studies in Exercise 19. If not, save the small portion of the spinal cord from this brain specimen. Otherwise, dispose of all the organic debris in the appropriate laboratory containers and clean the dissecting instruments and tray before leaving the laboratory.

Spinal Cord, Spinal Nerves, and the Autonomic Nervous System

EXERCISE
19

OBJECTIVES

1. To identify on a spinal cord model or appropriate diagram gray and white matter, anterior median fissure, posterior median sulcus, central canal, dorsal, ventral, and lateral horns of the gray matter, ventral and dorsal roots, dorsal root ganglia, and posterior, lateral, and anterior funiculi, and to cite the neuron type found in these areas (where applicable).

2. To name two major areas where the spinal cord is enlarged, and to explain the reasons for this anatomic characteristic.

3. To define *conus medullaris*, *cauda equina*, and *filum terminale*.

4. To locate on a diagram the fiber tracts in the spinal cord white matter and to state their functional importance.

5. To list two major functions of the spinal cord.

6. To name the meningeal coverings of the spinal cord and state their function.

7. To describe the origin, fiber composition, and distribution of the spinal nerves, differentiating between ventral and dorsal roots, the spinal nerve proper, and ventral and dorsal rami, and to discuss the result of transecting these structures.

8. To discuss the distribution of the dorsal rami and ventral rami of the spinal nerves.

9. To identify the four major nerve plexuses, the major nerves of each, and their distribution.

10. To identify on a dissected pig the musculocutaneous, radial, median, and ulnar nerves of the upper limb and the femoral, saphenous, sciatic, common peroneal, and tibial nerves of the lower limb.

11. To identify the site of origin and the function of the sympathetic and parasympathetic divisions of the autonomic nervous system, and to state how the autonomic nervous system differs from the somatic nervous system.

MATERIALS

Spinal cord model (cross section)
Laboratory charts of the spinal cord and spinal nerves and sympathetic chain.
Red and blue pencils
Preserved cow spinal cords with meninges and nerve roots intact (or spinal cord segment saved from the brain dissection in Exercise 18)
Dissecting tray and instruments
Dissecting microscope
Histologic slide of spinal cord (cross section)
Compound microscope
Pig specimen from previous dissections

ANATOMY OF THE SPINAL CORD

The cylindrical spinal cord, a continuation of the brain stem, is an association and communication center. It plays a major role in spinal reflex activity and provides neural pathways to and from higher nervous centers. Enclosed within the vertebral canal of the spinal column, the spinal cord extends from the foramen magnum of the skull to the first or second lumbar vertebra, where it terminates in the cone-shaped **conus medullaris** (Figure 19.1). Like the brain, it is cushioned and protected by the meninges. The meningeal coverings extend beyond the conus medullaris into the coccygeal canal as the **filum terminale.** Because of this

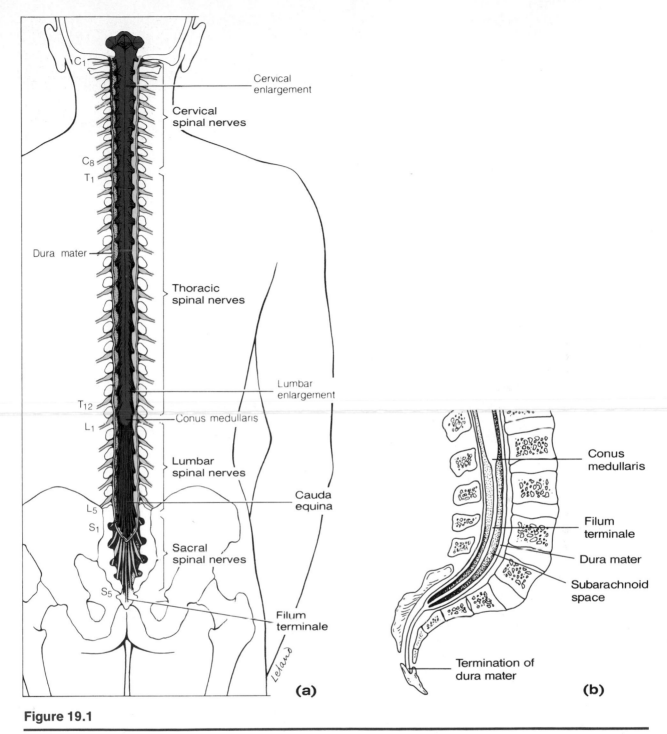

Figure 19.1

Anatomy of the human spinal cord: (a) dorsal view; (b) lateral view, showing the extent of the filum terminale and technique of lumbar puncture for obtaining a sample of cerebrospinal fluid for testing.

anatomic characteristic, cerebrospinal fluid can easily be obtained by a *lumbar* puncture (or lumbar tap) from the subarachnoid space below L_3 without endangering the delicate spinal cord.

In humans, 31 pairs of spinal nerves arise from the spinal cord and pass through intervertebral foramina to serve the body area at their approximate level of emergence. The cord is about the size of a thumb in circumference for most of its length,

but there are obvious enlargements in the cervical and lumbar areas where the nerves serving the upper and lower limbs issue from the cord.

Because the cord does not extend to the end of the spinal column, the spinal nerves emerging from the inferior end of the cord must travel through the vertebral canal for some distance before exiting at the appropriate intervertebral foramina. This collection of spinal nerves traversing the inferior end of

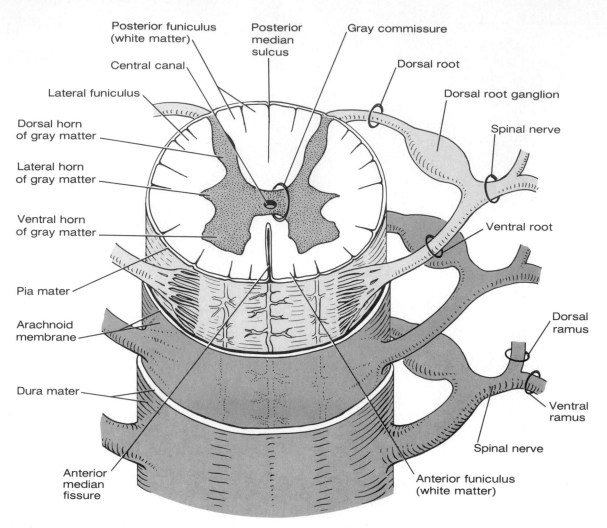

Figure 19.2

Anatomy of the human spinal cord (cross section).

the vertebral canal is called the **cauda equina** because of its similarity to a horse's tail (the literal translation of *cauda equina*).

Obtain a model of a cross section of a spinal cord and identify its structures as they are described next.

Gray Matter

In cross section, the gray matter of the spinal cord looks like a butterfly or the letter *H* (Figure 19.2). The two posterior projections are called the **dorsal** (or posterior) **horns**; the two anterior projections are the **ventral** (or anterior) **horns**. The tips of the ventral horns are broader and less tapered than those of the dorsal horns. In the thoracic and lumbar regions of the cord, there is also a lateral outpocketing of gray matter on each side referred

to as the **lateral horn**. The central area of gray matter connecting the two vertical regions is the **gray commissure**. The gray commissure surrounds the **central canal** of the cord, which contains cerebrospinal fluid.

Neurons with specific functions can be localized in the gray matter. The dorsal horns, for instance, contain internuncial neurons and sensory fibers that enter the cord from the body periphery via the **dorsal root**. The nerve cell bodies of these sensory neurons are found in an enlarged area of the dorsal root called the **dorsal root ganglion**. The ventral horns contain nerve cell bodies of motor neurons of the somatic nervous system (voluntary system), which send their axons out via the **ventral root** of the cord to enter the adjacent spinal nerve. The **spinal nerves** are formed from the fusion of the dorsal and ventral roots. The lateral horns, where present, contain nerve cell bodies of motor neurons of the autonomic nervous system (sympathetic division). Their axons also leave the cord via the ventral roots, with those of the motor neurons of the ventral horns.

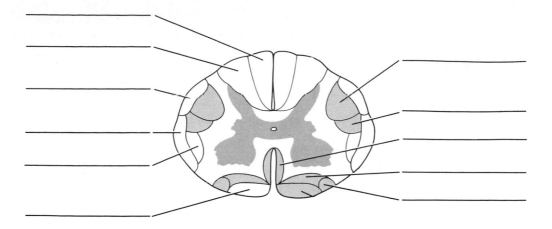

Figure 19.3

Cross section of spinal cord showing the various tracts.

White Matter

The white matter of the spinal cord is nearly bisected by fissures (see Figure 19.2). The more open anterior fissure is the **anterior median fissure**, and the posterior one is the **posterior median sulcus**. The white matter is composed of myelinated fibers—some running to higher centers, some traveling from the brain to the cord, and some conducting impulses from one side of the cord to the other.

Because of the irregular shape of the gray matter, the white matter on each side of the cord can be divided into three primary regions: the **posterior**, **lateral**, and **anterior funiculi**. Each of the funiculi contains a number of fiber **tracts** composed of axons with the same origin, terminus, and function. Tracts conducting sensory impulses to the brain are called ascending, or sensory, tracts; those carrying impulses from the brain to the skeletal muscles are descending, or motor, tracts.

 With the help of your textbook or a laboratory chart of the spinal cord and nerves, label Figure 19.3 with the tract names that follow. Since each tract is represented on both sides of the cord, you can label only one side of the diagram or use one side for sensory tracts and the other for motor tracts. *Color ascending tracts red and descending tracts blue.* Then fill in the functional importance of each tract beside its name. As you work, try to be aware of how the naming of the tracts is related to their anatomic distribution.

Fasciculus gracilis _____

Fasciculus cuneatus _____

Dorsal spinocerebellar _____

Ventral spinocerebellar _____

Lateral spinothalamic _____

Ventral spinothalamic _____

Lateral corticospinal _____

Ventral corticospinal _____

Rubrospinal _____

Tectospinal _____

Olivospinal _____

Vestibulospinal _____

Spinal Cord Dissection

 1. Obtain a dissecting tray and instruments and a segment of preserved spinal cord (from a cow or saved from the brain specimen used in Exercise 18). Identify the tough outer meninx (dura mater) and the weblike arachnoid membrane.

What name is given to the third meninx, and

where is it found? _____

_____

Peel back the dura mater and observe the fibers making up the dorsal and ventral roots. If possible, identify a dorsal root ganglion.

2. Cut a thin cross section of the cord and identify the dorsal and ventral horns of the gray matter with the naked eye or with the aid of a dissecting microscope.

How can you be certain that you are correctly identifying the dorsal and ventral horns? _____

_____

_____

Also identify the central canal, white matter, anterior median fissure, posterior median sulcus, and posterior, anterior, and lateral funiculi.

3. Obtain a prepared slide of the spinal cord (cross section) and a compound microscope. Examine the slide carefully under low power. Observe the shape of the central canal.

Is it basically circular or oval? _____

_____

What would you expect to find in this canal in the living animal? _____

_____

Does the anterior median fissure or the posterior median sulcus touch the gray matter? If so, which?_____

Can any nerve cell bodies be seen? _____

Where? _____

_____

What type of neurons would these most likely be—motor, internuncial, or sensory? _____

_____

SPINAL NERVES AND NERVE PLEXUSES

The 31 pairs of human spinal nerves arise from the fusions of the ventral and dorsal roots of the spinal cord. (Figure 19.4 shows how the nerves are named according to their point of issue.) Since the ventral roots contain the myelinated axons of motor neurons located in the cord and the dorsal roots carry sensory fibers entering the cord, all spinal nerves are **mixed nerves**. The first pair of spinal nerves leaves the vertebral canal between the base of the occiput and the atlas, but all the rest exit via the intervertebral foramina *above* the vertebra for which they are named.

Almost immediately after emerging, each nerve divides into **dorsal** and **ventral rami**. (Thus each spinal nerve is only about 1 or 2 cm long.) The rami, like the spinal nerves, contain both motor and sensory fibers. The smaller dorsal rami serve the skin and musculature of the posterior body trunk at their approximate level of emergence. The ventral rami of spinal nerves, T_2–T_{12}, pass anteriorly as the **intercostal nerves** to supply the muscles of intercostal spaces and the skin and muscles of the anterior and lateral trunk. The ventral rami of all other spinal nerves form complex networks of nerves called **plexuses**. These plexuses serve the motor and sensory needs of the muscles and skin of the limbs. The ventral rami unite in the plexuses and then diverge again to form peripheral nerves, which contain fibers from more than one spinal nerve. The four major nerve plexuses and their major peripheral nerves are illustrated in Figures 19.4 and 19.5 and are described below. Their names and site of origin should be committed to memory.

The **cervical plexus** arises from the ventral rami of C_1 through C_4 to supply muscles of the shoulder and neck. The major motor branch of this plexus is the **phrenic nerve**, which passes into the thoracic cavity in front of the first rib to innervate the diaphragm. The primary danger of a broken neck is that the phrenic nerve may be severed, leading to paralysis of the diaphragm and cessation of breathing.

The **brachial plexus** is large and complex, arising from the ventral rami of C_5 through C_8 and T_1. The plexus becomes subdivided into four major peripheral nerves. The large **radial nerve** passes down the posterolateral surface of the arm and forearm, supplying all the extensor muscles of the arm and hand and the skin along its course. The radial nerve is often injured in the axillary region by the pressure of a crutch or by hanging one's arm over the back of a chair. The **median nerve** passes down the anteromedial surface of the arm to supply most of the flexor muscles in the forearm and several muscles in the hand (plus the skin of the lateral surface of the palm of the hand). The **musculocutaneous nerve** supplies the flexor muscles of the upper arm and skin of the lateral surface of the forearm. The **ulnar nerve** travels down the posteromedial surface of the arm. It courses around the medial epicondyle of the humerus to supply the flexor carpi ulnaris, the ulnar head of the flexor digitorum profundus of the forearm, and all intrinsic muscles of the hand not served by the median nerve. It supplies the skin of the medial third of the hand, both its anterior and posterior surfaces. Trauma to the ulnar nerve, which often occurs when the elbow is hit, produces a smarting sensation commonly referred to as "hitting the funny bone."

The **lumbosacral plexus**, which serves the pelvic region of the trunk and the lower limbs, is actually a complex of two plexuses, the lumbar plexus and the sacral plexus (see Figure 19.5). The **lumbar**

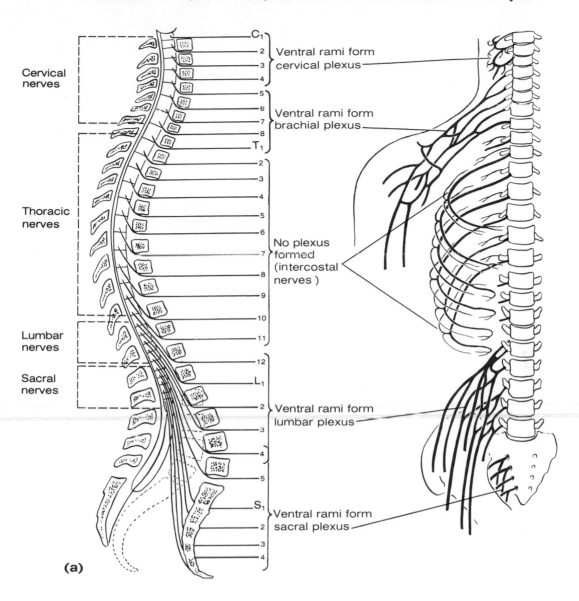

Cervical nerves

C₁
2
3
4

Ventral rami form cervical plexus

5
6
7
8
T₁

Ventral rami form brachial plexus

Thoracic nerves

2
3
4
5
6
7
8
9
10
11

No plexus formed (intercostal nerves)

Lumbar nerves

12
L₁

Sacral nerves

2
3
4
5

Ventral rami form lumbar plexus

S₁
2
3
4

Ventral rami form sacral plexus

(a)

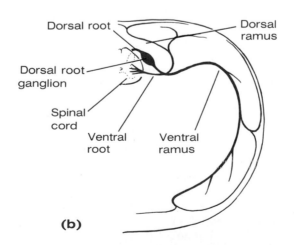

Dorsal root

Dorsal ramus

Dorsal root ganglion

Spinal cord

Ventral root

Ventral ramus

(b)

Figure 19.4

Human spinal nerves: (a) relationship of spinal nerves to vertebrae (areas of plexuses formed by the anterior rami are indicated); (b) relative distribution of the ventral and dorsal rami (cross section of left trunk).

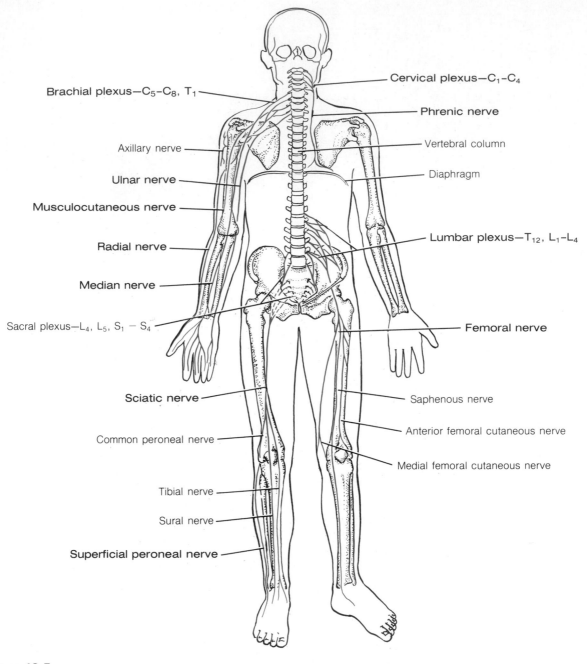

Brachial plexus—C_5-C_8, T_1

Cervical plexus—C_1-C_4

Phrenic nerve

Axillary nerve

Vertebral column

Ulnar nerve

Diaphragm

Musculocutaneous nerve

Radial nerve

Lumbar plexus—T_{12}, L_1-L_4

Median nerve

Sacral plexus—L_4, L_5, S_1 — S_4

Femoral nerve

Sciatic nerve

Saphenous nerve

Common peroneal nerve

Anterior femoral cutaneous nerve

Medial femoral cutaneous nerve

Tibial nerve

Sural nerve

Superficial peroneal nerve

Figure 19.5

Nerve plexuses and the major nerves arising from each.

plexus arises from ventral rami from T_{12} and L_1 through L_4. Its nerves serve the lower abdomino-pelvic region, the buttocks, and the anterior thighs. The largest nerve of this plexus is the **femoral nerve**, which passes beneath the inguinal ligament to innervate the anterior thigh muscles. The cutaneous branches of the femoral nerve (median and anterior femoral cutaneous and the saphenous nerves) supply the skin of the anteromedial surface of the entire leg.

Arising from L_4 through L_5 and S_1 through S_4, the nerves of the **sacral plexus** supply the lower trunk, the posterior surface of the legs, and the feet.

The major peripheral nerve of this plexus is the **sciatic nerve**, which is the largest nerve in the body. The sciatic nerve leaves the pelvis through the greater sciatic notch and travels down the posterior thigh, serving its flexor muscles and skin. In the popliteal region, the sciatic nerve divides into the **common peroneal nerve** and the **tibial nerve**, which together supply the balance of the leg muscles and skin, both directly and via several branches.

Identify each of the four major nerve plexuses (and its major nerves) in Figure 19.5 on a large laboratory chart. Trace the course of the nerves.

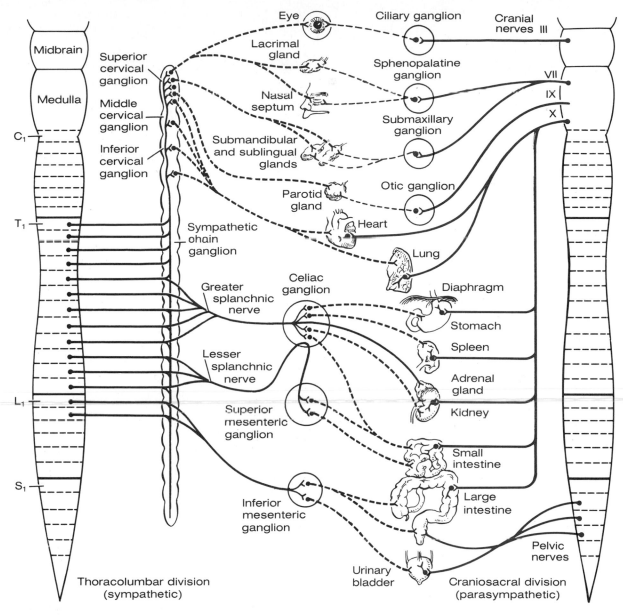

Figure 19.6

The autonomic nervous system. Solid lines indicate preganglionic nerve fibers; dotted lines indicate postganglionic nerve fibers.

THE AUTONOMIC NERVOUS SYSTEM

The **autonomic nervous system** is the subdivision of the PNS that regulates body activities that are generally not under conscious control. It is composed of a special group of motor neurons serving cardiac muscle (the heart), smooth muscle (found in the walls of the visceral organs and blood vessels), and internal glands. Because these structures function without conscious control, this system has often been referred to as the involuntary nervous system.

There is a basic anatomic difference between the motor pathways of the **somatic** (voluntary) nervous system, which innervates the skeletal muscles, and those of the autonomic nervous system. In the somatic division, the nerve cell bodies of the motor neurons reside in the CNS (spinal cord or brain), and their axons, sheathed in spinal nerves, extend all the way to the skeletal muscles they serve. However, the autonomic nervous system consists of chains of two motor neurons. The first motor neuron of each pair, called the *preganglionic neuron*, resides in the brain or cord. Its axon leaves the CNS to synapse with the second motor neuron (*postganglionic neuron*), which is located in a ganglion outside the CNS. The axon of the postganglionic neuron then extends to the organ it serves.

The autonomic nervous system has two major functional subdivisions (Figure 19.6). The sympa-

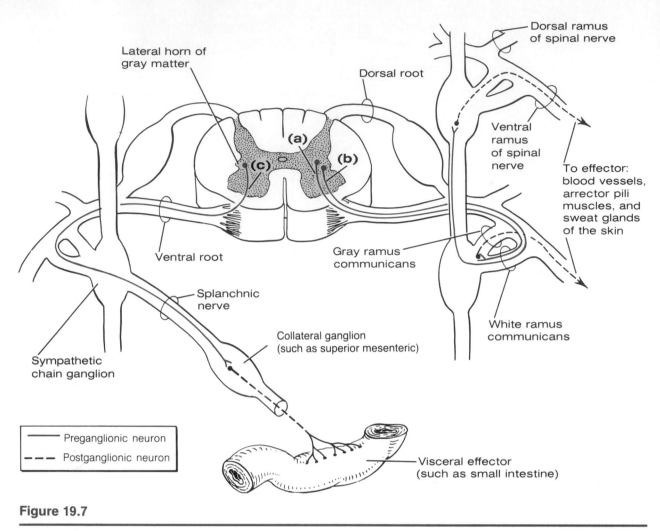

Figure 19.7

Sympathetic pathways: (a) synapse in a sympathetic chain ganglion at the same level; (b) synapse in a sympathetic chain ganglion at a different level; (c) synapse in a collateral ganglion.

thetic and parasympathetic divisions serve the same organs but cause opposing or antagonistic effects.

Sympathetic Division

The preganglionic neurons of the sympathetic (or thoracolumbar) division are in the lateral horns of the gray matter of the spinal cord from T_1 through L_2. The preganglionic axons leave the cord via the ventral root (in conjunction with the axons of the somatic motor neurons), enter the spinal nerve, and then travel briefly in the ventral ramus (Figure 19.7). From the ventral ramus, they pass through a small branch called the **white ramus communicans** to enter a ganglion in the **sympathetic chain**, or **trunk**, which lies alongside the vertebral column.

Having reached the ganglion, an axon may take one of three courses (see Figure 19.7). First, it may synapse with a postganglionic neuron in *that* sympathetic chain ganglion. Second, the axon may travel upward or downward through the sympathetic chain

to synapse with a postganglionic neuron at another level. In either of these two instances, the postganglionic axons then reenter the ventral or dorsal ramus of a spinal nerve via a **gray ramus communicans** and travel in the ramus to innervate skin structures (sweat glands, arrector pili muscles attached to hair follicles, and the smooth muscles of blood vessel walls). Third, the axon may pass through the ganglion without synapsing and form part of the **splanchnic nerves,** which travel to the viscera to synapse with a postganglionic neuron in a **collateral ganglion.** (The major collateral ganglia—the celiac, superior mesenteric, and inferior mesenteric ganglia—supply the abdominal and pelvic visceral organs.) The postganglionic axon then leaves the collateral ganglion and travels to a nearby visceral organ.

Parasympathetic Division

The preganglionic neurons of the parasympathetic (or craniosacral) division are located in brain

Organ or function	Parasympathetic effect	Sympathetic effect
Heart		
Bronchioles of lungs		
Digestive tract activity		
Urinary bladder		
Iris of the eye		
Blood vessels of skeletal muscles		
Blood vessels of visceral organs		
Penis		
Sweat glands		
Adrenal medulla		
Pancreas		

nuclei of cranial nerves III, VII, IX, X and in the S_1 through S_3 (or S_4) level of the spinal cord. The axons of the preganglionic neurons of the cranial region travel in their respective cranial nerves to the immediate area of the head and neck organs to be stimulated. There they synapse with the postganglionic neuron in a **terminal ganglion.** The postganglionic neuron then sends out a very short axon to the organ it serves. In the sacral region, the preganglionic axons leave the ventral roots of the spinal cord and collectively form the **pelvic nerve,** which travels to the pelvic cavity. In the pelvic cavity, the preganglionic axons synapse with the postganglionic neurons in terminal ganglia located on or close to the organs served.

 Locate the sympathetic chain on the spinal nerve chart.

Autonomic Functioning

As noted earlier, body organs served by the autonomic nervous system receive fibers from both the sympathetic and parasympathetic divisions. The only exceptions are the structures of the skin, some glands, and the adrenal medulla, all of which receive sympathetic innervation only. When both divisions serve an organ, they have antagonistic effects. This is because their postganglionic axons release different neurotransmitters. The parasympathetic fibers, called **cholinergic fibers**, release acetylcholine; the sympathetic postganglionic fibers,

called **adrenergic fibers,** release norepinephrine; and the preganglionic fibers of both divisions release acetylcholine.

The parasympathetic division is often referred to as the housekeeping system because it maintains the visceral organs in a state most suitable for normal functions and internal homeostasis; that is, it promotes normal digestion and elimination. In contrast, activation of the sympathetic division is referred to as the "fight or flight" response because it readies the body to cope with situations that threaten homeostasis. Under such conditions, the sympathetic nervous system induces an increase in heart rate, dilation of the bronchioles of the lungs, an increase in blood sugar levels, and many other effects that help the individual cope with stress.

Several body organs are listed above. Using your textbook as a reference, list the effect of the sympathetic and parasympathetic divisions on each.

Dissection of Pig Spinal Nerves

The pig has 33 pairs of spinal nerves (as compared to 31 in humans). Of these, 8 are cervical, 14 thoracic, 7 lumbar, and 4 sacral.

A complete dissection of the spinal nerves of the pig would be extraordinarily time-consuming and exacting and is not warranted in a basic anatomy and physiology course. However, it is desirable to have some dissection work to complement the anatomic chart study. Thus you will perform a partial dissection of the brachial plexus and some of the major nerves to provide some

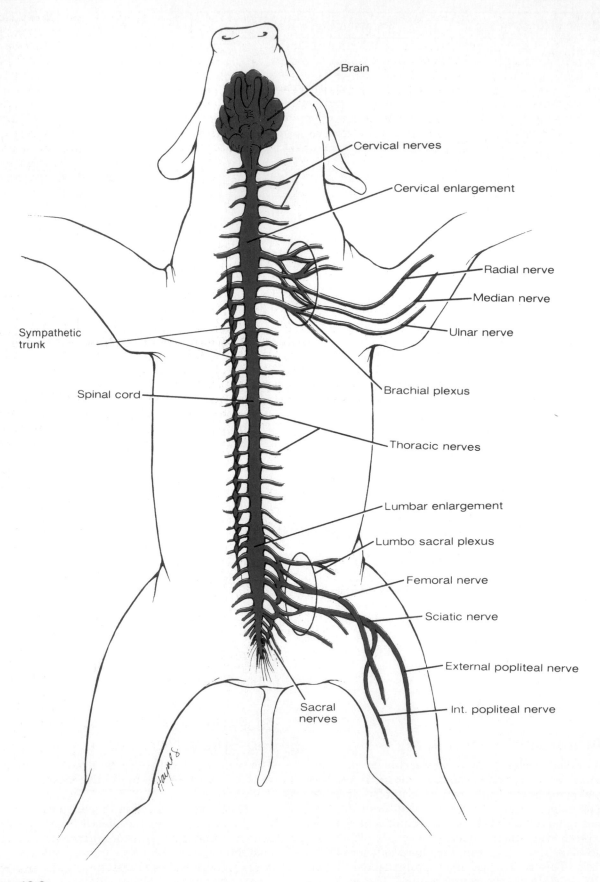

Figure 19.8

Brachial plexus and lumbosacral plexus of the fetal pig.

"hands on" experience. Refer to Figure 19.8 as you proceed.

NERVES OF THE BRACHIAL PLEXUS

1.   Place the pig dorsal side down on the dissecting tray to make these initial observations on the nerves of the brachial plexus. Reflect the cut ends of the left pectoralis muscles to expose the large brachial plexus in the axiliary region. Carefully clean the exposed nerves as far back toward their points of origin as possible. The brachial plexus, which consists of tough interconnected nerves emerging from the last three cervical and first thoracic vertebrae, should now be visible.

2.   The **median nerve** is closely associated with the brachial artery and vein, which course through the arm. It supplies most of the ventral muscles of the forearm. The **radial nerve** is a large nerve seen just superior to the median nerve. Follow it into the triceps brachii muscle of the arm. The **ulnar nerve** is the most posterior of the large brachial plexus nerves and supplies forearm and hand muscles that are not innervated by the median nerve.

3.   Although not actually part of the brachial plexus, the **sympathetic trunk** (chain of sympathetic nervous system ganglia) may be easily identified in the thoracic region. Pull the heart and lungs ventrally to expose the descending aorta. Carefully examine the dorsolateral surface of the aorta as it travels through the thorax. The sympathetic trunk appears as a slender white cord with segmental enlargements (the ganglia) in it lying alongside the aorta.

NERVES OF THE LUMBOSACRAL PLEXUS

1.   To locate the femoral nerve as it emerges from the pelvic cavity to enter the ventromedial aspect of the thigh, first identify the femoral artery and vein, with which it is closely associated. Follow it into the muscles and skin of the anterior thigh, which it supplies.

2.   Turn the pig ventral side down so that you can see the posterior aspect of the lower limb. Reflect the ends of the transected biceps femoris muscle to view the large cordlike **sciatic nerve.** The sciatic nerve arises from the sacral plexus and serves the dorsal thigh muscles and all of the muscles of the leg and foot via its various branches. Follow the nerve as it travels down the posterior thigh lateral to the semimembranous muscle. Note that it divides into its two major branches just superiorly to the gastrocnemius muscle of the calf. These branches, the internal and external popliteal nerves, innervate the leg.

3.   When you have finished making your observations, wrap the pig in paper towels soaked in embalming solution, place it in the plastic bag, and return it to the storage area. Clean all dissecting tools and equipment used before leaving the laboratory.

Neurophysiology of Nerve Impulses

OBJECTIVES

1. To list the two major physiologic properties of neurons.

2. To describe the polarized and depolarized state of the nerve cell membrane and to describe the events that lead to conduction of a nerve impulse—that is, depolarization by a threshold stimulus.

3. To explain the events of repolarization and the results of activation of the sodium-potassium pump.

4. To define *action potential* and *refractory period.*

5. To list various substances and factors that can stimulate neurons.

6. To recognize that neurotransmitters may be either stimulatory or inhibitory in nature.

7. To state the site of action of the blocking agents ether and curare.

8. To define *motor point.*

9. To explain how stimulation of motor points might be a useful treatment for an individual with nerve damage.

10. To demonstrate stimulation of motor points in the laboratory.

MATERIALS

Supply area 1:
Bullfrogs
Dissecting instruments and tray
Ringer's solution
Thread
Glass rods or probes
Glass plates or slides
Ring stand and clamp
Stimulator; platinum electrodes
Filter paper
0.01% hydrochloric acid (HCl) solution
Sodium chloride (NaCl) crystals
Heat-resistant mitts
Bunsen burner
Absorbent cotton
Ether
Pipettes
1-cc syringe with small-gauge needle
0.5% tubocurarine solution

Supply area 2:
Indifferent (plate) electrode (or ECG electrode)
Rubber straps
Insulated copper wire to be used as electrode lead or exploring electrode
Electrode paste
Penknife
Saline solution

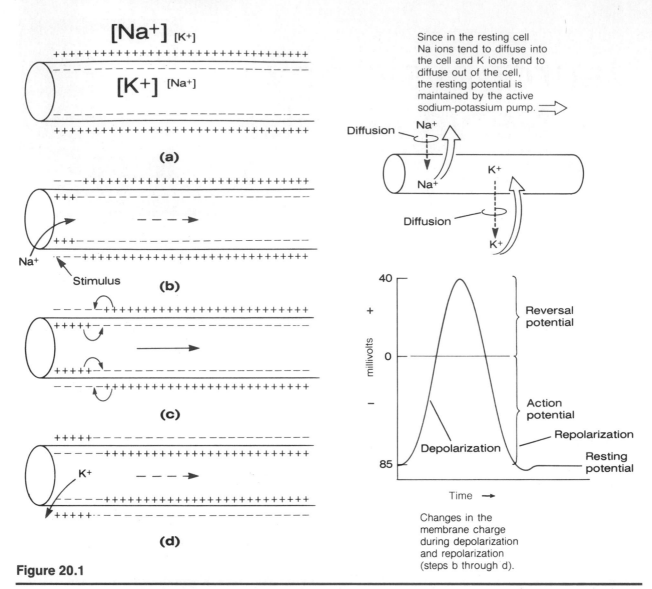

Figure 20.1

The nerve impulse. (a) Resting membrane potential (–85 mv). There is an excess of positive ions outside the cell, with Na$^+$ the predominant extracellular fluid ion and K$^+$ the predominant intracellular ion. The cell membrane has a low permeability to Na$^+$. (b) Depolarization—reversal of the resting potential. Application of a stimulus changes the membrane permeability, and Na$^+$ ions are allowed to diffuse rapidly into the cell. (c) Generation of the action potential or nerve impulse. If the stimulus is of adequate intensity, the action potential spreads rapidly along the entire length of the membrane. (d) Repolarization—reestablishment of the resting potential. The negative charge on the internal cell membrane surface and the positive charge on its external surface is reestablished by diffusion of K$^+$ ions out of the cell, proceeding in the same direction as in depolarization. The original ionic concentrations of the resting state are restored by the sodium-potassium pump.

THE NERVE IMPULSE

Neurons have two major physiologic properties: **irritability,** or the ability to respond to stimuli and convert them into nerve impulses, and **conductivity,** the ability to transmit the impulse to other neurons, muscles, or glands. Both phenomena reflect electrical changes across the neuron cell membrane that result in a separation of charges. or **membrane potential.**

In a resting neuron (as in resting muscle cells, which were discussed in Exercise 16), the exterior surface of the membrane has a higher net concentration of positive ions and is therefore more positively charged than the inner surface, as shown in Figure 20.1(a). This difference in electrical charge on the two sides of the membrane is referred to as the **resting potential,** and a neuron in this state is said to be **polarized.** In the resting state, the predominant intracellular ion is potassium (K$^+$), and sodium ions (Na$^+$) are found in greater concentration in the extracellular fluids. The resting potential is maintained by a very active sodium pump, which transports Na$^+$ out of the cell and K$^+$ into the cell.

When the neuron is activated by a stimulus

of adequate intensity—a **threshold stimulus**—its membrane briefly becomes more permeable to sodium. Sodium ions rush into the cell, increasing the number of positive ions inside the cell and reversing the polarity (Figure 20.1(b)). Thus the interior of the membrane becomes positive at that point and the exterior surface becomes negative—a phenomenon called **depolarization.**

Depolarization generates the membrane's **action potential.*** Once the action potential has been initiated, it is a self-propagating phenomenon that spreads rapidly along the entire length of the neuron, resulting in the depolarization of the entire membrane (Figure 20.1(c)). It is never partially transmitted; that is, it is an all-or-none response. This propagation of the action potential is the **nerve impulse.** When the nerve impulse reaches the axonal end bulbs, they release a neurotransmitter that acts either to stimulate or to inhibit the next neuron in the transmission chain. (Note that only stimulatory transmitters are considered here.)

Within a millisecond after the inward flux of sodium, the membrane permeability is again altered. As a result, Na^+ permeability decreases, K^+ permeability increases, and K^+ rushes out of the cell. Since K^+ ions are positively charged, their movement out of the cell reverses the membrane potential again, so that the external membrane surface is again positive relative to the internal membrane (Figure 20.1(d)). This event, called **repolarization,** reestablishes the resting potential. During the time of repolarization, the neuron is insensitive to further stimulation; thus this period is referred to as the **refractory period.**

Once repolarization has been completed, the neuron can quickly respond again to a stimulus, since only minute amounts of sodium and potassium ions have changed places. In fact, thousands of impulses can be generated before ionic imbalances prevent the neuron from transmitting impulses. Eventually, however, it is necessary to restore the original ionic concentrations on the two sides of the membrane.

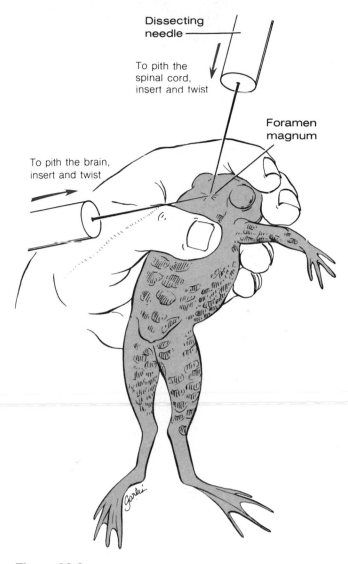

Labels: Dissecting needle · To pith the spinal cord, insert and twist · Foramen magnum · To pith the brain, insert and twist

Figure 20.2

Procedure for pithing a frog.

PHYSIOLOGY OF NERVE FIBERS

In this experiment you will investigate the functioning of nerve fibers by subjecting the sciatic nerve of a frog to various types of stimuli and blocking agents. (Work in groups of two to four to lighten the work load.) Stimulation of the nerve and generation of the action potential will be indicated by the contraction of the gastrocnemius muscle. Because you will make no mechanical recording (unless your instructor asks you to), you must keep complete and accurate records of all experimental procedures and results.

 1. Obtain a bullfrog and dissecting instruments and tray. Destroy the central nervous system of the frog by double pithing it. Hold the frog firmly in the palm of your hand with its ventral surface down. Flex the head, holding it securely between your index and middle fingers, as shown in Figure 20.2. Insert a dissecting needle anteriorly into the foramen magnum (the small depression on the posterior surface of the frog's head), and quickly twist the needle to destroy the brain. Then insert the needle posteriorly into the spinal cavity to destroy the spinal cord.

---

*If the stimulus is of less than threshold intensity, depolarization is limited to a small area of the membrane, and no action potential is generated.

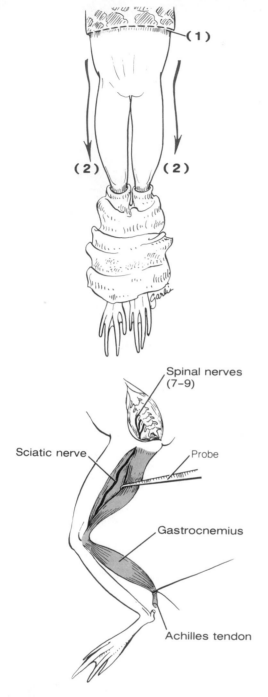

Figure 20.3

Removal of the sciatic nerve and gastrocnemius muscle.
(1) Cut through the frog's skin around the circumference
of the trunk. (2) Pull the skin down over the trunk and
legs. (3) Make a longitudinal cut through the abdominal
musculature and expose the roots of the sciatic nerve
(arising from spinal nerves 7–9). Ligate the nerve and cut
the roots proximal to the ligature. (4) Use a glass probe to
expose the sciatic nerve beneath the posterior thigh
muscles. (5) Ligate the Achilles tendon and cut it free
distal to the ligature. (6) Release the gastrocnemius
muscle from the connective tissue of the knee region.

2. Prepare the sciatic nerve as illustrated in
Figure 20.3. Place the pithed frog on the dissecting tray, dorsal side down. Make a cut through
the skin around the circumference of the frog
approximately halfway down the trunk, and then
pull the skin down over the muscles of the legs.
Open the abdominal cavity and push the abdominal organs to one side to expose the origin of
the glistening white sciatic nerve, which arises
from the last three spinal nerves. Once the
sciatic nerve has been exposed, it should be kept
continually moist with Ringer's solution.

3. Slip a piece of thread moistened with Ringer's solution under the sciatic nerve close to
its origin at the vertebral column. Make a single
ligature (tie it firmly with the thread), and then
cut through the nerve roots to free the proximal
end of the sciatic nerve from its attachments.
Using a glass rod or probe, carefully separate
the posterior thigh muscles to locate and then
free the sciatic nerve, which runs down the
posterior aspect of the thigh.

4. Tie a piece of thread around the Achilles
tendon of the gastrocnemius muscle, and then
cut through the tendon distal to the ligature to
free the gastrocnemius muscle from the heel. Using a scalpel, very carefully release the gastrocnemius muscle from the connective tissue in the
knee region. At this point you should have completely freed both the gastrocnemius muscle
and the sciatic nerve, which innervates it.

5. With glass rods, transfer the muscle to a
glass plate or slide, and then attach the slide to
a ring stand with a clamp. Allow the end of the
sciatic nerve to hang over the edge of the glass
slide, so that it is easily accessible for stimulation. Remember to keep the nerve moist at all
times.

6. You are now ready to investigate the response of the sciatic nerve to various stimuli,
beginning with electrical stimulation. Using the
stimulator and platinum electrodes, stimulate the
sciatic nerve with single shocks, gradually
increasing the intensity of the stimulus until the
threshold stimulus is determined.

(The muscle as a whole will just barely contract

at this stimulus.) Record this stimulus: _____ v

Continue to increase the voltage until you find
the point beyond which no further increase in the
strength of muscle contraction occurs—that is,
the point at which the maximal contraction of the

muscle is obtained. Record this voltage: _____ v

Delivering multiple or repeated shocks to the
sciatic nerve causes volleys of impulses in the

nerve. Shock the muscle with multiple stimuli. Observe the response of the muscle. How does this response compare with the response to the

single electrical shocks? _____

_____

_____

7. To investigate mechanical stimulation, pinch the free end of the nerve by firmly pressing it

between two glass rods. What is the result? _____

8. Chemical stimulation can be tested by applying a small piece of filter paper saturated with 0.01% hydrochloric acid (HCl) solution to the free

end of the nerve. What is the result? _____

Drop a few grains of salt (NaCl) on the free end

of the nerve. What is the result? _____

9. Now test thermal stimulation. Wearing the heat-resistant mitts, heat a glass rod for a few moments over a Bunsen burner. Then touch the rod to the free end of the nerve. What is the

result? _____

What do these muscle reactions say about the

irritability and conductivity of neurons? _____

_____

10. The next part of this experiment investigates the effects of various blocking factors. Ether is one such blocking agent. Since ether is extremely volatile and explosive, be sure that laboratory fans are on and that *all Bunsen burners are off* during this procedure.

Clamp a second glass slide to the ring stand just slightly below the first slide. With glass rods, gently position the sciatic nerve on this second slide, allowing a small portion of the nerve's distal end to extend over the edge. Place a piece of absorbent cotton soaked with ether under the midsection of the nerve on the slide, probing it into position with a glass rod. Stimulate the distal end of the nerve at 2-min intervals, using a voltage slightly above the threshold stimulus, until the muscle fails to respond. (If the cotton dries before this, rewet it with ether using a pipette.) How long did it take for anesthesia

to occur? _____ sec

Once anesthesia has occurred, stimulate the nerve beyond the anesthetized area, between the ether-

soaked pad and the muscle. What is the result?

_____

Remove the ether-soaked pad and flush the nerve fibers with saline. Again stimulate the nerve at its distal end at 2-min intervals. How long does it take

for recovery? _____

Does ether exert its blocking effect on the nerve

fibers or on the muscle cells? _____

Explain your reasoning. _____

_____

If sufficient frogs are available and time allows, you may do the following classic experiment. In the 1800s Claude Bernard described an investigation into the effect of curare on nerve-muscle interaction. Curare was commonly used by some South American Indian tribes to tip their arrows. Victims struck with these arrows were paralyzed, but the paralysis was not accompanied by loss of sensation.

1. Prepare another frog as in steps 1 through 3 of the previous dissection. However, in this case the frog should be positioned ventral side down on a frog board. In exposing the sciatic nerve, take care not to damage the blood vessels in the thigh region, as the success of the experiment depends on the maintenance of the blood supply to the muscles of the leg.

2. Slip a length of thread under the nerve, and then tie the thread tightly around the thigh muscles to cut off circulation to the leg. (The sciatic nerve should be above the thread and not in the ligatured tissue.) Expose and gently tie the sciatic nerve so that it can be lifted away from the leg musculature for stimulation. Expose and ligature the sciatic nerve of the left leg in the same manner, but this time do *not* ligate the thigh muscles.

3. Using a syringe and needle, slowly inject 1 cc of 0.5% tubocurarine into the dorsal lymph sac of the frog.* The dorsal lymph sacs are located dorsally at the level of the scapulae, so introduce the needle of the syringe just beneath the skin between the scapulae and toward one side of the spinal column. *Handle the tubocurarine very carefully, because it is extremely poisonous.* Do not get any on your skin.

4. Wait 15 min after injection of the tubocurarine, and then stimulate electrically the left sciatic nerve.

_____

*To obtain 1 cc of the tubocurarine, inject 1 cc of air into the vial through the rubber membrane, and then draw up 1 cc of the chemical into the syringe.

Be careful not to touch any of the other tissues with the electrode. Gradually increasing the voltage, deliver single shocks until the threshold stimulus has been determined for this specimen. Threshold

stimulus: _____v

Now stimulate the right sciatic nerve with the same voltage intensity. Is there any difference in the

reaction of the two muscles? _____If so,

explain. _____

_____

If you did not find any difference, wait an additional 10 to 15 min and restimulate both sciatic nerves.

What is the result? _____

5. To determine the site at which tubocurarine acts, directly stimulate each gastrocnemius muscle. What

is the result? _____

_____

Explain the difference between the responses of the

right and left sciatic nerves. _____

_____

_____

Explain the results when the muscles were stimu-

lated directly. _____

_____

At what site does tubocurarine (or curare) act? ____

_____

MOTOR POINTS

Motor points are specific points on the body's surface where motor nerves supplying particular skeletal muscles may be stimulated with relative ease. When a motor point is stimulated, the muscle or muscles that it supplies contract; thus motor points may be located and mapped out in the laboratory by using an exploring electrode.

 1. Work in pairs or groups of four to conduct this investigation. One student, who will be the subject, should sit comfortably with her or his arms resting on the laboratory bench.

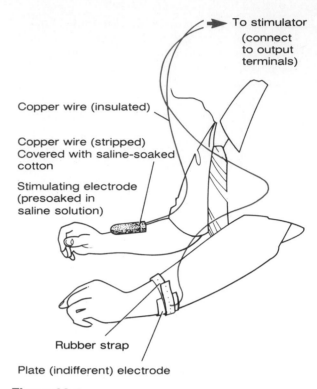

To stimulator (connect to output terminals)

Copper wire (insulated)

Copper wire (stripped) Covered with saline-soaked cotton

Stimulating electrode (presoaked in saline solution)

Rubber strap

Plate (indifferent) electrode

Figure 20.4

Setup for stimulation and localization of motor points.

2. Attach the indifferent electrode (the plate electrode) to the positive terminal of the stimulator with an insulated copper wire. Apply the electrode paste to the plate and rest the subject's left forearm on the plate. Attach the electrode to the arm with a rubber strap.

3. A simple "wick" electrode can be used as the exploring or stimulating electrode. If necessary, make this electrode by cutting off with a penknife approximately 2 in. of the insulated covering of a copper wire at the distal end. Wrap the exposed 2-in. length of wire with a piece of absorbent cotton soaked in saline solution. (The saline-soaked "wick" diffuses the current so that the subject will feel no unpleasant shocks.) Attach the distal end of the wick electrode to the negative pole of the stimulator. The experimental setup is shown in Figure 20.4.

4. Set the stimulator to deliver a tetanizing current—that is, one that produces a prolonged contraction. Do this by setting the frequency at 15–20 per second, the duration at 0.5 msec, and the voltage at 40 v. (If the subject complains of the sensation of shock, lower the voltage.)

5. With the exploring electrode, examine the right forearm (the arm not resting on the plate electrode) for motor points. Remember, the subject's stimulated muscle will contract involuntarily. Find at least four motor points on the

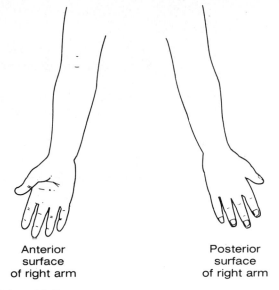

Anterior
surface
of right arm

Posterior
surface
of right arm

Figure 20.5

Anterior and posterior surfaces of the right arm for plotting localized motor points.

anterior forearm and three on the back of the forearm and hand. One of the specific motor points of the posterior arm that you should locate is the motor point of the ulnar nerve at the elbow. If you find no motor points, try again with a higher voltage.

6. On the adjacent figures of the anterior and posterior arm surfaces in Figure 20.5 indicate the approximate location of the motor points you have stimulated, and number each one. On the corresponding numbered lines that follow, describe the specific muscle activity you have seen (for example, pronation of the forearm or flexion of specific digits). Then, using an appropriate reference, attempt to name the muscle or muscles and motor nerve stimulated at each motor point. Verify your conclusions with your instructor.

Anterior surface of arm:

(1) _____

(2) _____

(3) _____

(4) _____

Posterior surface of arm:

(1) _____

(2) _____

(3) _____

Do the motor points you have mapped always lie

directly over the stimulated muscle? _____

Explain. _____

_____

General Sensation

OBJECTIVES

1. To recognize various types of general sensory receptors as studied in the laboratory, and to describe the function and location of each type.

2. To define *exteroceptor, interoceptor,* and *proprioceptor.*

3. To demonstrate and relate differences in relative density and distribution of tactile and thermoreceptors in the skin.

4. To define *tactile localization* and describe how this ability varies in different areas of the body.

5. To explain the tactile two-point discrimination test, and to state its anatomic basis.

6. To define *referred pain* and attempt to explain it.

7. To define *adaptation, negative afterimage,* and *projection.*

MATERIALS

Compound microscope
Histologic slides of pacinian corpuscles (longitudinal section), Meissner's corpuscles, Golgi tendon organs, and muscle spindles
Temperature cylinders (or pointed metal rods of approximately 1-in. diameter and 2- to 3-in. length, with insulated blunt end
Large beaker of ice water; chipped ice
Hot water bath set at 45 C
Thermometer
Fine-point, felt-tipped markers (black, red, and blue)
Small metric ruler
Von Frey's hairs (or 1-in. lengths of horsehair glued to a matchstick) or sharp pencil
Towel
Caliper or esthesiometer
Four coins (nickels or quarters)
Three large finger bowls or 1000-ml beakers

It can be said without reservation that people are irritable creatures. Hold a sizzling steak before them and their mouths water. Flash your high beams in their eyes on the highway and they cuss. Stroke their arms gently and they smile. These "irritants" (the steak, the light, and the soft touch) and many others are stimuli that continually assault the individual.

The body's sense organs, which include its sensory receptors, react to stimuli or changes within the body and in the external environment. The minute sensory receptors of the general senses react to touch, pressure, pain, heat, cold, and changes in position and are distributed throughout the body. In contrast to the widely distributed general receptors, the special senses are large, complex sensory organs or small, localized groups of receptors. The special senses include sight, hearing, equilibrium, smell, and taste.

Sensory receptors may be classified according to their location in the body and the source of the stimulus. **Exteroceptors** are found close to the body surface, and they react to stimuli in the external environment. Exteroceptors include the cutaneous receptors and the highly specialized receptor structures of two of the special senses (the vision apparatus of the eye and the hearing and equilibrium receptors of the ear). Interoceptors respond to stimuli arising within the body. **Interoceptors** are found in the internal organs as well as in the olfactory mucosa of the nose and the taste buds of the mouth. A subdivision of the interoceptors, the **proprioceptors,** are located in the skeletal muscles and their tendons. They provide information on the position and degree of stretch of the muscles and tendons. The receptors of the special sense organs are complex and deserve considerable study. Thus the special senses (vision, hearing, equilibrium, taste, and smell) are covered separately in Exercises 24 through 26. Only the anatomically simpler cutaneous sensory receptors and proprioceptors will be studied in this exercise.

STRUCTURE OF SENSORY RECEPTORS

You cannot become aware of changes in the environment unless your sensory neurons and their receptors are operating properly. Sensory receptors are considered to be modified dendritic endings (or specialized cells associated with the dendrites) that are sensitive to certain environmental stimuli. They react to such stimuli by initiating a nerve impulse. Several histologically distinct types of receptors in the skin have been identified; their structures are depicted in Figure 21.1.

Many references link each type of receptor to specific stimuli, but there is still considerable controversy about the precise qualitative function of each receptor. It may be that the responses of all these receptors overlap considerably. Certainly, intense stimulation of any of them is always interpreted as pain.

The least specialized of the cutaneous receptors are the **pain receptors,** which are essentially bare dendritic endings of sensory neurons. (The pain receptors are widespread in the skin and make up a sizable portion of the visceral interoceptors.) The other cutaneous receptors are more complex, with the dendritic endings encapsulated in other types of cells. **Meissner's corpuscles,** commonly referred to as the touch receptors because they respond to light pressure, are located just beneath the epidermis in the dermal papillary layer of the skin. **Krause's end bulbs** and **Ruffini's corpuscles** are presumed to be thermoreceptors that respond to cold and heat respectively. As you inspect Figure 21.1, note that all of these receptor types are quite similar, with the possible exception of the **pacinian corpuscles** (deep pressure receptors), which are anatomically more unique.

1.   Obtain histologic slides of pacinian and Meissner's corpuscles. Locate a Meissner's corpuscle in the dermal layer of the skin under low power and then switch to the oil immersion lens for a detailed study. Note that the naked dendritic fibers within the capsule are aligned parallel to the skin surface. Compare your observations to Figure 21.1(b).

2.   Next observe the pacinian corpuscles. If possible, note the slender naked dendrite ending in the center

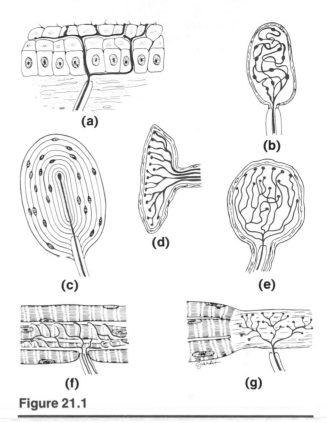

(a)

(b)

(d)

(c)

(e)

(f)

(g)

Figure 21.1

Types of sensory receptors: (a) naked nerve endings (pain receptors); (b) Meissner's corpuscle (touch receptor); (c) pacinian corpuscle (deep pressure receptor); (d) Ruffini's corpuscle (heat receptor); (e) Krause's end bulb (cold receptor); (f) muscle spindle; (g) Golgi tendon organ (proprioceptors).

of the capsule and the heavy capsule of connective tissue surrounding it (which looks rather like an onion cut lengthwise). Also, note how much larger the pacinian corpuscles are than the Meissner's corpuscles. Compare your observations to the view shown in Figure 21.1(c).

3.   Obtain slides of muscle spindles and Golgi tendon organs, the two major types of proprioceptors (see Figure 21.1(f) and 21.1(g)). In the slide of muscle spindles, note that minute extensions of the dendrites of the sensory neuron coil around specialized slender skeletal muscle cells called intrafusal cells, or fibers. The Golgi tendon organs are composed of dendrites that ramify through the tendon tissue close to the muscle tendon attachment. Stretching of the muscle or tendon excites both types of receptors, which then transmit impulses that ultimately reach the cerebellum for interpretation. Compare your observations to Figure 21.1(f) and (g).

RECEPTOR PHYSIOLOGY

Sensory receptors act as transducers, changing the environmental stimulus into afferent nerve impulses.

Since the action potential generated in all nerves is essentially identical, the stimulus is identified entirely by the area of the brain's sensory cortex that is stimulated (which, of course, differs for the various afferent nerves).

Four qualities of cutaneous sensations have traditionally been recognized: tactile (touch), heat, cold, and pain. Mapping these sensations on the skin has revealed that the specialized sensory receptors have discrete locations and are characterized by clustering at certain points—**punctate distribution**—rather than by uniform distribution.

The simple pain receptors, extremely important in protecting the body, are the most numerous. Touch and temperature receptors tend to be clustered where greater sensitivity is desirable, as on the hands and face. It is surprising to learn that rather large areas of the skin are quite insensitive to touch because of a relative lack of touch receptors.

There are several simple experiments you can conduct to investigate the location and physiology of the cutaneous receptors. In each of the following activities, work in pairs with one person as the subject and the other as the experimenter. After you have completed an experiment, switch roles and go through the procedures again so that all class members obtain individual results. Keep an accurate account of each test that you perform.

Density and Location of Temperature and Touch Receptors

 1. Place one temperature cylinder in a beaker of ice water and the other in a water bath controlled at 45 C. Draw a square (2 cm on each side) with a felt marker on the ventral surface of the subject's forearm. During the following tests, the subject's eyes are to remain closed; the subject should tell the examiner when a stimulus is detected.

2. Working in a systematic manner from one side of the marked square to the other, gently touch the Von Frey's hairs or a sharp pencil tip to different points within the square. The hairs should be applied with a pressure that just causes them to bend; use the same pressure for each contact. Do not apply deep pressure; only the more superficially located Meissner's corpuscles are to be stimulated by the touch stimulus (as opposed to the pacinian corpuscles located in the subcutaneous tissue). Mark with a black dot all points at which the touch is perceived.

3. Remove the temperature rod from the ice water and quickly wipe it dry. Repeat the procedure outlined above, noting all points of cold

perception with a blue dot.* Perception should be for temperature, not simply touch.

4. Remove the second temperature rod from the 45 C water bath and repeat the procedure once again, marking all points of heat perception with a red dot.*

5. After each student has acted as the subject, produce a "map" of receptor areas in the squares provided here.

Student 1 _____

Student 2 _____

How does the density of the heat receptors correspond to that of the touch receptors? _____

_____

To that of the cold receptors? _____

_____

On the basis of your observations, which type of receptor appears to be most abundant (at least in the area tested)? _____

Two-Point Discrimination Test

The density of the touch receptors varies significantly in different areas of the body. In general, areas that have the greatest density of tactile receptors have a heightened ability to "feel." These areas correspond to areas that receive the greatest motor innervation; thus they are areas of fine motor control.

On the basis of this information, which areas of the body do you *predict* will have the greatest density of touch receptors? _____

_____

* The temperature rods will have to be returned to the water baths approximately every 2 minutes to maintain the desired testing temperatures.

Using a caliper or esthesiometer and a metric ruler, test the ability of the subject to differentiate two distinct sensations when the skin is touched simultaneously at two points. Beginning with the face, start with the caliper arms completely together. Gradually increase the distance between the arms, testing the subject's skin after each adjustment. Continue with this testing procedure until the subject reports that two points of contact can be felt. This measurement, the smallest distance at which two points of contact can be felt, is the **two-point threshold.** Repeat this procedure on the back and palm of the hand, fingertips, lips, back of the neck, and back of the calf. Record your results in the accompanying chart.

Body area tested	Two-point threshold (millimeters)
Face	
Back of hand	
Palm of hand	
Fingertips	
Lips	
Back of neck	
Back of calf	

Tactile Localization

Tactile localization is the ability to determine which portion of the skin has been touched. The tactile receptor field of the body periphery has a corresponding "touch" field in the brain's somatosensory association area. Some body areas are well represented with touch receptors, which allows tactile stimuli to be localized with great accuracy, but the density of the touch receptors in other body areas allows only a crude discrimination.

1. The subject's eyes should be closed during the testing. The experimenter touches the palm of the subject's hand with a pointed black felt-tipped marker. The subject should then try to touch the exact point with his or her own marker, which should be of a different color. Measure the error of localization in millimeters. Repeat the test in the same spot twice more, recording the error of localization for each test. Average the results of the three determinations and record it in the chart below. Does the

ability to localize the stimulus improve the second time? _____

The third time? _____ Explain. _____

_____

_____

2. Repeat the above precedure on a fingertip, the ventral forearm, the ventral surface of the upper arm, and the upper back between the shoulder blades. Record the averaged results in the chart.

Body area tested	Average area of localization (millimeters)
Palm of hand	
Fingertip	
Ventral forearm	
Ventral surface of upper arm	
Upper back	

Adaptation of Touch Receptors

The number of impulses transmitted by sensory receptors often changes both with the intensity of the stimulus and with the length of time the stimulus is applied. In many cases, when a stimulus is applied for a prolonged period, the rate of receptor discharge slows and conscious awareness of the stimulus is decreased or lost until some type of stimulus change occurs. This phenomenon is referred to as **adaptation.** The touch receptors adapt particularly rapidly, which is highly desirable. Who, for instance, would want to be continually aware of the pressure of clothing on their skin?

1. The subject's eyes should be closed. Place a coin on the anterior surface of the subject's forearm, and determine how long the sensation persists for the subject. Record the duration of the

sensation: _____ sec

2. Repeat the test, placing the coin at a different forearm location. How long does the sensation persist at the second location? _____ sec

After awareness of the sensation has been lost at the second site, stack three more coins atop the first one.

Does the pressure sensation return? _____

If so, for how long is the subject aware of the pressure in this instance? _____ sec. Are the same receptors being stimulated when the four coins, rather than the one coin, are used? _____

Explain. _____

_____

3.   To further illustrate the adaptation of touch receptors—in this case, the touch receptors of the hair follicles—gently and slowly bend one hair shaft with a pen or pencil until it springs back (away from the pencil) to its original position. Is the tactile sensation greater when the hair is being slowly bent

or when it springs back? _____

_____

Why is the adaptation of the touch receptors in the hair follicles particularly important to a woman who wears her hair in a ponytail? If the answer is not immediately apparent, consider the opposite phenomenon: what would happen, in terms of sensory input from her hair follicles, if these receptors

did not exhibit adaptation? _____

_____

_____

Adaptation of Temperature Receptors

Adaptation of the temperature receptors can be tested through very unsophisticated methods.

 1.   Obtain three large finger bowls or 1000-ml beakers and fill the first with 45 C water. Have the subject immerse her or his left hand into the water and report the sensation. Keep the left hand immersed for 1 min and then also immerse the right hand in the same bowl.

What is the sensation of the left hand when it is

first immersed? _____

What is the sensation of the left hand after 1 min as compared to the sensation in the right hand just immersed? _____

Had adaptation occurred in the left hand? _____

2.   Rinse both hands in tap water, dry them, and wait 5 min before conducting the next test. Just before beginning the test, refill the finger bowl with fresh 45 C water, fill a second with ice water, and fill a third with water at room temperature.

3.   Place the left hand in the ice water and the right hand in the 45 C water. What is the sensation in each hand after 2 min as compared to the sensation perceived when the hands were first

immersed? _____

_____

Which hand seemed to adapt more quickly? _____

_____

4.   After reporting these observations, the subject should then place both hands simultaneously into the finger bowl containing the water at room temperature. Record the sensation in the left hand:

_____

The right hand: _____

The sensations that the subject experienced when both hands were put into room-temperature water are called **negative afterimages.** They are explained by the fact that sensations of heat and cold depend on the rapidity of heat loss or gain by the skin and differences in the temperature gradient.

Referred Pain

Experiments on pain receptor localization and adaptation are commonly conducted in the laboratory. However, there are certain problems in conducting such experiments. The pain receptors are densely distributed in the skin, and they adapt very little, if at all. (This lack of adaptability is due to the protective function of the receptors. The sensation of pain often indicates tissue damage or trauma to body structures.) Thus no attempt will be made in this Exercise to localize the pain receptors or to prove their nonadaptability, since both would cause needless discomfort to the subject and would not add any additional insight.

However, the phenomenon of referred pain is easily demonstrated in the laboratory, and such experiments provide information that may be useful in explaining common examples of this phenomenon. Referred pain is a sensory experience in which pain is perceived as arising in one area of the body when in fact another, often quite remote area is receiving the painful stimulus. Thus the pain is said to be "referred" to a different area. The phenomenon of **projection,** the process by which the brain refers sensations to their usual point of stimulation, provides the most simple explanation of such experiences. Many of us have experienced referred pain as a radiating pain in the forehead after quickly swallowing an ice-cold drink. Referred pain is important in many types of clinical diagnosis, since damage to many visceral organs results in this phenomenon. For example, inadequate oxygenation of the heart muscle often results in pain being referred to the chest wall and shoulder (angina pectoris), and the reflux of gastric juice into the esophagus causes a sensation of intense discomfort in the thorax referred to as heartburn. In addition, amputees often report phantom limb pain—feelings of pain that appear to be coming from a part of the body that is no longer there.

Immerse the subject's elbow in a finger bowl containing ice water. Record the quality (such as discomfort, tingling, or pain) and the quality progression of the sensations for 2 min. Also record the location of the perceived sensations. The ulnar nerve, which serves the medial two fingers and the medial surface of the hand, is involved in the phenomenon of referred pain experienced during this test. How does the localization of this referred pain correspond to the areas served by the ulnar nerve?

_____

_____

_____

_____

Time of observation	Quality of sensation	Localization of sensation
On immersion		
After 1 min		
After 2 min		

Human Reflex Physiology

OBJECTIVES

1. To define *reflex*.

2. To name, identify, and describe the function of each element of a reflex arc.

3. To state why reflex testing is an important part of every physical examination.

4. To describe and discuss several types of reflex activities as observed in the laboratory; to note the functional or clinical importance of each; and to categorize each as a somatic or autonomic reflex action.

5. To explain why cord-mediated reflexes are generally much faster than those involving input from the higher brain centers.

6. To differentiate between reaction time and reflex activity.

MATERIALS

Reflex hammer
Lancets
Alcohol swabs
Cot (if available)
Absorbent cotton
Tongue depressor
Metric and 12-in. ruler
Flashlight
100- or 250-ml beaker
10- or 25-ml graduated cylinder
Lemon juice
Wide-range pH paper

THE REFLEX ARC

Reflexes are rapid, predictable, and involuntary motor responses to stimuli; they are mediated over neural pathways called **reflex arcs**.

The essential components of all reflex arcs are a sensory receptor endorgan, which reacts to a stimulus; an effector organ, which is the muscle or gland stimulated; and afferent and efferent neurons, which provide the communication pathway between the two. The simple patellar or knee-jerk reflex shown in Figure 22.1(a) is an example of a simple, two-neuron, monosynaptic reflex arc, and it will be demonstrated in the laboratory.

Most reflexes are more complex, involving the participation of one or more internuncial neurons in the reflex arc pathway. A three-neuron reflex arc (flexor reflex) is diagrammed in Figure 22.1(b). It consists of five structures: receptor, afferent neuron, internuncial neuron, efferent neuron, and effector. Since delay or inhibition of the reflex may occur at the synapses, the larger the number of synapses encountered in a reflex pathway, the greater the time required to effect the reflex.

Reflexes of many types may be considered programed into the neural anatomy. For example, many spinal reflexes, such as the flexor reflex, occur without the involvement of higher brain centers. These reflexes work equally well in decerebrate animals (those in which the brain has been destroyed), as long as the spinal cord is functional. Conversely, other reflexes require the involvement of functional brain tissue, since many different inputs must be evaluated before the appropriate reflex is determined. The pupillary responses to light and the vestibular reflexes are in this category. In addition, although many spinal reflexes do not require the involvement of higher centers, the brain is frequently "advised" of spinal cord reflex activity and may alter it by facilitating or inhibiting the reflexes.

If the spinal cord is suddenly transected, all functions mediated by the cord are immediately

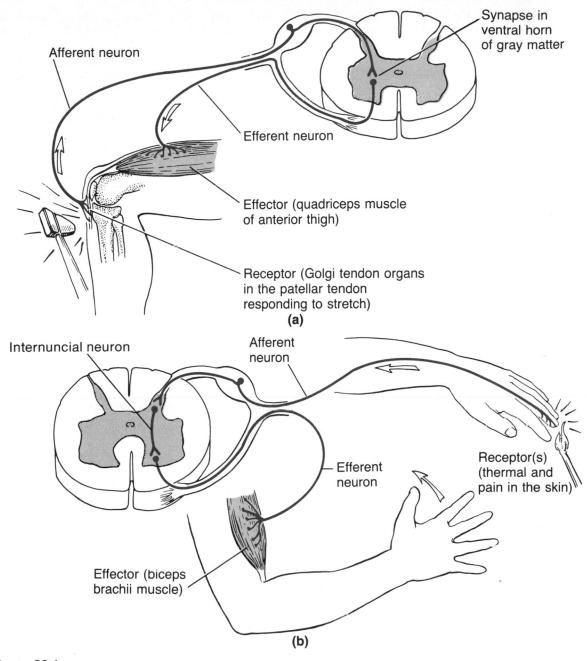

Figure 22.1

Simple reflex arcs: (a) two-neuron reflex arc; (b) three-neuron reflex arc.

depressed, a phenomenon called spinal shock. The reason for this appears to be that continual facilitating signals from the higher centers are necessary for normal cord activity. After a few weeks in spinal shock, the spinal cord neurons normally regain their excitability, and most cord-mediated reflexes reappear (often, sadly in a hyperexcitable state that leads to the spasticity of certain muscles).

Reflex testing is an important diagnostic tool to the doctor in assessing the condition of the nervous system. Exaggeration, distortion, or absence of the various reflex responses may indicate degeneration or pathology of portions of the nervous system, often before other signs are apparent.

Reflexes can be categorized into one of two large groups: the somatic reflexes and the autonomic reflexes. **Autonomic** (or visceral) **reflexes** are mediated through the autonomic nervous system, and are not subject to conscious control. These reflexes result in the activation of smooth muscles, cardiac muscle, and the glands of the body; they involve the regulation of such body functions as digestion, elimination, blood pressure, salivation, and sweating. **Somatic reflexes** include all those reflexes that involve the stimulation of skeletal muscles by the somatic division of the nervous system. An example of such a reflex is the rapid withdrawal of a hand from a hot object.

SOMATIC REFLEXES

There are several types of somatic reflexes, including the stretch, crossed extensor, superficial cord, corneal, and gag reflexes. Some require only spinal cord activity, and others require brain involvement as well.

Deep Tendon or Stretch Reflexes

The deep tendon reflexes are initiated by tapping a tendon, which stretches the muscle the tendon is attached to. This stimulates both the Golgi tendon receptors and muscle spindle receptors, and results in the reflex contraction of the muscle or muscles. Stretch reflexes are important postural reflexes, normally acting to maintain posture, balance, and locomotion. The deep tendon reflexes tend to be hypoactive or absent in cases of peripheral nerve damage or ventral horn disease and hyperactive in corticospinal tract lesions. They are absent in deep sedation and coma.

 1.   Test the knee-jerk, or **patellar, reflex** by seating a subject on the laboratory bench with legs hanging free (or with knees crossed). Tap the patellar tendon sharply with the reflex hammer just below the knee to elicit the knee-jerk response. Test both knees and record your observations.

_____

_____

_____

Which muscles contracted? _____

_____

What nerve is carrying the afferent and efferent

impulses? _____

2.   Test the effect of mental distraction on the patellar reflex by having the subject add a column of three-digit numbers while you test the reflex again. Is the response greater than or less than the

first response? _____

What are your conclusions about the effect of mental

distraction on reflex activity? _____

_____

_____

3.   Now test the effect of muscular activity occurring simultaneously in other areas of the body. Have the subject clasp the edge of the laboratory bench and vigorously attempt to pull it upward with both hands. At the same time, test the patellar reflex again. Is the response more or less vigorous than the

first response? _____

4.   Fatigue also influences the reflex response. The subject should jog in position until she or he is very fatigued (really fatigued—no slackers). Test the patellar reflex again and record the results (that is,

is it more or less vigorous?) _____

_____

Would you say that nervous system activity or muscle function is responsible for the changes you

have just observed? _____

_____

Explain your reasoning. _____

_____

_____

_____

5.   To demonstrate the **Achilles**, or ankle-jerk, **reflex**, kneel on a chair with the feet dangling over the seat edge. Dorsiflex the foot to increase the tension on the gastrocnemius muscle. Have your partner sharply tap the Achilles tendon with the reflex hammer.

What is the result? _____

_____

Does the contraction of the gastrocnemius normally

result in the activity you have observed? _____

Crossed Extensor Reflex

The crossed extensor reflex is more complex than the stretch reflex, consisting of a flexor or withdrawal reflex followed by extension of the opposite limb.

This reflex is quite obvious when, for example, a stranger suddenly and strongly grips one's arm. The immediate response is to withdraw the clutched arm and push the intruder away with the other arm. The reflex is more difficult to demonstrate in a laboratory because it is anticipated, and under these conditions the extensor part of the reflex may be inhibited.

The subject should sit with eyes closed and with the back of one hand resting on the laboratory bench. Obtain a lancet and suddenly prick the subject's index finger. What are the results?

_____

_____

_____

Wipe the subject's pricked finger with an alcohol swab and press for a few seconds to stop the bleeding. Did the extensor part of this reflex seem to be slow compared to the other reflexes you

have observed? _____

What are the reasons for this? _____

_____

_____

The reflexes that have been demonstrated so far—the stretch and crossed extensor reflexes—are examples of reflexes in which the reflex pathway is initiated and completed at the cord level.

Superficial Cord Reflexes

The superficial cord reflexes (abdominal, cremaster, and plantar reflexes) result from pain and temperature changes. They are initiated by stimulation of receptors in the skin and mucosae. The superficial cord reflexes depend *both* on functional upper-motor pathways and on the cord-level reflex arc.

1. To test the **abdominal reflex**, the subject should lie in the supine position on a cot or the laboratory bench. Since the skin of the abdominal wall will have to be exposed, the subject should wear a blouse or shirt that can be pulled up and aside to expose the umbilical area. Gently stroke the skin on one side of the abdomen slightly above the umbilicus and record the

response._____

_____

_____

_____

Then stroke the skin on the same side, but this time below and to the side of the umbilicus. What is the

result? _____

_____

_____

In each case, the skin of the umbilicus should have moved toward the point of stimulation. In cases of upper motor neuron lesions involving the pyramidal or corticospinal tract, the reflexes are absent.

How does the speed of this superficial type of reflex compare with that of deep tendon reflexes investi-

gated earlier? _____

_____

2. The **plantar**, or **Babinski, reflex**, an important neurologic test, is elicited by stimulating the cutaneous receptors in the sole of the foot. In adults, stimulation of these receptors causes the toes to flex and move closer together; however, damage to the pyramidal (or corticospinal) tract, produces a *positive* Babinski response in which the toes flare and the great toe moves in an upward direction. (In newborn infants, the positive Babinski response is seen and is due to incomplete myelination of the nervous system.)

Have the subject remove a shoe and lie on the cot or laboratory bench with knees slightly bent and thighs rotated so that the lateral side of the foot is resting on the cot. Alternatively, the subject may sit up and rest the lateral surface of the foot on a chair. Draw the handle of the reflex hammer firmly down the medial side of the exposed sole from the heel to the base of the toes.

What is the response? _____

_____

Is this a positive or negative Babinski response?

_____

Corneal Reflex

The corneal reflex is mediated through the trigeminal nerve (cranial nerve V). The absence of this reflex is an ominous sign, as it often indicates damage to the brain stem, resulting from compression of the brain or other trauma.

Stand to one side of the subject; the subject should look away from you toward the opposite wall. Wait a few seconds and then quickly, *but gently*, touch the subject's cornea (on the side toward you) with a wisp of absorbent cotton. What is the

reaction?_____

_____

What is the function of this reflex? _____

_____

_____

Was the sensation of touch *or* pain experienced?

_____

_____

Why? _____

_____

_____

Gag Reflex

The gag reflex tests the somatic motor responses of cranial nerves IX and X. When the oral mucosa on the side of the uvula is stroked, each side of the mucosa should rise, and the amount of elevation should be equal.*

For this experiment, select a subject who does not have a "queasy" stomach, because regurgitation is a possibility. Stroke the subject's oral mucosa on each side of the uvula with a tongue

depressor. What are the results? _____

_____

_____

AUTONOMIC REFLEXES

The autonomic reflexes include the pupillary, ciliospinal, and salivary reflexes, as well as a multitude of other reflexes that are difficult to observe in a laboratory situation.

Pupillary Reflexes

There are two types of pupillary reflexes: the pupillary light reflex and the consensual reflex. In both, the retina of the eye is the receptor and the smooth muscle of the iris is the effector. Many central nervous system centers are involved in the integration of these responses, but cranial nerve III is responsible for conducting impulses from and to the eye. Absence of the normal pupillary reflexes is generally a late indication of severe trauma or deterioration of the vital brain stem tissue due to metabolic imbalance.

*The uvula is the fleshy tab hanging from the roof of the mouth just above the root of the tongue.

1. Conduct the reflex testing in an area where the lighting is relatively dim. Before beginning, measure and record the size of the subject's

pupils. Right pupil: _____ mm; left

pupil: _____ mm.

2. Stand to the left of the subject to conduct the testing. The subject should shield her right eye by holding her hand vertically between the eye and the right side of the nose.

3. Shine a flashlight into the subject's left eye.

What is the pupillary response?_____

_____

Measure the size of the left pupil: _____ mm

4. Observe the right pupil. Has the same type of change (called a consensual response) occur-

red in the right eye? _____

Measure the size of the right pupil: _____ mm

The consensual response, or any reflex observed on one side of the body when the other side has been stimulated, is called a **contralateral response**. The pupillary light response or any reflex occurring on the same side stimulated is referred to as an **ipsilateral response**.

When a contralateral response occurs, what does this indicate about the pathways involved?

_____

_____

_____

Was the sympathetic *or* the parasympathetic division of the autonomic nervous system active during the testing of these reflexes?

_____

What is the function of these pupillary responses?

_____

_____

_____

Ciliospinal Reflex

The ciliospinal reflex is another example of reflex activity in which pupillary responses can be observed. This response may initially seem a little bizarre, especially in view of the consensual reflex just demonstrated. While observing the subject's eyes, gently stroke the skin on the left side of the back of the subject's neck, close to the hairline.

What is the reaction of the left pupil?_____

The reaction of the right pupil?_____

If you see no reaction, repeat the test using a gentle pinch in the same area.

   The response you should have noted—pupillary dilation—is consistent with the pupillary changes occurring when the sympathetic nervous system is stimulated. Such a response may also be elicited in a single pupil when more impulses from the sympathetic nervous system reach it for any reason. For example, when the left side of the subject's neck was stimulated, sympathetic impulses to the left iris increased, resulting in the ipsilateral reaction of the left pupil.

On the basis of your observations, would you say that the sympathetic innervation of the two irises

is closely integrated? _____

Why or why not? _____

_____

_____

Salivary Reflex

Unlike the other reflexes, in which the effectors were smooth or skeletal muscles, the effectors of the salivary reflex are glands. The salivary glands secrete varying amounts according to reflex stimulation.

1. Obtain a small beaker, a graduated cylinder, lemon juice, and wide-range pH paper. After refraining from swallowing for 2 minutes, the subject is to expectorate the accumulated saliva into a small beaker. Measure the volume of the expectorated saliva and determine its pH.

Volume:_____ cc;   pH: _____

2. Now place 2 or 3 drops of lemon juice on the subject's tongue. Allow the lemon juice to mix with the saliva for 5 to 10 seconds, and then determine the pH of the subject's saliva by touching a piece of pH paper to the tip of his

tongue. pH: _____

As before, the subject is to refrain from swallowing for 2 minutes. After the 2 minutes is up, again collect and measure the volume of the saliva and determine its pH.

Volume:_____ cc;   pH: _____

3. How does the volume of saliva collected after the application of the lemon juice compare

with the volume of the first saliva sample?_____

How does the final saliva pH reading compare

to initial reading? _____

To that obtained 10 seconds after the application

of lemon juice? _____

What division of the autonomic nervous system

mediates the reflex release of saliva?_____

_____

Other Autonomic Reflexes

Very few autonomic reflexes were investigated in this exercise because they are difficult to illustrate in a laboratory situation. To rectify this omission, several autonomic reflexes are listed below. Name the organ involved, receptors stimulated, and resulting action involved in each case. (Use an appropriate reference as necessary.)

Micturition (urination):

Organ/receptors: _____

Result: _____

_____

Hering-Breuer:

Organ/receptors:_____

Result: _____

_____

Defecation:

Organ/receptors:_____

Result: _____

_____

Carotid sinus:

Organ/receptors:_____

Result: _____

_____

REACTION TIME OF UNLEARNED RESPONSES

The time required for reaction to a stimulus depends on many factors—the sensitivity of the receptors, the velocity of nerve conduction, the number of neurons and synapses involved, and the speed of effector activation, to name just a few. The type of response to be elicited is also important. If the response involves the reflex arc, the synapses are facilitated and the response time will be short. If, on the other hand, the response can be categorized as an unlearned response, then a far larger number of neural pathways and many types of higher intellectual activities—including choice and decision making—will be involved, and the time for response will be considerably lengthened.

There are various ways of testing the reaction time of unlearned responses. The tests range in difficulty from simple to ultrasophisticated. Since the objective here is to demonstrate the major time difference between reflexes and unlearned responses, the simple approach will suffice.

1. Using a reflex hammer, elicit the patellar reflex in your partner. Note the reaction time needed for this reflex to occur.

2. Now test the reaction time for unlearned responses. The subject should stretch his hand out, with the thumb and index finger extended. Hold a 12-in. ruler so that its end is exactly 1 in. above the subject's outstretched hand. The ruler should be in the vertical position with the numbers reading from the bottom up. When the ruler is dropped, the subject should be able to grasp it between thumb and index finger as it passes, without having to change position. Have the subject catch the ruler five times, varying the time between each trial. The relative speed of reaction can be determined by reading the number on the ruler at the point of the subject's fingertips. (Thus if the number at the fingertips is 6 in., the subject was unable to catch the ruler until 7 in. of length had passed through his fingers; 6 in. of ruler length plus 1 in. to account for the distance of the ruler above the hand.) Record the number of inches that pass through the subject's fingertips for each trial:

Trial 1: _____ in.;   Trial 2: _____ in.;

Trial 3: _____ in.;   Trial 4: _____ in.;

Trial 5: _____ in.

3. Perform the test again, but this time say a simple word each time you release the ruler. Designate a specific word as a signal for the subject to catch the ruler; on all other words, the subject is to allow the ruler to pass through his fingers. Trials in which the subject erroneously catches the ruler are to be disregarded. Record the distance the ruler travels for five successful trials:

Trial 1: _____ in.;   Trial 2: _____ in.;

Trial 3: _____ in.;   Trial 4: _____ in.;

Trial 5: _____ in.

Did the addition of a specific word to the stimulus

increase or decrease the reaction time? _____

4. Perform the testing once again to investigate the subject's reaction to word association. As you drop the ruler, say a word—for example, *hot*. The subject is to respond with a word he associates with the stimulus word—for example, *cold*—catching the ruler as he responds. If he is unable to make the word association, he must allow the ruler to pass through his fingers. Record the distance the ruler travels for five successful trials, as well as the number of times the ruler is not caught by the subject.

Trial 1: _____ in.;   Trial 2: _____ in.;

Trial 3: _____ in.;   Trial 4: _____ in.;

Trial 5: _____ in.

The number of times the subject was unable to catch

the ruler: _____

You should have noted quite a large variation in reaction time in this series of trials. Why is this so?

_____

_____

Electro-encephalography

OBJECTIVES

1. To define *electroencephalogram*.

2. To describe or recognize typical tracings of the various brain wave patterns (alpha, beta, theta, and delta waves) and to note the conditions under which each is most likely to be predominant.

3. To state the source of brain waves.

4. To discuss the clinical significance of the electroencephalogram.

5. To define *alpha block*.

6. To monitor electroencephalography and recognize the alpha rhythm.

7. To describe the effect of a sudden sound, mental concentration, and alkalosis on brain wave patterns.

MATERIALS

Oscilloscope and EEG lead-selector box or polygraph and high-gain preamplifier
Cot
Electrode gel
EEG electrodes and leads
Collodion gel

BRAIN WAVE PATTERNS AND THE ELECTROENCEPHALOGRAM

Any physiologic investigation of the brain can emphasize and expose only a very minute portion of its function. The higher functions of the brain, such as consciousness and logical reasoning, are extremely difficult to investigate. It is obviously much easier to do experiments on the brain's input-output functions, some of which can be detected with appropriate recording equipment. Still, the ability to record brain activity does not necessarily guarantee an understanding of the brain.

The **electroencephalogram** (EEG), a record of the electrical activity of the brain, can be obtained through electrodes placed at various points on the skin or scalp of the head. This electrical activity, which is recorded as waves (Figure 23.1), is not completely understood at present but may be regarded as action potentials generated by brain neurons.

Certain characteristics of brain waves are known. They have a frequency of 1 to 60 cycles per second (cps) and a dominant rhythm of 10 cps, a maximal amplitude (voltage) in normal conscious adults of about 100 microvolts, and an average amplitude of 10 to 50 microvolts. They vary in frequency in different brain areas, occipital waves having a lower frequency than those associated with the frontal and parietal lobes. In addition, brain waves are known to change with age, sensory stimuli, brain pathology or disease, and the physiochemical state of the body. (Glucose deprivation, oxygen poisoning, and sedatives all interfere with the rhythmic activity of brain output by disturbing the metabolism of the neurons.) Since spontaneous waves are always present, even during unconsciousness, the absence of brain waves (a "flat" EEG) is taken as evidence of clinical death. The EEG is used clinically to aid in the diagnosis and localization of a number of brain lesions, including epileptic lesions, infections, abscesses, and brain tumors.

The first of the brain waves to be described by scientists were the **alpha waves** (or alpha rhythm).

203

(a)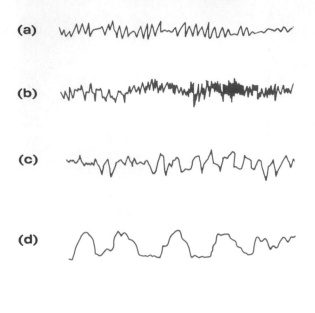

(b)

(c)

(d)

| 1 Sec |

Figure 23.1

EEG patterns: (a) alpha waves; (b) beta waves; (c) theta waves; (d) delta waves.

Alpha waves have an average frequency range of 8 to 13 cps and are produced when the individual is in a relaxed state with the eyes closed. **Alpha block,** the suppression of the alpha rhythm, occurs if the eyes are opened or if the individual begins to concentrate on some mental problem or visual stimulus. Under these conditions, the waves decrease in amplitude but increase in frequency. Under conditions of fright or excitement, the frequency increases still more.

Beta waves, closely related to alpha waves, are faster (15 to 30 cps) and have a lower amplitude. They are typical of the attentive or alert state.

Very large (high-amplitude) waves with a frequency of 4 cps that are seen in deep sleep are **delta waves. Theta waves** are large, abnormally contoured waves with a frequency of 4 to 7 cps. Although theta waves are normal in children, they represent emotional problems or some sort of neural imbalance in adults.

Sleeping individuals and patients in a stupor have EEGs that are slower (or lower frequency) than the alpha rhythm of normal adults. Fright, epileptic seizures, and various types of drug intoxication are associated with comparatively faster cortical activity. Thus impairment of cortical function is indicated by neuronal activity that is either too fast or too slow; unconsciousness occurs at both extremes of the frequency range. In young infants, the frequency of the major wave pattern is quite slow (0.5 to 2 cps), but the frequency of the major waves increases with age, and at about age 12 the EEG recording is similar to that of an adult.

OBSERVING BRAIN WAVE PATTERNS

If one electrode (the active electrode) is placed over a particular cortical area and another (the indifferent electrode) is placed over an inactive part of the head, such as the earlobe, all of the activity of the cortex underlying the active electrode will, theoretically, be recorded. (The inactive area provides a zero reference point or a base line, and the EEG represents the difference between "activities" occurring under the two electrodes.)

 1. Connect the EEG selector box to the oscilloscope preamplifier, or connect the high-gain preamplifier to the polygraph channel amplifier. Adjust the horizontal sweep and sensitivity-according to the directions given in the manual or by your instructor.

2. Prepare the subject. The subject should lie undisturbed with eyes closed in a quiet, dimly lit area. (Someone who is able to relax easily makes a good subject.) Apply a small amount of electrode gel to the subject's forehead above the left eye and on the left earlobe. Press an electrode to each prepared area and secure them by applying a film of collodion gel to the electrode surface and the adjacent skin. Allow the collodion to dry before continuing.

3. Connect the active frontal lead (forehead) to the EEG selector box outlet marked "L Frontal" and the lead from the indifferent electrode (earlobe) to the ground outlet (or to the appropriate input terminal on the high-gain preamplifier).

4. Turn the oscilloscope or polygraph on, and observe the EEG pattern of the relaxed subject for a period of 5 min. If the subject is truly relaxed, you should see a typical alpha-wave pattern. (If the subject is unable to relax and the alpha-wave pattern does not appear in this time interval, test another subject.) Discourage all muscle movement during the monitoring period.

5. Abruptly and loudly clap your hands. The subject's eyes should open and alpha block should occur. Observe the immediate brain wave pattern. How do the frequency and amplitude of

the brain waves change? _____

_____

Would you characterize this as beta rhythm? _____

Why? _____

6. Allow the subject about 5 min to achieve complete relaxation once again, and then ask the subject to compute a number problem that requires concentration (for example, add 3 and 36, subtract 7, multiply by 2, add 50 and so on). Observe the brain wave pattern during the period of mental computation.

7. Once again allow the subject to relax until the alpha rhythm is observed, and then ask him or her to hyperventilate for 3 min. (Be sure to tell the subject when to stop.) Hyperventilation rapidly flushes carbon dioxide out of the lungs, decreasing carbon dioxide levels in the blood and producing respiratory alkalosis. Record the changes noted in the rhythm and amplitude of the brain waves during the period of hyperventila-

tion. Record your observations: _____

_____

_____

_____

UNIT 9

THE SPECIAL SENSES

Special Senses: Vision

EXERCISE
24

OBJECTIVES

1. To describe the structure and function of the accessory visual structures.

2. To identify the structural components of the eye when provided with a model, an appropriate diagram, or a preserved sheep or cow eye, and list the function(s) of each.

3. To describe the cellular makeup of the retina.

4. To discuss the mechanism of image formation on the retina.

5. To trace the visual pathway to the optic cortex and note the effects of damage to various parts of this pathway.

6. To define the following terms:

   refraction   myopia
   accommodation   hypermetropia
   convergence   cataract
   astigmatism   glaucoma
   emmetropia   conjunctivitis

7. To discuss the importance of the pupillary and convergence reflexes.

8. To explain the difference between the rods and cones with respect to visual perception and retinal localization.

9. To state the importance of an ophthalmoscopic examination.

MATERIALS

Dissectible eye model
Chart of eye anatomy
Preserved cow or sheep eye
Dissecting pan and instruments
Laboratory lamp
Histologic section of an eye showing retinal layers
Snellen eye chart (floor marked with chalk to indicate 20-ft distance from posted Snellen chart)
Ishihara's color-blindness plates
Ophthalmoscope
Compound microscope

1-inch-diameter disks of colored paper (white, red, blue, green)

White, red, blue, and green chalk

Metric ruler

Common straight pins

Related film: *The Eyes and Seeing* (color, sound, 16 mm, 20 minutes. Encyclopaedia Britannica Films)

ANATOMY OF THE EYE

External Anatomy and Accessory Structures

The adult human eye is a sphere measuring about 1 inch (2.5 cm) in diameter. Only about one-sixth of the eye's anterior surface is observable, the remainder is enclosed and protected by a cushion of fat and the walls of the bony orbit.

Six **external,** or **extrinsic, eye muscles** attached to the exterior surface of each eyeball control eye movement and make it possible for the eye to follow a moving object. The names and positioning of these extrinsic muscles are noted in Figure 24.1. Their actions are given in the following chart.

The anterior surface of each eye is protected by the **eyelids,** or **palpebrae.** (See Figure 24.2). The medial and lateral junctions of the upper and lower eyelids are referred to as the *medial* and *lateral canthus* (respectively). A mucous membrane, the **conjunctiva,** lines the internal surface of the eyelids and continues over the anterior surface of the eyeball. The conjunctiva secretes mucus, which aids in

Name	Innervation (cranial nerve)	Action
Lateral rectus	VI	Moves eye horizontally (laterally)
Medial rectus	III	Moves eye horizontally (medially)
Superior rectus	III	Elevates eye
Inferior oblique	III	Elevates eye and turns it laterally
Inferior rectus	III	Depresses eye
Superior oblique	IV	Depresses eye and turns it laterally

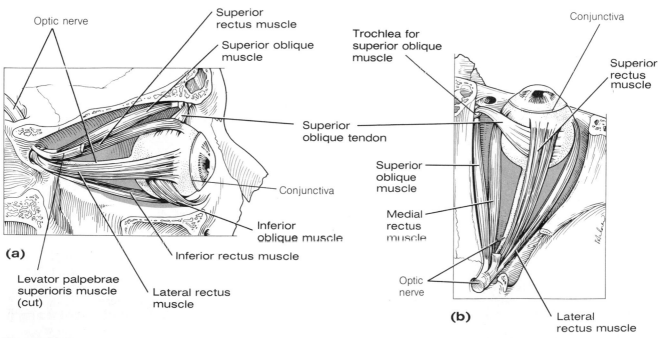

Figure 24.1

Extrinsic muscles of the eye: (a) lateral view of right eye; (b) superior view of right eye.

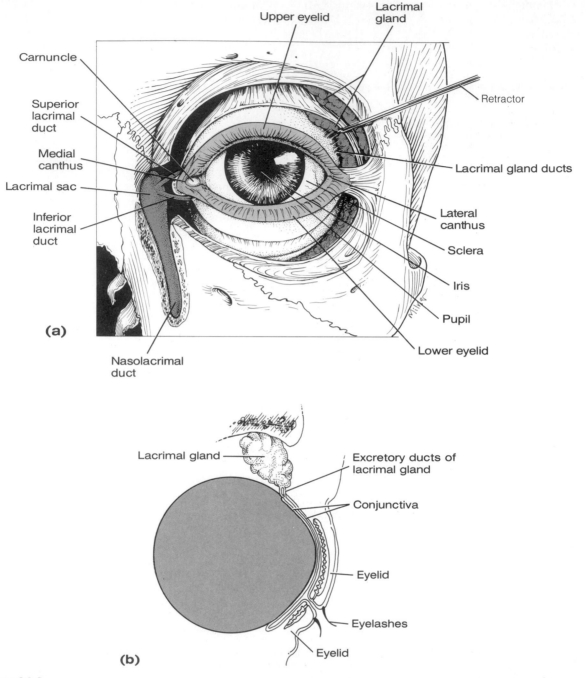

Figure 24.2

External anatomy of the eye and accessry structures: (a) anterior view; (b) sagittal section.

lubricating the eyeball. An inflammation of the conjunctiva, often accompanied by redness of the eye, is called **conjunctivitis.**

Projecting from the border of each eyelid is a row of short hairs, the **eyelashes.** The **meibomian glands,** a type of sebaceous gland, are associated with the eyelash hair follicles and aid in the lubrication of the eyeball by producing an oily secretion. An inflammation of one of these glands is called a **sty.**

The **lacrimal apparatus** consists of the **lacrimal gland, lacrimal duct, lacrimal sac,** and the **nasolacri-** mal duct. The lacrimal glands are situated superior to the lateral aspect of each eye. They continually liberate a dilute salt solution (tears) that flows onto the anterior surface of the eyeball through several small ducts. The tears flush across the eyeball into the lacrimal ducts medially, then into the lacrimal sac, and finally into the nasolacrimal duct, which empties into the nasal cavity. The lacrimal secretion also contains **lysozyme,** an antibacterial enzyme. Because it constantly flushes the eyeball, lysozyme cleanses and protects the eye surface as it moistens

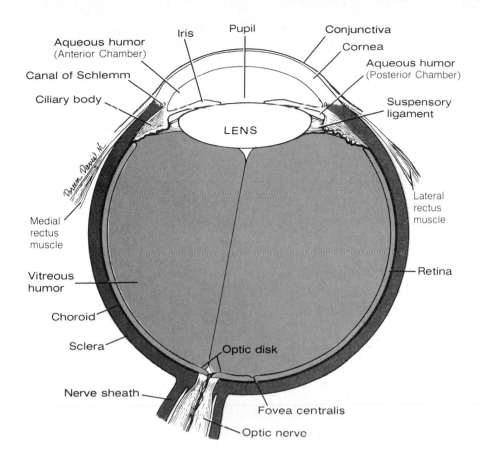

Figure 24.3

Internal anatomy of the eye (transverse section).

and lubricates it. Eyes tend to become dry as we age, due to decreased lacrimation, and thus are more vulnerable to bacterial invasion and irritation.

Observe the eyes of another student and identify as many of the accessory structures as possible. Ask the student to look to the left. What extrinsic eye muscles are responsible for this action?

Right eye _____

Left eye _____

Internal Anatomy of the Eye

 Obtain a dissectible eye model and identify its internal structures as they are described as follows. As you work, also refer to Figure 24.3.

Anatomically, the wall of the eye is constructed of three tunics, or coats. The outermost tunic, the protective **sclera,** is primarily thick white connective tissue and is observable anteriorly as the "white of the eye." Its anteriormost portion is modified structurally to form the transparent **cornea,** through which light enters the eye.

The middle tunic is the **choroid,** a richly vascular nutritive layer that contains a dark pigment. Anteriorly, the choroid is modified to form the **ciliary body,** or **ciliary muscle,** to which the lens is attached, and then the pigmented **iris.** The iris is incomplete, resulting in a rounded opening, the **pupil,** through which light passes.

The iris is composed of circularly and radially arranged smooth muscle fibers and acts as a reflexly activated diaphragm in regulating the amount of light entering the eye. In close vision and bright light, the circular muscles of the iris contract, and the pupil becomes more constricted. In distant vision and in dim light, the radial fibers contract to enlarge (dilate) the pupil and allow more light to enter the eye.

The innermost sensory tunic of the eye is the **retina,** which extends anteriorly only to the ciliary body. The retina contains the photoreceptor cells, the **rods** and **cones,** which act as transducers to convert light energy into nerve impulses that are transmitted to the optic cortex of the brain. Vision is the result. The photoreceptor cells are distributed over the entire retina, except where the optic nerve leaves the eyeball. This site is called the **optic disc,** or blind spot. Lateral to each blind spot is an area called the **macula lutea** (yellow spot), an area of high cone density. In its center is the **fovea centralis,** a minute pit about ½ mm in diameter, which contains only cones and is the area of greatest visual acuity. Focusing for discriminative vision occurs in the fovea centralis.

Light entering the eye is focused on the retina by the **lens,** a flexible crystalline structure held vertically in the eye's interior by **suspensory ligaments** attached to the ciliary body. Activity of the ciliary body changes lens thickness to allow light to be properly focused on the retina. In the elderly the lens becomes increasingly hard and opaque. **Cataracts,** which are the result of this process, cause vision to become hazy or entirely obstructed.

The lens divides the eye into two chambers: the aqueous chamber anterior to the lens, which contains a clear watery fluid called the **aqueous humor,** and the vitreous chamber posterior to the lens, filled with a gellike substance, the **vitreous humor,** or **vitreous body.** The aqueous chamber is further divided into **anterior** and **posterior chambers,** located before and after the iris, respectively. The aqueous humor is continually formed by the **ciliary processes** of the ciliary body. It helps to maintain the intraocular pressure of the eye and provides nutrients for the avascular lens and cornea. The aqueous humor is resorbed into the **canal of Schlemm.** Any interference with this drainage increases intraorbital pressure, resulting in pain and possible blindness, a condition called **glaucoma.** Vitreous humor also prevents the eyeball from collapsing.

DISSECTION OF THE COW (SHEEP) EYE

 1. Obtain a preserved cow or sheep eye, dissecting instruments, and a dissecting pan.

2. Examine the external surface of the eye, noting the thick cushion of adipose tissue. Identify the optic nerve (cranial nerve II) as it leaves the eyeball, the remnants of the extrinsic eye muscles, the conjunctiva, the sclera, and the cornea. The normally transparent cornea is opalescent or opaque if the eye has been preserved. Refer to Figure 24.4 as you work.

3. Trim away most of the fat and connective tissue, but leave the optic nerve intact. Holding the eye with the cornea facing downward, make an incision with a sharp scalpel into the sclera about ¼ inch above the cornea. Complete the incision around the circumference of the eyeball parallel to the corneal edge.

4. Carefully lift the posterior part of the eyeball away from the lens. (Conditions being proper, the vitreous body should remain with the posterior part of the eyeball.)

5. Examine the anterior portion of the eye and identify the following structures:

Ciliary body: black pigmented body that appears to be a halo encircling the lens.
Lens: biconvex structure that appears opaque in preserved specimens.
Suspensory ligaments: delicate fibers attaching the lens to the ciliary body.

Carefully remove the lens and identify the adjacent structures:

Iris: anterior continuation of the ciliary body penetrated by the pupil.
Cornea: more convex anteriormost portion of the sclera; normally transparent but cloudy in preserved specimens.

6. Examine the posterior portion of the eyeball. Remove the vitreous humor, and identify the following structures:

Retina: appears as a delicate white, probably crumpled membrane that separates easily from the pigmented choroid.

Note its point of attachment. What is this point

called? _____

Pigmented choroid coat: appears iridescent in the cow or sheep eye owing to a special reflecting surface called the **tapetum lucidum.** This specialized surface reflects the light within the eye and is found in the eyes of animals that live under conditions of low-intensity light. It is not found in humans.

MICROSCOPIC ANATOMY OF THE RETINA

The retina consists of two types of cells: a pigmented epithelial layer, which abuts the choroid, and an inner cell layer composed of neurons, which is in contact with the vitreous humor (Figure 24.5). The inner nervous layer is composed of three major neuronal populations. These are, from outer to inner aspect, the **photoreceptor layer** (rods and cones), the **bipolar cells,** and the **ganglion cell neuron layer.**

The rods are the specialized receptors for dim light. Visual interpretation of their activity is in gray tones. The cones are color receptors that permit high levels of visual acuity, but they function only under conditions of high light intensity; thus, for example, no color vision is possible in moonlight. Only cones are found in the fovea centralis, and their number decreases as the retinal periphery is approached.

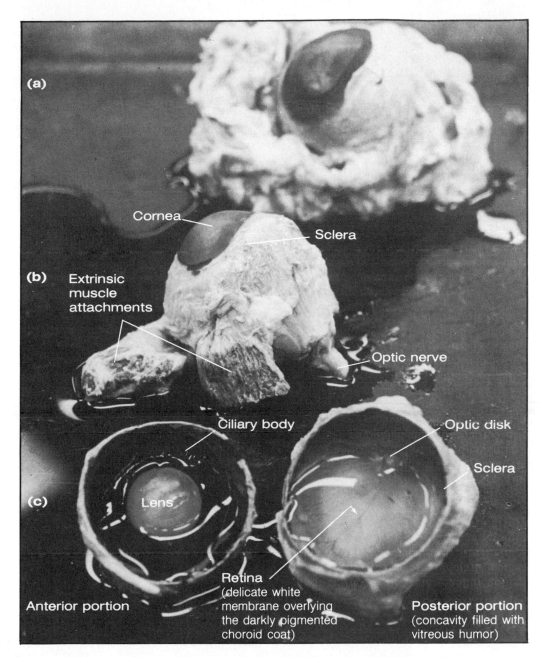

Figure 24.4

Anatomy of the cow eye: (a) cow eye (entire) removed from orbit (note the large amount of fat cushioning the eyeball); (b) cow eye (entire) with fat removed to show the extrinsic muscle attachments and optic nerve; (c) cow eye cut along the coronal plane to reveal internal structures. (Photograph courtesy of Jack Scanlon, Holyoke Community College Art Graphics Studio.)

Conversely, only rods are found in the periphery, and their number decreases in density near the macula.

Light must pass through the ganglion cell neuron layer and the bipolar neuron layer to reach and excite the rods and cones, which then initiate nervous impulses that pass to the bipolar neurons. These in turn stimulate the ganglion cells, whose axons leave the retina in the tight bundle of fibers known as the optic nerve. The retinal layer is thickest (approximately 4 μm) where the optic nerve attaches to the eyeball because an increasing number of ganglion cells converge at this point. It thins as it approaches the ciliary body.

Obtain a histologic slide of a longitudinal section of the eye. Identify the retinal layers by comparing it to Figure 24.5.

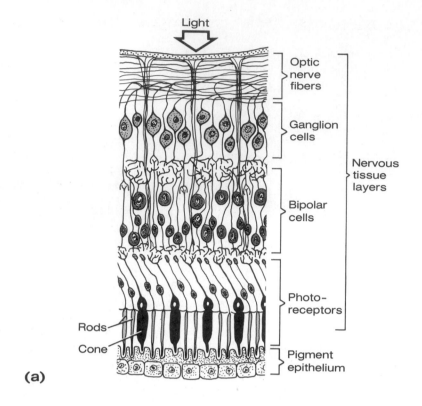

(a)

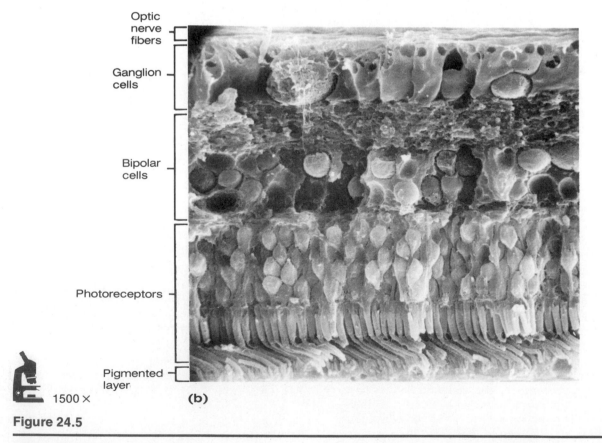

1500 ×   **(b)**

Figure 24.5

Microscopic anatomy of the cellular layers of the retina: (a) diagrammatic view; (b) scanning electron micrograph. (*From Tissues and Organs: A Text-Atlas of Scanning Electron Microscopy* by Richard G. Kessel and Randy H. Kardon. W. H. Freeman and Company. Copyright © 1979.)

Left eye   Right eye

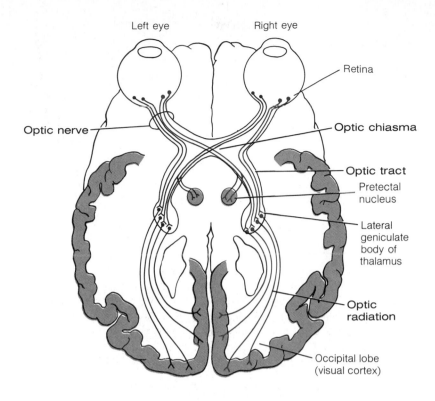

Figure 24.6

Visual pathway to the brain. (Note that fibers from the lateral portion of each retinal field do not cross at the optic chiasma.)

VISUAL PATHWAYS TO THE BRAIN

The axons of the ganglion cells of the retina converge at the posterior aspect of the eyeball and exit from the eye as the optic nerve. At the **optic chiasma,** the fibers from the medial side of each eye cross over to the opposite side. The fiber tracts thus formed are called the **optic tracts.** Each optic tract thus contains fibers from the lateral side of the eye on the same side and from the medial side of the opposite eye. The optic tract fibers synapse with neurons in the **lateral geniculate nucleus** of the thalamus, whose axons form the **optic radiation,** terminating in the **optic cortex** in the occipital lobe of the brain. Here they synapse with the cortical cells, and visual interpretation occurs.

After examining Figure 24.6, determine what the effects of lesions in the following areas would have on vision:

In the right optic nerve _____

_____

Through the optic chiasma _____

_____

In the left optic tract _____

_____

In the right cerebral cortex (visual area) _____

_____

VISUAL TESTS AND EXPERIMENTS

Demonstration of the Blind Spot

1. Hold Figure 24.7 about 18 inches from your eyes. Close your left eye, and focus your right eye on the X. Move the figure slowly toward your face, keeping the right eye focused on the X. When the dot focuses on the blind spot, which lacks photoreceptors, it will disappear.

2. Have your laboratory partner record in metric units the distance at which this occurs. The dot will reappear as the figure is moved closer. Distance at which the dot disappears:

Right eye _____

X

Figure 24.7

Blind spot test figure.

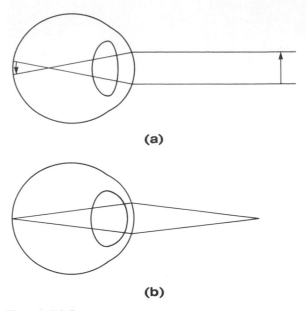

(a)

(b)

Figure 24.8

Relative convexity of the lens in focusing for close and distant vision: (a) light rays from a distant object are nearly parallel and can be focused with normal lens convexity; (b) diverging light rays from close objects require increased convexity of the lens to be focused sharply on the retina.

Repeat the test for the left eye, this time closing the right eye and focusing the left eye on the dot. Record the distance at which the X disappears:

Left eye _____

Refraction, Tests for Visual Acuity, and Astigmatism

When light rays pass from one medium to another, their velocity, or speed of transmission, changes, and the rays are bent or refracted. When passing from a less dense medium (air) to a more dense medium (water or the cornea), the rate of transmission slows, and the light rays are bent. Thus the light rays in the visual field are refracted as they encounter the cornea, lens, and vitreous body of the eye.

The refractive index (bending power) of the cornea and vitreous humor are constant; however, the refractive index of the lens can be varied by changing its shape, that is, by making it more or less convex, thus causing the light to be properly converged and focused on the retina. The greater the lens convexity, or bulge, the more the light will be bent. Conversely, the less the lens convexity, the flatter it is, the less it bends the light. In general, light from a distant source (over 20 feet) approaches the eye as parallel rays (Figure 24.8), and no change in lens convexity is necessary for it to focus properly on the retina. However, light from a close source tends to diverge, and the convexity of the lens must increase to make close vision possible. To achieve this, the ciliary muscle contracts, decreasing the tension on the suspensory ligaments attached to the lens. This ability of the eye to focus differentially for objects of near vision (less than 20 feet) is called **accommodation.** It should be noted that the image formed on the retina as a result of the refractory activity of the lens (see Figure 24.8(a)) is a **real image** (reversed from left to right, inverted, and smaller than the object).

The normal or **emmetropic** eye is able to accommodate properly; however, visual problems may result from lenses that are too strong or too "lazy" (overconverging and underconverging, respectively) or from structural problems such as an eyeball that is too long or too short to provide for proper focusing by the lens, or a cornea or lens with improper curvatures.

Individuals in whom the image normally focuses in front of the retina are said to have **myopia,** or "nearsightedness"; they can see close objects without difficulty, but distant objects are blurred or seen indistinctly. Correction requires a concave lens, which causes the light reaching the eye to diverge.

If the image focuses behind the retina, the individual is said to have **hypermetropia** or farsightedness. Such persons have no problems with distant vision but need glasses with convex lenses to augment the converging power of the lens for close vision.

Irregularities in the curvatures of the lens and/or the cornea lead to a blurred vision problem called **astigmatism.** Circularly ground lenses, which compensate for inequalities in the curvatures of the refracting surfaces, are prescribed to correct the condition.

TEST FOR NEAR-POINT ACCOMMODATION
The elasticity of the lens decreases dramatically with age, resulting in difficulty in focusing for near or close vision. This condition is called **presbyopia** —literally, old vision. Lens elasticity can be tested by measuring the **near point of accommodation.** The near point of vision is about 7.5 cm (or 3 inches) at age 10, 9 cm (or 3.5 inches) at age 20, 17 cm (or 6.75 inches) at age 40, and 83 cm (or 33 inches) at the age of 60.

To determine your near point of accommodation, hold a common straight pin at arm's length in front of one eye. Slowly move the pin toward that eye until the pin image becomes distorted. Have your lab partner measure the distance from your eye to the pin at this point, and record the distance below. Repeat the procedure for the other eye.

Near point for right eye _____

Near point for left eye _____

TEST FOR VISUAL ACUITY Visual acuity, or sharpness of vision, is generally tested with a Snellen eye chart, which consists of letters of various sizes printed on a white card. This test is based on the fact that letters of a certain size can be seen clearly by eyes with normal vision at a specific distance. The distance at which the normal, or emmetropic, eye can read a line of letters is printed at the end of that line.

1. Have your partner stand 20 feet from the posted Snellen eye chart and cover one eye with a card or hand. As your partner reads each consecutive line aloud, check for accuracy. (If this individual wears glasses, the test should be taken twice—first with glasses off and then with glasses on.)

2. Record the number for the line of smallest-sized letters read. If it is 20, the person's vision for that eye is 20/20. If it is 40, vision is 20/40, or about half the normal acuity. (Such an individual is myopic). If it is 15, visual acuity is 20/15. (This is better than normal since this person can read letters at 20 feet that are only discernible by the normal eye at 15 feet from the chart.)

3. Have your partner test and record your visual acuity.

Visual acuity right eye _____

Visual acuity left eye _____

TEST FOR ASTIGMATISM The astigmatism chart (Figure 24.9) is designed to test for defects in the refracting surface of the lens and/or cornea.

View the chart first with one eye and then with the other, focusing on the center of the chart. If all the radiating lines appear equally dark and distinct, there is no distortion of your refracting surfaces. If some of the lines are blurred or appear less dark than others, at least some degree of astigmatism is present.

Is astigmatism present in your left eye? _____

Right eye? _____

Test for Color Blindness

Ishihara's color plates are designed to test for deficiencies in the cones or color photoreceptor

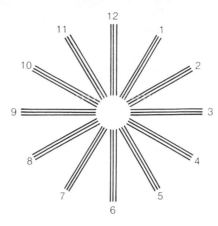

Figure 24.9

Astigmatism testing chart.

cells. Studies suggest that there are three cone types, each containing a different photoreceptor pigment. One type primarily absorbs the red wavelengths of the visible light spectrum, another the blue wavelengths, and a third the green wavelengths. Nervous impulses reaching the brain from these different photoreceptor types are then interpreted (seen) as red, blue, and green, respectively.

The interpretation of the intermediate colors of the visible light spectrum is a result of overlapping input from more than one cone type.

1. View the various color plates in bright light or sunlight while holding them about 30 inches away and at right angles to your line of vision. Report to your laboratory partner what you see in each plate. (Take no more than 3 seconds for your decision.)

2. Your partner is to write down your responses and then check their accuracy with the correct answers given at the front of the color plate book. Is there any indication that you have some degree

of color blindness? _____

If so, what type? _____

Repeat the procedure to test your partner's color vision.

Relative Positioning of Rods and Cones on the Retina

The test subject in a demonstration of the relative positioning of rods and cones should have shown no color vision problems during the test for color blindness. Students may work in pairs, or two students may perform the test for the entire class.

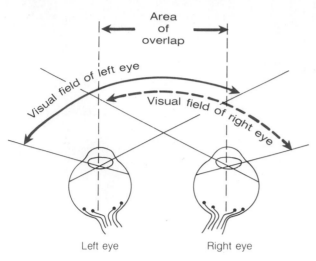

Figure 24.10

Overlapping of the visual fields.

White, red, blue and green paper disks and chalk will be needed for this demonstration.

1.   Position a test subject about 1 foot away from the blackboard.

2.   Make a white chalk circle on the board immediately in front of the subject's right eye. Have her close her left eye and stare fixedly at the circle with the right eye throughout the test.

3.   To map the extent of the rod field, begin to move a white paper disk into the field of vision from various sides of the visual field (beginning at least 2 feet away from the white chalk circle) and plot with white chalk dots the points at which the disk first becomes visible to the test subject.

4.   Repeat the procedure, using red, green, and blue paper disks and like-colored chalk to map the cone fields. Color (not object) identification is required.

5.   After the test has been completed for all four disks, connect all dots of the same color. It should become apparent that each of the color fields has a different radius and that the rod and cone distribution on the retina is not uniform. (Normal fields have white outermost, followed by blue, red, and green as the fovea is approached.)

6.   Record the rod and cone color field distribution observed in the laboratory review section using appropriately colored pencils.

Tests for Binocular Vision

Humans, cats, predatory birds, and most primates are endowed with **binocular,** or two-eyed, vision. Although both eyes look in approximately the same direction, they see slightly different views. Their visual fields, each about 170 degrees, overlap to a considerable extent; thus there is two-eyed vision at the overlap area (Figure 24.10).

In contrast, the eyes of many animals (rabbits, pigeons, and others) are more on the sides of their head. Such animals see in two different directions and thus have a panoramic field of view (**panoramic vision**).

Although both types of vision have their good points, we now know that binocular vision provides an accurate means of locating objects in space. This three-dimensional vision is called **stereopsis.** The slight differences between the views seen by the two eyes are fused by the higher centers of the visual cortex to give us depth perception. Because of the manner in which the visual cortex resolves these two different views into a single image, it is often re-referred to as the "cyclopean eye of the binocular animal."

 1.   To demonstrate that a slightly different view is seen by each eye, perform the following simple experiment: hold a pencil at arm's length directly in front of the left eye. Hold another pencil directly beneath it and move it back and forth toward and away from you after closing one eye. Repeat the test with the other eye closed. Note that with only the left eye open the pencil stays in the same plane as the fixed pencil, but when viewed with the right eye, the moving pencil moves laterally away from the plane of the fixed pencil.

To demonstrate the importance of two-eyed binocular vision for depth perception, perform this second simple experiment.

2.   Have your laboratory partner hold a test tube erect about arm's length in front of you. With both eyes open, quickly insert a pencil into the test tube. Remove the pencil, bring it back close to your body, close one eye, and quickly and without hesitation insert the pencil into the test tube. Repeat with the other eye closed.

Was it as easy to dunk the pencil with one eye closed

as with both eyes open? _____

Ophthalmoscopic Examination of the Eye (Optional)

The opthalmoscope is an instrument used to examine the fundus, or eyeball interior, to determine visually the condition of the retina, optic disk, and internal blood vessels. Certain pathologic conditions such as diabetes, arteriosclerosis, and degenerative changes of the optic nerve and retina can be detected by such an examination. The ophthalmoscope consists of a set of lenses mounted on a rotating disk, a light source, and

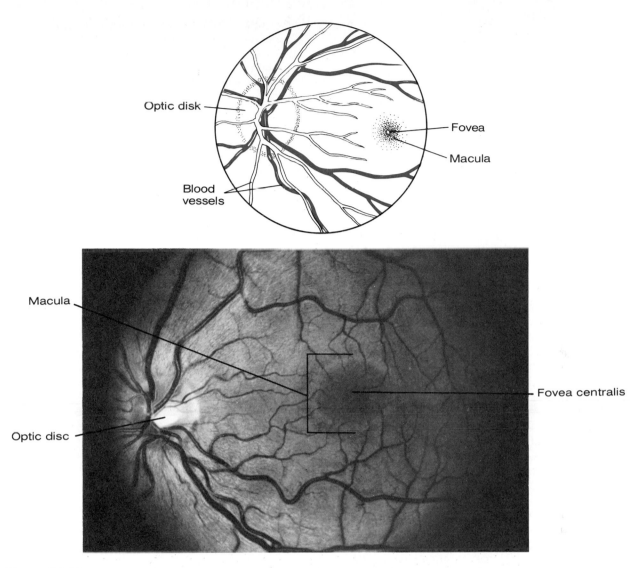

Figure 24.11

Posterior portion of the left retina. (Photograph taken with slit-lamp camera.)

mirrors that reflect the light so that the eye interior can be illuminated and viewed. When the ophthalmoscope is correctly set, the fundus should appear as in Figure 24.11.

 1. Conduct the examination in a dimly lit or darkened room with the test subject comfortably seated and gazing straight ahead. To examine the right eye, sit facing the subject, and hold the instrument in your right hand. Use your right eye to view the eye interior. To view the left eye, use your left eye, and hold the instrument in your left hand.

2. Begin the examination with the 0 lens in position. Hold the instrument so that the lens disk may be rotated with the index finger. Hold the ophthalmoscope about 6 inches from the subject's eye and direct the light into the pupil at a slight angle—through the pupil edge rather than directly through its center. You will see a red circular area that is the illuminated eye interior.

3. Move in more closely (about 2 inches from the cornea) as you continue to observe the area and slowly begin to rotate the lens disk to lens numbers of higher value (the black numbers) until you have the optic disk in sharp focus. (The focal length of the lenses decreases as the stronger lenses are used.) When the examination is proceeding correctly, the subject can often see images of retinal vessels in his own eye which appears rather like cracked glass. If you are not able to achieve a sharp focus or see the optic disk, move medially or laterally, and begin again.

4. Examine the optic disk for color, elevation, and sharpness of outline. Observe the blood vessels radiating from near its center. Locate the

macula, which is lateral to the optic disk. It is a darker area in which blood vessels are absent and the fovea appears to be a slightly lighter area in its center.

Tests of Eye Reflexes

Both intrinsic (internal) and extrinsic (external) muscles are necessary for proper eye functioning. The intrinsic muscles, controlled by the autonomic nervous system, are those of the ciliary body (which alters the lens curvature in focusing) and the radial and circular muscles of the iris (which control pupillary size and thus regulate the amount of light entering the eye). The extrinsic muscles are the rectus and oblique muscles, which are attached to the eyeball exterior. These muscles control eye movement and make it possible to keep moving objects focused on the fovea centralis. They are also responsible for **convergence,** or medial eye movements, which are essential for near vision. When convergence occurs, both eyes are directed toward the near object viewed. The extrinsic eye muscles are controlled by the somatic nervous system.

Involuntary activity of both of these muscle types is brought about by reflex actions that can be observed in the following experiments.

PHOTOPUPILLARY REFLEX   A sudden illumination of the retina by a bright light causes the pupil to contract reflexly in direct proportion to the light intensity. This protective device prevents damage to the delicate photoreceptor cells.

Obtain a laboratory lamp or penlight. Have your laboratory partner sit with eyes closed and hands over eyes. Turn on the light and position it so that it shines on the subject's right hand. After 1 minute, ask your partner to open his right eye. Quickly observe the pupil of that eye. What

happened to the pupil? _____

_____

Shut off the light and ask your partner to open the opposite eye. What are your observations?

_____

_____

ACCOMMODATION PUPILLARY REFLEX
Have your partner gaze for approximately 1 minute at a distant object in the lab—*not* toward the windows or another light source. Observe his pupils, then hold some printed material 6 to 10 inches from his face.

How does pupil size change as your partner focuses

on the printed material? _____

_____

Explain the value of this reflex. _____

_____

_____

_____

CONVERGENCE REFLEX   Repeat the previous experiment, this time using a pen or pencil as the close object to be focused on. Note the position of your partner's eyeballs both while he is gazing at the distant and at the close object. Do they change position as the object of focus is changed?

_____

In what way? _____

_____

_____

_____

Explain the importance of the convergence reflex.

_____

_____

_____

_____

_____

Special Senses: Hearing and Equilibrium

OBJECTIVES

1. To identify the anatomic structures of the external, middle, and internal ear by labeling a diagram appropriately.

2. To describe the anatomy of the organ of hearing (organ of Corti in the cochlea) and explain its function in sound reception.

3. To describe the anatomy of the equilibrium organs of the inner ear (ampullae crystallaris and maculae), and to explain their relative function in the maintenance of equilibrium.

4. To define or explain *central deafness, conduction deafness,* and *nystagmus.*

5. To state the purpose of the Rinne and Romberg tests.

6. To explain how one is able to localize the source of sounds.

7. To describe the effects of acceleration on the semicircular canals.

8. To explain the role of vision in the maintenance of equilibrium.

MATERIALS

Three-dimensional dissectible ear model and/or chart of ear anatomy
Histologic slides of the cochlea of the ear
Compound microscope
Tuning forks (range of frequencies)
Rubber mallet
Absorbent cotton
Otoscope (if available)
Alcohol swabs
12-inch ruler
Related film: *The Ears and Hearing* (black and white, sound, 16 mm, 10 minutes, Encyclopaedia Britannica Films)

ANATOMY OF THE EAR

Gross Anatomy

The ear is a complex structure containing sensory receptors for hearing and equilibrium. The ear is divided into three major areas: the **outer,** or **external ear,** the **middle ear,** and the **inner ear** (Figure 25.1). The outer and middle ear structures serve the needs of the sense of hearing *only,* while the inner ear structures function both in equilibrium and hearing reception.

Obtain a dissectible ear model and identify its structures described as follows. Refer to Figure 25.1 as you work.

The outer ear is composed primarily of the external **pinna,** or **auricle,** and the **external auditory canal.** The pinna is the skin-covered cartilaginous structure encircling the auditory canal opening. In many animals, it collects and directs sound waves into the auditory canal. In humans this function of the pinna is largely lost.

The external auditory canal is a short, narrow (about 1 inch long by ¼ inch wide) chamber carved into the temporal bone. In its skin-lined walls are wax-secreting glands called **ceruminous glands.** The sound waves that enter the external auditory canal eventually encounter the **tympanic membrane,** or **eardrum,** which vibrates at exactly the same frequency as the sound wave(s) hitting it. The membranous eardrum separates the outer from the middle ear.

The middle ear is essentially a small chamber—the **tympanic cavity**—found within the temporal bone. The cavity is spanned by three small bones, the **ossicles*** (hammer, anvil, and stirrup), which articulate to form a lever system that transmits the vibratory motion of the eardrum to the fluids of the

---

*The ossicles are often commonly referred to by their Latin names, that is, **malleus, incus,** and **stapes,** respectively.

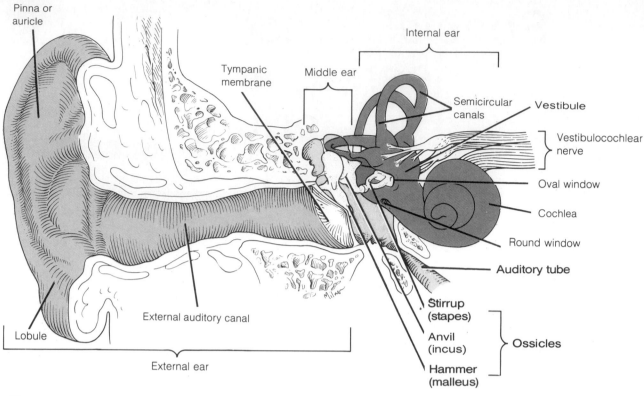

Figure 25.1

Anatomy of the ear.

inner ear. When the eardrum moves, the hammer attached to it moves with it and transfers the vibratory movement to the anvil. The anvil in turn passes it on to the stirrup, which drives its footplate against a small membrane, the **oval window** of the inner ear. Although the ossicles work as a chain of levers, their combined action is more like that of a hydraulic press. (The same total pressure is delivered to a much smaller area.) As a result of their activity, the amplitude of the sound waves hitting the eardrum is reduced, but the pressure on the oval window is increased, since the area of oval window membrane is much smaller than that of the tympanic membrane.

Connecting the middle ear chamber with the nasopharynx is the **auditory,** or **eustachian, tube.** Normally this tube is flattened and closed, but swallowing or yawning can cause it to open temporarily to equalize the pressure of the middle ear cavity with external air pressure. This is an important function, since the eardrum does not vibrate properly unless the pressure on both of its surfaces is the same. Because the mucosal membranes of the middle ear cavity and the nasopharynx are continuous through the eustachian tube, **otitis media,** or inflammation of the middle ear, is a fairly common condition, especially among youngsters prone to sore throats.

The inner ear consists of a system of bony and rather tortuous chambers called the **osseous,** or **bony, labyrinth,** which is filled with an aqueous fluid called **perilymph** (Figure 25.2). Suspended in the perilymph

is the **membranous labyrinth,** a system structured much like the osseous labyrinth. The interior of the membranous labyrinth is filled with a more viscous fluid called **endolymph.** The three subdivisions of the bony labyrinth are the **cochlea,** the **vestibule,** and the **semicircular canals,** with the vestibule situated between.

The snaillike cochlea (see Figures 25.2 and 25.3) contains the sensory receptors for hearing. The cochlear membranous labyrinth, the **cochlear duct,** is a soft wormlike tube about 1½ inches long. It winds through the full two and three-quarter turns of the cochlea and separates the cochlear cavity into upper and lower perilymph-containing chambers, the **scala vestibuli** and **scala tympani,** respectively. The scala vestibuli terminates at the oval window. The scala tympani is bounded by another membranous area, the **round window.** The cochlear duct, itself filled with endolymph, supports the **organ of Corti,** which contains the receptors for hearing—the sensory hair cells and auditory nerve endings.

Otoscopic Examination of the Ear (Optional)

1. Obtain an otoscope. To examine the eardrum of your partner's ear, grasp the ear pinna firmly and pull it up, back, and slightly laterally. (If your partner experiences pain or discomfort when the pinna is manipulated,

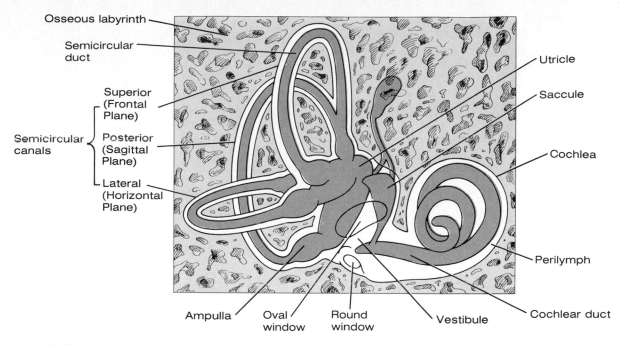

Figure 25.2

Inner ear: right membranous labyrinth shown within the bony labyrinth.

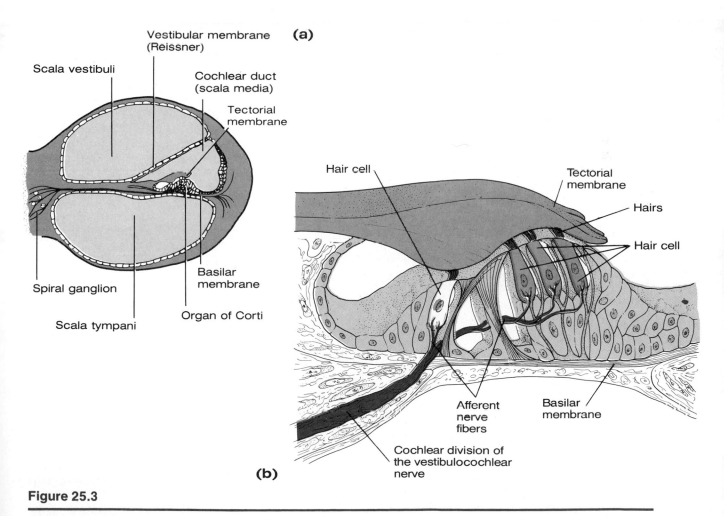

Figure 25.3

Organ of Corti: (a) cross section through one turn of the cochlea showing the position of the organ of Corti; (b) enlarged view of the organ of Corti.

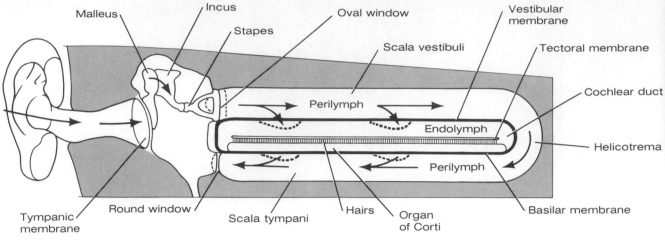

Figure 25.4

Fluid movement in the cochlea following stirrup thrust on the oval window. Compressional wave created causes the round window to bulge into the middle ear.

an inflammation or infection of the external ear may be present. If this occurs, do not attempt to examine the ear canal.)

2. Turn on the light switch of the otoscope and carefully insert the speculum of the otoscope into the external auditory canal in a downward and forward direction only far enough to permit examination of the shape and color and vascular network of the tympanic membrane or eardrum. The healthy tympanic membrane is pearly white.

3. Note during the examination if there is any discharge or redness in the canal and identify earwax.

4. Thoroughly clean the speculum after the examination with an alcohol swab before returning it to the supply area.

Microscopic Anatomy of the Receptors

ORGAN OF CORTI AND THE MECHANISM OF HEARING   The anatomic details of the organ of Corti are shown in Figure 25.3. The hair (auditory receptor) cells rest on the **basilar membrane,** which forms the floor of the cochlear duct, and their cilia project into a gelatinous membrane, the **tectorial membrane,** which overlies them. The roof of the cochlear duct is called the **vestibular membrane,** or **Reissner's membrane.** The endolymph-filled chamber of the cochlear duct is called the **scala media.**

Obtain a prepared microscope slide of the cochlea and identify the areas noted in Figure 25.3(a).

The mechanism of hearing begins as sound waves pass through the external auditory canal and through the middle ear, into the inner ear, where the vibration eventually reaches the organ of Corti, which contains the receptors for hearing. Many theories have attempted to explain how the organ of Corti actually responds to sound.

The popular "traveling wave" hypothesis of Von Békésy suggests that vibrations at the oval window initiate traveling waves that peak, creating secondary circular eddies at the region of maximal amplitude and thus stimulating the hair cells of the organ of Corti in that region. Since the area at which the traveling waves peak is a high-pressure area, the vestibular membrane is compressed at this point. The vestibular membrane in turn compresses the endolymph and the basilar membrane of the cochlear duct. The resulting pressure on the perilymph in the scala tympani causes the membrane of the round window to bulge outward into the middle ear chamber, thus acting as a relief valve for the compressional wave (Figure 25.4). Von Békésy found that high-frequency waves (high-pitched sounds) peaked close to the oval window and that low-frequency waves (low-pitched sounds) peaked farther up the basilar membrane near the apex of the cochlea. Although the mechanism of sound reception by the organ of Corti is incompletely understood, we do know that hair cells on the basilar membrane are uniquely stimulated by sounds of various frequencies and amplitude and that once stimulated they depolarize and begin the chain of nervous impulses to the auditory centers of the temporal lobe cortex. This series of events results in the phenomenon we call hearing.

EQUILIBRIUM APPARATUS AND MECHANISMS OF EQUILIBRIUM   The equilibrium apparatus of the inner ear is in the vestibular and semicircular canal portions of the body labyrinth. Their chambers are filled with perilymph, in which membranous labyrinth structures are suspended. (The

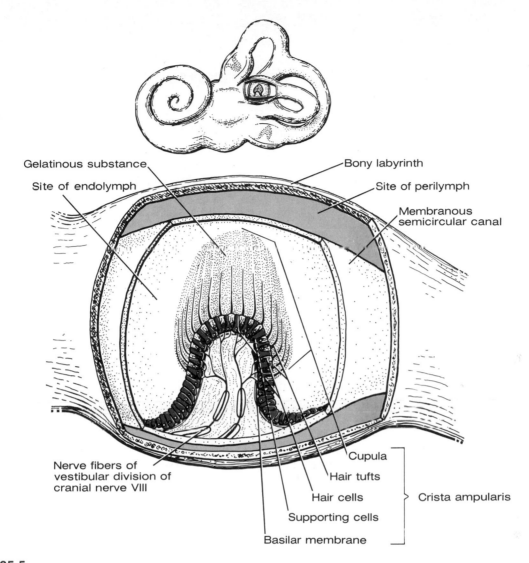

Gelatinous substance

Site of endolymph

Bony labyrinth

Site of perilymph

Membranous
semicircular canal

Nerve fibers of
vestibular division of
cranial nerve VIII

Cupula

Hair tufts

Hair cells

Supporting cells

Basilar membrane

Crista ampularis

Figure 25.5

Crista ampullaris in the semicircular canal.

vestibule contains the saclike **utricle** and **saccule,** and the semicircular chambers contain **membranous semicircular canals.)** Like the cochlear duct, these membranes are filled with endolymph and contain receptor cells that are activated by the disturbance of their cilia.

The semicircular canals are centrally involved in the **mechanism of dynamic equilibrium.** They are about ½ inch in circumference and are oriented in three planes—horizontal, frontal, and sagittal. At the base of each membranous canal is an enlarged region, the **ampulla,** which communicates with the utricle of the vestibule. Within each ampulla is a receptor region called a **crista ampullaris,** which consists of a tuft of hair cells covered with a gelatinous cap, or **cupula** (Figure 25.5). When your head position changes in an angular direction, as when twirling on the dance floor or when taking a rough boat ride, the endolymph in the canal lags behind and moves in the opposite direction, pushing

the cupula—like a swinging door—in a direction opposite to that of the angular motion. This movement initiates the action potential of the hair cells. These impulses are then transmitted up the vestibular division of the eighth cranial nerve to the brain. Then, when the angular motion stops suddenly, the endolymph flows backward, reversing the direction of the cupula's movement and again initiating electrical impulses in the hair cells. (It is this backflow that accounts for the reversed motion sensation you feel when you stop suddenly after twirling.) If you begin to move at a constant rate of motion, the cupula gradually returns to its original position. The hair cells, no longer bent, send no new signals, and you lose the sensation of spinning. Thus the response of these dynamic equilibrium receptors is a reaction to *changes* in angular motion rather than to motion itself.

The vestibule contains receptors (**maculae**) that are essential to the **mechanism of static equilibrium.**

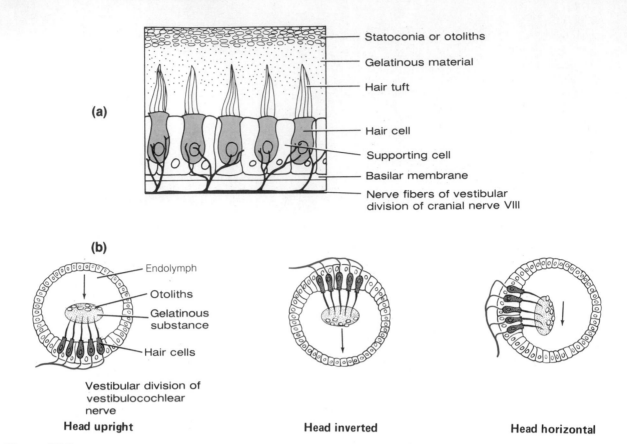

(a)
- Statoconia or otoliths
- Gelatinous material
- Hair tuft
- Hair cell
- Supporting cell
- Basilar membrane
- Nerve fibers of vestibular division of cranial nerve VIII

(b)
- Endolymph
- Otoliths
- Gelatinous substance
- Hair cells

Vestibular division of vestibulocochlear nerve

Head upright **Head inverted** **Head horizontal**

Figure 25.6

Structure and function of static equilibrium receptors (maculae): (a) diagrammatic view of a portion of a macula; (b) stimulation of the maculae by movement of the otoliths in the gelatinous material that creates a pull on the hair cells. Arrows indicate direction of gravitational pull.

The maculae respond to gravitational pull, thus providing information on which way is up or down, and to linear or straightforward changes in speed. They are located on the walls of the saccule and utricle. A gelatinous material, containing small grains of calcium carbonate (**otoliths** or **statoconia**), overrides the hair cells in each macula. As the head moves, the otoliths move in response to variations in gravitational pull and deflect different hair cells, thus triggering nerve impulses along the vestibular nerve (Figure 25.6).

Although the receptors of the semicircular canals and the vestibule are responsible for dynamic and static equilibrium, respectively, they rarely act individually. Complex interaction of many of the receptors is often the rule. Also, the information these equilibrium, or balance, senses provide is significantly augmented by the proprioceptors and sight, as some of the following laboratory experiments demonstrate.

LABORATORY TESTS

Hearing Tests

Perform the following hearing tests in a quiet area.

ACUITY TEST Have your lab partner pack one ear with cotton and sit quietly with eyes closed. Hold a watch very close to her unpacked ear and then slowly move it away from the ear until she signals that the ticking is no longer audible. Record the distance in inches at which ticking is inaudible.

Right ear _____ Left ear _____

Is the threshold of audibility sharp or indefinite?

_____

SOUND LOCALIZATION Ask your partner to close both eyes. Hold a watch at an audible distance (about 6 inches) from her ear, and move it to various locations (front, back, sides, and above her head). Have her locate the position by pointing in each instance. Can the sound be localized equally well at all positions? _____

If not, at what position(s) was the sound less easily located? _____

_____

FREQUENCY RANGE OF HEARING Obtain three tuning forks: one with a low frequency (75 to 100 cps), one with a frequency of approximately 1000 cps, and one with a frequency of 4000 to 5000 cps. Strike the lowest-frequency fork with a rubber mallet, and hold it close to your partner's ear. Repeat with the other two forks.

Which fork was heard most clearly and comfortably? _____cps. Which was heard least well? _____cps.

WEBER TEST TO DETERMINE CONDUCTIVE AND PERCEPTIVE DEAFNESS Strike a tuning fork with a mallet, and place the handle of the tuning fork medially on your forehead. Is the tone equally loud in both ears, or is it louder in one ear? _____

If it is equally loud in both ears, you have equal hearing or loss of hearing in both ears. If perceptive (sensineural) deafness is present in one ear, the tone will be heard in the unaffected ear but not in the ear with sensineural deafness. If conduction deafness is present, the sound will be heard more strongly in the ear in which there is a hearing loss. Conduction deafness can be simulated by plugging one ear with cotton to interfere with the conduction of sound to the inner ear.

RINNE TEST FOR COMPARING BONE AND AIR-CONDUCTION HEARING

1. Strike the tuning fork, and place its handle on your partner's mastoid process.
2. When the sound is no longer audible to your partner, hold the still-vibrating prongs close to his auditory canal. If your partner hears the fork again when it is moved to that position (by air conduction), hearing is not impaired, and the test result is recorded as positive (+). (Record below.)
3. Repeat the test, but this time test air-conduction hearing first.
4. After the tone is no longer heard by air conduction, hold the handle of the tuning fork on the bony mastoid process. If the subject hears the tone again by bone conduction after hearing by air conduction is lost, there is some conductive deafness and the result is recorded as negative (−).
5. Repeat the sequence for the opposite ear.

   Right ear _____ Left ear _____

   Does the subject hear better by bone or by air conduction? _____

   _____

Equilibrium Experiments

The function of the semicircular canals and vestibule are not routinely tested in the laboratory, but the following simple tests should serve to illustrate normal equilibrium apparatus functioning.

BALANCE TEST Have your partner walk a straight line, placing one foot directly in front of the other.

Is she able to walk without undue wobbling from side to side? _____

Did she experience any dizziness? _____

The ability to walk with balance and without dizziness, unless subject to rotational forces, indicates normal function of the equilibrium apparatus.

Was nystagmus* present? _____

BANARY TEST (INDUCTION OF NYSTAGMUS AND VERTIGO †) This experiment evaluates the semicircular canals and should be conducted as a group effort to protect the test subject(s) from possible injury. The following precautionary notes should be read before beginning:

- The subject(s) chosen should not be easily inclined to dizziness on rotational or turning movements.
- Rotation should be stopped immediately if the subject complains of feeling nauseous.
- Since the subject(s) will experience vertigo and loss of balance as a result of the rotation, several classmates should be prepared to catch, hold, or support the subject(s) as necessary until the symptoms pass.

1. Instruct the subject to sit on a rotating chair or stool and to hold on to the arms or seat of the chair, feet on stool rungs.
2. Rotate the chair approximately 10 revolutions in 20 seconds and then suddenly stop the rotation.
3. Note the direction of the subject's resultant nystagmus and ask him to describe his feelings of movement, indicating speed and direction sensation. (If the semicircular canals are operating normally, the subject will experience a sensation that the stool is still rotating immediately after it has stopped and *will* demonstrate nystagmus.)

* **Nystagmus** is the involuntary rolling of the eyes in any direction or the trailing of the eyes slowly in one direction, followed by their rapid movement in the opposite direction. It is normal after rotation; abnormal otherwise. The direction of nystagmus is that of its quick phase on acceleration.

†Vertigo is a sensation of dizziness and rotational movement when such movement is not occurring or has ceased.

4. Conduct each test on a different subject. Tabulate the results on the following charts.

ROMBERG TEST The Romberg test determines the integrity of the dorsal white column of the spinal cord, which transmits impulses to the brain from the proprioceptors involved with posture.

1. Have your partner stand with her back to the blackboard.
2. Draw one line parallel to each side of her body. She should stand erect, eyes open, and staring straight ahead for 2 minutes while you observe her movements. Did you see any gross swaying movements? _____

Series I: Nystagmus Observations

No.	Head position	Direction of rotation	Nature of nystagmus observed	Apparent motion to the subject (direction and speed)
1	Erect	Clockwise	_____	_____
2	Erect	Counterclockwise	_____	_____
3	Head flexed forward on chest	Clockwise	_____	_____
4	Head resting on right shoulder	Clockwise	_____	_____
5	Head resting on right shoulder	Counterclockwise	_____	_____
6	Head resting on left shoulder	Clockwise	_____	_____
7	Head resting on left shoulder	Counterclockwise	_____	_____

Series II: Postural and Ambulatory Movements After Rotation*

No.	Head position	Direction of rotation	Direction of ambulation	Direction of fall	Apparent movement of objects to subject
1	Erect	Clockwise	_____	_____	_____
2	Erect	Counterclockwise	_____	_____	_____
3	Head forward on chest	Clockwise	_____	_____	_____
4	Head forward on chest	Counterclockwise	_____	_____	_____
5	Head on right shoulder	Clockwise	_____	_____	_____
6	Head on right shoulder	Counterclockwise	_____	_____	_____

*Essentially the same testing procedures are used with the following additional instructions to the subject: after rotation, the subject should stand up immediately and walk straight away from the chair. Record results on the chart.

3. Repeat the test. This time the subject's eyes should be closed. Note and record the degree

    of side-to-side movement. _____

    _____

    _____

4. Repeat the test with the subject's eyes closed. This time the subject should be positioned with her left shoulder toward, but not touching, the board so you may observe and record

    the degree of front-to-back swaying. _____

    _____

    _____

Do you think the equilibrium apparatus of the inner ear was operating equally well in all these

tests? _____

The proprioceptors? _____

Why was the observed degree of swaying greater

when the eyes were closed? _____

_____

_____

What conclusions can you draw regarding the factors necessary for maintaining body equilibrium

and balance? _____

_____

_____

_____

_____

ROLE OF VISION IN MAINTAINING EQUILIBRIUM   To further demonstrate the role of vision in maintaining equilibrium, perform the following experiment. Ask your lab partner to record observations and act as a "spotter." Stand erect, eyes open. Raise your left foot approximately 1 foot off the floor and hold for 1 minute. Observations:

_____

_____

_____

Rest for 1 or 2 minutes and repeat with the opposite foot raised, but with the eyes closed. Observations:

_____

_____

_____

_____

Special Senses: Taste and Olfaction

OBJECTIVES

1. To describe the structure and function of the taste receptors.

2. To describe the location and cellular composition of the olfactory epithelium.

3. To name the four basic types of taste sensation and list the chemical substances that elicit these sensations.

4. To point out on a diagram of the tongue the predominant location of the basic types of taste receptors (salty, sweet, sour, bitter).

5. To explain the interdependence between the senses of smell and taste.

6. To name two factors other than olfaction that influence taste appreciation of foods.

7. To define *olfactory adaptation*.

MATERIALS

Prepared histologic slides: the tongue showing taste buds; the nasal epithelium (longitudinal section)
Compound microscope
Paper towels
Small mirror
Granulated sugar
Cotton-tipped swabs
Absorbent cotton
Paper cups
Prepared vials of 10% NaCl, 0.1% quinine or Epsom salt solution, 5% sucrose solution, and 1% acetic acid
Prepared dropper bottles of oil of cloves, oil of peppermint, and oil of wintergreen
Like-sized food cubes of cheese, apple, raw potato, dried prunes, banana, raw carrot, and hard-cooked egg white
Chipped ice

The receptors for taste and olfaction are classified as **chemoreceptors** because they respond to chemicals or volatile substances in solution. Although four relatively specific types of taste receptors have been identified, the olfactory receptors are considered sensitive to a much wider range of chemical sensations. The sense of smell is one of the least understood of the special senses.

LOCALIZATION AND ANATOMY OF TASTE BUDS

The **taste buds,** specific receptors for the sense of taste, are widely but not uniformly distributed in the oral cavity. Most are located on the dorsal surface of the tongue (as described next). A few are found on the soft palate, epiglottis, and inner surface of the cheeks.

The dorsal tongue surface is covered with small projections, or **papillae,** of three major types: sharp filiform papillae and the rounded fungiform and circumvallate papillae. The taste buds are located primarily on the sides of the circumvallate papillae (arranged in a V-formation on the posterior surface of the tongue) and on the more widely distributed fungiform papillae, which look rather like minute mushrooms and are widely distributed on the tongue. (See Figure 26.1.)

Use a mirror to examine your tongue. Can you pick

out the various papillae types? _____

If so, which? _____

_____

Each taste bud consists of a globular arrangement of two types of modified epithelial cells: the **gustatory,** or actual receptor cells, and the **sustentacular,** or supporting cells. Several nerve fibers enter

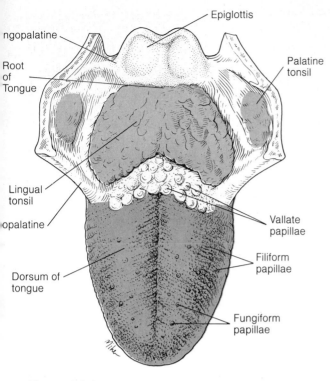

Figure 26.1

Dorsal surface of the tongue, showing major structures.

each taste bud and supply sensory nerve endings to each of the receptor cells whose microvilli emerge through the epithelial surface through an opening called the **taste pore.** When the microvilli contact specific chemicals in the solution, the receptor cells depolarize. The afferent fibers from the taste buds to the sensory cortex in the postcentral gyrus of the brain are carried in three cranial nerves: the facial nerve (VII) serves the anterior two-thirds of the tongue; the glossopharyngeal nerve (IX) serves the posterior third of the tongue; and the vagus nerve (X) carries a few fibers from the pharyngeal region. Obtain a prepared slide of a tongue cross section, and use Figure 26.2 as a guide to aid you in locating the taste buds on the papillae. Make a detailed drawing of one taste bud in the space provided here. Label the taste pore and sensory hairs if observed.

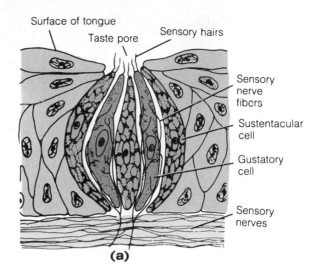

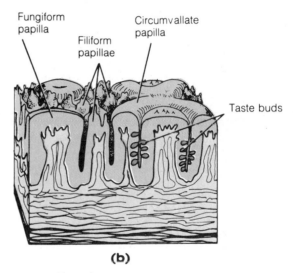

Figure 26.2

Taste bud anatomy and localization: (a) diagrammatic view of a taste bud; (b) enlarged diagrammatic view of the tongue papillae, showing positioning of taste buds.

LOCALIZATION AND ANATOMY OF THE OLFACTORY RECEPTORS

The **olfactory epithelium** (organ of smell) occupies an area of about 2.5 cm in the roof of each nasal cavity. Since the air entering the human nasal cavity must make a hairpin turn to enter the respiratory passages below, the nasal epithelium is in a rather poor position for performing its function. This is why sniffing, which brings more air into contact with the receptors, intensifies the sense of smell.

The specialized receptor cells in the olfactory epithelium are surrounded by sustentacular, or nonsensory, epithelial cells. The receptor cells are believed to be bipolar neurons whose olfactory hairs (actually dendrites) extend outward from the epithe-

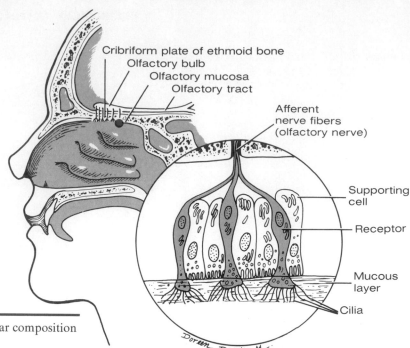

Cribriform plate of ethmoid bone
Olfactory bulb
Olfactory mucosa
Olfactory tract

Afferent
nerve fibers
(olfactory nerve)

Supporting
cell

Receptor

Mucous
layer

Cilia

Figure 26.3

Diagrammatic representation of the cellular composition of olfactory epithelium.

lium. Emerging from their basal ends are axonal nerve fibers that penetrate the cribriform plate of the ethmoid bone and proceed as the olfactory nerves to synapse in the olfactory bulbs lying on either side of the crista galli of the ethmoid bone. Impulses from neurons of the olfactory bulbs are then conveyed to the olfactory portion of the cortex (rhinencephalon).

Obtain a longitudinal section of olfactory epithelium. Examine it closely, comparing it to Figure 26.3.

LABORATORY EXPERIMENTS

Stimulation of Taste Buds

Dry the dorsal surface of your tongue with a paper towel, and place a few sugar crystals on it. Do *not* close your mouth. Time how long it takes to taste

the sugar. _____ sec

Why couldn't you taste the sugar immediately? ____

_____

_____

Plotting Taste Bud Distribution

There are four basic taste sensations, which correspond to the stimulation of four major types of taste buds. Although all taste buds are believed to respond in some degree to all four classes of chemical stimuli, each type responds optimally to only one. This characteristic makes it possible to map the tongue to show the relative density of each type of taste bud.

The sweet receptors respond to a number of seemingly unrelated compounds such as sugars (fructose, sucrose, glucose), saccharine, and some amino acids. Some believe the common factor is the hydroxyl (OH^-) group. Sour receptors respond to hydrogen ions (H^+) or the acidity of the solution, bitter receptors to alkaloids, and salty receptors to metallic ions in solution.

1. Prepare to make a taste sensation map of your lab partner's tongue by obtaining the following: cotton-tipped swabs, one vial each of NaCl, quinine or Epsom salt solution, sucrose solution, acetic acid, paper cups, and a flask of distilled or tap water.
2. Before each test, the subject should rinse her mouth thoroughly with water and lightly dry her tongue with a paper towel.
3. Moisten a swab with 5% sucrose solution and touch it to the center, back, tip, and sides of the dorsal surface of the subject's tongue.
4. Map the location of the sweet receptors with an O on the tongue outline below.

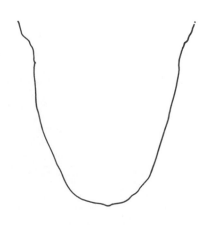

5. Repeat the procedure to map the location of the bitter receptors (use the symbol B) with quinine (or Epsom salt solution), salt receptors (symbol +) with NaCl, and the sour receptors (symbol −) with acetic acid. *Use a fresh swab for each test.*

What area of the dorsum of the tongue seems to lack taste receptors? _____

_____

How closely does your localization of the different taste receptors coincide with the information in your textbook? _____

_____

Combined Effects of Smell, Texture, and Temperature on Taste

1. Ask the subject to sit with eyes closed and to stuff her nostrils with absorbent cotton.

2. Obtain samples of the food items listed in the chart below.

3. In each test, place a cube of food in the subject's mouth and ask her to identify the food by using the following sequence of activities:

• First, by manipulation of the food with the tongue
• Second, by chewing the food
• If a positive identification is not made with these techniques along with the taste sense, ask the subject to remove the cotton and

continue chewing with the nostrils open to determine if a positive identification can be made.

At no time should the subject be allowed to see the foods being tested. Record the results on the chart by checking the appropriate column. (Use an out-of-sequence order of food testing.)

Was the sense of smell equally important in all cases? _____

Where did it seem to be important and why? _____

_____

_____

Effect of Olfactory Stimulation on Taste

There is no question that what is commonly referred to as the sense of taste depends heavily on stimulation of the olfactory receptors, particularly in the case of strongly odoriferous substances. The following experiment should illustrate this fact.

1. Obtain vials of oil of wintergreen, peppermint, and cloves and some fresh cotton-tipped swabs. Ask the subject to sit, so that he cannot see which vial is being used, and to dry his tongue and close his nostrils.
2. Apply a drop of one of the oils to his tongue.

Can he distinguish the flavor? _____

3. Have the subject open his nostrils and record the change in sensation he reports. _____

Method of Identification

Food	Texture only	Chewing with nostrils packed	Chewing with nostrils open	Identification not made
Cheese	_____	_____	_____	_____
Apple	_____	_____	_____	_____
Raw potato	_____	_____	_____	_____
Banana	_____	_____	_____	_____
Dried prunes	_____	_____	_____	_____
Raw carrot	_____	_____	_____	_____
Hard-cooked egg white	_____	_____	_____	_____

4. Have the subject rinse his mouth well and dry his tongue.
5. Prepare two swabs, each with one of the two remaining oils.
6. Hold one swab under the subject's open nostrils, while touching the second swab to

his tongue. Record his sensations. _____

_____

_____

_____

Which sense, taste or smell, appears to be more important in the proper identification of a strongly

flavored volatile substance? _____

In addition to the effect that olfaction and textural factors play in determining our taste sensations, the temperature of foods may also determine if the food is appreciated or even tasted. To illustrate this, have your partner hold some chipped ice on her tongue for approximately a minute and then close her eyes. Immediately place any of the foods previously identified in her mouth.

Olfactory Adaptation

Obtain some absorbent cotton and two of the following oils (oil of wintergreen, peppermint, or cloves). Press one nostril shut or stuff it with cotton. Hold the bottle of oil under the open nostril and exhale through the mouth. Record the time required for the odor to disappear (for olfactory adaptation to occur).

_____ sec

Repeat the procedure with the other nostril.

_____ sec

Immediately test another oil with the nostril that has just experienced olfactory adaptation. What are

the results? _____

_____

_____

What conclusions can you draw? _____

_____

_____

_____

_____

Anatomy and Basic Function of the Endocrine Glands

EXERCISE 27

OBJECTIVES

1. To identify and name the major endocrine glands and tissues of the body when provided with an appropriate diagram.

2. To list the hormones produced by the endocrine glands and discuss the general function of each.

3. To indicate the means by which hormones contribute to body homeostasis by giving appropriate examples of hormonal actions.

4. To cite the mechanism by which the endocrine glands are stimulated to release their hormonal products.

5. To describe the physiologic relationship between the hypothalamus and the pituitary gland.

6. To describe a major pathologic consequence of hypersecretion and hyposecretion of each hormone considered.

7. To correctly identify the histologic structure of the anterior and posterior pituitary, thyroid, parathyroid, adrenal cortex and medulla, and pancreas by microscopic inspection or when presented with an appropriate photomicrograph or diagram. (Optional)

8. To name and point out the specialized hormone-secreting cells in the above tissues as studied in the laboratory. (Optional)

MATERIALS

Human torso model
Anatomic chart of the human endocrine system
Compound microscopes
Colored pencils
Histologic slides of the anterior-posterior pituitary (differential staining), thyroid gland, parathyroid glands, adrenal gland, pancreas tissue
Dissection animal, dissection tray, and instruments (if desired by the instructor)
Related film: *The Hormones: Small but Mighty* (black and white, sound 16 mm, 29 minutes, Indiana University, Audio-Visual Center, Bloomington, IN 47401)

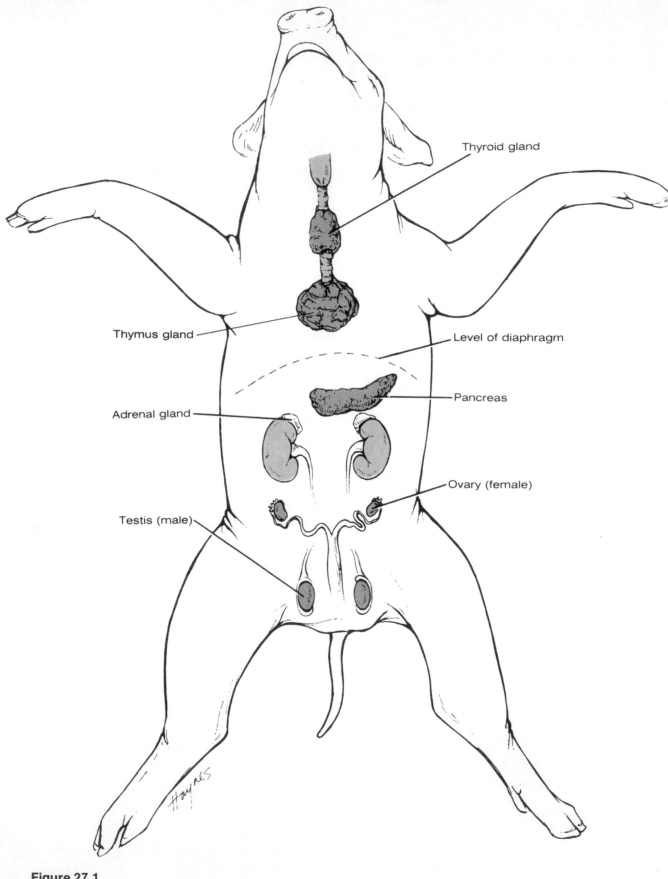

Figure 27.1

Localization of endocrine organs in a fetal pig.

The endocrine system is the second major controlling system of the body. Acting with the nervous system, it helps to coordinate and integrate the activity of the body's cells. However, the nervous system employs electrochemical impulses to bring about rapid control, while the more slowly acting endocrine system employs chemical "messengers," or **hormones,** which are released into the blood to be transported throughout the body.

The term *hormone* comes from a Greek word meaning to arouse. The body's hormones, which are steroids or protein derivatives, arouse the body's tissues and cells by stimulating changes in their metabolic activity. These changes lead to growth and development and to the physiologic homeostasis of many body systems. Hormones affect the body cells primarily by altering (increasing or decreasing) a metabolic process rather than by initiating a new one. Although all hormones are blood-borne, a given hormone affects only the biochemical activity of a specific organ or organs. Organs that respond to a particular hormone are referred to as the *target organs* of that hormone. The ability of the target tissue to respond seems to depend on the ability of the hormone to bind with specific receptors (proteins), occurring on the cell membranes or within the cells.

Although the function of some hormone-producing glands (the anterior pituitary, thyroid, adrenals, parathyroids) is purely endocrine, the function of others (the pancreas and gonads) is mixed—both endocrine and exocrine. Both types of glands are derived from epithelium, but the endocrine, or ductless glands release their product (always hormonal) directly into the blood. The exocrine glands release their products at the body's surface or outside an epithelial membrane via ducts. In addition, there are varied numbers of hormone-producing cells within organs such as the intestine, stomach, kidney, and placenta, whose functions are primarily nonendocrine.

You will use anatomic charts and models to localize the endocrine organs in this exercise and wait to identify the endocrine glands of the laboratory animals along with their other major organ systems in subsequent units. Partial dissection would cause the laboratory animals to become desiccated. However, if your instructor wishes you to perform an animal dissection to identify the endocrine organs. Figure 27.1 may be used as a guide for this purpose. Your instructor will provide methods of making incisions that minimize desiccation and damage to other organs.

GROSS ANATOMY AND BASIC FUNCTION OF THE ENDOCRINE GLANDS

As the endocrine organs are described, *locate and identify them by name* on Figure 27.2. When you have completed the descriptive material, also locate the organs on the anatomic charts or torso.

Pituitary Gland

The pituitary gland, or hypophysis, is located in the concavity of the sella turcica of the sphenoid bone. It consists of two functional areas, the **adenohypophysis,** or **anterior pituitary,** and the **neurohypophysis,** or **posterior pituitary,** and is attached to the hypothalamus by a stalk called the **infundibulum.**

ADENOHYPOPHYSEAL HORMONES   The adenohypophyseal hormones consist of the following tropic hormones or hormone groups: the **gonadotropins** regulate the hormonal activity of the gonads (ovaries and testes). In the female, **follicle-stimulating hormone (FSH)** stimulates graafian follicle development and the production of **estrogen.** In the male, it promotes sperm cell production (spermatogenesis). A second gonadotropic hormone, **luteinizing hormone (LH),** triggers ovulation in the female in conjunction with FSH and induces the conversion of the ruptured graafian follicle (created by the ovulatory event) to a corpus luteum. It then stimulates production of **progesterone** and some estrogen by the newly formed corpus luteum. In men LH is referred to as **interstitial cell-stimulating hormone (ICSH),** and it stimulates **testosterone** production by the interstitial cells of the testes.

The third gonadotropin is **luteotropic hormone (LTH),** or **prolactin.** After childbirth it promotes and maintains lactation by the mammary glands. (In female rats LTH also plays a role in prolonging the life and function of the corpus luteum in conjunction with LH . This has not yet been demonstrated in the human female.) Its function in the male is unknown.

Adrenocorticotropic hormone (ACTH) regulates the endocrine activity of the adrenal cortex portion of the adrenal gland. **Thyrotropic hormone (TSH)** influences the growth and activity of the thyroid gland.

These adenohypophyseal hormones are all **tropic** hormones. In each case, after a tropic hormone has been released by the anterior pituitary, it stimulates its target organ, which is also an endocrine gland, to secrete its hormones, which then exert their effects on other body organs and tissues. Two other major hormones produced by the anterior pituitary are not directly involved in the regulation of other endocrine glands of the body. **Somatotropin (STH),** or **growth hormone,** is a general metabolic hormone that plays an important role in determining body size. It is essential for normal retention of body protein and affects many tissues of the body. Its major effects, however, are exerted on the growth of muscle and the long bones of the body.

Melanocyte-stimulating hormone (MSH) does not appear to be of major significance in humans, but it darkens the skin of reptiles and amphibians and may similarly affect the melanocytes that produce pigment in human skin.

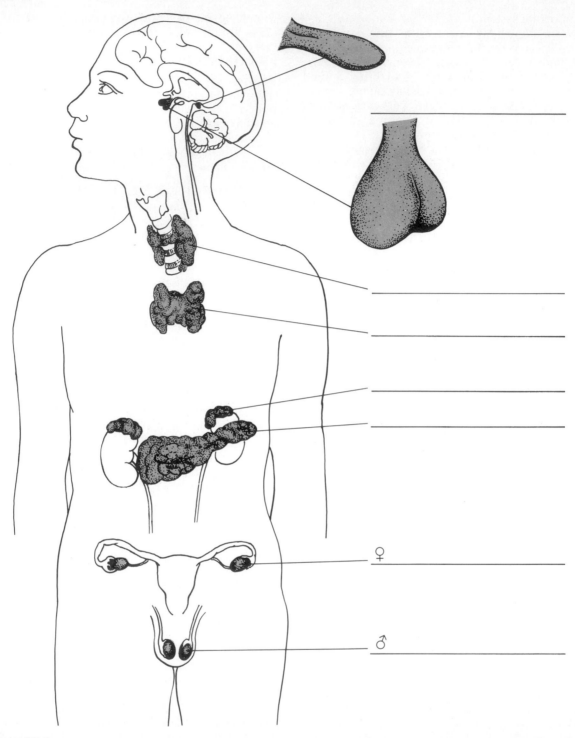

Figure 27.2

Human endocrine organs.

The anterior pituitary controls the activity of so many other endocrine glands that it has often been called the master endocrine gland. Its removal dramatically interferes with body metabolism: the gonads and adrenal and thyroid glands atrophy, and changes from subsequent inadequate hormone production become obvious. However, the anterior pituitary is not autonomous in its control, since the release of the anterior pituitary hormones is controlled by neurosecretions, or releasing factors, produced by the hypothalamus and liberated into the hypophyseal portal system, which serves the circulatory needs of the anterior pituitary (Figure 27.3).

NEUROHYPOPHYSEAL HORMONES The neurohypophysis or posterior pituitary is not an

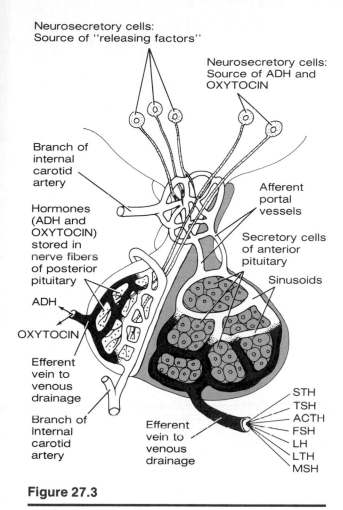

Neurosecretory cells:
Source of "releasing factors"

Neurosecretory cells:
Source of ADH and
OXYTOCIN

Branch of
internal
carotid
artery

Afferent
portal
vessels

Hormones
(ADH and
OXYTOCIN)
stored in
nerve fibers
of posterior
pituitary

Secretory cells
of anterior
pituitary

Sinusoids

ADH

OXYTOCIN

Efferent
vein to
venous
drainage

Branch of
internal
carotid
artery

Efferent
vein to
venous
drainage

STH
TSH
ACTH
FSH
LH
LTH
MSH

Figure 27.3

Neural and vascular relationships between the hypothalamus and the anterior and posterior lobes of the pituitary.

endocrine gland in a strict sense, since it does not synthesize the hormones it releases. (This relationship is also indicated in Figure 27.3.) Instead, it acts as a storage area for two hormones transported to it from the hypothalamus. The first of these hormones is **oxytocin,** which stimulates powerful uterine contractions during birth and coitus and also causes milk ejection in the lactating mother. The second, **antidiuretic hormone (ADH),** causes the kidney tubule cells to resorb water from the urinary filtrate, thereby reducing urine production and conserving body water. It also plays a role in increasing blood pressure because of its vasoconstrictor effect on the arterioles. Hyposecretion of ADH results in dehydration from excessive urine output, a condition called **diabetes insipidus.** Individuals with this condition experience an insatiable thirst.

Thyroid Gland

The thyroid gland is composed of two lobes joined by a central mass, or isthmus. It is located

in the throat, just inferior to the larynx. It produces two major hormones, thyroxine and thyrocalcitonin.

Because of the primary function of **thyroxine** (actually two physiologically active substances known as T_3 and T_4) is to control the rate of body metabolism and cellular oxidation, it affects every cell in the body. In addition, it is an important regulator of tissue growth and development, especially in the reproductive and nervous systems. Hyposecretion of thyroxine leads to a condition of mental and physical sluggishness, which is called **myxedema** in the adult.

Thyrocalcitonin (or calcitonin) decreases blood calcium levels by causing calcium to be deposited in the bones. It acts antagonistically to **parathormone (PTH),** the hormonal product of the parathyroid glands.

- Try to palpate your thyroid gland by placing your fingers against your windpipe. As you swallow, the thyroid gland will move up and down on the sides and front of the windpipe.

Parathyroid Glands

The parathyroid glands are found on the posterior surface of the thyroid gland. Typically, there are two small oval glands on each lobe. They secrete **parathormone (PTH),** the most important regulator of calcium-phosphate ion homeostasis of the blood. When blood calcium levels decrease below a certain critical level, the parathyroids release PTH, which causes bone to release calcium from the matrix and causes the kidney to decrease resorption of phosphate from the filtrate. If blood calcium levels fall too low, **tetany** results and may be fatal.

Adrenal Glands

The two bean-shaped adrenal, or suprarenal, glands are located atop or close to a kidney. Anatomically, the adrenal medulla, like the neurohypophysis, develops from neural tissue, and the **adrenal medulla** is directly controlled by sympathetic nervous system neurons. The cells of the medulla respond to this stimulation by releasing **epinephrine** (80%) or **norepinephrine** (20%), which act in conjunction with the sympathetic nervous system to elicit the "flight or fight" response to stress.

The **adrenal cortex** produces three major groups of steroid hormones, collectively called the **corticosteroids**—the mineralocorticoids, the glucocorticoids, and the gonadocorticoids. The **mineralocorticoids,** chiefly **aldosterone,** regulate water and electrolyte balance in the extracellular fluids primarily by regulating sodium ion resorption by kidney tubules. The **glucocorticoids** (cortisone, hydrocortisone, and corticosterone) enable the body to resist long-term stress, primarily by increasing blood glucose levels. Because of their anti-inflammatory properties, they are often administered to decrease tissue edema and counteract vascular dilation. The **gonadocorticoids,** or **sex hormones,** produced by the adrenal cortex are chiefly androgens (male sex hormones), but some estrogens (female sex hormones) are also formed. The gonadocorticoids are produced throughout life in relatively insignificant amounts; however, hypersecretion of these hormones produces abnormal hairiness (**hirsutism**), and masculinization occurs.

Pancreas

The pancreas, which functions as both an endocrine and exocrine gland, produces digestive enzymes as well as **insulin** and **glucagon,** important hormones concerned with the regulation of blood sugar levels. The pancreas is found close to the stomach.

Elevated blood glucose levels stimulate release of insulin, which decreases blood sugar levels, primarily by accelerating the transport of glucose into the body cells, where it is oxidized for energy or converted to glycogen or fat for storage. Hyposecretion of insulin leads to **diabetes mellitus,** which is characterized by the inability of body cells to utilize glucose and the subsequent loss of glucose in the urine. Alterations of protein and fat metabolism also occur, but these are probably secondary to derangements in carbohydrate metabolism.

Glucagon acts antagonistically to insulin. Its release is stimulated by low blood glucose levels, and its action is basically hyperglycemic. Its primary target organ is the liver, which it stimulates to break down its glycogen stores to glucose and subsequently to release the glucose to the blood.

The Gonads

The female gonads, or ovaries, are paired, almond-sized organs located in the pelvic cavity. In addition to producing the female sex cells (ova), the ovaries produce two steroid hormone groups, the **estrogens** and **progesterone.** The endocrine and exocrine functions of the ovaries do not begin until the onset of puberty, when the anterior pituitary gonadotropic hormones prod the ovary into action that produces rhythmic ovarian cycles in which ova develop and hormonal levels rise and fall. The estrogens are responsible for the development of the secondary sex characteristics of the female at puberty (primarily maturation of the reproductive organs and development of the breasts) and act with progesterone to bring about cyclic changes of the uterine lining that occur during the menstrual cycle. The estrogens also help maintain pregnancy and prepare the mammary glands for lactation.

Progesterone, as already noted, acts with estrogen to bring about the menstrual cycle. During pregnancy it maintains the uterine musculature in a quiescent state and helps to prepare the breast tissue for lactation.

The paired oval testes of the male are suspended in a pouchlike sac, the scrotum, outside the pelvic cavity. In addition to the male sex cells, sperm, the testes also produce the male sex hormone, **testosterone.** Testosterone promotes the maturation of the reproductive system accessory structures, brings about the development of the secondary sex characteristics, and is responsible for the male sexual drive, or libido. Both the endocrine and exocrine functions of the testes begin at puberty under the influence of the anterior pituitary gonadotropins.

Two glands not enumerated earlier as major endocrine glands should also be briefly considered here, the thymus and the pineal gland.

Thymus

The thymus is a bilobed gland situated in the upper thorax, posterior to the sternum and overlying the heart and lungs. Conspicuous in the infant, it begins to atrophy at puberty, and by adulthood it is no longer functional as an endocrine-producing gland. Researchers now believe that the thymus produces a hormone called **thymosin** and that during early life the thymus acts as an incubator for the maturation and specialization of a unique population of white blood cells called T-lymphocytes. The T-lymphocytes are responsible for the cellular immunity aspect of body defense; that is, rejection of foreign grafts, tumors, or virus-infected cells.

Pineal Body

The pineal body, or **epiphysis cerebri,** is a small cone-shaped gland located in the roof of the third ventricle of the brain. Like the thymus gland, it has a limited functional life and begins to atrophy and become fibrous at an early age—presumably around the age of 7. The endocrine role of the pineal body is still controversial (early philosophers regarded it as the seat of the soul), but many investigators suggest that it is the formation site of three hormones—**melatonin, adrenoglomerulotropin,** and **serotonin.**

Melatonin appears to exert some inhibitory effect on the reproductive system (especially on the ovaries) that prevents precocious sexual maturation. Some evidence suggests that adrenoglomerulotropin stimulates the adrenal cortex to release aldosterone, but this has not been sufficiently substantiated. Serotonin, a neurotransmitter, is produced by many regions of brain tissue and is involved in normal neural transmission of impulses. In the pineal body, serotonin is known to be a chemical precursor of melatonin.

The endocrine functions of the intestine, stomach, and kidney will be considered in conjunction with their respective organ systems.

 Once you are satisfied that you can name and locate the endocrine organs and have labeled Figure 27.2, use an appropriate reference to define the following pathologic conditions resulting from hypersecretion or hyposecretion of the various hormones (which have not been described here).

acromegaly _____

_____

_____

_____

_____

Addison's disease _____

_____

_____

_____

_____

cretinism _____

_____

_____

Cushing's syndrome _____

_____

_____

diabetes insipidus _____

_____

_____

diabetes mellitus _____

_____

_____

pituitary dwarfism _____

_____

_____

eunuchism _____

_____

exophthalmic goiter _____

_____

_____

giantism _____

_____

_____

hirsutism _____

_____

_____

hypercalcemia _____

_____

_____

myxedema _____

_____

_____

Paget's disease _____

_____

_____

Simmond's disease _____

_____

_____

tetany _____

_____

MICROSCOPIC ANATOMY OF SELECTED ENDOCRINE GLANDS (OPTIONAL)

To prepare for the histologic study of the endocrine glands, obtain a microscope and one each of the slides listed in the list of materials. We will study only organs in which it is possible to identify the endocrine-producing cells.

Thyroid Gland

1. Scan the thyroid under low power, noting the spherical sacs (follicles), containing a pink-stained material (colloid). Stored thyroxine is attached to the protein colloidal material in the follicles as **thyroglobulin** and is released gradually to the blood.
2. Observe the tissue under high power. Note that the walls of the follicles are formed by simple cuboidal epithelial cells that synthesize the follicular products. The **parafollicular cells** you see between the follicles are responsible for calcitonin production.
3. Sketch two or three follicles in the space below. Label the colloid, cuboidal epithelial cells (follicular cells), and parafollicular cells.

When the thyroid gland is actively secreting, the follicles appear small, and the colloidal material has a ruffled border. When the thyroid is hypoactive or inactive, the follicles are large and plump and the follicular epithelium appears to be squamous-like. What is the physiologic state of the tissue you have

been viewing? _____

_____

_____

Parathyroid Glands

1. Observe the parathyroid tissue under low power to localize its two major cell types, the **principal,** or **chief, cells** and the **oxyphil cells.** The oxyphil cells, which are thought to bear the major responsibility for the synthesis of PTH, occur singly or in groups of a few cells. The chief cells appear in masses.
2. Sketch a small portion of the parathyroid tissue in the space provided. Label the chief cells, oxyphil cells, and the connective tissue matrix.

Pancreas

1. Observe pancreas tissue under low power to identify the roughly circular **islets of Langerhans,** the endocrine portions of the pancreas. The islets are scattered amid the more numerous acinar cells and stain differentially (usually lighter), which makes their identification possible.
2. Focus on an islet and examine its cells under high power. Note that the islet cells are densely packed and have no definite arrangement. In contrast, the cuboidal acinar cells are arranged around secretory ducts. Unless special stains are used, it will not be possible to distinguish the **alpha cells,** which produce glucagon, from the **beta cells,** which synthesize insulin. With these specific stains, the beta cells are larger and stain blue, and the alpha cells are smaller and appear to contain pink granules. What is the product of the

acinar cells? _____

_____

3. Draw a section of the pancreas in the space provided. Label the islets and the acinar cells. Differentiate between the alpha and beta cells if possible.

Pituitary Gland

1. Observe the general structure of the pituitary gland under low power to differentiate between the glandular anterior pituitary and the neural posterior pituitary.
2. Using the h.p. lens, focus on the nests of cells of the anterior pituitary. It is possible to identify the specialized cell types that secrete the specific hormones when differential stains are used. Locate the pink-staining **acidophil cells,** which produce growth hormone and LTH, and the **basophil cells,** whose deep-blue granules are responsible for the production of the other four anterior pituitary hormones (TSH, ACTH, FSH, and LH). **Chromophobes,** the third cellular population, do not take up the stain and appear rather dull and colorless. The role of the chromophobes is controversial, but they apparently are not directly involved in hormone production.
3. Switch your focus to the posterior pituitary. Observe the nerve fibers (axons of hypophyseal neurons) that compose most of this portion of

the pituitary. Also note the **pituicytes,** which are randomly distributed among the nerve fibers and which resemble the connective tissue neuroglia of the CNS.

What two hormones are stored here? _____

_____

What is their source? _____

_____

4. Make sketches of the anterior pituitary and posterior pituitary tissue. Use appropriately colored pencils to indicate the acidophils, basophils, and chromophobes. Label these cells as well as the nerve fibers and pituicytes of the posterior pituitary.

their clumped arrangement and that they are relatively large ovoid-shaped cells.

What hormonal products are produced by the

medulla? _____ _____

and _____

3. Draw a representative area of each of the adrenal regions, indicating in your sketch the differences in relative cell size and arrangement.

Zona glomerulosa   Zona fasciculata

Anterior pituitary   Posterior pituitary

Adrenal Gland

1. Hold the slide of the adrenal gland up to the light to distinguish outer cortical and inner medulla areas. Then scan the cortex under low power to distinguish the differences in cell appearance and arrangement in the three cortical areas. Identify the following cortical areas:

- Connective tissue capsule of the adrenal gland
- The outermost zona glomerulosa, where mineralocorticoid secretion occurs and where the tightly packed cells are rather irregularly arranged
- The deeper intermediate zona fasciculata, which produces glucocorticoids (This is the thickest part of the cortex. Its cells are arranged in irregular cords.)
- The innermost cortical zone, the zona reticularis, abutting the medulla, which produces sex hormones and some glucocorticoids (The cells here stain deeply and form interconnections with each other.)

2. Switch focus to view the lightly stained cells of the adrenal medulla under high power. Note

Zona reticularis   Adrenal medulla

Ovary

Because you will consider the ovary in greater histologic detail when you study the reproductive system, the objective in this laboratory exercise is just to identify the endocrine-producing parts of the ovary.

1. Scan an ovary slide under low power, and look for a graafian follicle—a circular arrangement of cells enclosing a central cavity. This structure synthesizes estrogens.
2. Examine the graafian follicle under high power, identifying the follicular cells that produce estrogen, the antrum (fluid-filled cavity), and developing ovum (if present). The ovum will be the largest cell in the follicle.

3. Draw a graafian follicle below, labeling the antrum, follicle cells, and ovum.

4. Switch to low power, and scan the slide to find a corpus luteum, a large amorphous-looking area that produces progesterone.

Testis

1. Examine a section of a testis under low power. Identify the seminiferous tubules, which produce sperm, and the **interstitial cells of Leydig,** which produce testosterone. The interstitial cells are scattered between the seminiferous tubules in the connective tissue matrix.

2. Draw a representative area of the testis in the space provided. Label the seminiferous tubules and area of the interstitial cells.

Experiments on Hormonal Action

OBJECTIVES

1. To describe and explain the effect of pituitary hormones on the ovary.

2. To describe and explain the effects of hyperinsulinism.

3. To describe and explain the effect of epinephrine on the heart.

4. To understand the physiologic (and clinical) importance of metabolic rate measurement.

5. To investigate the effect of hypo-, hyper-, and euthyroid conditions on oxygen consumption and metabolic rate.

6. To assemble the necessary apparatus and properly use a manometer to obtain experimental results.

7. To calculate metabolic rate in terms of ml O_2/kg/hr.

MATERIALS

Female frog (*Rana pipiens*)
Syringe
20- to 25-gauge needle
Pituitary extract
Physiologic saline
Battery jar
Wax marking pencils

500 or 600 ml beakers

10% glucose solution
Commercial insulin in dropper bottle (40 units/ml)
Small (1½–2″) freshwater fish (bluegill, sunfish, or goldfish in order of preference)

1:10,000 epinephrine (Adrenalin) solution in dropper bottles
Dissecting pan, instruments, and pins
Frog Ringer's solution in dropper bottle

Glass dessicator; manometer, 20 ml glass syringe, two-hole cork, and T-valve (1 for each 3 to 4 students)
Soda lime (dessicant)
Hardware cloth squares
Petrolatum
Rubber tubing; tubing clamps; scissors
3-in. pieces of glass tubing
Animal balances
Young rats of the same sex, obtained 2 weeks prior to the laboratory session and treated as follows for 14 days:

> **Group 1: control group**—fed normal rat chow and water
> **Group 2: experimental group A**—fed normal rat chow, and drinking water containing 0.02% propylthiouracil
> **Group 3: experimental group B**—fed rat chow, containing dessicated thyroid (2% by weight), and normal drinking water

Chart set up on chalkboard so that each student group can record their computed metabolic rate figures under the appropriate headings

---

The endocrine system exerts many complex and interrelated effects on the body as a whole as well as on specific organs and tissues. Most scientific knowledge about this system is contemporary, and new information is constantly being presented. Most experiments on the endocrine system require relatively large laboratory animals, are time consuming (requiring days to weeks of observation), and often involve technically difficult surgical procedures to remove the glands or parts of them, all of which makes it difficult to conduct more general types of laboratory experiments. Nevertheless, the four technically unsophisticated experiments presented here should illustrate how dramatically hormones affect the body functioning.

To conserve laboratory specimens, the experiments can be conducted by groups of four students. The use of larger working groups should not detract from benefits gained, since the major value of these experiments lies in observation.

EFFECT OF PITUITARY HORMONES ON THE OVARY

As discussed in Exercise 27, a group of anterior pituitary hormones called the gonadotropins regulate the ovarian cycles of the female. Although amphibians normally ovulate seasonally, many can be stimulated to ovulate "on demand" by the injection of an extract of pituitary hormones.

 1.   Aspirate 1 to 2 ml of the pituitary extract into a syringe. Inject the extract subcutaneously into the anterior lymph sac of the frog you have selected to be the experimental animal. To inject into the lymph sac, hold the frog with its ventral surface superiorly. Open the frog's mouth, and insert the needle into the floor of the mouth. Avoid piercing the tongue. The needle enters the lymph sac as it pierces the tissue. With a wax marker, label a large battery jar "experimental," and place the frog in it.

2.   Aspirate 1 to 2 ml of normal saline into a syringe and inject it into the anterior lymph sac of a second frog—the control animal. (Make sure you inject the same volume of fluid into both frogs.) Place this frog into a second battery jar, marked "control."

3.   After 1½ hours*, check each frog for the presence of eggs in the cloaca. Hold the frog firmly with one hand and exert pressure on its abdomen toward the cloaca (in the direction of the legs). Any eggs in the oviduct will be forced out and will appear at the cloacal opening. If no eggs are present, make arrangements with your laboratory instructor to return to the lab on the next day to check your frogs for the presence of eggs.

In which of the prepared frogs was ovulation induced? _____

Specifically, what hormones in the pituitary extract cause ovulation to occur? _____

*The student should use this time to conduct the experiment on the effects of hyperinsulinism (page 244) or that on the effects of thyroxin on metabolic rate (p. 245).

EFFECTS OF HYPERINSULINISM

Many people with diabetes mellitus need injections of insulin to maintain a proper physiologic balance between the glycogen stored in liver and muscle tissue and the glucose in the blood. Adequate levels of blood glucose are essential for the proper functioning of the nervous system; thus, the administration of insulin must be carefully controlled. If blood glucose levels fall precipitously, the patient will go into insulin shock.

A small fish will be used to demonstrate the effects of hyperinsulinism. Since the action of insulin on the fish parallels that in the human, this experiment should provide valid information concerning its administration to humans.

 1.   Prepare two beakers. Using a wax marker, mark one A and the other B. To beaker A, add 200 ml of water and 10 to 15 drops of commercial insulin. To beaker B, add 200 ml of 10% glucose solution.

2.   Place a small fish in beaker A and observe its actions carefully as the insulin diffuses into its bloodstream through the capillary circulation of its gills.

Approximately how long did it take for the fish to become comatose? _____

_____

What types of activity did you observe in the fish before it became comatose? _____

_____

3.   When the fish is comatose, carefully transfer it to beaker B and observe its actions. What happens to the fish after it is transferred to beaker B? _____

_____

_____

Approximately how long did it take for this recovery?

_____

4.   After all observations have been made and recorded, carefully return the fish to the aquarium.

EFFECT OF EPINEPHRINE ON THE HEART

As noted in Exercise 27, the adrenal medulla and the sympathetic nervous system are closely interrelated, specifically because the cells of the adrenal medulla and the postganglionic neurons of the sympathetic nervous system both release catecholamines. This experiment demonstrates the effects of epinephrine on the frog heart.

1.   Destroy the nervous system of a frog. (The frog used in the experiment on pituitary hormone effects may be used if your test results were positive. Otherwise, obtain another frog.) Insert one blade of a scissors into its mouth as far as possible and quickly cut off the top of its head posterior to the eyes. Then, identify the spinal cavity and insert a dissecting needle into it to destroy the spinal cord.

2.   Place the frog dorsal side down on a dissecting pan, and carefully open its ventral body cavity by making a vertical incision with the scissors.

3.   Identify the beating heart, and carefully cut through the saclike pericardium to expose the heart tissue.

4.   Visually count the heart rate for 1 minute, and record below. Keep the heart moistened with frog Ringer's solution during this interval. Beats per

minute: _____

5.   Flush the heart with epinephrine solution. Record the heart beat rate per minute for 5 consecutive minutes.

minute 1 _____

minute 2 _____

minute 3 _____

minute 4 _____

minute 5 _____

What was the effect of epinephrine on the heart

beat rate? _____

Was it long-lived? _____

6.   Dispose of the frog in an appropriate container, and clean the dissecting pan and instruments before returning them to the supply area.

EFFECT OF THYROXINE ON METABOLIC RATE

Metabolism is a broad term referring to all chemical reactions that are necessary to maintain life. It involves enzymatically controlled processes in which substances are broken down to simpler substances, *catabolism*, and processes in which larger molecules or structures are built from smaller ones, *anabolism*. During catabolic reactions, energy is released as chemical bonds are broken. Some of the liberated energy is captured to make ATP, the energy-rich molecule used by body cells to energize all their activities; the balance is lost in the form of thermal energy or heat. Maintenance of body temperature is critically related to the heat-liberating aspects of metabolism.

Various foodstuffs make different contributions to the process of metabolism. For example, carbohydrates, particularly glucose, are generally broken down or oxidized to make ATP, whereas fats are utilized to form cell membranes, myelin sheaths, and to insulate the body with a fatty cushion. (Fats are used secondarily for producing ATP when there are inadequate carbohydrates in the diet.) Proteins and amino acids tend to be carefully conserved by body cells and understandably so, since most structural elements of the body are proteinaceous in nature.

Thyroxine, produced by the thyroid gland, is the single most important hormone influencing the rate of cellular metabolism and body heat production. Under conditions of excess thyroxine production (hyperthyroidism), an individual's basal metabolic rate (BMR) increases; heat production and oxygen consumption increase; and the individual tends to lose weight, become heat-intolerant, and irritable. Conversely, hypothyroid individuals become obese, are cold-intolerant because of their low BMR, and become mentally and physically sluggish. Many factors other than thyroxine levels contribute to metabolic rate (for example, body size and weight, age, and activity level), but the focus of the following experiments is to investigate how differences in thyroxine concentration impact on metabolism.

Three groups of laboratory rats will be used. The *control group* animals are assumed to be euthyoid and to have normal metabolic rates for their relative body weights. *Experimental group A* animals have received water containing the chemical propylthiouracil, which counteracts or antagonizes the effects of thyroxine in the body. *Experimental group B* animals have been fed rat chow containing dried thyroid tissue, which contains thyroxine. The rate of oxygen consumption (an indirect means of determining metabolic rate) in the animals of the three groups will be measured and compared to investigate the effects of hyperthyroid, hypothyroid, and euthyroid conditions.

Oxygen consumption will be measured with a simple respirometer-manometer apparatus. Each animal will be placed in a closed chamber, containing soda lime. As carbon dioxide is evolved and expired, it will be absorbed by the soda lime; therefore, pressure changes observed will indicate the volume of oxygen consumed by the animal during the testing interval. Students will work in groups of 3 to 4 to assemble the apparatus, make preliminary weight measurements on the animals, and record the data.

Preparation of the Respirator-Manometer Apparatus

1. Obtain a dessicator, a two-hole rubber stopper, 20-ml glass syringe, hardware cloth square, T-valve, scissors, rubber tubing, two short pieces of glass tubing, and petrolatum, and bring them to your laboratory bench. The apparatus will be assembled as illustrated in Figure 28.1.

2. Shake soda lime into the bottom of the dessicator to thoroughly cover the glass bottom, then place the hardware cloth on the ledge of the dessicator above the soda lime. The hardware cloth should be well above the soda lime, so that the animal will not be able to touch it. Soda lime is quite caustic and can cause chemical burns.

3. Lubricate the ends of the two pieces of glass tubing with petrolatum, and twist them into the holes in the rubber stopper until their distal ends can be seen protruding from the opposite side. *Do not plug the tubing with petrolatum.* Place the stopper into the dessicator cover, and set the cover on the dessicator temporarily.

4. Cut off a short (3-in.) piece of rubber tubing, and attach it to the top of one piece of glass tubing extending from the stopper. Cut and attach a 12–14 in. piece of rubber tubing to the other glass tubing. Insert the T-valve stem into the distal end of the longer tubing length.

5. Cut another short piece of rubber tubing; attach one end to the T-valve and the other to the nib of the 20-ml syringe. Remove the plunger of the syringe and grease its sides generously with petrolatum. Insert the plunger back into the syringe barrel and work it up and down to evenly disperse the petrolatum on the inner walls of the syringe, then pull the plunger out to the 20-ml marking.

6. Cut a piece of rubber tubing long enough to reach from the third arm of the T-valve to one arm of the manometer. (The manometer should be partially filled with water so that a U-shaped water column is seen.) Attach the tubing to the T-valve and the manometer arm.

7. Remove the dessicator cover and generously grease its bottom edge with petrolatum. Place the cover back on the dessicator and firmly move it from side to side to spread the lubricant evenly.

8. Test the system for leaks as follows: firmly clamp the *short* length of rubber tubing extending from the stopper. Now gently push in on the plunger of the syringe. If the system is properly sealed, the fluid in the manometer will move away from the rubber tubing attached to its arm. If there is an air leak, the manometer fluid level will not change, or it will change and then return to its original level. If either of these events occurs, check all glass to glass or glass to rubber tubing connections. Smear additional petrolatum on suspect areas and test again. The apparatus must be airtight before experimentation can begin.

9. After ensuring that there are no leaks in the system, unclamp the short rubber tubing, and remove the dessicator cover.

Preparation of the Animal

1. Obtain one of the animals as directed by your instructor. Handling it gently, weigh the animal to the nearest 0.1 g on the animal balance.
2. Place the animal carefully on the hardware cloth in the dessicator. The objective is to measure oxygen usage at basal levels; so you do not want to prod the rat into high levels of activity, which would produce errors in your measurements.
3. Record the animal's group (control, experimental group A or B) and its weight in kilograms (that is, weight in grams/1000) on the data sheet on p. 248.

Equilibration of the Chamber

1. Place the lid on the dessicator, and move it slightly from side to side to seal it firmly.
2. Leave the short tubing unclamped for 7 to 10 minutes to allow for temperature equilibration in the chamber. (Since the animal's body heat will warm the air in the container, the air will expand initially. This must be allowed to occur before any measurements of oxygen consumption are taken. Otherwise, it would appear that the animal is evolving oxygen rather than consuming it.)

Determination of Oxygen Consumption of the Animal

1. Once again clamp the short rubber tubing extending from the dessicator lid stopper.
2. Check the manometer to make sure that the fluid level is the same in both arms. (If not, manipulate the syringe plunger to make the

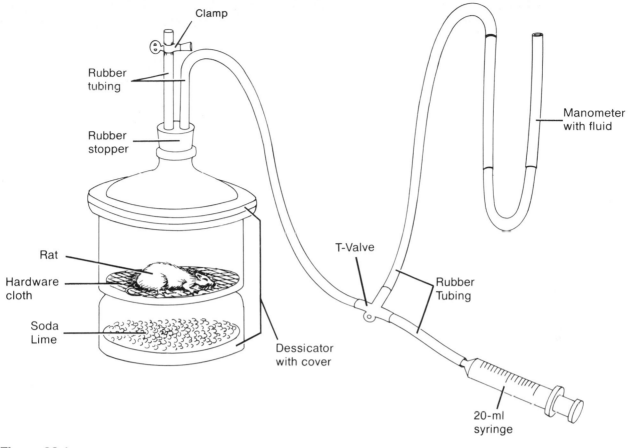

Figure 28.1

Respirator-manometer apparatus.

fluid levels even.) Record the time and the position of the bottom of the plunger (use ml marking) in the syringe.

3. Observe the manometer fluid levels at 1-minute intervals. Each time make the necessary adjustment to bring the fluid levels even in the manometer by carefully pushing the syringe plunger further into the barrel. Determine the amount of oxygen used per minute intervals by computing the difference in air volumes within the syringe. For example, if after the first minute, you push the plunger from the 20- to the 17-ml marking, the oxygen consumption is 3 ml/min. Then, if the plunger is pushed from 17 ml to 15 ml at the second minute reading, the oxygen usage during minute 2 would be 2 ml.

4. Continue taking readings (and recording oxygen consumption per minute intervals on the data sheet) for 10 consecutive minutes or until the syringe plunger has been pushed nearly to the 0 ml mark. Then unclamp the short rubber tubing, remove the dessicator cover, and allow the apparatus to stand open for 2 to 3 minutes to flush out the stale air.

5. Repeat the recording procedures for another 10-minute interval. *Make sure that you equilibrate the temperature within the chamber before beginning this second recording series.*

6. After you have recorded the animal's oxygen consumption for 2 10-minute intervals, unclamp the short rubber tubing, remove the dessicator lid, and carefully return the rat to its cage.

Computation of Metabolic Rate

Metabolic rate calculations are generally reported in terms of $KCal/m^2/hr$ and require that corrections be made to present the data in terms of standardized pressure and temperature conditions. These more complex calculations will not be used here, since the object is simply to arrive at some generalized conclusions concerning the effect of thyroxine on metabolic rate.

STUDENT DATA SHEET

Animal used from group: _____

14-day prior treatment of animal: _____

Body weight in grams: _____ /1000 = body weight in kg: _____

O_2 consumption/min: Test 1 O_2 consumption/min: Test 2

     Syringe reading _____ ml   Syringe reading _____ ml

min 1 _____   min 1 _____

min 2 _____   min 2 _____

min 3 _____   min 3 _____

min 4 _____   min 4 _____

min 5 _____   min 5 _____

min 6 _____   min 6 _____

min 7 _____   min 7 _____

min 8 _____   min 8 _____

min 9 _____   min 9 _____

min 10 _____   min 10 _____

_____ Total O_2/10 min   _____Total O_2/10 min

Average ml O_2 consumed/10 min: _____

Milliters O_2 consumed/hr: _____

Metabolic rate: _____ ml O_2/kg/hr

Class results averaged:

   Metabolic rate of control animals: _____ ml O_2/kg/hr

   Metabolic rate of experimental group A animals

   (propylthiouracil-treated): _____ ml O_2/kg/hr

   Metabolic rate of experimental group B animals

   (dessicated thyroid-treated): _____ ml O_2/kg/hr

1. Obtain the average figure for milliliters of oxygen consumed per 10-minute intervals by adding up the minute interval consumption figures for each 10-minute testing series and dividing the total by 2.

$$\frac{\text{Total ml } O_2 \text{ test series } 1 \ + \ \text{total ml } O_2 \text{ test series } 2}{2}$$

2. Determine oxygen consumption per hour using the following formula, and record the figure on the data sheet:

$$\frac{\text{Average ml } O_2 \text{ consumed}}{10 \text{ min}} \ \text{x} \ \frac{60 \text{ min}}{\text{hr}} = \text{ml } O_2/\text{hr}$$

3. To determine the metabolic rate in milliliters of oxygen consumed per kilogram of body weight per hour so that the results of all experiments can be compared, divide the figure just obtained in procedure 2 by the animal's weight in kilograms.

$$\text{Metabolic rate} = \frac{\text{ml } O_2/\text{hr}}{\text{wt in kg}} = \underline{\hspace{1cm}} \text{ml } O_2/\text{kg}/\text{hr}$$

Record the metabolic rate on the data sheet and also in the appropriate space on the chart on the chalkboard.

4. Once all groups have recorded their final metabolic rate figures on the chalkboard, average the results of each animal grouping to obtain the mean for each experimental group. Also record this information on your data sheet.

UNIT 11

THE CIRCULATORY SYSTEM

Blood

EXERCISE
29

OBJECTIVES

1. To name the two major components of blood and state their average percentages in whole blood.

2. To cite the composition of plasma and state the functional importance of many of its constituents.

3. To define *formed elements* and list the cell types composing them, cite their relative percentages, and describe their major functions.

4. To identify red blood cells, basophils, eosinophils, monocytes, lymphocytes, and neutrophils when provided with a microscopic preparation or appropriate diagram.

5. To conduct the following blood test determinations in the laboratory, state their norms, and the importance of each.

   hematocrit
   hemoglobin determination
   clotting time
   sedimentation rate
   differential white blood cell count
   total white blood cell count
   total red blood cell count
   ABO and Rh blood typing

6. To discuss the reason for transfusion reactions resulting from the administration of mismatched blood.

7. To define *anemia, polycythemia, leukopenia,* and *leukocytosis* and to cite a possible reason for each condition.

MATERIALS

Compound microscope
Immersion oil
Clean microscope slides
Lancets
Alcohol swabs
Wright's stain in dropper bottle
Distilled water in dropper bottle
Three-dimensional model of blood cells (if available)

Plasma (obtained from a clinical agency or prepared by centrifuging whole blood)
Wide-range pH paper
Test tubes
Test tube racks

Since many blood tests are to be conducted in this exercise, it seems advisable to set up a number of appropriately labeled supply areas for the various tests. These are designated here.

General supply area:
Heparinized blood (obtained from a clinical agency if desired by the instructor)
Blood lancets
Alcohol swabs
Absorbent cotton balls
Millimeter ruler
Heparinized capillary tubes
Pipette cleaning solutions—(1) 10% household bleach solution, (2) distilled water, (3) 70% ethyl alcohol, (4) acetone

Blood cell count supply area:
Hemacytometer
RBC and WBC dilution fluids (Hayem's solution for RBCs; 1% acetic acid for WBCs)
RBC and WBC dilution pipettes and tubing
Mechanical hand counters

Hematocrit supply area:
Heparinized capillary tubes
Microhematocrit centrifuge and reading gauge (if the reading gauge is not available, millimeter ruler may be used)
Seal-ease (Clay Adams Co.) or modeling clay

Hemoglobin determination supply area:
Tallquist hemoglobin scales and test paper
Sahli hemoglobin determination kit (or another type of hemoglobinometer such as Spencer or American Optical)
1% HCl
Slender glass stirring rod

Sedimentation rate supply area:
Landau Sed-rate pipettes with tubing and rack
Wide-mouthed bottle of 5% sodium citrate
Mechanical suction device

Coagulation time supply area:
Capillary tubes (nonheparinized)
Triangular file

Blood typing supply area:
Blood typing sera (anti-A, anti-B, and anti-Rh [D])
Rh typing box
Wax marker
Toothpicks
Clean microscope slides

COMPOSITION OF BLOOD

The blood circulating to and from the body cells within the vessels of the vascular system is a rather viscous substance that varies from bright scarlet to a dull brick red, depending on the amount of oxygen it is carrying. The circulatory system of the average adult contains about 5.5 liters of blood.

Blood is classified as a type of connective tissue, since it is composed of a nonliving fluid matrix (the **plasma**), in which the living cells (**formed elements**) are suspended. The fibers typical of a connective tissue matrix become visible in blood only when clotting occurs. They then appear as fibrin threads, which form the structural basis for clot formation.

Over 100 different substances are dissolved or suspended in the plasma, which is over 90% water. These include nutrients, gases, hormones, various wastes and metabolites, many types of functional proteins, and mineral salts. The composition of plasma varies continuously as cells remove or add substances to the blood.

Three types of formed elements are present in blood. The most numerous are the **erythrocytes,** or red blood cells (RBCs), which are literally sacs of hemoglobin molecules that transport the bulk of the oxygen carried in the blood (and a small percentage of the carbon dioxide). **Leukocytes,** or white blood cells (WBCs), are part of the body's immunity system, and the **thrombocytes,** or platelets, function in hemostasis (blood clot formation). Formed elements constitute 40% to 45% of whole blood, plasma the remaining 55% to 60%.

Physical Characteristics of Plasma

Go to the supply area and pour a few milliliters of plasma into a test tube. Return to your laboratory bench to make the following simple observations.

pH OF PLASMA   Test the pH of the plasma with wide-range pH paper. Record the pH observed.

COLOR AND CLARITY OF PLASMA   Hold the test tube up to a source of natural light and note its color and degree of transparency. Is it clear, translucent, or opaque?

Color _____

Degree of transparency _____

CONSISTENCY Dip your finger and thumb into the plasma and then press them firmly together for a few seconds. Gently pull them apart. How would you describe the consistency of plasma? Slippery, watery, sticky, or granular?

Record your observations. _____

_____

Formed Elements of Blood

Conduct your observations of blood cells on a slide prepared from your own blood.

1. Obtain two glass slides, dropper bottles of Wright's stain and distilled water, two or three lancets, cotton balls, and alcohol swabs. Bring this equipment to the laboratory bench.

2. Clean the slides thoroughly and dry. Open the alcohol swab packet and scrub your third or fourth finger with the swab. (Because the pricked finger may be a little sore later, it is better to prepare a finger on the hand used less often.) Allow the alcohol to dry completely, and then open the lancet packet and grasp the lancet by its blunt end. Quickly jab the pointed end into the prepared finger to produce a free flow of blood. It is *not* a good idea to squeeze or "milk" the finger, as this forces out tissue fluid as well as blood. If the blood is not flowing freely, another puncture should be made. <u>Under no circumstances is a lancet to be used for more than one puncture.</u> Dispose of the lancets in proper containers *immediately* after use.

3. Wipe away the first drop of blood with a cotton ball, and allow another large drop of blood to form. Touch the blood to one of the cleaned slides approximately ½ inch from the end. Then quickly (to prevent clotting) use the second slide to form a blood smear as shown in Figure 29.1. The blood smear when properly prepared is uniformly thin. If it appears streaked, the blood probably began to clot or coagulate before the smear was made, and another slide should be prepared.

4. Dry the slide by waving it in the air. When it is completely dry, it will look dull. Place it on a paper towel, and flood it with Wright's stain. Count the number of drops of stain used. Allow the stain to remain on the slide for 3 or 4 minutes and then flood the slide with an equal number of drops of distilled water. Allow the water and Wright's stain mixture to remain on the slide for 4 or 5 minutes or until a metallic green film or scum is apparent on the fluid surface. Blow on the slide gently every minute or so to keep the water and stain mixed during this interval.

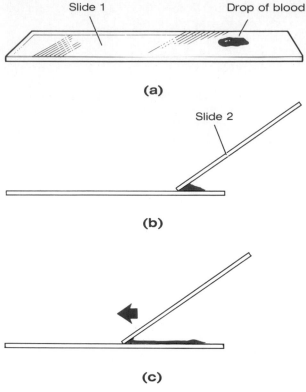

(a)

(b)

(c)

Figure 29.1

Procedure for making a blood smear: (a) place drop of blood on slide 1 approximately 1/2 inch from one end; (b) hold slide 2 at a 30° to 45° angle to slide 1 (it should touch the drop of blood); allow blood to spread along entire bottom edge of angled slide; (c) smoothly advance slide 2 to end of slide 1 (blood should run out before reaching the end of slide 1).

5. Rinse the slide with a stream of distilled water. Then flood it with distilled water, and allow it to lie flat until the slide becomes translucent and takes on a pink cast. Then stand the slide on its long edge on the paper towel, and allow it to dry completely.

6. Once the slide is dry, you can begin your observations. Obtain a microscope and scan the slide under low power to find the most evenly stained area. Use the oil immersion lens to identify the formed elements. Read the following descriptions of cell types, find each one on Figure 29.2, and color the diagram to correspond to the description. Then observe the slide carefully to identify each cell type.

ERYTHROCYTES Erythrocytes, or red blood cells, which range in size from 5 to 10 μm in diameter (averaging 7.5 μm), are cells whose color varies from brick red to pale pink, depending on the effectiveness of the stain. They have a distinctive biconcave disk shape and appear paler in the center than at the edge.

Red blood cells differ from the other blood cells because they are anucleate when mature

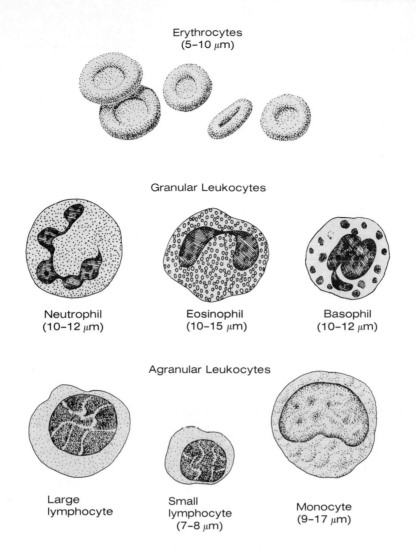

Erythrocytes
(5–10 μm)

Granular Leukocytes

Neutrophil
(10–12 μm)

Eosinophil
(10–15 μm)

Basophil
(10–12 μm)

Agranular Leukocytes

Large
lymphocyte

Small
lymphocyte
(7–8 μm)

Monocyte
(9–17 μm)

Figure 29.2

Diagrammatic representation of blood cells (relative size relationships indicated).

and circulating in the blood. As a result, they are unable to reproduce and have a limited life span of 100 to 120 days, after which they begin to fragment and are destroyed by the spleen and other reticuloendothelial tissues of the body. In various anemias, the red blood cells may appear pale (an indication of decreased hemoglobin content) or may be nucleated (an indication that the bone marrow is turning out cells prematurely).

As you observe the slide, note that the red blood cells are by far the most numerous blood cells seen in the field. Their number averages 4.5 million to 5.0 million cells per cubic millimeter of blood (for women and men, respectively).

LEUKOCYTES Leukocytes, or white blood cells, are nucleated cells and are much less numerous than the red blood cells, averaging from 5000 to 9000 cells per cubic millimeter. Basically, white blood cells are protective, pathogen-destroying cells that are transported to all parts of the body in the blood or lymph. Important to their protective function is their ability to move in and

out of blood vessels, a process called **diapedesis,** and wander through body tissues by ameboid motion to reach sites of inflammation or tissue destruction. They are classified into two major groups, depending on their formation site and whether or not they contain conspicuous granules in their cytoplasm.

Granulocytes, which are formed in the bone marrow from the same stem cell (hemocytoblast) as red blood cells, comprise the first group. The granules in their cytoplasm stain differentially with Wright's stain, and they have peculiarly lobed nuclei, which often consist of expanded nuclear regions connected by thin strands of nucleoplasm. Other information about the three types of granulocytes is noted as follows:

Neutrophil: The most abundant of the white blood cells (55% to 65% of the leukocyte population); nucleus consists of 3 to 7 lobes and the cytoplasm contains fine cytoplasmic granules, which are generally indistinguishable; function as active phagocytes whose number increases exponentially during acute infections.

Eosinophil: Represent 1% to 3% of the leukocyte population; nucleus is generally figure 8 or bilobed in shape; contain large cytoplasmic granules that stain red-orange with Wright's stain. Precise function is unknown, but seen to increase during allergy and parasite infections and may selectively phagocytize antigen-antibody complexes.

Basophil: Least abundant leukocyte type representing less than 1% of the population; large U- or S-shaped nucleus with two or more indentations. Cytoplasm contains coarse sparse granules that stain deep purple with Wright's stain. The granules are believed to contain histamine and heparin, which are discharged on exposure to antigens and help to mediate the inflammatory response.

The second group, **agranulocytes,** or **agranular leukocytes,** contain no observable cytoplasmic granules. They arise from stem cells in bone marrow but tend to reside in lymphoid tissue. They exist in two forms. Their nuclei tend to be closer to the norm, that is, spherical, oval, or kidney-shaped. Specific characteristics of these cells are listed below.

Lymphocyte: The smallest of the leukocytes, approximately the size of a red blood cell. The dark-staining nucleus is generally spherical or slightly indented. Sparse cytoplasm appears as a thin rim around the nucleus. Concerned with immunologic responses in body; one population, B-lymphocytes, produces antibodies that are released to blood; second population, T-lymphocytes, plays surveillance role and destroys grafts, tumors, and virus-infected cells. Activate B-lymphocytes. Represent 20% to 35% of the WBC population.

Monocyte: The largest of the leukocytes; approximately twice the size of red blood cells. Represent 3% to 8% of the leukocyte population. Dark blue nucleus is generally kidney-shaped; abundant cytoplasm stains grey-blue. Function as active phagocytes (the "long-term cleanup team"), which increase dramatically in number during chronic infections such as tuberculosis.

THROMBOCYTES Thrombocytes, or platelets, are cell fragments of large multinucleate cells (**megakaryocytes**) formed in the bone marrow. They appear as darkly staining, irregularly shaped bodies interspersed among the other blood cells. The normal platelet count in blood is 500,000 per cubic millimeter. Platelets are instrumental in the clotting process that occurs in plasma when blood vessels are ruptured.

After you have identified these cell types on your slide, observe three-dimensional models of blood cells if these are available. *Do not dispose of your slide,* as it will be used later for the differential white blood cell count.

HEMATOLOGIC TESTS

When a person enters a hospital as a patient, several hematologic studies are routinely done to determine general level of health as well as the presence of pathologic conditions. You will be conducting the most common of these studies in this exercise.

Materials such as cotton balls, lancets, and alcohol swabs are used in nearly all of the following diagnostic tests. These supplies are at the general supply area and should be properly disposed of immediately after use. Other necessary supplies and equipment are at specific supply areas marked according to the test with which they are used. Since nearly all of the tests require a finger stab to obtain blood, it might be wise to quickly read through the various tests to determine in which instances more than one preparation can be done from the same finger stab. For example, the hematocrit capillary tubes and sedimentation rate samples might be prepared at the same time the chamber is being prepared for the total blood cell counts. A little preplanning will save you the discomfort of a multiply punctured finger.

An alternative to using blood obtained from the finger stab technique is using heparinized blood samples supplied by your instructor. The purpose of using heparinized tubes is to prevent the blood from clotting. Thus blood collected and stored in such tubes will be suitable for all tests except coagulation time testing.

Differential White Blood Cell Count

To make a differential white blood cell count, 100 WBCs are counted and classified according to type. Such a count is routine in a physical examination and in sickness, since any abnormality or significant elevation in the normal percentages of the specific types of WBCs may indicate the possible source of pathology. Use the slide prepared for the identification of the blood cells (pp. 252–253) for the count.

1. Begin at the edge of the smear and move the slide in a systematic manner on the microscope stage—either up and down or from side to side as indicated in Figure 29.3.
2. Record each type of white blood cell you observe by making a count on the chart on p. 255 (for example, ***卌 ǁ*** = 7 cells) until you have observed and recorded a total of 100 WBCs. Compare the percentage of each (number observed ÷ 100), and record the percentages on the data sheet at the end of this exercise.

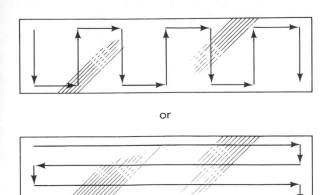

Figure 29.3

Alternative methods of moving the slide for a differential WBC count.

Cell type	Number observed
Neutrophils	
Eosinophils	
Basophils	
Lymphocytes	
Monocytes	

How does your differential white blood cell count correlate with the percentages given for each type

on pp. 252–254? _____

_____

Total White and Red Blood Cell Counts

To conduct a total or direct WBC or RBC count, dilute a known volume of blood with a fluid that prevents blood coagulation and place the mixture into a counting chamber of known volume—a hemacytometer. The cells are then counted by microscopic inspection. Make corrections for dilution and chamber volume to obtain the result in cells per cubic millimeter. The white and red blood cell diluents differ. They are isotonic for the cell type to be counted and cause dissolution of the cell types not being counted. In other words, if the WBC diluent is used, the RBCs are hemolyzed and thus do not interfere with the WBC counting process. Thus, the diluents must not be intermixed or used haphazardly. Fig-

ure 29.4 depicts the apparatus and procedure used for the counts. (Note that this technique is rather outdated, since most clinical agencies now have computerized equipment for performing blood counts.)

TOTAL WHITE BLOOD CELL COUNT Since white blood cells are an important part of the body's defense system, it is essential to note any abnormalities in them. **Leukocytosis,** an abnormally high WBC count, may indicate bacterial or viral infection, metabolic disease, hemorrhage, or poisoning by drugs or chemicals. A decrease in the white cell number (**leukopenia**) below 5000/mm^3 may indicate typhoid fever, measles, infectious hepatitis or cirrhosis, tuberculosis, or excessive antibiotic or x-ray therapy. A person with leukopenia lacks the usual protective mechanisms.

1.   Obtain a hemacytometer and cover glass, WBC diluent, WBC diluting pipette (white bead in bulb) and tubing, cotton balls, lancets, alcohol swabs, and a hand counter.

2.   Cleanse your finger thoroughly, and obtain a second drop of blood as described on page 252.

3.   Attach the tubing (with attached mouthpiece) to the pipette, place the tip of the pipette in the blood drop, and fill the WBC pipette exactly to the 0.5 ml mark by drawing air through the pipette via the mouthpiece. _Hold the pipette horizontally during the filling process._ Wipe the tip of the pipette with a cotton ball and then immerse it into WBC diluent and draw it up into the pipette until the mixture is exactly at the 11.0 mark. Rotate the pipette as the fluid is being aspirated to ensure proper mixing. The dilution of the blood is now 1:20.

4.   Still holding the pipette in the horizontal position, remove the tubing from the pipette with a gentle rocking motion. Place your thumb at one end and the second finger at the opposite end of the pipette. Mix the pipette contents for 2 to 3 minutes, moving the pipette in a back and forth arclike movement approximately parallel with the bench top.

5.   Place a cover slip on the hemacytometer, and charge it. Discard the first 2 to 3 drops—which is only diluent—from the pipette onto a paper towel. When your finger is on the distal end of the pipette, the mixture will remain in the pipette. When you remove your finger, the mixture will be discharged. Place 1 drop of the mixture at each end of the cover slip at the junction of the hemacytometer and cover slip. _Do this quickly;_ otherwise the hemacytometer will overfill, and the mixture will flow under the cover slip by capillary action. You now have two charged chambers. If the chambers are properly charged, no fluid will appear in the moat.

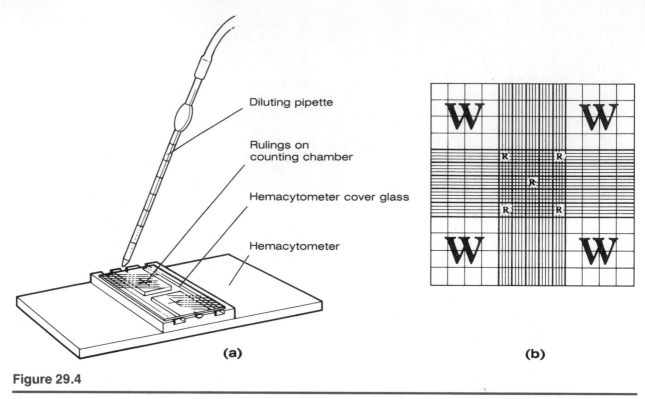

(a)   **(b)**

Figure 29.4

Apparatus and preparation for blood cell counts: (a) procedure for introducing diluted blood to the hemacytometer; (b) areas of the hemacytometer used for red and white blood cell counts.

6.   Place the hemacytometer on the microscope stage and allow it to remain undisturbed for 2 to 3 minutes to allow the cells to become evenly distributed.

7.   Bring the grid lines into focus under low power (10× objective lens) and check for uniformity of white blood cell distribution. If it is uneven, switch to the other chamber for observation.

8.   Move the slide into position so that one of the W areas can be seen (see Figure 29.4), and proceed to count all WBCs observed in the four corner W areas. Use a hand counter to facilitate the counting process. *To prevent overcounts of cells at boundary lines, count only those that touch the left and upper boundary lines of the W area but not those touching the right and lower boundaries.* Record the number of WBCs counted and multiply this number by 50 to obtain the number of WBCs per cubic milli meter.*

No. WBCs: _____ WBC/mm^3: _____

---

*The factor of 50 is obtained by multiplying the volume dilution factor (20) by the volume correction factor (2.5): 20 × 2.5 = 50. The 2.5 volume correction factor is obtained in the following manner: Each W area on the grid is exactly 1 mm^2 × 0.1 mm deep; therefore, the volume of each W section is 0.1 mm^3. Since WBCs in four W areas are counted (a total of 0.4 mm^3), this volume must be multiplied by 2.5 (correction factor) to obtain the number of cells in 1 mm^3.

Record this information on the data sheet on page 261. How does your result compare with the

norms for a white blood cell count? _____

9.   Clean the pipette by drawing up the bleach, distilled water, alcohol, and acetone in succession. *If you use mouth suction, watch the fluid column rise to prevent aspiration of the cleaning solutions.* Discharge the contents each time before drawing up the next solution. Return the cleaned pipette to the supply area. Clean the hemacytometer and cover slip in preparation for the red blood cell count.

TOTAL RED BLOOD CELL COUNT   The red blood cell count, like the white blood cell count, determines the total number of this cell type per unit volume of blood. Since RBCs are absolutely necessary for oxygen transport, the doctor should investigate any excessive change in their number immediately. An increase in the number of RBCs (**polycythemia**) may result from bone marrow cancer or from living at high altitudes where less oxygen is available. A decrease in the number of RBCs results in anemia. (The term *anemia* simply indicates a decreased oxygen-carrying capacity of blood that may result from a decrease in RBC number or size or a decreased hemoglobin content of the RBCs.) A decrease in RBCs may result suddenly from hemorrhage or more gradually from conditions that increase RBC destruction or decrease RBC production.

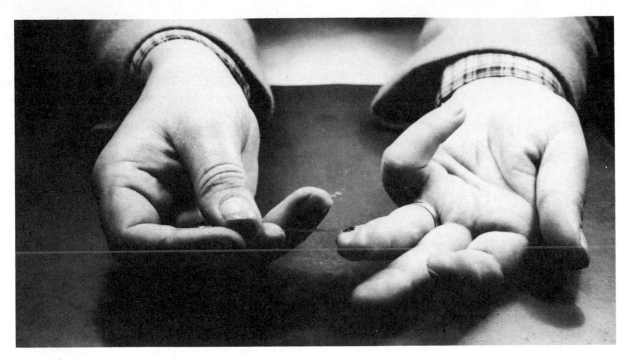

Figure 29.5

Filling a capillary tube with blood. (Photo courtesy of Ann Allworth.)

 1.   Obtain a RBC diluting pipette (red bead in bulb), RBC diluent, and other supplies as required for the WBC count. Note that the RBC pipette has a much larger bulb. Follow the same procedure for obtaining blood and charging the chamber as for the WBC count, except use Hayem's solution diluent, and draw it up to the 101 mark after 0.5 ml of blood have been aspirated. This provides a dilution of 1:200.

2.   Allow the hemacytometer to rest on the microscope stage as before. Using the h.p. lens, count all RBCs in the R-marked areas (see Figure 29.4). Again, at the boundary lines, count only those cells touching the left and upper lines.

3.   Record the number of cells counted, and multiply that number by 10,000 to compute the RBC/mm³. (Note: the dilution factor (200) × volume correction factor (50) = 10,000).

No. RBCs: _____   RBCs/mm³: _____

How does your result compare with normal RBC

values? _____

Hematocrit

The hematocrit, or packed cell volume (PCV), is routinely determined when anemia is suspected.

Centrifuging whole blood causes the formed elements to spin to the bottom of the tube, with plasma forming the top layer. Since the blood cell population is primarily RBCs, the PCV is generally considered equivalent to the RBC volume, and this is the only value reported. However, the relative percentage of WBCs can be differentiated, and both WBC and plasma volume are reported here. Normal PCV values for the male and female, respectively, are 47.0 ± 7 and 42.0 ± 5.

 The PCV is determined by the micromethod, so only a drop of blood is needed. If possible, all members of the class should prepare their capillary tubes at the same time so the centrifuge can be properly balanced and run only once.

1.   Obtain two heparinized capillary tubes. Sealease or modeling clay, a lancet, alcohol swabs, and some cotton balls.

2.   Cleanse the finger, and allow the blood to flow freely. Wipe away the first few drops and, holding the red-line-marked end of the capillary tube to the blood drop, allow the tube to fill at least three-fourths full by capillary action (Figure 29.5). If the blood is not flowing freely, the end of the capillary tube will not be completely submerged in the blood during filling, air will enter, and you will have to make another sample.

3. Plug the blood-containing end by pressing it into the Seal-ease or clay. Prepare a second tube in the same manner.

4. Place the prepared tubes opposite one another in the radial grooves of the centrifuge with the sealed ends abutting the rubber gasket at the centrifuge periphery. This loading procedure balances the centrifuge and prevents blood from spraying everywhere by centrifugal force. Make a note of the numbers of the grooves your tubes are in. When all the tubes have been loaded, make sure the centrifuge is properly balanced, and secure the centrifuge cover. Turn the centrifuge on, and set the timer for 4 or 5 minutes.

5. Determine the percentage of RBCs, WBCs, and plasma by using the microhematocrit reader. The RBCs are the bottom layer, the plasma is the top layer, and the WBCs are the buff-colored layer between the two. If the reader is not available, use a millimeter ruler to measure the length of the filled capillary tube occupied by each element, and compute its percentage by using the following formula:

$$\frac{\text{Height of the column composed by the element (mm)}}{\text{Height of the original column of whole blood (mm)}} \times 100$$

Record your calculations below and on the data sheet.

% RBC _____ % WBC _____ % plasma _____

Usually WBCs constitute 1% of the total blood volume. How do your blood values compare to this figure and to the normal percentages for RBCs and plasma? (See page 251.)

_____

Generally, the PCV is considered a more accurate test for determining the RBC composition of the blood than the total RBC count, and a RBC hematocrit within the normal range is generally indicative of a normal RBC number.

Hemoglobin Concentration Determination

As noted earlier, a person can be anemic even with a normal RBC count. Since hemoglobin is the RBC protein responsible for oxygen transport, perhaps the most accurate way of measuring the oxygen-carrying capacity of the blood is to determine its hemoglobin content. Oxygen, which combines reversibly with the heme (iron-containing portion) of the hemoglobin molecule, is picked up by the blood cells in the lungs and unloaded in the tissues. Thus, the more hemoglobin molecules the RBCs contain, the more oxygen they will be able to transport. Normal blood contains 12 to 16 g hemoglobin per 100 ml blood. Hemoglobin content in men is slightly higher (13 to 16 g) than in women (12 to 14 g).

Several techniques have been developed to estimate the hemoglobin content of blood, ranging from the old, rather inaccurate Tallquist method to expensive colorimeters, which are precisely calibrated and yield highly accurate results. Both the Tallquist method and a procedure using the Sahli-Adams hemoglobinometer are used in the hemoglobin determinations here, so you can compare their results. (If another type of hemoglobinometer is available, it may be used in lieu of the Sahli equipment. Simply follow the instructions provided with the instrument, and record your results noting the equipment used.)

Obtain a Tallquist hemoglobin scale, a Sahli hemoglobin determination kit, lancets, alcohol swabs, cotton balls, slender glass stirring rod, dropper bottle of 1% HCl, and a dropper bottle of distilled water. Read through the directions for both procedures before beginning. Obtain blood for both tests from the same finger prick.

TALLQUIST METHOD   In the Tallquist method, you will compare the color of drawn blood with a color chart.

1. Prepare the finger as previously described. Make sure the alcohol evaporates before puncturing for best results. Place 1 good-sized drop of blood on the special absorbent paper provided with the color chart. The blood stain should be larger than the holes on the color chart.

2. As soon as the blood has dried and loses its glossy appearance, match its color under natural light with the color standards by moving the specimen under the comparison chart until the blood stain appears at the various apertures. (The blood should not be allowed to dry to a brown color, as this will result in an inaccurate reading.) Since the colors on the chart represent 10% variations in hemoglobin content, it may be necessary to estimate the percentage if the color of your blood sample is intermediate between two color standards.

3. Record your results in the percentage of hemoglobin concentration and in grams per 100 ml blood here and on the data sheet.

% hemoglobin concentration _____

g/100 ml _____

SAHLI-ADAMS METHOD   In the Sahli-Adams method, hemoglobin is converted to a brown acid (hematin) by mixing the blood with hydrochloric acid (HCl). This technique provides a more accurate estimation of the hemoglobin content because the hemoglobin is released from the blood cells.

1. Fill the graduated tube to the 2 g mark with 1% HCl. Produce a free flow of blood from a finger stab, and draw the blood into the pipette to the 20 mm^3 marking.

2. Wipe the tip of the pipette with a cotton ball, and blow the blood from the pipette into the acid solution in the tube.

3. Draw the mixture up into the pipette, and blow it back into the tube gently two or three times to ensure proper mixing.

4. Place the tube in the hemoglobinometer and turn it so that the graduated scale will not interfere with your color observations later.

5. Allow the mixture to stand for 10 minutes and then compare the color of your test sample with the color standard. Be sure you have a good light source behind the color standard before attempting to match the colors.

6. If the test sample is darker than the standard, add distilled water, a drop at a time, mixing with a glass stirring rod after each addition, until the colors of the test sample and the standard are identical. When the colors appear identical to you, read the two scales on the test tube to record the percentage of hemoglobin and the grams of hemoglobin per 100 ml blood. Record here and on the data sheet.

% hemoglobin concentration _____

g/100 ml _____

Are your values within the normal range? _____

How closely do these values compare with those obtained by using the Tallquist method? _____

7. Clean the pipette with the cleaning solutions before returning it and the other equipment to the supply area.

Generally speaking, the relationship between the PCV and grams of hemoglobin per 100 ml blood is 3:1. How do your values compare?

_____

Record this value (obtained from your data) on the data sheet.

Blood Indices and Corpuscular (or Blood) Constants

Most health agencies perform tests to determine absolute values of red blood cell count, hematocrit, and hemoglobin concentration, and also re-port the interrelationships in terms of ratios or indices. To be meaningful the indices must be calculated on very accurate RBC counts and hematocrit and hemoglobin determinations. When this is so, the indices aid in the diagnosis of the various types of anemias, allowing them to be classified as macrocytic, normocytic, or micro-cytic, relative to the size of the RBCs, and hypo-chromic or hyperchromic, according to the amount of hemoglobin contained in the RBCs. These indices will not be computed here, but are defined as follows:

Mean corpuscular volume (**MCV**): the volume of the average RBC in a given sample of blood. The normal volume range is 80–94μ^3. Higher values indicate enlarged (macrocytic) RBCs, whereas lower values indicate smaller than normal (microcytic) RBCs.

Mean corpuscular hemoglobin (**MCH**): the amount of hemoglobin (by weight) in the average RBC in a given sample of blood. Normal value is 27–32 $\mu\mu g$.

Mean corpuscular hemoglobin concentration (**MCHC**): the weight to volume ratio of hemo-globin in the average RBC in a given sam-ple of blood. Normal value is 33%–38%. The MCHC relates the blood samples to norms of hemoglobin content relative to RBC size. RBCs that have lower than normal MCHC values are pale (hypochromic) and considered to be hemoglobin-deficient.

Sedimentation Rate

The speed at which red blood cells settle to the bottom of a vertical tube when allowed to stand is called the sedimentation rate. The normal rate for adults is 0 to 6 mm/hr (averaging 3 mm/hr) and for children 0 to 8 mm/hr (averaging 4 mm/hr). Sedimentation of RBCs apparently proceeds in three stages: rouleaux formation, rapid settling, and final packing. Rouleaux forma-tion (alignment of RBCs like a stack of pennies) does not occur with abnormally shaped red blood cells (as in sickle cell anemia); therefore, the sedimentation rate is decreased. The size and number of RBCs affect the packing phase. In anemia the sedimentation rate increases; in polycythemia the rate decreases. The sedimenta-tion rate is greater than normal during men-ses and pregnancy, and very high sedimentation rates may be indicative of infectious conditions or tissue destruction occurring somewhere in the body. Although this test is nonspecific, it alerts the diagnostician to the need for further tests to pinpoint the site of pathology. The Landau micromethod, which uses just one drop of blood, is used here.

1. Obtain lancets, cotton balls, alcohol swabs, and Landau Sed-rate pipette and tubing, the Landau rack, a wide-mouthed bottle of 5% sodium citrate, a mechanical suction device, and a milli-meter ruler.

2. Use the mechanical suction device to draw up the sodium citrate to the first (most distal) marking encircling the pipette. Prepare the finger for puncture, and produce a free flow of blood.

3. Wipe off the first drop, and then draw the blood into the pipette until the mixture reaches the second encircling line. Keep the pipette tip immersed in the blood to avoid air bubbles.

4. Thoroughly mix the blood with the citrate (an anticoagulant) by drawing the mixture into the bulb and then forcing it back down into the lumen. Repeat this mixing procedure six times, and then adjust the top level of the mixture as close to the zero marking as possible. *If any air bubbles are introduced during the mixing process, discard the sample and begin again.*

5. Seal the tip of the pipette by holding it tightly against the tip of your index finger, and then carefully remove the suction device from the upper end of the pipette.

6. Stand the pipette in an exactly vertical position on the Sed-rack with its lower end resting on the base of the rack. Record the time, and allow it to stand for exactly 1 hour.

time _____

7. After 1 hour, measure the number of millimeters of visible clear plasma (which indicates the amount of settling of the RBCs), and record this figure here and on the data sheet.

sedimentation rate _____ mm/hr

Coagulation Time

Blood clotting, or coagulation, is a protective device that minimizes blood loss when blood vessels are ruptured. This process requires the interaction of many substances normally present in the plasma (clotting factors, or procoagulents) as well as some released by platelets and injured tissues. Basically hemostasis proceeds as follows: The injured tissues and platelets release **thromboplastin,** which triggers the clotting mechanism, or cascade. Thromboplastin interacts with other blood protein clotting factors and calcium ions to convert **prothrombin** (present in the plasma) to **thrombin.** Thrombin then acts enzymatically to polymerize the soluble **fibrinogen** proteins (present in plasma) into insoluble **fibrin,** which forms a meshwork of strands that traps the RBCs and forms the basis of the clot. Normally, blood removed from the body clots within 2 to 6 minutes.

1. Obtain a *nonheparinized* capillary tube, a lancet, cotton balls, a triangular file, and alcohol swabs.

2. Clean and prick the finger to produce a free flow of blood.

3. Place one end of the capillary tube in the blood drop, and hold the opposite end at a lower level to collect the sample.

4. Lay the capillary tube on a paper towel and record the time. _____

5. At 30-sec intervals, make a small nick on the tube close to one end with the triangular file, and then carefully break the tube. Slowly separate the ends to see if a gellike thread of fibrin spans the gap. When this occurs, record the time.

Time for coagulation to occur _____

Are your results within the normal time range?

_____

Blood Typing

Blood typing is a system of blood classification based on the presence of specific proteins (combined with polysaccharides) on the outer surface of the RBC plasma membrane. Such proteins are called **antigens,** or **agglutinogens,** and are genetically determined. In many cases, these antigens are accompanied by other proteins found in the plasma, **antibodies** or **agglutinins,** which react with RBCs bearing different antigens, causing them to become clumped, agglutinated, and eventually hemolyzed. It is because of this phenomenon that a person's blood must be carefully typed before a whole blood or packed cell transfusion.

Several blood typing systems exist, based on the various possible antigens, but the factors most commonly typed for are the antigens of the ABO and Rh blood groups that are most commonly involved in transfusion reactions. Other blood factors, such as Kell, Lewis, M, and N, are not routinely typed for unless the individual is expected to require multiple transfusions. The basis of the ABO typing is shown on p. 261.

Individuals whose red blood cells carry the Rh antigen are considered to be Rh positive (approximately 85% of the U.S. population); those lacking the antigen are Rh negative. Unlike ABO blood groups neither the blood of the Rh-positive nor Rh-negative individuals carries preformed anti-Rh antibodies. This is understandable in the case of the Rh-positive individual. However, Rh-negative persons who receive transfusions of Rh-positive blood become sensitized by the Rh antigens of the donor RBCs and their systems begin to produce anti-Rh antibodies. On subsequent exposures to Rh-positive blood, typical transfusion reactions occur, resulting in the clumping and hemolysis of the donor blood cells.

1. Obtain two clean microscope slides, a wax marking pencil, anti-A, anti-B, and anti-Rh typing sera, toothpicks, lancets, alcohol swabs, and the Rh typing box.

ABO blood type	Antigens present on RBC membranes	Antibodies present in plasma	% of U.S. Cau- casian) population
A	A	Anti-B	42
B	B	Anti-A	10
AB	A and B	None	3
O	Neither	Anti-A and anti-B	45

2.   Divide slide 1 into two equal halves with the wax marking pencil. Label the lower left hand corner "anti-A" and the lower right hand corner "anti-B." Mark the bottom of slide 2 "anti-Rh."

3.   Place one drop of anti-A serum on the *left* side of slide 1. Place one drop of anti-B serum on the *right* side of slide 1. Place one drop of anti-Rh serum in the center of slide 2.

4.   Cleanse your finger with an alcohol swab, pierce the finger with a lancet, and wipe away the first drop of blood. Obtain 3 drops of freely flowing blood, placing one drop on each side of slide 1 and a drop on slide 2.

5.   Quickly mix each blood-antiserum sample with a *fresh* toothpick.

6.   Place slide 2 on the Rh typing box and rock gently back and forth. (A slightly higher temperature is required for precise Rh typing than for ABO typing.)

7.   After 2 minutes, observe all three blood samples for evidence of clumping. The agglutination that occurs in the positive test for the Rh factor is very fine and difficult to perceive; thus if there is any question, observe the slide under the microscope. Record your observations below:

	Observed (+)	Not observed (−)
Presence of clumping with anti-A		
Presence of clumping with anti-B		
Presence of clumping with anti-Rh		

8.   Interpret your results in light of the following information: Slide 1: If clumping occurs on both sides, your ABO blood group is AB. If clumping occurs only in the blood sample mixed with anti-A serum, you are ABO type A. If clumping occurs only in the blood sample mixed with anti-B serum, you

are type B. If clumping was not observed with either serum, you are type O. Slide 2: If clumping was observed, you are Rh positive. If not, you are Rh negative.

9.   Record your blood type on the data sheet.

Hematologic Test Data Sheet

Differential WBC count:

_____ % granulocytes _____ % agranulocytes

_____ % neutrophils _____ % lymphocytes

_____ % eosinophils _____ % monocytes

_____ % basophils

Total WBC count _____ WBCs/mm^3

Total RBC count _____ RBCs/mm^3

Hematocrit (PCV):

_____ RBC % of blood volume

_____ WBC % of blood volume   } not

_____ % Plasma % of blood volume   } generally reported

Hemoglobin content:

Tallquist method _____ %

_____ g/100 ml blood

Sahli-Adams method: _____ %

_____ g/100 ml blood

Ratio (PCV/grams Hb per 100 ml blood): _____

Sedimentation rate _____ mm/hr

Coagulation time _____

Blood typing:

ABO group _____   Rh factor _____

Anatomy of the Heart

EXERCISE 30

OBJECTIVES

1. To describe the location of the heart.

2. To name and locate the major anatomic areas and structures of the heart when provided with an appropriate model, diagram, or dissected sheep heart, and to explain the function of each.

3. To trace the pathway of blood through the heart.

4. To explain why the heart is called a double pump, and to compare the pulmonary and systemic circuits.

5. To explain the operation of the atrioventricular and semilunar valves.

6. To name the functional blood supply of the heart.

MATERIALS

Torso model or laboratory chart showing heart anatomy

Red and blue pencils

Preserved sheep heart, pericardial sacs intact (if possible)

Dissecting pan and instruments

Pointed glass rods for probes

Three-dimensional model of cardiac muscle

Heart model (three-dimensional)

X-ray of the human thorax for observation of the position of the heart *in situ*

When they hear the term *circulatory system,* most people immediately think of the heart. People have long recognized the vital importance of the heart, and for centuries have alluded to it in song and poetry.

From a more scientific standpoint, it is important to understand why the heart and circulatory system play such a vital role in human physiology. The major function of the circulatory system is transportation. Using blood as the transport vehicle, the system carries oxygen, digested foods, cell wastes, electrolytes, and many other substances vital to the body's homeostasis to and from the body cells. The system's propulsive force is the contracting heart, which can be compared to a muscular pump equipped with one-way valves. As the heart contracts, it forces blood into a closed system of large and small plumbing tubes (blood vessels) within which the blood is confined and circulated. This exercise deals with the structure of the heart or circulatory pump. The anatomy of the blood vessels is considered separately in Exercise 32.

GROSS ANATOMY OF THE HUMAN HEART

The heart is a cone-shaped organ approximately the size of a fist and is located within the mediastinum of the thorax. It is flanked laterally by the lungs, posteriorly by the vertebral column, and anteriorly by the sternum. Its more pointed **apex** extends slightly to the left and rests on the diaphragm, approximately at the level of the fifth intercostal space. Its broader **base,** from which the great vessels emerge, lies beneath the second rib. *In situ,* the right ventricle of the heart forms most of its anterior surface.

If an x-ray of a human thorax is available, verify the relationships described above.

Figure 30.1 shows two views of the heart—an external anterior view and a frontal section. As anatomic areas are described in the text, consult the figure. When you have pinpointed all the structures, observe the human heart model, and reidentify the same structures without reference to the figure.

The heart is enclosed within a double sac of serous membrane called pericardium. The thin **visceral pericardium,** or **epicardium,** which is closely applied to the heart muscle, reflects downward at the base of the heart to form the other, loosely applied **parietal pericardium,** which is attached at the apex to the diaphragm. Serous fluid produced by the pericardial membranes allows the heart to beat in a relatively frictionless environment. Inflammation of the pericardium, **pericarditis,** causes painful adhesions between the pericardial layers. These adhesions interfere with heart movements.

The walls of the heart are composed of cardiac muscle—the **myocardium**—which is reinforced internally by a dense fibrous connective tissue network. This network—the skeleton of the heart—is more elaborate and thicker in certain areas, for example, around the valves and at the base of the great vessels leaving the heart (see Figure 30.2).

Heart Chambers

The heart is divided into four chambers: two **atria** and two **ventricles,** each lined with a thin serous lining called the **endocardium.** The septum that divides the heart longitudinally is referred to as the **interatrial** or **interventricular septum,** depending on which chambers it partitions. The superiorly located atria are receiving chambers and are relatively ineffective as pumps. Blood flows into the atria under relatively low pressure from the veins of the body. The right atrium receives oxygen-poor blood from the body via the **superior** and **inferior vena cavae.** Four **pulmonary veins** deliver oxygen-rich blood from the lungs to the left atrium.

The inferior thick-walled ventricles, which form the bulk of the heart, are the discharging chambers, which force blood out of the heart into the large arteries that emerge from its base. The right ventricle pumps blood into the **pulmonary trunk,** which carries the blood to the lungs to be oxygenated. The left ventricle discharges blood into the **aorta,** from which all systemic arteries of the body diverge to supply the body tissues. Discussions of the heart's pumping action usually refer to ventricular activity.

Pulmonary and Systemic Circulations

The heart functions as a double pump. The right side serves as the **pulmonary circuit** pump, shunting the oxygen-poor blood entering its chambers to the lungs to unload carbon dioxide and pick up oxygen, and then back to the left side of the heart. The function of this circuit is strictly to provide for gas exchange. The second circuit, which pumps oxygen-rich blood from the left heart through the body tissues and back to the right heart is called the **systemic circulation.** It supplies the functional blood supply to all body tissues.

Trace the pathway of blood through the heart by adding arrows to the frontal section diagram (see Figure 30.1). Use red arrows for the oxygen-rich blood and blue arrows for the oxygen-poor blood.

Heart Valves

Four valves enforce a one-way blood flow through the heart chambers. The **atrioventricular valves,** located between the atrial and ventricular chambers on each side, prevent backflow into the atria when the ventricles begin to contract. The left atrioventricular valve, also called the **mitral** or **bicuspid valve,** consists of two cusps, or flaps, of endocardium. The right atrioventricular valve, the **tricuspid valve,** has three cusps (see Figure 30.2). Tiny white collagenic cords called the **chordae tendineae** (literally, heart strings) anchor the cusps to the ventricular walls. The chordae tendineae originate from small bundles of cardiac muscle, **papillary muscles,** that project from the myocardial wall. When blood is flowing passively into the atria and then into the ventricles during **diastole** (the period of ventricular relaxation), the atrioventricular valve flaps hang passively into the ventricular chambers. When the ventricles begin to contract (**systole**) and blood in their chambers is compressed, the intraventricular blood pressure begins to rise, causing the valve flaps to be reflected superiorly and close the valve. The chordae tendineae anchor the flaps in a closed position preventing backflow into the atria during ventricular contraction. If unanchored, the flaps would blow upward into the atria rather like an umbrella being turned inside out by a strong wind.

The second set of valves, the **semilunar valves,** each composed of three pocketlike cusps, guards the bases of the two large arteries leaving the ventricular chambers. These are referred to as the **pulmonary** and **aortic semilunar valves.** The valve cusps are forced open and flatten against the walls of artery as the ventricles discharge their blood into the large arteries during systole. However, when the ventricles relax, blood begins to flow backward toward the heart and the cusps fill with blood, closing the valves and preventing arterial blood from reentering the heart.

Cardiac Circulation

Even though the heart chambers are almost continually bathed with blood, this contained blood does not nourish the myocardium. The functional blood supply of the heart is provided by the right and left **coronary arteries,** which issue from the base of the aorta just above the aortic semilunar valve and encircle the heart in the **atrioventricular groove** at the junction of the atria and ventricles. They then ramify over the heart's surface, the right coronary artery supplying the

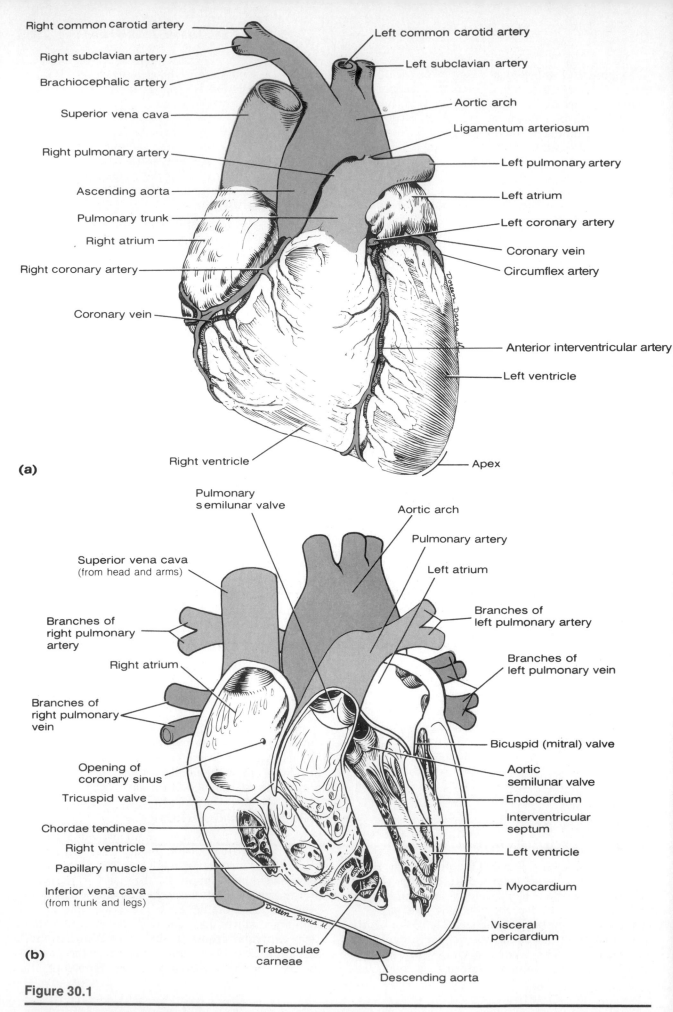

(a)

(b)

Figure 30.1

Anatomy of the human heart: (a) external anterior view; (b) frontal section.

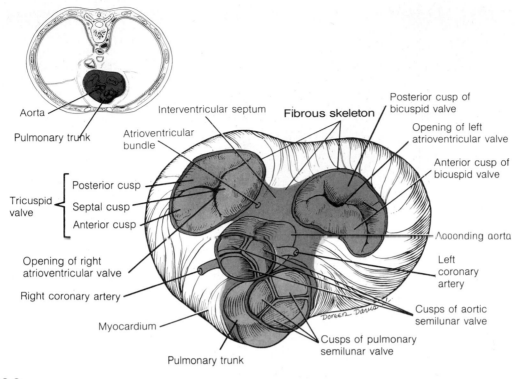

Figure 30.2

Heart valves (superior view).

posterior surface (right side) of the heart and the left coronary artery supplying its anterior and the left side of its posterior surface via its two major branches, the *anterior interventricular artery* and the *circumflex artery*. The coronary arteries and their branches are compressed during systole and fill when the heart is relaxed. The myocardium is drained by the **coronary veins,** which empty into the **coronary sinus,** which in turn empties into the right atrium.

DISSECTION OF THE SHEEP HEART

Dissection of the sheep heart is valuable because of the similarity of its size and structure to the human heart. Also, a dissection experience allows the student to view structures in a way not possible with models and diagrams. Refer to Figure 30.3 as you proceed with the dissection.

1. Obtain a preserved sheep heart, a dissection tray, and dissecting instruments. Rinse the sheep heart in cold water to remove excessive preservatives and to flush out any trapped blood clots. Now you are ready to make your observations.

2. Observe the texture of the pericardium. Also, note its point of attachment to the heart. Where

is it attached? _____

_____

3. If the pericardial sac is still intact, slit open the parietal pericardium and cut it from its attachments. Observe the visceral pericardium (epicardium). Using a sharp scalpel, carefully pull a little of this serous membrane away from the myocardium. How does its position, thickness, and apposition to the heart differ from that of the

parietal pericardium? _____

_____

_____

_____

4. Examine the external surface of the heart. Notice the accumulation of adipose tissue, which in many cases marks the separation of the chambers and the location of the coronary arteries that nourish the myocardium. Carefully scrape away some of the fat with a scalpel to expose the coronary blood vessels.

5. Identify the base and apex of the heart, and then identify the two wrinkled **auricles,** earlike flaps of tissue projecting from the atrial chambers. The

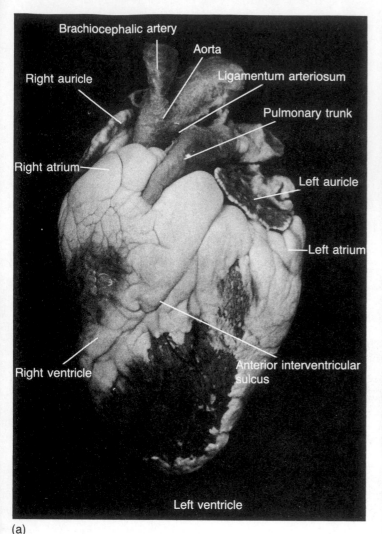

(a)

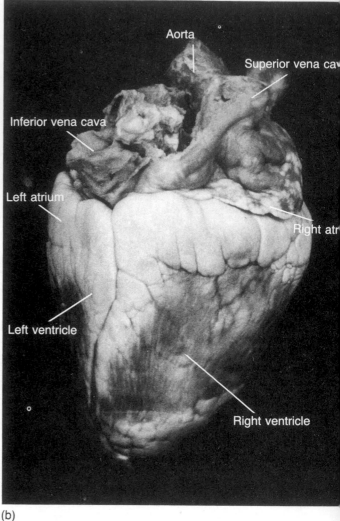

(b)

Figure 30.3

Anatomy of the sheep heart. Photos: (a) anterior view; (b) posterior view. (Photos courtesy of Ann Allworth.)

balance of the heart muscle is ventricular tissue. To identify the left ventricle, compress the ventricular chambers on each side of the longitudinal fissures carrying the coronary blood vessels. The side that feels more solid and thicker is the left ventricle. The right ventricle feels much thinner and somewhat flabby on compression. This difference reflects the greater demand placed on the left ventricle, which must pump blood through the much longer systemic circulation, a pathway with much higher resistance than the pulmonary circulation served by the right ventricle. Hold the heart in its anatomic position (Figure 30.3(a)), with the anterior surface uppermost. In this position the left ventricle composes the entire apex and the left side of the heart.

6. Identify the pulmonary trunk and the aorta extending from the base of the heart. The pulmonary trunk is the most anterior, and you may see its division into the right and left pulmonary arteries if

it has not been cut too closely to the heart. The aorta, which is thicker walled and branches almost immediately, is located just beneath the pulmonary trunk. The first branch of the sheep aorta, the **brachiocephalic artery,** is identifiable unless the aorta has been cut immediately as it leaves the heart. The brachiocephalic artery later splits to form the carotid and subclavian arteries, which supply the right side of the head and right forelimb, respectively. Carefully clear away some of the fat between the pulmonary trunk and the aorta to expose the **ligamentum arteriosum,** a cordlike remnant of the **ductus arteriosus.** (In the fetus the ductus arteriosus allows blood to pass directly from the pulmonary trunk to the aorta, thus bypassing the nonfunctional fetal lungs.)

7. Cut through the wall of the aorta until you see the aortic semilunar valve. Identify the two openings to the coronary arteries just above the valve. Place a probe into one of these holes to see if you can fol-

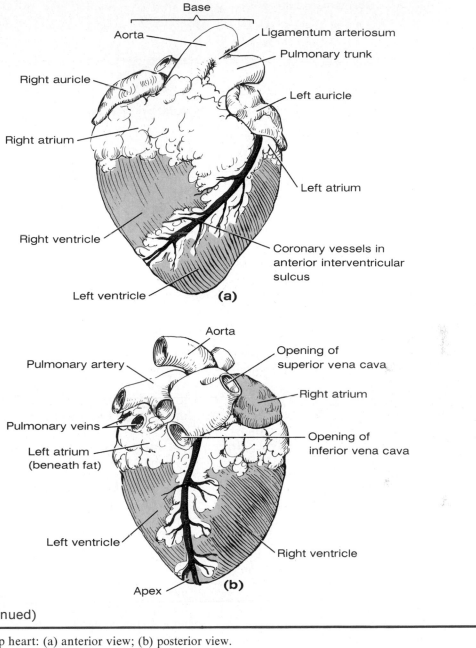

Figure 30.3 (continued)

Anatomy of the sheep heart: (a) anterior view; (b) posterior view.

low the course of a coronary artery between the right and left ventricles.

8. Turn the heart to view its posterior surface. The heart will appear as shown in Figure 30.3(b). The right and left ventricles appear equal-sized in this view. Identify the four thin-walled pulmonary veins entering the left atrium. (It may or may not be possible to locate the pulmonary veins from this vantage point, depending on how they were cut as the heart was removed.) Identify the superior and inferior vena cavae entering the right atrium. Compare the approximate diameter of the superior vena cava with the diameter of the aorta.

Which is larger? _____

Which has thicker walls? _____

Why do you suppose these differences exist? _____

_____

_____

_____

9. Insert a probe into the superior vena cava and use scissors to cut through its wall so that you can view the interior of the right atrium. Do not extend your cut entirely through the right atrium or into the ventricle. Observe the right atrioventricular valve.

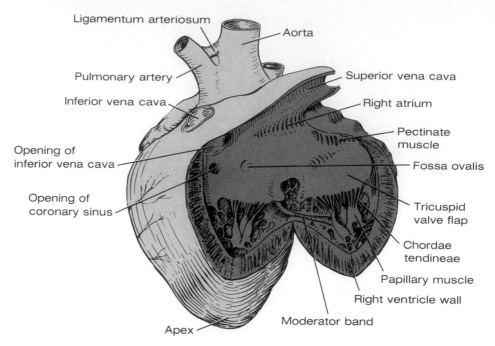

Ligamentum arteriosum

Aorta

Pulmonary artery

Superior vena cava

Inferior vena cava

Right atrium

Pectinate muscle

Opening of inferior vena cava

Fossa ovalis

Opening of coronary sinus

Tricuspid valve flap

Chordae tendineae

Papillary muscle

Right ventricle wall

Moderator band

Apex

Figure 30.4

Right side of the sheep heart opened and reflected to reveal internal structures (diagrammatic view).

How many flaps does it have? _____

Pour some water into the right atrium and allow it to flow into the ventricle. Slowly and gently squeeze the right ventricle to watch the closing action of this valve. (If you squeeze too vigorously, you'll get a face full of water!)

10.   Return to the pulmonary trunk and cut through its anterior wall until you can see the pulmonary semilunar valve. Pour some water into the base of the pulmonary trunk to observe the closing action of this valve. How does its action differ from that

of the atrioventricular valve? _____

_____

_____

After observing valve action, continue the cut from the pulmonary trunk through the right ventricle, paralleling the anterior lateral sulcus until you reach the interventricular septum on the dorsal aspect of the heart. Return to the superior vena cava, and continue the cut made in its wall through the right atrium and right atrioventricular valve until this cut meets the lateral cut just made through the right ventricle (Figure 30.4).

11.   Reflect the cut edges of the superior vena cava, right atrium, and right ventricle to obtain the view seen in Figure 30.4. Observe the comblike ridges of muscle throughout most of the right atrium. This is

called **pectinate muscle** (pectin = comb). Identify the large opening of the inferior vena cava on the ventral atrial wall and follow it to its external opening with a probe. Notice that the atrial walls in the vicinity of the vena cavae are smooth and do not have the roughened appearance (pectinate musculature) of the other regions of the atrial walls. Just below the inferior vena caval opening, identify the opening of the **coronary sinus,** which returns venous blood of the coronary circulation to the heart.

12.   Identify the papillary muscles in the right ventricle, and follow their attached chordae tendineae to the flaps of the tricuspid valve. Also identify the **moderator band,** which spans the lumen of the right ventricle from the interventricular septum to the ventricular wall. This band reinforces the walls of the right ventricle and prevents its possible overdistention. Is the moderator band

present in the human heart? _____

Note the pitted and ridged appearance (**trabeculae carneae**) of the inner ventricular muscle.

13.   Make a longitudinal incision through the aorta and continue it into the left ventricle. Notice how much thicker the myocardium of the left ventricle is as compared to that of the right ventricle. Compare the *shape* of the left ventricular cavity to

the shape of the right ventricular cavity. _____

_____

_____

_____

Are the papillary muscles and chordae tendineae observed in the right ventricle also present in the left

ventricle? _____ Count the number of

cusps in the left atrioventricular valve. How does this compare with the number seen in the right

atrioventricular valve? _____

_____ _____

_____

How do the sheep valves compare with their human

counterparts? _____

_____

14.   Continue your incision from the left ventricle superiorly into the left atrium. Reflect the cut edges of the atrial wall, and attempt to locate the entry points of the pulmonary veins into the left atrium. Follow them to the heart exterior with a probe. Note how thin-walled these vessels are. Locate an oval depression, the **fossa ovalis,** in the interatrial septum. This depression marks the site of an opening in the fetal heart, the **foramen ovale,** which allows blood to pass from the right to the left atrium, thus bypassing the fetal lungs.

15.   Properly dispose of the organic debris, and clean the dissecting tray and instruments.

Electro-cardiography

OBJECTIVES

1. To enumerate and localize the elements of the Purkinje, or nodal, system of the heart, and to describe the initiation and conduction of impulses through this system and the myocardium.

2. To interpret the ECG in terms of depolarization and repolarization events occurring in the myocardium, and to identify the P, QRS, and T waves on an ECG recording.

3. To calculate the heart rate, QRS interval, P-R interval, and S-T interval from an ECG obtained during the laboratory period.

4. To define *tachycardia, bradycardia,* and *fibrillation.*

MATERIALS

ECG recording apparatus
Electrode paste; alcohol swabs
Cot

THE PURKINJE SYSTEM

Heart contraction results from a series of electrical potential changes (depolarization wave) that travel through the heart preliminary to each beat. The ability of cardiac muscle to beat is intrinsic—it does not depend on impulses from the nervous system to initiate its contraction and will continue to contract rhythmically even if all nerve connections are severed. However, two types of controlling systems exert their effects on heart activity. One of these involves nerves of the autonomic nervous system, which increase or decrease the heart beat rate depending on which division is activated. The second system is the **Purkinje,** or nodal, system of the heart, which is a built-in conduction system of specialized myocardial tissue. The Purkinje system ensures that heart muscle depolarizes in an orderly and sequential manner (from atria to ventricles) and that the heart beats as a coordinated unit. Various impairments of this system result in out-of-phase contractions of the atria and ventricles or fibrillation (a shuddering of the heart muscle), which decreases or obliterates the effectiveness of the heart as a pump.

The components of the Purkinje system include the **SA (sinoatrial) node,** located in the right atrium just inferior to the entrance of the superior vena cava, the **AV (atrioventricular) node** in the lower atrial septum at the junction of the atria and ventricles, the **bundle of His** and right and left **bundle branches,** located in the interventricular septum, and the **Purkinje fibers,** which ramify within the muscle bundles of the ventricular walls. The Purkinje fiber network is much denser and more elaborate in the left ventricle because of the larger size of this chamber (Figure 31.1).

The SA node, which has the highest rate of discharge, provides the stimulus for contraction. Since it sets the rate of depolarization for the heart as a whole, the SA node is often referred to as the *pacemaker.* From the SA node, the impulse spreads throughout the atria and to the AV node. This electrical wave is immediately followed by atrial contraction. At the AV node, the impulse is momentarily delayed (approximately 0.1 sec), allowing the atria to complete their contraction. It then passes through the bundle of His, the right and left bundle branches, and the Purkinje fibers, finally resulting in ventricular contraction. Note that the atria and ventricles are separated from one another by a region of electrically inert connective tissue, and the depolarization wave can be transmitted to the ventricles only via the AV node. Thus, any damage to the AV node partially or totally in-

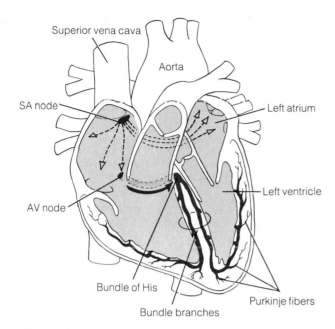

Figure 31.1

The Purkinje system of the heart. Dotted-line arrows indicate transmission of the impulse from the SA node through the atria. Solid arrow indicates transmission of the impulse from the AV node to the bundle of His.

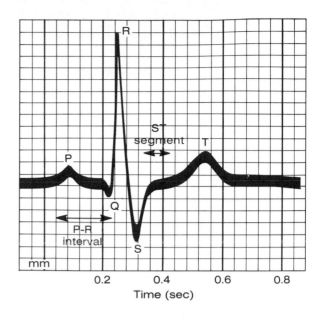

Figure 31.2

Normal ECG TRACING.

sulates the ventricles from the influence of the SA node. Although ventricular muscle fibers are capable of spontaneous depolarization, the Purkinje system increases the rate of ventricular depolarization, thus synchronizing heart activity.

ELECTROCARDIOGRAPHY

The conduction of impulses through the heart generates electrical currents that eventually spread throughout the body. These impulses can be detected on the body surface and recorded with an instrument called an electrocardiograph. The graphic recording of the electrical changes (depolarization and repolarization) occurring during the cardiac cycle is called an **electrocardiogram (ECG)** (Figure 31.2). The typical ECG consists of a series of three recognizable waves called *deflection waves*. The first wave, the **P wave,** is a small wave that indicates the depolarization of the atria immediately before atrial contraction. The large **QRS complex,** resulting from ventricular depolarization, has a complicated shape (primarily because of the variability in size of the two ventricles and the time differences required for these chambers to depolarize). It precedes ventricular contraction. The **T wave** results from currents propagated during ventricular repolarization. The repolarization of the atria, which occurs during the QRS interval, is generally obscured by the large QRS complex.

Abnormalities of the deflection waves and changes in the time intervals of the ECG may be useful in detecting myocardial infarcts or problems with the conduction system of the heart. The P-R interval represents the time between the beginning of atrial depolarization and ventricular depolarization; thus it includes the period during which the depolarization wave passes to the AV node, atrial systole, and the passage of the excitation wave to the balance of the conducting system. Generally, the P-R interval is about 0.16 to 0.18 second. A longer interval may suggest a partial AV heart block due to damage to the AV node itself or weak SA nodal impulses. In total heart block, no impulses are transmitted through the AV node, and the atria and ventricles beat independently of one another—the atria at the SA node rate and the ventricles at their intrinsic rate, which is considerably slower. If the QRS interval (normally 0.06 to 0.10 second) is prolonged, it may indicate a right or left bundle branch block in which one ventricle is contracting later than the other. The Q-T interval reflects the duration of ventricular systole. With a heart rate of 70/min, this interval is normally 0.31 to 0.41 second. As the rate increases, this interval becomes shorter; conversely, when the heart rate drops, the interval is longer. A heart rate over 100/min is referred to as **tachycardia;** a rate below 60/min is **bradycardia.** Although neither condition is pathologic, prolonged tachycardia may progress to **fibrillation,** a condition of rapid uncoordinated heart contractions. Many types of ECG deviations may result from **myocardial infarcts** (regions of dead myo-

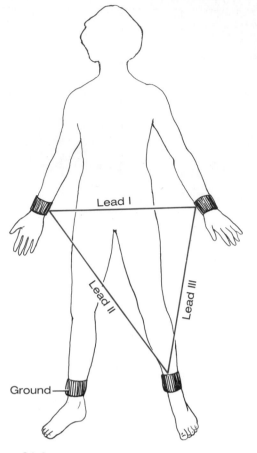

Figure 31.3

ECG recording positions for the standard limb leads.

cardial tissue that do not depolarize); however, elevations or depressions of the ST segment are the most common abnormalities.

Many types of polygraphs or ECG recorders are available. Your instructor will provide directions on how to set up and use the available apparatus.

Many different electrode positionings are used to obtain information for diagnostic purposes. For this investigation, four electrodes are used (Figure 31.3), and results are obtained from the three standard leads (also shown in Figure 31.3). With each of the leads, the electrical changes between two of the electrodes is determined; thus, as shown in the figure, the potential difference between the left and right arm (LA-RA) is determined with lead I, between the right arm and left leg (RA-LL) with lead II, and between the left arm and left leg (LA-LL) with lead III.

Preparation of Subject

Place electrode paste on four electrode plates, and position the electrodes at the sites described below after first scrubbing each designated area with an alcohol swab. Attach an electrode to the anterior surface of each arm about 3 to 4 inches above the wrist, and secure them with rubber

straps. In the same manner, attach an electrode to each leg, approximately 3 to 4 inches above the medial malleolus (inner aspect of the ankle).

Attach the appropriate tips of the patient cable to the electrodes. The cable leads are marked RA (right arm), LA (left arm), LL (left leg), and RL (right leg = ground).

Recording the ECG

The ECG will be recorded first under baseline (resting) conditions and then under conditions of fairly strenuous activity. Finally, recordings will be made while the subject holds his breath. The activity and breath-holding recordings will be compared to the baseline recordings, and the student will be asked to determine the underlying reasons for the observed differences in the recordings. (Hopefully, this exercise will lead to an increased understanding of the interrelatedness of body systems.)

BASE LINE RECORDINGS

1. Position the subject comfortably in a supine position on a cot.

2. Set the paper speed at 25 mm/ sec and the lead selector switch to the position corresponding to recording from lead I. Record the at-rest subject's ECG from lead I for 2 to 3 minutes or until the recording stabilizes. You will need a tracing long enough to provide each student in the group with a representative segment. The subject should try to relax and not move unnecessarily, as the skeletal muscle action potentials will also be picked up and recorded.

3. Stop the recording, and mark it "lead I."

4. Repeat the recording procedure for leads II and III.

5. Each student should take a representative segment of one of the lead recordings and label the record with the name of the subject and the lead used. Identify and label the P, QRS, and T waves. The calculations you perform for your recording should be based on the following information: since the paper speed was 25 mm/ sec, each millimeter of paper corresponds to a time interval of 0.04 sec. Thus, if an interval requires 4 mm of paper, its duration is 4 mm × 0.04 sec/mm = 0.16 sec. Compute the heart rate. Measure the distance (mm) from the beginning of one QRS complex to the beginning of the next QRS complex, and plug this value into the equation below to find the time for one heartbeat.

_____ mm × 0.04 sec/mm = ___ _____

sec per heartbeat

Now find beats per minute, or heart rate, by using the figure just computed (x) in the following equation:

$$\text{Beats/min} = \frac{1}{\underline{x}\ (\text{sec/beat})} \times 60\ (\text{sec/min})$$

Beats/min computed _____

Is the value obtained within normal limits? _____

Measure the QRS interval and compute its duration. _____

Measure the ST segment and compute its duration. _____

Measure the P-R interval and compute its duration. _____

Are the values computed within normal limits?

_____

6. Attach segments of the ECG recordings from leads I to III in the space below. Note the paper speed, lead, and the subject's name on each tracing. On the recording on which you based your previous computations, add your calculations for the duration of the QRS, P-R intervals, and the ST segment above the appropriate area of tracing. Also record the heart rate on that tracing.

"RUNNING IN PLACE" RECORDING

1. Make sure the electrodes are securely attached to prevent electrode movement while recording the ECG.
2. Set the paper speed at 25 mm/sec, and prepare to make the recording using lead I.
3. Record the ECG while the subject is running in place for 3 minutes.
4. Stop the recording and compute the beats/

min during the third minute of running and record below.

_____ beats/min while running in

place.
5. Compare this recording with the previous recording from lead I. Which intervals are

shorter in this recording? _____

"BREATH-HOLDING" RECORDING

1. Position the subject comfortably in the sitting position.
2. Using lead I and a paper speed of 25 mm/sec, *begin* the recording. After approximately 10 seconds have passed, notify the subject to begin breath holding and mark the record to indicate the onset of the 1-minute breath-holding interval.
3. Stop the recording after 1 minute (and remind the subject to breathe). Compute the beats/min during the 1-minute experimental period.

Beats/min during breath holding _____

_____

4. Compare this recording with the recording (Lead I) obtained under baseline conditions.

What differences are seen? _____

_____

Attempt to *explain* the physiologic reason for the differences you have seen. (Hint: a good place to start might be to check "hypoventilation" or the role of the *respiratory* system in acid-base balance of the blood.)

_____

_____

_____

Anatomy of Blood Vessels

OBJECTIVES

1. To describe the tunics of arterial and venous walls and state the function of each layer.

2. To correlate differences observed in artery, vein, and capillary structures with the functions these vessels perform.

3. To recognize a cross-sectional view of an artery and vein when provided with a microscopic view or appropriate diagram.

4. To list and/or identify the major arteries arising from the aorta, and to indicate the body region supplied by each.

5. To list and/or identify the major veins draining into the superior and inferior vena cavae, and to indicate the body regions drained.

6. To point out and/or discuss the unique features of special circulations in the body:

   - The pattern of blood flow in the portal system and its importance
   - The vascular supply of the brain and the importance of the circle of Willis

   - Components of the pulmonary circulation
   - Structures unique to the fetal circulation and the importance of each

7. To point out anatomic differences between the vascular system of the human and the laboratory dissection specimen.

MATERIALS

Anatomic charts of human arteries and veins (or a three-dimensional model of the human circulatory system)

Anatomic charts of the following specialized circulations: pulmonary circulation, hepatic portal circulation, arterial supply and circle of Willis of the brain (or a brain model showing this circulation), fetal circulation

Compound microscope

Prepared microscope slides showing cross sections of an artery and vein

Dissection animals

Dissecting pans and instruments

Blood circulates within the blood vessels, which constitute a closed transport system. As the heart contracts, blood is propelled into the large arteries leaving the heart. It moves into successively smaller arteries and then to the arterioles, which feed the capillary beds in the tissues. Capillary beds are drained by the venules, which in turn empty into veins that ultimately converge on the great veins entering the heart. Thus arteries, carrying blood away from the heart, and veins, which drain the tissues and return blood to the heart, function simply as conducting vessels or conduits. Only the tiny capillaries that connect the arterioles and venules and ramify throughout the tissues directly serve the needs of the body's cells. It is through the capillary walls that exchanges between tissue cells and blood occur.

Respiratory gases, nutrients, and wastes move along diffusion gradients; thus, oxygen and nutrients diffuse from the blood to the tissue cells, and carbon dioxide and metabolic wastes move from the cells to the blood.

In this exercise you will examine the microscopic structure of blood vessels and the major arteries and veins of the systemic circulation and other special circulations.

MICROSCOPIC STRUCTURE OF THE BLOOD VESSELS

Except for the microscopic capillaries, the walls of blood vessels are constructed of three coats, or tunics (Figure 32.1). The **tunica intima,** or

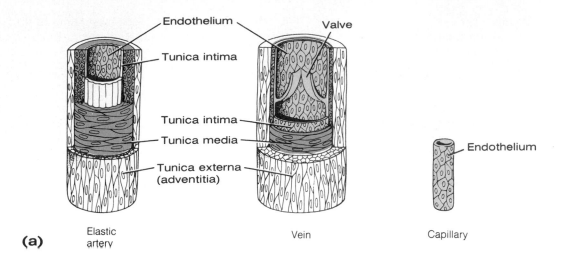

(a)   Elastic artery   Vein   Capillary

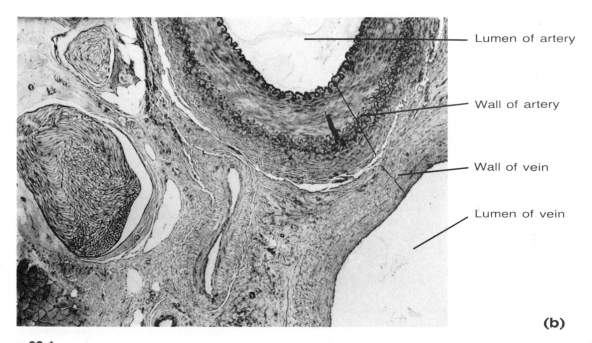

(b)

Figure 32.1

Structure of arteries, veins, and capillaries. Photomicrograph shows portion of an artery and a vein.

interna, which lines the lumen of a vessel, is a single thin layer of endothelium that is continuous with the endocardium of the heart. Its cells fit closely together, forming an extremely smooth blood vessel lining that helps to decrease resistance to blood flow.

The **tunica media** is the more bulky middle coat and is composed primarily of smooth muscle and elastic tissue. The smooth muscle, under the control of the sympathetic nervous system, plays an active role in reducing or increasing the diameter of the vessel, which in turn increases or decreases the peripheral resistance and blood pressure.

The **tunica externa,** or **adventitia,** the outermost tunic, is composed of areolar or fibrous connective tissue. Its function is basically supportive and protective.

In general, the walls of arteries are much thicker than those of veins. The tunica media in particular tends to be considerably heavier and contains substantially more smooth muscle and elastic tissue. This anatomic difference reflects a functional difference in the two types of vessels. Arteries, which are closer to the pumping action of the heart, must be able to expand as an increased volume of blood is propelled into them and then recoil passively as the blood flows off into the circulation during diastole. Their walls must be sufficiently strong and resilient to withstand such pressure fluctuations. Since these larger arteries have such large amounts of elastic tissue in their media, they are often referred to as *elastic arteries.* Smaller arteries, further along in the circulatory pathway, are exposed to less extreme pressure fluctuations. They have somewhat less elastic tissue but still have substantial amounts of smooth muscle in their media. For this reason, they are called *muscular arteries.*

In contrast, the veins, which are far removed from the heart in the circulatory pathway, are not subjected to such pressure fluctuations and are essentially low-pressure vessels. Thus, veins may be thinner-walled without jeopardy. However, the low-pressure condition itself requires structural modifications to ensure that venous return equals cardiac output; thus, the lumens of veins tend to be substantially larger than those of corresponding arteries.

Since blood returning to the heart often flows against gravity, there are other aids to venous return. The valves of the larger veins function in much the same manner as the semilunar valves of the heart to prevent backflow of blood. Skeletal muscle activity also promotes venous return; as the skeletal muscles surrounding the veins contract and relax, the blood is "milked" through the veins toward the heart. (Anyone who has had to stand relatively still for an extended time will be happy to show you their swollen ankles, caused by blood pooling in their feet during the period of muscle inactivity!) Finally, pressure changes that occur in the thorax during breathing also facilitate the return of blood to the heart.

To demonstrate the efficiency of the venous valves in preventing backflow of blood, perform the following simple experiment. Allow one hand to hang by your side until the blood vessels on the dorsal aspect become distended. Place two fingertips against one of the distended veins and, pressing firmly, move the superior finger proximally along the vein and then release this finger. The vein will remain flattened and collapsed despite gravity. Then remove the distal fingertip and observe the rapid filling of the vein.

The transparent walls of the tiny capillaries are only one cell layer thick, consisting of just the endothelium or tunica intima. Because of this exceptional thinness, exchanges are easily made between the blood and tissue cells.

1. Obtain a cross-sectional view preparation of blood vessels and a microscope.

2. Scan the section to identify a thick-walled artery. Very often, but not always, its lumen will appear scalloped due to the constriction of its walls by the elastic tissue of the media.

3. Identify a vein. Its lumen may appear elongated or irregularly shaped and collapsed, and its walls will be considerably thinner. Note the difference in the relative amount of elastic fibrils in the media of the two vessels. Also, note the thinness of the intima layer, which is composed of flat squamous type cells.

4. Make a drawing of your observations of the two vessel types below, and label the tunics. Try to indicate the proper size relationships relative to wall thickness and the tunic widths.

Artery   Vein

MAJOR SYSTEMIC ARTERIES OF THE BODY

The aorta is the largest artery of the body. Extending upward as the ascending aorta from the left ventricle, it arches posteriorly and to the left (aortic arch) and courses downward as the descending aorta through the thoracic cavity. It penetrates the diaphragm to enter the abdominal cavity just anterior to the vertebral column. Figure 32.2 depicts the course of the aorta and its major branches. As you locate the arteries on the figure, be aware of ways in which you can make your memorization task easier—in many cases the name of the artery reflects the body region traversed (axillary, subclavian, brachial, popliteal), the organ served (renal, hepatic), or the bone followed (tibial, femoral, radial, ulnar). Once you have identified these arteries on the figure, attempt to locate and name them (without a reference) on a large anatomic chart or a three-dimensional model of the circulatory system vessels. (Arteries described here are shown in bold type on the figure, but other less important arteries are also named. Ask your instructor which arteries you are required to identify.)

Ascending Aorta

The only branches of the ascending aorta are the **right** and the **left coronary arteries,** which supply the myocardium.

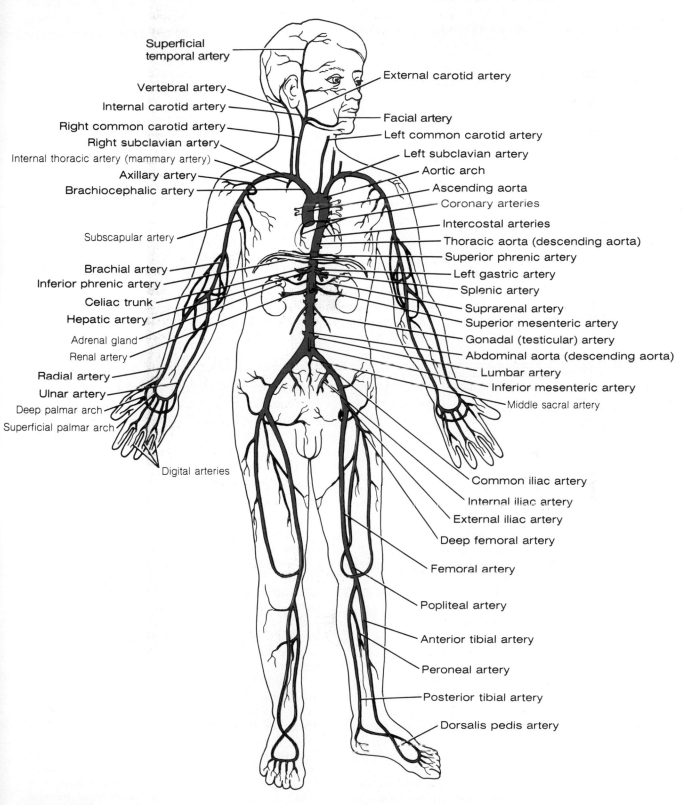

Figure 32.2

Major systemic arteries of the body.

Aortic Arch

The **brachiocephalic,** or **innominate, artery** is the first branch of the aortic arch. It persists briefly before dividing into the right **common carotid artery** and the right **subclavian artery.** The common carotid divides to form the **internal carotid artery,** which serves the brain, and the **external carotid artery,** which supplies the extracranial tissues of the neck and head. The subclavian artery gives off three branches to the head and neck, the most important being the **vertebral artery,** which runs up the posterior neck to supply a portion of the brain. In the axillary region, the subclavian artery becomes the **axillary artery** and then the **brachial artery** as it enters the arm. At the elbow, the brachial artery divides into the **radial** and **ulnar arteries,** which follow the same-named bones to supply the forearm and hand.

The **left common carotid artery,** the second branch of the aortic arch, supplies the left side of the head and neck in the same manner the right common carotid serves the right side. The third branch is the **left subclavian artery,** which supplies the left upper extremity and subdivides as described for the right subclavian artery.

Descending Aorta

The 9 or 10 pairs of **intercostal arteries** that supply the muscles of the thoracic wall are small branches of the descending aorta, as are the **phrenic arteries,** which supply the diaphragm. Other more major branches of the descending artery supply the abdominal region. The **celiac trunk** is an unpaired artery that subdivides into three branches: the **left gastric** artery supplying the stomach, the **splenic artery** supplying the spleen, and the **hepatic artery,** which provides the functional blood supply of the liver. The largest branch of the descending aorta, the **superior mesenteric artery,** supplies most of the small intestine and the first half of the large intestine. The small paired **suprarenal arteries** emerge at approximately the same level as the superior mesenteric artery and run laterally to supply the adrenal glands. The paired **renal arteries** supply the kidneys, and the **gonadal arteries,** arising from the ventral surface of the aorta slightly below the renal arteries, run inferiorly to serve the gonads. They are called **ovarian arteries** in the female and **testicular** (or **internal spermatic**) **arteries** in the male. Since these vessels must travel through the inguinal canal to supply the testes in the male, they are considerably longer in the male than in the female. The small unpaired artery supplying the second half of the large intestine is the **inferior mesenteric artery.**

In the pelvic region, the descending aorta divides into the two large **common iliac arteries.** Each of these vessels extends for about 2 inches into the pelvis before it divides into the **internal iliac artery,** which supplies the pelvic organs (bladder, rectum, and some reproductive structures) and the **external iliac artery,** which continues into the thigh, where its name changes to the **femoral artery.** A branch of the femoral artery, the **deep femoral artery,** supplies the posterior thigh region. In the knee region, the femoral artery briefly becomes the **popliteal artery;** its subdivisions—the **anterior** and **posterior tibial arteries**—supply the lower leg, ankle, and foot. The posterior tibial gives off one main branch, the **peroneal artery,** which supplies the lateral portion of the calf (peroneal muscles). The anterior tibial artery terminates at the **dorsalis pedis** artery, which supplies the dorsum of the foot. The dorsalis pedis is often palpated in patients with circulation problems of the leg to determine the circulatory efficiency to the limb as a whole.

MAJOR SYSTEMIC VEINS
OF THE BODY

Arteries are generally located in deep, well-protected body areas. However, many veins follow a more superficial course and are often easily visualized and palpated on the body surface (Figure 32.3). Most deep veins parallel the course of the major arteries; thus in many cases the naming of the veins and arteries is identical except for the designation of the vessels as veins. Although the major systemic arteries branch off the aorta, the veins tend to converge on the vena cavae, which enter the right atrium of the heart. Veins draining the head and upper extremities empty into the **superior vena cava,** and those draining the lower body empty into the **inferior vena cava.**

Veins Draining into the Superior Vena Cava

Veins draining into the superior vena cava are named from the superior vena cava distally; **but remember that the flow of blood is in the opposite direction.**

The **azygos vein,** which drains the thorax, enters the dorsal aspect of the superior vena cava immediately before it enters the right atrium. The **right** and **left brachiocephalic veins** drain the head, neck, and upper extremities and unite to form the superior vena cava. (Note that although there is only one brachiocephalic artery, there are two brachiocephalic veins.)

Branches of the brachiocephalic veins include the **internal jugular veins,** large veins that drain the dural sinuses of the brain; the **vertebral veins,** which drain the posterior aspect of the head; and the **subclavian veins,** which receive the venous blood from the upper extremity. The small **external jugular vein** joins the subclavian vein near its origin to return the venous drainage of the extracranial tissues of the head and neck. As the subclavian vein traverses the axilla, it becomes the **axillary vein** and then the **brachial vein** as it enters the arm to course down the posterior aspect of the humerus. The brachial vein then divides to form the deep **radial** and **ulnar veins** of the forearm. The superficially located venous drainage of the arm includes the **cephalic vein,** which

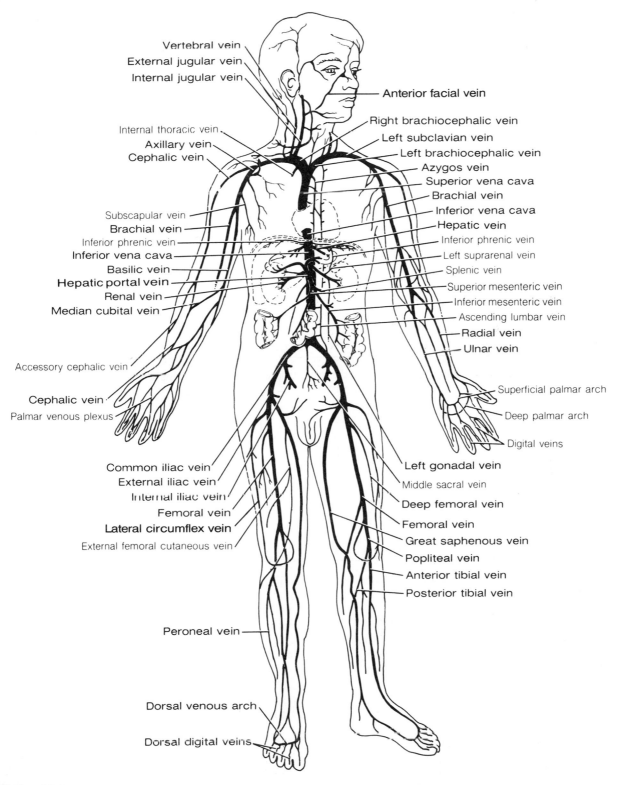

Figure 32.3

Major systemic veins of the body.

courses along the lateral aspect of the arm and empties into the axillary vein; the **basilic vein,** found on the medial aspect of the arm and entering the brachial vein; and the **median cubital vein,** which runs between the cephalic and basilic veins in the anterior aspect of the elbow (this vein is often the site of choice for removing blood for testing purposes).

Veins Draining into the Inferior Vena Cava

The inferior vena cava, which is much longer than the superior vena cava, returns blood to the heart from all body regions below the diaphragm. It begins in the lower abdominal region with the union of the paired **common iliac veins,** which drain venous blood from the legs and pelvis. Each common iliac vein in turn is formed by the union of the **internal iliac vein,** draining the pelvis, and the **external iliac vein,** which receives venous blood from the leg. Veins of the leg include the **anterior** and **posterior tibial veins,** which serve the calf and foot. The posterior tibial vein becomes the **popliteal vein** in the knee region and the **femoral vein** in the thigh. The femoral vein empties into the external iliac vein in the inguinal region. The **great saphenous vein,** a superficial vein, is the longest vein of the body. Beginning in the foot with the **dorsal venous arch,** it extends up the medial side of the leg, knee, and thigh to empty into the femoral vein. Moving superiorly into the abdominal cavity, the inferior vena cava receives blood from the **right gonadal vein** (testicular or spermatic vein in the male; ovarian vein in the female), which drains the right gonad. (The left testicular or ovarian vein drains into the left renal vein.) The **right** and **left renal veins** drain the kidneys, and the **right** and **left hepatic veins** drain the liver. The unpaired veins draining the digestive tract organs empty into a special vessel, the **hepatic portal vein,** which carries this blood through the liver before it enters the systemic venous system. (The hepatic portal system is discussed separately on page 281.)

 Identify the important arteries and veins on the large anatomic chart or model without referring to the figures.

SPECIAL CIRCULATIONS

Pulmonary Circulation

The pulmonary circulation (discussed previously in relation to heart anatomy on page 263) differs in many ways from the systemic circulation, since it does not serve the metabolic needs of body tissues (in this case, lung tissue). It functions instead to bring the blood into close contact with the alveoli of the lungs to permit gaseous exchanges that rid the blood of excess carbon dioxide and replenish its supply of vital oxygen. The arteries of the pulmonary circulation are structurally much like veins; thus they create a low pressure bed in the lungs. (If the arterial pressure in the systemic circulation is 120/80, the pressure in the pulmonary artery is likely to be approximately 70/10.) The functional blood supply of the lungs is provided by the **bronchial arteries,** which diverge from the thoracic portion of the descending aorta.

Pulmonary circulation begins with the large **pulmonary trunk,** which leaves the right ventricle and divides into the **right** and **left pulmonary arteries** about 2 inches above its origin (Figure 32.4). The right and left pulmonary arteries plunge into the lungs, where they subdivide into the **lobar arteries** (three on the right and two on the left), which accompany the main bronchi into the lobes of the lungs. Within the lungs, the lobar arteries branch extensively to form the arterioles, which finally terminate in the capillary networks surrounding the alveolar sacs of the lungs. Diffusion of the respiratory gases occurs across the walls of the alveoli and **pulmonary capillaries.** The pulmonary capillary beds are drained by venules, which converge to form sequentially larger veins and finally the four **pulmonary veins** (two leaving each lung), which return the blood to the left atrium of the heart.

Using the terms provided in Figure 32.4, _label_ all structures provided with leader lines.

How is the pulmonary artery unlike any of the systemic arteries? _____

_____

_____

How are the pulmonary veins different from systemic veins? _____

_____

_____

Arterial Supply of the Brain and the Circle of Willis

A continuous blood supply to the brain is crucial, since deprivation for even a few minutes causes irreparable damage to the delicate brain tissue. The brain is supplied by two pairs of arteries arising from the region of the aortic arch—the **internal carotid arteries** and the **vertebral arteries.** Figure 32.5 is a diagram of the brain's arterial supply. The internal carotid and vertebral arteries are

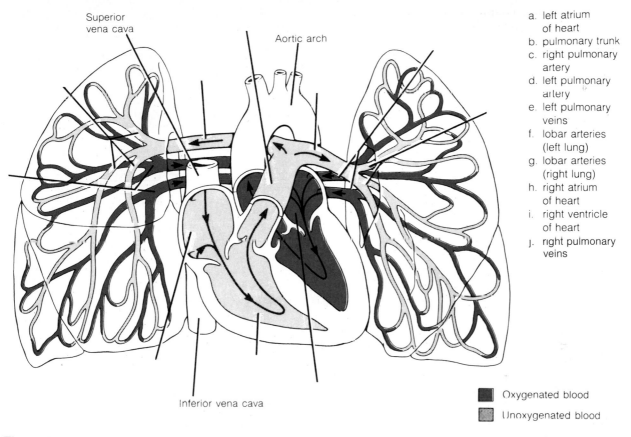

Superior
vena cava

Aortic arch

a. left atrium
   of heart
b. pulmonary trunk
c. right pulmonary
   artery
d. left pulmonary
   artery
e. left pulmonary
   veins
f. lobar arteries
   (left lung)
g. lobar arteries
   (right lung)
h. right atrium
   of heart
i. right ventricle
   of heart
j. right pulmonary
   veins

Inferior vena cava

■ Oxygenated blood

▨ Unoxygenated blood

Figure 32.4

The pulmonary circulation.

labeled. As you read the description of the blood supply below, _complete the labeling_ of this diagram.

The internal carotid arteries, branches of the common carotid arteries, follow a deep course through the neck and along the pharynx, entering the skull through the carotid canals of the temporal bone. Within the cranium, each divides into the **anterior** and **middle cerebral** arteries, which supply the bulk of the cerebrum. The internal carotid arteries also contribute to the formation of the **circle of Willis,** an arterial circle at the base of the brain surrounding the pituitary gland and the optic chiasma, by forming a **posterior communicating artery** on each side. The circle is completed by the **anterior communicating artery,** a short shunt connecting the right and left anterior cerebral arteries.

The paired vertebral arteries diverge from the subclavian arteries and pass superiorly through the foramina of the transverse process of the cervical vertebrae to enter the skull through the foramen magnum. Within the skull, the vertebral arteries unite to form a single **basilar artery,** which continues superiorly along the ventral aspect of the brain stem, giving off branches to the pons, cerebellum, and inner ear. At the base of the cerebrum, the artery divides to form the **posterior cerebral arteries,** which supply portions of the temporal and occipital lobes of the cerebrum and also become part of the circle of Willis by joining with the posterior communicating arteries.

The uniting of the blood supply of the internal carotid arteries and the vertebral arteries via the circle of Willis is a protective device that theoretically provides an alternate set of pathways for blood to reach the brain tissue in the case of arterial occlusion or impaired blood flow anywhere in the system. (In actuality, the communicating arteries are tiny, and in many cases the communicating system is defective.)

Hepatic Portal Circulation

Blood vessels of the hepatic portal circulation drain the digestive viscera, spleen, and pancreas and deliver this blood to the liver for processing via the **hepatic portal vein.** If a meal has recently been eaten, the hepatic portal blood will contain a high concentration of nutrient substances. Since the liver is a key body organ involved in maintainance of proper sugar, fatty acid, and amino acid concentrations in the blood, this system ensures that these substances pass through the liver before entering the systemic circulation. As blood percolates through the liver sinusoids, some of the nutrients are removed to be stored or processed in various ways for release

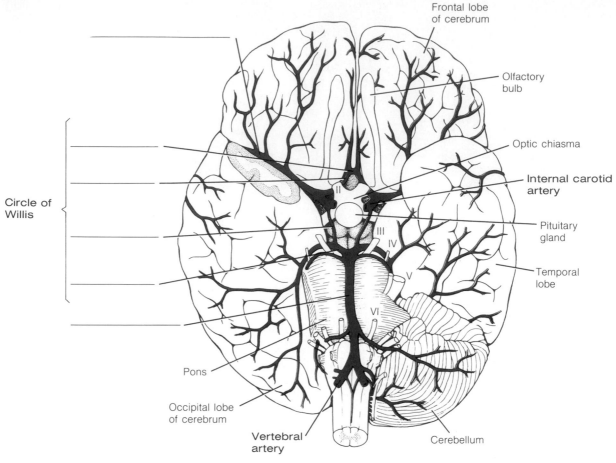

Frontal lobe
of cerebrum

Olfactory
bulb

Optic chiasma

**Internal carotid
artery**

Pituitary
gland

Temporal
lobe

Circle of
Willis

II

III

IV

V

VI

Pons

Occipital lobe
of cerebrum

Vertebral
artery

Cerebellum

Figure 32.5

Arterial supply of the brain.

to the general circulation. The liver in turn is drained by the hepatic veins that enter the inferior vena cava.

The **inferior mesenteric vein,** draining the transverse and terminal portions of the large intestine, drains into the **splenic vein,** which drains the spleen, pancreas, and greater curvature of the stomach. The splenic vein and the **superior mesenteric vein,** which receives blood from the small intestine and the ascending colon, join to form the hepatic portal vein. The **gastric vein,** which drains the lesser curvature of the stomach, drains directly into the hepatic portal vein.

Identify the vessels named above, and *label* them on Figure 32.6.

Fetal Circulation

In a developing fetus, the lungs and digestive systems are not yet functional, and all nutrient, excretory, and gaseous exchanges must occur through the placenta. Thus nutrients and oxygen move across placental barriers from the mother's

blood into fetal blood, and carbon dioxide and other metabolic wastes move from the fetal blood supply to the mother's blood.

Fetal blood travels through the umbilical cord which contains three blood vessels: two smaller **umbilical arteries** and one large **umbilical vein.** The umbilical vein carries blood rich in nutrients and oxygen to the fetus; the umbilical arteries carry carbon dioxide and waste-laden blood from the fetus to the placenta. The umbilical arteries meet the umbilical vein at the umbilicus (navel, or belly button) and wrap around the vein within the cord en route to their placental attachments. Newly oxygenated blood flows in the umbilical vein superiorly toward the fetal heart. En route, some of this blood perfuses the liver, but a larger proportion is ducted through the relatively nonfunctional liver via a vessel called the **ductus venosus** to enter the inferior vena cava, which carries the blood to the right atrium of the heart.

Since fetal lungs are nonfunctional and collapsed, two shunting mechanisms ensure that the lungs are almost entirely bypassed. Much of the blood entering the right atrium is shunted into

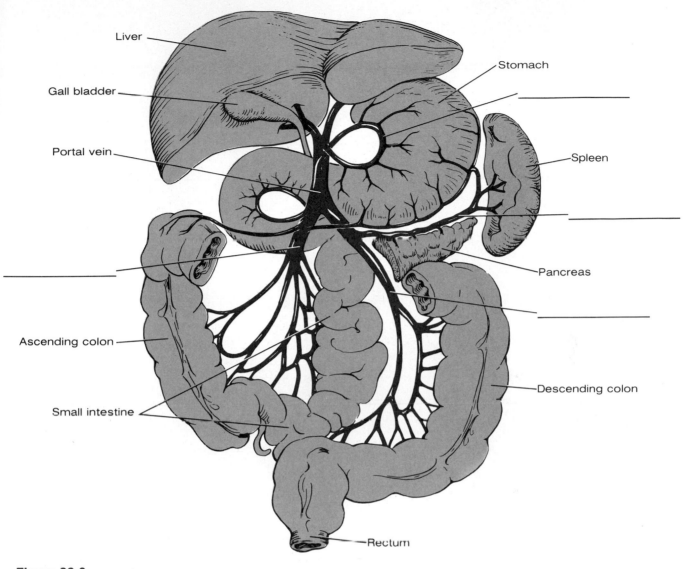

Figure 32.6

The hepatic portal circulation.

the left atrium through a flaplike opening in the interatrial septum—the **foramen ovale.** The left ventricle then pumps the blood out the aorta to the systemic circulation. Blood that does enter the right ventricle and is pumped out of the pulmonary trunk encounters a second shunt, the **ductus arteriosus,** a short vessel connecting the pulmonary trunk and the aorta. Because the collapsed lungs present an extremely high-resistance pathway, blood preferentially enters the systemic circulation through the ductus arteriosus.

The aorta carries blood to the tissues of the body; this blood ultimately finds its way back to the placenta via the umbilical arteries. The only fetal vessel that carries highly oxygenated blood is the umbilical vein; all other vessels contain varying degrees of oxygenated and deoxygenated blood.

At birth, or shortly after, the foramen ovale closes and becomes the **fossa ovalis,** and the ductus arteriosus collapses and is converted to the fibrous **ligamentum arteriosum.** Lack of blood flow through the umbilical vessels leads to their eventual obliteration, and the circulatory pattern becomes that of the adult. Remnants of the umbilical arteries exist as the **umbilical ligament** on the inner surface of the anterior abdominal wall, of the umbilical vein as the **round ligament** of the liver, and of the ductus venosus as a fibrous band called the **ligamentum venosus** on the inferior surface of the liver.

The pathway of fetal blood flow is indicated with arrows on Figure 32.7. Appropriately _label_ all specialized fetal circulatory structures provided with leader lines.

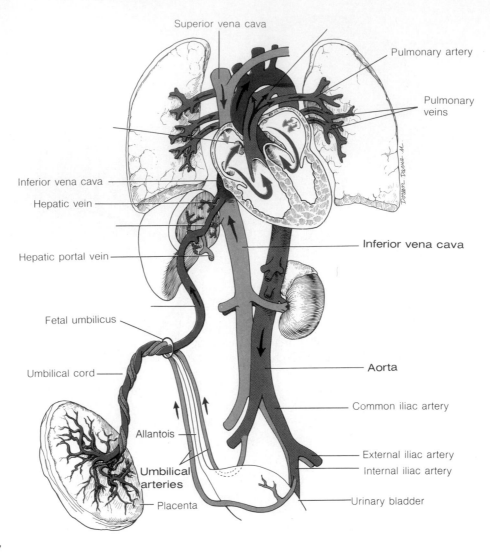

Figure 32.7

The fetal circulation.

DISSECTION OF FETAL PIG BLOOD VESSELS

Opening the Ventral Body Cavity

 To identify the blood vessels of the circulatory system of the pig, it is necessary to open the ventral body cavity. The outline drawing provided in the inset on Figure 32.8 indicates where to make the incisions.

1.   Place the pig dorsal side down on the dissecting pan and secure its forelimbs and hindlimbs with cord. Using scissors, make two longitudinal incisions through the ventral body wall beginning just superiorly and lateral to the midline of the pubic bone. Cut to and then around each side of the umbilical cord until the cuts meet medially. Con-

tinue anteriorly to the rib cage with a single midline incision.

2.   Angle the scissors slightly (½ inch) to the right or left of the sternum, and continue the cut through the rib cartilages, just lateral to the body midline, to the base of the throat.

3.   Make two lateral cuts on either side of the ventral body surface, anterior and posterior to the diaphragm. Leave the diaphragm intact. Spread the thoracic walls laterally to expose the thoracic organs.

4.   If the body cavity contains a dark fluid, flush it out with tap water before continuing.

THORACIC CAVITY ORGANS

Thymus: a large brownish elongated mass of tissue seen extending downward over the heart and superior thoracic organs

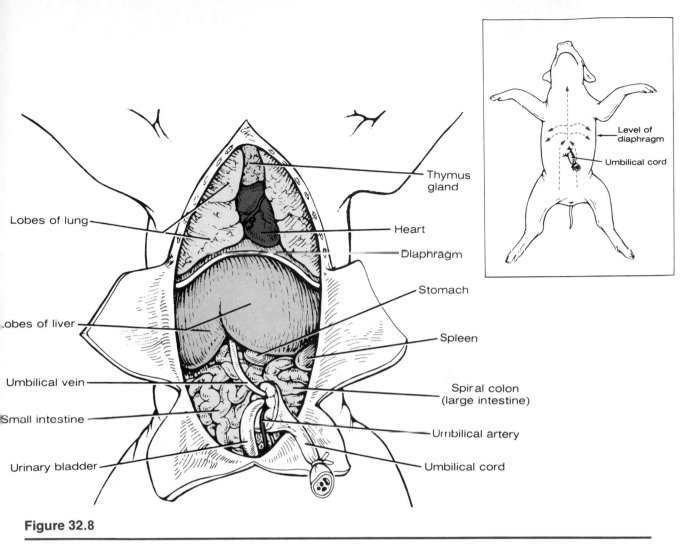

Figure 32.8

Ventral body cavity organs of the fetal pig. (Inset indicates incision lines for opening ventral body cavity.)

Heart: in the mediastinum enclosed by the pericardium

Lungs: flanking the heart

Thyroid gland: an oval reddish organ (gland) seen at the base of the throat

ABDOMINAL CAVITY ORGANS

Liver: a large, brown, lobed organ posterior to the diaphragm

Stomach: lying to the left and nearly covered by the liver

Spleen: a brown organ curving around the lateral aspect of the stomach

Small intestine: continuing posteriorly from the stomach

Large intestine: a large composite mass of coils lying within the coils of the small intestine

Urinary bladder: the large saclike structure seen attached to and entering the umbilical cord in the lower abdominal wall

Blood Vessels of the Body Cavity and Lower Extremities

1. Carefully clear away any thymus tissue or fat obscuring the heart and the large vessels associated with the heart. Before identifying the blood vessels, try to locate the phrenic nerve (from the cervical plexus), which innervates the diaphragm. The phrenic nerves can be seen ventral to the root of the lung on each side, passing to the diaphragm. Also attempt to locate the vagus nerve (cranial nerve X) passing laterally along the trachea and dorsal to the root of the lung.

2. Slit the parietal pericardium and reflect it superiorly; then cut it away from its heart attachments. Review the structures of the heart. Note its pointed inferior end (apex) and its superior broader

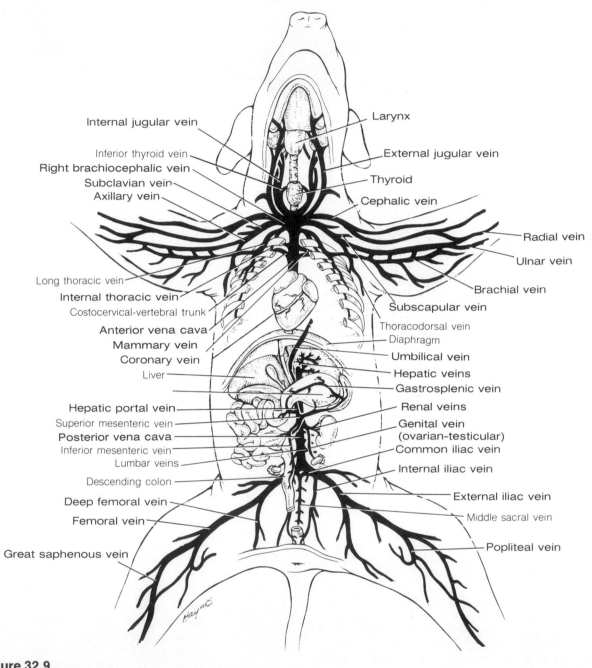

Figure 32.9

Venous system of the fetal pig.

base. Identify the two atria, which appear darker than the inferior ventricles. Identify the coronary arteries in the sulcus on the ventral surface of the heart; these should be injected with red latex. (As an aid to blood vessel identification, laboratory specimens prepared for dissection have the arteries injected with red latex, the veins are injected with blue latex. Exceptions to this will be noted as they are encountered.)

3.   Identify the two large vena cavae—the anterior and posterior vena cavae—entering the right atrium. (These vessels are homologous to the superior and inferior vena cavae, respectively, in the human.) The caval veins drain the same relative body areas as in humans. The anterior vena cava is the largest dark-colored vessel entering the base of the heart. Also identify the pulmonary trunk (usually injected with blue latex) extending anteriorly from the right ventricle and the right and left pulmonary arteries. Trace the pulmonary arteries until they enter the lungs. Locate the pulmonary veins entering the left atrium and the ascending aorta arising from the left ventricle running dorsally to the anterior vena cava and to the left of the body midline. Also identify the ductus arteriosus, a short vessel connecting the aorta to the pulmonary trunk, approximately at the point where the pulmonary trunk divides to form the pulmonary arteries. This is a fetal structure that shunts blood from the pulmonary artery to the aorta, bypassing the nonfunctional fetal lungs. After birth it is occluded and becomes the fibrous ligamentum arteriosum.

VENOUS SUPPLY   As you study the venous system of the fetal pig, keep in mind that the vessels are named for the region drained, not for the point of union with other veins. Refer to Figures 32.9 and 32.10 as you proceed with the dissection, but note that not all vessels shown on the figures are discussed.

1.   Reidentify the posterior vena cava and trace it to its passage through the diaphragm. Note as you follow its course that the intercostal veins drain into a much smaller vein lying to the left of the posterior vena cava (and partially beneath the aorta)—the **hemiazygos vein.** The hemiazygos vein does not enter the posterior vena cava (as does the corresponding azygos vein in humans) but instead generally enters independently into the right atrium, immediately posterior to the point of entry of the posterior vena cava. (In some cases the hemiazygos appears as the first branch off the posterior vena cava and is thus an extremely variable vessel in the pig.) Follow the hemiazygos vein to its entrance into the heart.

2.   Trace the posterior vena cava into the liver. Using a scalpel, scrape away the liver tissue surrounding the posterior vena cava to expose the hepatic

veins (three to four in the pig) that enter it. At this point, also identify the **umbilical vein** (a fetal vessel only), which carries oxygen and nutrient-rich blood from the placenta to the fetus. Trace it into the liver, where it becomes a large thin-walled duct, the **ductus venosus.** The ductus venosus, which also receives blood from the hepatic portal vein, empties into the posterior vena cava. Both the ductus venosus and the umbilical vein atrophy and degenerate after birth.

3.   Displace the intestines to the left side of the body cavity, and proceed posteriorly to identify the following veins in order. All of these veins empty into the posterior vena cava and drain the same organs served by the same-named arteries. Variations in the connections of the veins to be localized are not uncommon. If you observe deviations, call them to the attention of your instructor.

Adrenolumbar veins: from the adrenal glands and body wall. (Generally the right adrenolumbar vein drains into the posterior vena cava, and the left drains into the left renal vein posteriorly, but you may observe variations.)
Renal veins: from the kidneys (It is common to find two renal veins on the right side.)
Genital veins (testicular or **ovarian veins):** these veins are small and generally poorly injected; the right member drains into the posterior vena cava while the left genital vein empties into the left renal vein
Iliolumbar veins: drain the muscles of the back
Common iliac veins: form the posterior vena cava by their union

The common iliac veins are formed in turn by the union of the internal and external iliac veins. The more medial internal iliac veins receive branches from the pelvic organs and gluteal region. The external iliac vein receives blood from the lower extremity. As the external iliac vein enters the thigh, by running beneath the inguinal ligament, it receives the deep femoral vein, which serves the thigh and the external genital region and then becomes the femoral vein, which receives blood from the thigh, leg, and foot. Follow the femoral vein down the thigh to identify the great saphenous vein, a superficial vein that courses up the inner aspect of the calf and across the inferior portion of the gracilis muscle (accompanied by the great saphenous artery and nerve) to enter the femoral vein. The femoral vein is formed by the union of this vein and the popliteal vein, which can be found deep in the thigh.

4.   Return to the abdominal cavity to trace the portal drainage (Figure 32.10). Locate the hepatic portal vein by reidentifying the ductus venosus and tracing it to the point of entry of the hepatic portal vein. Follow the hepatic portal vein posteriorly to the viscera. The hepatic portal system is usually

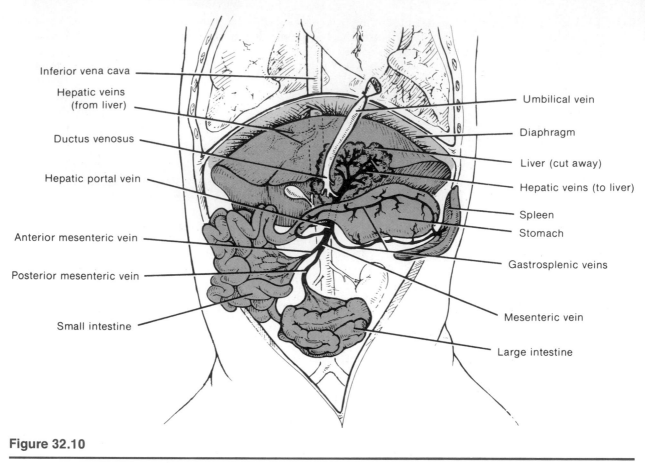

Figure 32.10

Hepatic portal circulation of the fetal pig.

poorly injected, but with a little patience, you can trace it. In the pig, the hepatic portal vein is formed by the union of the **gastrosplenic** and **mesenteric veins.** (In humans, the portal vein is formed by the union of the splenic and superior mesenteric veins.) Carefully tear away the greater omentum, and loosen the mesenteric attachment of the pancreas. If possible, locate the following vessels, which empty into the hepatic portal vein:

Gastrosplenic vein: formed from the union of veins draining the stomach and spleen

Anterior mesenteric vein: lift the small intestine to expose the mesentery with its fan-shaped network of veins draining the small intestine; the point of union of these veins into a single vessel is the beginning of the anterior mesenteric vein. Trace the anterior mesenteric vein to the point where it joins with the posterior mesenteric vein, a small vessel draining the large intestine.

Mesenteric vein: formed by the union of the anterior and posterior mesenteric veins; runs anteriorly to join with the gastrosplenic vein to form the hepatic portal vein

ARTERIAL SUPPLY Refer to Figure 32.11 as you continue the dissection.

1. Return to the thorax, lift the left lung, and follow the course of the ascending aorta through the thoracic cavity. The esophagus overlies it along its course. Note the paired intercostal arteries that branch laterally from the aorta in the thoracic region.

2. Follow the aorta through the diaphragm into the abdominal cavity. Carefully pull the peritoneum away from its ventral surface to make the identification of the following vessels possible:

Celiac trunk: the first abdominal branch of the aorta seen diverging from the aorta immediately as it enters the abdominal cavity; supplies the stomach, liver, gall bladder, pancreas, and spleen. Trace as many of its branches to these organs as possible.

Superior (or **anterior**) **mesenteric artery:** emerges from the ventral surface of the abdominal aorta approximately ½ inch posterior to the celiac trunk and supplies the small intestine and

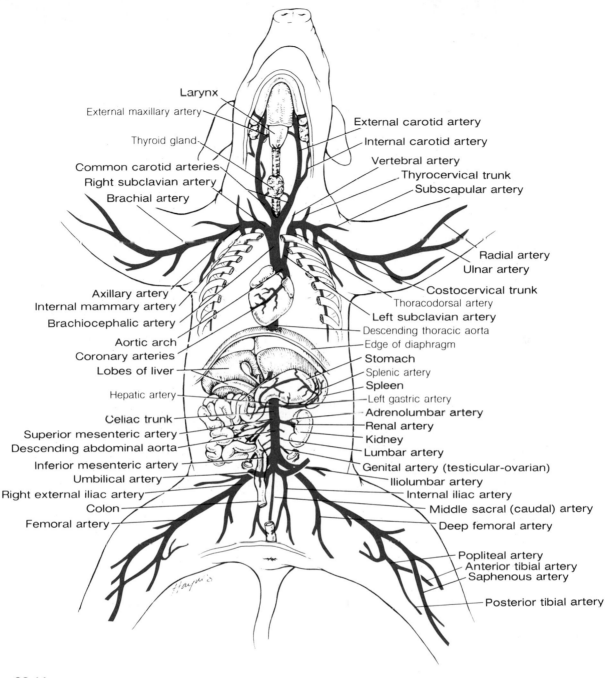

Larynx
External maxillary artery
Thyroid gland
Common carotid arteries
Right subclavian artery
Brachial artery
External carotid artery
Internal carotid artery
Vertebral artery
Thyrocervical trunk
Subscapular artery
Radial artery
Ulnar artery
Axillary artery
Internal mammary artery
Brachiocephalic artery
Aortic arch
Coronary arteries
Lobes of liver
Hepatic artery
Celiac trunk
Superior mesenteric artery
Descending abdominal aorta
Inferior mesenteric artery
Umbilical artery
Right external iliac artery
Colon
Femoral artery
Costocervical trunk
Thoracodorsal artery
Left subclavian artery
Descending thoracic aorta
Edge of diaphragm
Stomach
Splenic artery
Spleen
Left gastric artery
Adrenolumbar artery
Renal artery
Kidney
Lumbar artery
Genital artery (testicular-ovarian)
Iliolumbar artery
Internal iliac artery
Middle sacral (caudal) artery
Deep femoral artery
Popliteal artery
Anterior tibial artery
Saphenous artery
Posterior tibial artery

Figure 32.11

Arterial system of the fetal pig.

most of the large intestine. Spread the mesenteric attachment of the small intestine to observe the branches of this artery as they run to supply the small intestine.

Adrenolumbar arteries: paired arteries seen diverging from the aorta slightly posterior to the superior mesenteric artery and running to supply the muscles of the body wall and the adrenal glands

Renal arteries: large paired arteries arising from the aorta about 1 inch posterior to the adre-

nolumbar arteries and coursing to the kidneys, which they supply. Generally, the right renal artery is superiorly located. Branching of the renal arteries frequently occurs as they approach and enter the kidneys.

Genital arteries (testicular or ovarian): slender paired arteries serving the needs of the gonads

Inferior (or posterior) mesenteric artery: an unpaired thin vessel arising from the ventral surface of the aorta posterior to the genital arteries; supplies the last half of the large intestine

Iliolumbar arteries: paired, rather large arteries that supply the body musculature in this region of the trunk

External iliac arteries: large arteries that issue from the lateral surface of the aorta and continue through the body wall, passing under the inguinal ligament to enter the thigh

3. The descending aorta ends posteriorly by dividing into three arteries, the two lateral **umbilical arteries** and the midline **middle sacral artery,** which continues posteriorly through the sacrum to serve the tail. The thick umbilical arteries, which carry oxygen-poor blood from the fetus to the placenta, pass through the ventral abdominal wall to enter the umbilical cord. (After birth, the umbilical arteries degenerate.) The **internal iliac arteries** arise from the umbilical arteries and proceed into the pelvis to serve organs located there.

4. Return to an external iliac artery and trace it into the thigh, where it becomes the femoral artery, which supplies the leg and the deep femoral artery that supplies the medial proximal thigh. In the region of the knee, the femoral artery gives off a superficial branch, the saphenous artery, and then descends deep to the knee to become the popliteal artery. The branches of the popliteal artery (**anterior** and **posterior tibial arteries**) supply the distal portion of the leg and foot.

Blood Vessels of the Head, Neck, and Upper Extremities

VENOUS SUPPLY Refer to Figure 32.9 as you continue the dissection.

1. Return to the thoracic region, and reidentify the anterior vena cava as it enters the right atrium. Trace it anteriorly to identify its branches:

Costocervical trunks: paired arteries entering the anterior vena cava just before it enters the heart; receive blood from three veins, the most important of which is the vertebral vein, which drains the cervical vertebral area

Internal mammary veins: draining into the anterior vena cava at the level of the third rib; drain the ribs and mammary glands

Right and left brachiocephalic (innominate) veins: form the anterior vena cava by their union

2. Reflect the pectoral muscles, and trace a brachiocephalic vein laterally. Identify the three large veins that join to form it, the external jugular vein, the internal jugular vein, and the subclavian vein. The internal jugular vein drains the deeper regions of the head and lies adjacent to the trachea, common carotid artery, and vagus nerve in the neck region. The external jugular vein, seen lateral to the internal jugular vein, drains the superficial tissues of the head. It is formed by the union of the external and internal maxillary veins in the general area of the submaxillary gland. Several other veins drain into the external and internal jugular veins (inferior cervical, lingual, occipital, and others); these will not be discussed here but are shown on the figure. The thoracic duct (lymphatic system duct) enters the left external jugular or subclavian vein from behind the peritoneum. If possible, identify the thoracic duct, which will appear as an elongated beaded vessel, at this time. If the venous system has been well injected, some of the blue latex may have gotten past the valves and entered the first portion of this duct.

3. The third vessel contributing to the formation of the brachiocephalic vein, the subclavian vein, drains much of the shoulder and all of the arm. Follow the course of this vessel as it moves laterally out of the thoracic cavity into the forelimb, where it becomes the axillary vein. The axillary vein gives off several branches, among them the subscapular vein, which drains the proximal part of the forelimb and shoulder. Other branches that receive venous drainage from the shoulder and the pectoral muscles are shown in the figure but need not be identified in this dissection.

4. Follow the axillary vein into the arm, where it becomes the brachial vein. This vein can be found in the medial side of the arm accompanying the brachial artery and nerve. Trace it to the point where it receives the radial and ulnar veins (which drain the forelimb) at the inner bend of the elbow. (In the pig, there are often two brachial veins, which anastamose frequently along their course. This condition is depicted in the figure.) Also locate the cephalic vein, a prominent vessel draining the more superficial areas of the forelimb. Follow it to its point of entry into the base of the external jugular vein.

ARTERIAL SUPPLY Refer to Figure 32.11 as you continue with the dissection.

1. Reidentify the aorta as it emerges from the left ventricle. As you noted in the dissection of the sheep heart, the first branches of the aorta are the coronary arteries, which supply the myocardium. The coronary arteries emerge from the base of the aorta and can be seen on the surface of the heart. Follow the aorta as it arches (aortic arch), and identify its major branches. In the pig, the aortic arch gives off two large vessels, the brachiocephalic artery and the left subclavian artery. The brachiocephalic artery has two major branches, the right subclavian artery (laterally) and the medial bicarotid trunk, a very short vessel that bifurcates to form the right and left common carotid arteries. In some cases, the two common carotid arteries may be seen arising directly from the brachiocephalic artery.

2. Follow the right common carotid artery along the right side of the trachea as it moves anteriorly, giving off branches to the neck muscles, thyroid gland, and trachea. At the superior border of the larynx, it branches to form the external and internal carotid arteries. The internal carotid is generally much smaller than the external carotid artery in the pig. The distribution of the carotid arteries parallels that in humans.

3. Follow the right subclavian artery laterally. The first vessel to issue from the subclavian is the **costocervical trunk,** which soon splits into three branches. (Any or all of these branches may also arise separately from the subclavian artery.) The first of these is the vertebral artery, which along with the internal carotid artery provides the arterial circulation of the brain. Follow the vertebral artery to the cervical vertebrae. The second and third branches supply the musculature of the dorsal aspect of the neck, the deep muscles of the shoulder, and some of the intercostal muscles of the rib cage. Other branches of the subclavian include the large **thyrocervical artery** (to the deep muscles of the neck and shoulder and the thyroid and parotid glands), and the **internal mammary artery** (serving the ventral thoracic wall). These vessels arise from the subclavian approximately opposite one another, distal to the costocervical trunk. As the subclavian artery passes in front of the first rib, it becomes the **axillary artery,** which provides several branches to the trunk and shoulder muscles (the **ventral thoracic artery,** to the pectoral and latissimus dorsi muscles; the **lateral thoracic artery,** to the teres and latissimus dorsi muscles; and the large **subscapular artery,** which serves the subscapular muscle region).

4. Follow the course of the axillary artery into the arm, where it becomes the **brachial artery** and travels with the median nerve down the length of the humerus. At the elbow, the brachial artery branches to produce the two major arteries serving the forearm and hand, the radial and ulnar arteries.

5. Once you have completed your dissection, properly clean your dissecting instruments and dissecting pan, and wrap and tag your pig for storage.

Human Cardiovascular Physiology—Blood Pressure and Pulse Determinations

EXERCISE

33

OBJECTIVES

1. To define *systole, diastole,* and *cardiac cycle.*

2. To state the normal length of the cardiac cycle, the relative pressure changes occurring within the atria and ventricles during the cycle, and the timing of valve closure.

3. To use the stethoscope to auscultate heart sounds and to relate heart sounds to cardiac cycle events.

4. To describe the clinical significance of heart sounds.

5. To demonstrate the thoracic locations where the first and second heart sounds are most accurately auscultated.

6. To define *murmur.*

7. To define *pulse, pulse deficit, blood pressure,* and *sounds of Korotkoff.*

8. To accurately determine the apical and radial pulse of a subject.

9. To accurately determine the blood pressure of a subject using a sphygmomanometer and to relate systolic and diastolic pressures to events of the cardiac cycle.

10. To note factors affecting and/or determining blood pressure.

MATERIALS

Stethoscope
Sphygmomanometer
Watch (or clock) with a second hand
Step stool
Alcohol swabs
Millimeter ruler
Ice
Small basin suitable for the immersion of one hand
Laboratory thermometer (°C)
Record or audiotape: "Interpreting Heart Sounds" (available on free loan from the local chapters of the American Heart Association)
Phonograph or tapedeck

Any comprehensive study of human cardiovascular physiology takes much more time than a single laboratory period. However, it is possible to investigate a few phenomena such as pulse determinations, auscultation of heart sounds, and blood pressure measurements, which reflect the heart in action and the function of blood vessels. The electrocardiogram is studied separately in Exercise 31. A discussion of the cardiac cycle will provide a basis for understanding and interpreting the various physiologic measurements to be taken.

CARDIAC CYCLE

In the healthy heart, the two atria contract simultaneously. As they begin to relax, simultaneous contraction of the ventricles occurs. According to general usage, the terms **systole** and **diastole** refer to events of ventricular contraction and relaxation, respectively. The **cardiac cycle** is equivalent to one complete heart beat, during which both atria and ventricles contract and then relax. It is marked by a succession of changes in blood volume and pressure within the heart. Figure 33.1 is a graphic representation of the events of the cardiac cycle for the left side of the heart. Although pressure changes in the right side are lower than those in the left, the same relationships apply.

We will begin the discussion of the cardiac cycle with the heart in complete relaxation. At this point, pressure in the heart is very low, blood is flowing passively into and through the atria into the ventricles from the pulmonary and systemic circulations,

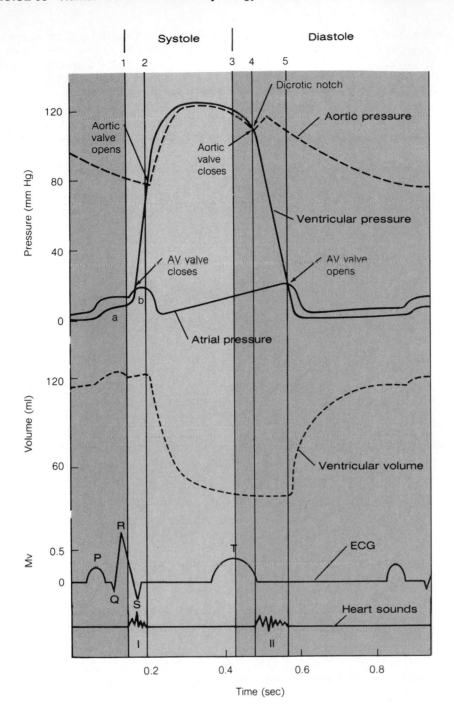

Figure 33.1

Graphic representation of pressure and volume changes in the left ventricle during one cardiac cycle. Depicts valvular events and correlates changes to heart sounds and ECG. (ECG is considered in Exercise 31.)

the semilunar valves are closed, and the AV valves open. As atrial contraction occurs, atrial pressure increases, forcing residual blood into the ventricles. Shortly after, ventricular systole begins, and the intraventricular pressure increases rapidly, closing the AV valves. When ventricular pressure exceeds that of the large arteries leaving the heart, the semilunar valves are forced open and blood is expelled through them from the ventricular chambers. During this phase, the aortic pressure reaches

approximately 120 mm Hg. During ventricular systole, the atria relax, and their chambers fill with blood, which results in a gradually increasing atrial pressure. At the end of ventricular systole, the ventricles relax, the semilunar valves snap shut, preventing backflow, and momentarily the ventricles are closed chambers. When the aortic semilunar valve snaps shut, there is a momentary increase in the aortic pressure caused by the elastic recoil of the aorta after valve closure. This event results in the pressure fluctuation called the **dicrotic notch** (see Figure 33.1). As the ventricles relax, the pressure within them begins to drop. When it is again less than atrial pressure, the AV valves are forced open and the ventricles again begin to fill with blood. Atrial and aortic pressures decrease, and the ventricles rapidly refill, completing the cycle.

Since the average heart beats approximately 72 beats per minute, the length of the cardiac cycle is about 0.8 seconds. Of this time period, atrial contraction occupies the first 0.1 second, which is followed by atrial relaxation and ventricular contraction for the next 0.3 second. The last 0.4 second is the quiescent, or ventricular relaxation, period. When the heart beats at a more rapid pace than normal, this period decreases.

Note that two different types of phenomena control the movement of blood through the heart: the alternate contraction and relaxation of the myocardium and the opening and closing of valves (which is entirely dependent on the pressure changes within the heart chambers).

Study the figure carefully to make sure you understand what has been discussed before continuing with the next portion of the exercise.

AUSCULTATION OF HEART SOUNDS

You can hear two distinct sounds during each cardiac cycle. These heart sounds are commonly described by the monosyllables "lup" and "dup," and the sequence is designated lup-dup, pause, lup-dup, pause, and so on. The first heart sound (lup) is associated with the closure of the AV valves at the beginning of systole. Likewise, the second heart sound (dup) is most commonly associated with the closure of the semilunar valves and corresponds with the end of systole. Figure 33.1 indicates the timing of heart sounds in the cardiac cycle.

Abnormal heart sounds are referred to as *murmurs* and often indicate valvular problems. In valves that do not close tightly, closure is followed by a swishing sound caused by the backflow of blood (regurgitation). Distinct sounds are also associated with the tortuous flow of blood through constricted or stenosed valves. Before auscultating your partner's heart sounds, listen to the recording, "Interpreting Heart Sounds," so that you may hear both normal and abnormal heart sounds.

In the following procedure, you will auscultate heart sounds with an ordinary stethoscope. A number of more sophisticated heart sound amplification systems are on the market, and your instructor may prefer to use such if it is available. If so, directions for the use of this apparatus will be provided by the instructor.

1. Obtain a stethoscope and some alcohol swabs. Heart sounds are best auscultated (listened to) if the subject's outer clothing is removed.

2. Clean the earpieces of the stethoscope with an alcohol swab and allow the alcohol to dry. Notice that the earpieces are angled. For comfort and best auscultation, the earpieces should be angled in a forward direction when placed into the ears.

3. Don the stethoscope, and place the bell of the stethoscope on your partner's thorax just to the sternal side of the left nipple at the fifth intercostal space, and listen carefully for heart sounds. The first sound will be a longer, louder (more booming) sound than the second, which is short and sharp. After listening for a couple of minutes, try to time the pause between the second and first heart sounds.

How long is this interval? _____ sec

How does it compare to the interval between the first

and second heart sounds? _____

_____

_____

4. To differentiate individual valve sounds somewhat more precisely, auscultate the heart sounds over specific thoracic regions. Refer to Figure 33.2 for the positioning of the stethoscope.

Auscultation of AV Valves

The mitral valve generally closes slightly before the tricuspid valve. You can hear it more clearly if you place the stethoscope over the apex of the heart, which is at the fifth intercostal space approximately in line with the middle region of the clavicle. Listen to the heart sounds at this region, and then move the stethoscope medially to the left margin of the sternum, where the triscuspid valve is most clearly auscultated. Are you able to detect the slight lag between the

closure of the mitral and tricuspid valves? _____

Auscultation of Semilunar Valves

Again there is a slight dysynchrony of valve closure; the aortic semilunar valve normally snaps shut just ahead of the pulmonary semilunar valve. If the subject inspires deeply but gently, the filling of the right ventricle will be delayed slightly (due to the compression of the pulmonary

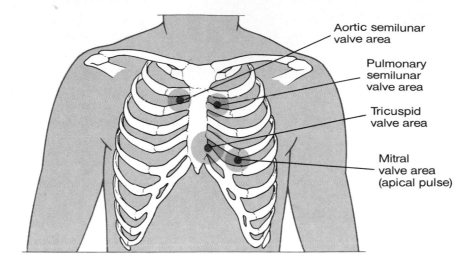

Figure 33.2

Areas of the thorax where valvular sounds can be best detected.

blood vessels by the increased intrapulmonary pressure), and the two sounds can be heard more distinctly. Position the stethoscope over the second intercostal space just to the right of the sternum. As you listen, have your partner take a deep breath. The aortic valve is best heard in this position. Then move the stethoscope to the left side of the sternum in the same line, and auscultate the pulmonary valve. Listen carefully and try to heart the "split" between the closure of these two valves in the second heart sound.

PALPATION OF THE PULSE

The term **pulse** refers to the alternating surges of pressure (expansion and then recoil) in an artery that occur with each contraction and relaxation of the left ventricle. Normally the pulse rate (pressure surges per minute) equals the heart rate (beats per minute), and the pulse averages 70 to 76 beats per minute in the resting state. Parameters other than pulse rate are also useful clinically. You may also assess the regularity or rhythmicity of the pulse and its amplitude and/or tension—does the blood vessel expand and recoil (sometimes visibly) with the pressure waves? Can you feel it strongly or is it difficult to detect? Is it regular like the ticking of a clock or does it seem to skip beats?

Superficial Pulse Points

The pulse may be felt easily on any superficial artery when the artery is compressed over a bone or firm tissue. Palpate the following pulse or pressure points on your partner by placing the fingertips of the first two or three fingers of one hand over the artery. It is generally helpful to compress the artery firmly as you begin your palpation, and then immediately ease up on the

pressure slightly. In each case, note the regularity of the pulse and assess the degree of tension or amplitude.

Carotid artery: place the fingertips at the side of the neck

Temporal artery: anterior to the ear, in the temple region

Brachial artery: in the antecubital fossa at the point where it bifurcates into the radial and ulnar arteries

Radial artery: at the lateral aspect of the wrist above the thumb

Facial artery: clench the teeth and palpate the pulse just anterior to the masseter muscle on the mandible (in line with the corner of the mouth)

Femoral artery: in the groin

Popliteal artery: at the back of the knee

Dorsalis pedis artery: on the dorsum of the foot

Which pulse point had the greatest amplitude?

_____

Which the least? _____

Can you offer any explanation for this? _____

_____

_____

Because of its easy accessibility, the pulse is most often taken on the radial artery. With your partner sitting quietly, practice counting the radial pulse for 1 minute. Make three counts and average the results.

count 1 _____ count 2 _____

count 3 _____ average _____

Apical-Radial Pulse

The correlation between the apical and radial pulse rates can be determined by simultaneously counting them. The apical pulse (actually the counting of heart beats) may be slightly more rapid because of a slight lag between the time the blood rushes from the heart into the large arteries where it can be palpated. However, any *large* difference between the values observed (referred to as a **pulse deficit**) may indicate cardiac impairment or arteriosclerosis. Apical pulse counts are routinely ordered for those with cardiac decompensation.

With the subject sitting quietly, one student should determine the apical pulse with a stethoscope while another simultaneously counts the radial pulse rate. The stethoscope should be positioned over the fifth left intercostal space. The person taking the radial pulse should determine the starting point for the count and give the stop-count signal exactly 1 minute later. Record your values below.

apical count _____ beats/min

radial count _____ pulses/min

pulse deficit _____ /min

BLOOD PRESSURE DETERMINATIONS

Blood pressure is defined as the pressure the blood exerts against any unit area of the blood vessel walls, and it is generally measured in the arteries. Because the heart alternately contracts and relaxes, the rhythmic flow of blood into the arteries causes the blood pressure to rise and fall during each beat. Thus you must take two blood pressure readings: the **systolic pressure,** which is the pressure in the arteries at the peak of ventricular ejection, and the **diastolic pressure,** which reflects the pressure during ventricular relaxation. Blood pressures are reported in millimeters of mercury (mm Hg), with the systolic pressure appearing first; 120/80 translates to 120 over 80, or a systolic pressure of 120 mm Hg and a diastolic pressure of 80 mm Hg. Normal blood pressure varies considerably from one person to another.

In this procedure, you will measure arterial and venous pressures by indirect means and under various conditions. You will investigate and demonstrate factors affecting blood pressure, the rapidity of blood pressure changes, and the large differences between arterial and venous pressures.

Indirect Measurement of Arterial Blood Pressure

The sphygmomanometer is an instrument used to obtain blood pressure readings by the ausculta-

tory method. It consists of an inflatable cuff with an attached pressure gauge. The cuff is placed around the upper arm and inflated to a pressure higher than systolic pressure to occlude circulation to the forearm. Cuff pressure is gradually released, and the examiner listens with a stethoscope for characteristic sounds called the **sounds of Korotkoff,** which indicate the resumption of blood flow into the arm. The pressure at which the first soft tapping sounds can be detected is recorded as the systolic pressure. As the pressure is reduced further, blood flow becomes more turbulent, and the sounds become louder. As pressure is reduced still further below the diastolic pressure, the artery is no longer compressed and blood flows freely and without turbulence. At this point, the sounds of Korotkoff can no longer be detected. The pressure at which the sounds disappear is recorded as the diastolic pressure.

 1. Work in pairs to obtain radial artery blood pressure readings. Obtain a stethoscope, alcohol swabs, and a sphygmomanometer. Clean the earpieces of the stethoscope, and check the cuff for the presence of trapped air by compressing it against the laboratory table.

2. The subject should sit in a comfortable position at the laboratory table with the arm supported on the bench (approximately at heart level if possible). Wrap the cuff around the subject's arm just above the elbow with the inflatable area on the medial arm surface. The cuff may be marked with an arrow; if so, the arrow should be positioned over the brachial artery (Figure 33.3). Secure the cuff by tucking the distal end under the wrapped portion or by bringing the Velcro areas into apposition.

3. Palpate the brachial pulse, don the stethoscope, and place its diaphragm over the pulse point. (*The cuff should not be kept inflated for more than 1 minute;* so if you have any trouble obtaining a reading within this time, deflate the cuff, wait 1 or 2 minutes, and try again.)

4. Inflate the cuff to approximately 160 mm Hg pressure, and slowly release the pressure valve. Watch the pressure gauge as you listen carefully for the first soft thudding sounds of the blood spurting through the partially occluded artery. Note this pressure (systolic pressure), and continue to release the pressure. You will notice first an increase, then a muffling of the sound. Note the pressure at which the sound becomes muffled or disappears as the diastolic pressure. Controversy exists over the point at which the diastolic pressure should be recorded: so in some cases you may see readings such as 120/80/78, which would indicate the systolic pressure, the pressure at which the sound muffles, and the pressure at which the sound disappears. It makes little difference here as to which of the two diastolic

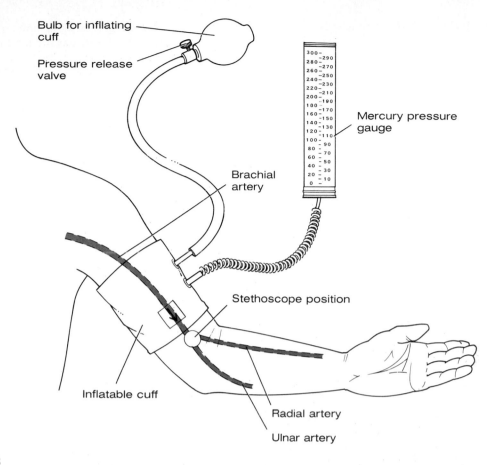

Bulb for inflating cuff

Pressure release valve

Brachial artery

Mercury pressure gauge

Stethoscope position

Inflatable cuff

Radial artery

Ulnar artery

Figure 33.3

Positioning of the sphygmomanometer cuff and stethoscope when auscultating blood pressure in the brachial artery.

pressures is recorded but be consistent. Make two blood pressure determinations, and record your results below.

First trial:

systolic pressure _____ diastolic pressure _____

Second trial:

systolic pressure _____ diastolic pressure _____

5.   Compute the **pulse pressure** for each trial. The pulse pressure is the difference between the systolic and diastolic pressures and indicates the amount of blood forced from the heart during systole, or the actual "working" pressure.

Pulse pressure:

first trial _____   second trial _____

Estimation of Venous Pressure

It is not possible to measure venous pressure with the sphygmomanometer, and the available meth-

ods produce estimates at best, because venous pressures are so much lower than arterial pressures. The pressure difference between veins and arteries becomes obvious when these vessels are cut. If a vein is cut, the blood flows evenly from the cut; a lacerated artery produces rapid spurts of blood.

1.   Ask your lab partner to stand with her right side toward the blackboard with arms hanging freely at the sides. Mark the approximate level of the right atrium on the board (just slightly higher than the point at which you auscultated the apical pulse; See Figure 33.2.)
2.   Observe the superficial veins on the dorsum of the subject's right hand as she alternately raises and lowers it, noting the collapsing and filling of the veins with changing internal pressures. Have the subject do this maneuver several times until you can determine the point at which the veins have just collapsed; then measure the distance in millimeters in the vertical plane from this point to the level of the right atrium (previously marked). Record this value. Distance of right arm from right atrium at point of venous collapse:

_____ mm

3.   Compute the venous pressure in millimeters of mercury with the following formula:

Venous pressure =

$$\frac{1.056 \text{ (specific gravity of blood)} \times \text{mm (measured)}}{13.6 \text{ (specific gravity of Hg)}}$$

Venous pressure computed: _____ mm Hg

Normal venous pressure varies from approximately 30 to 120 mm Hg; that of the hand ranges between 30 and 40 mm Hg. How does your

computed value compare? _____

4.   Because of the relative thinness of the venous walls, pressure within them is readily affected by external factors such as muscle activity, deep pressure, and pressure changes occurring in the thorax during breathing. The Valsalva maneuver, which increases intrathoracic pressure, can be used to demonstrate the effect of thoracic pressure changes on venous pressure.

To perform this maneuver take a deep breath, and then mimic the motions of exhaling forcibly without exhaling. In reaction to this, the glottis will close, and intrathoracic pressure will increase. (Most of us have performed this maneuver unknowingly in acts of defecation in which there is "straining at stool.") The subject should perform the Valsalva maneuver while the height of the hand at the point of venous collapse is measured under these conditions. Compute the venous pressure and record below. Distance of right arm from right atrium at point of venous collapse while performing the Valsalva maneuver _____ mm

Venous pressure computed _____ mm Hg

How does this value compare with the venous pressure measurement computed for the relaxed

state? _____

Explain: _____

_____

EFFECT OF VARIOUS FACTORS ON BLOOD PRESSURE AND HEART RATE

Arterial blood pressure can be shown to be directly proportional to cardiac output (amount of blood pumped out of the left ventricle per unit time) and peripheral resistance to blood flow. Peripheral resistance is increased by constriction of blood vessels (most importantly, the arterioles), increased blood viscosity or volume,

and by a loss of elasticity of the arteries (seen in arteriosclerosis). Thus, any factor that increases either the cardiac output or peripheral resistance causes an almost immediate reflex rise in blood pressure. A close examination of these relationships reveals that many factors alter blood pressure— age, weight, time of day, exercise, body position, emotional state, and various drugs, for example. The influence of a small number of these factors is investigated here.

The following tests require one student as subject and two as examiners (one taking the radial pulse and the other auscultating the brachial blood pressure). The sphygmomanometer cuff should be left on the subject's arm throughout the experiments (in a deflated state, of course), so that the blood pressure can be taken quickly at the proper times. In each case, take the measurements at least twice.

Posture
Monitor blood pressure and pulse under the conditions on chart 1 and record your values on that chart.

Exercise
Conduct the tests listed on chart 2 (deep knee bends and bench stepping), to determine the effect of exercise on blood pressure and pulse. If possible, obtain results for a well-conditioned subject and a poorly conditioned subject, and compare the results. Make base line measurements before beginning each test; then measure blood pressure and pulse immediately after exercise and at 1-minute intervals until blood pressure stabilizes. Allow the subject to rest for 5 minutes between tests. Record your values on chart 2.

When did you note a greater elevation of blood

pressure and pulse? _____

Explain. _____

_____

Did you note a sizeable difference between the after-exercise values for well-conditioned and

poorly conditioned individuals? _____

Explain. _____

_____

Did the diastolic pressure also increase? _____

Explain. _____

_____

Chart 1   Posture

	Trial 1 BP	Pulse	Trial 2 BP	Pulse
Sitting quietly	_____	_____	_____	_____
Reclining (after 2 to 3 min)	_____	_____	_____	_____
Immediately on standing from the reclining position	_____	_____	_____	_____
After standing for 3 min	_____	_____	_____	_____

Chart 2   Exercise

Well-conditioned individual	Base line BP	P	Immediately BP	P	1 min BP	P	Interval following test 2 min BP	P	3 min BP	P	4 min BP	P
10 deep knee bends	____	____	____	____	____	____	____	____	____	____	____	____
Bench step for 3 min at 30/min	____	____	____	____	____	____	____	____	____	____	____	____

Poorly conditioned individual	Base line BP	P	Immediately BP	P	1 min BP	P	Interval following test 2 min BP	P	3 min BP	P	4 min BP	P
10 deep knee bends	____	____	____	____	____	____	____	____	____	____	____	____
Bench step for 3 min at 30/min	____	____	____	____	____	____	____	____	____	____	____	____

_____

_____

Nicotine

To investigate the effects of nicotine on blood pressure and pulse, take a base line blood pressure and pulse of a smoker in a standing position. Ask the subject to light up and smoke in the usual manner. After 2 minutes, take blood pressure and pulse readings. Repeat the measurements for 2 consecutive minutes, and record your values on chart 3 (p. 300).

How do the effects of nicotine bring about the changes noted? (Hint: nicotine has vasoconstric-

tor effects.) _____

_____

_____

_____

Temperature

Heat acts as a vasodilator of the peripheral blood vessels, but cold has a vasoconstrictor effect. You

Chart 3 Nicotine

Base line BP P	After smoking for 2 min BP P	After smoking for 3 min BP P	After smoking for 4 min BP P
‾‾‾‾ ‾‾‾‾	‾‾‾‾ ‾‾‾‾	‾‾‾‾ ‾‾‾‾	‾‾‾‾ ‾‾‾‾

Chart 4 Temperature (cold)

Base line BP P	1 min BP P	2 min BP P	3 min BP P	4 min BP P	5 min BP P	6 min BP P	7 min BP P
‾‾ ‾‾	‾‾ ‾‾	‾‾ ‾‾	‾‾ ‾‾	‾‾ ‾‾	‾‾ ‾‾	‾‾ ‾‾	‾‾ ‾‾

will investigate the effects of cold by performing the **cold pressor test.**

Measure the blood pressure and pulse of the subject as he sits quietly. Obtain a basin, fill it with ice cubes, and add water. When the temperature of the ice bath has reached 5 C, immerse one of the subject's hands in the ice water. With the hand still immersed throughout the testing, take blood pressure and pulse readings at 1-minute intervals for a period of 5 to 7 minutes, and record the values in chart 4.

How did the blood pressure change during cold

exposure? _____

_____

Was there any change in pulse? _____

_____

Explain. _____

_____

_____

_____

What is the function of the vasoconstriction resulting

from the cold? _____

_____

_____

Frog Cardiovascular Physiology

OBJECTIVES

1. To list the properties of cardiac muscle as automaticity and rhythmicity and define each.

2. To explain the statement "Cardiac muscle has an intrinsic ability to beat."

3. To compare the intrinsic rate of contraction of the "pacemaker" of the frog heart (sinus venosus) to that of the atria and ventricle.

4. To compare the relative length of the refractory period of cardiac muscle with that of skeletal muscle, and explain why it is not possible to tetanize cardiac muscle.

5. To define *extrasystole.*

6. To explain at what point in the cardiac cycle (and on an ECG tracing) an extrasystole can be induced.

7. To describe the effect of the following on heart rate: cold, heat, vagal stimulation, atropine sulfate, epinephrine, and potassium, sodium, and calcium ions and to explain the mechanism by which each exerts its effect(s).

8. To define *vagal escape* and discuss the value of this phenomenon.

9. To define partial and *total heart block.*

10. To describe how heart block was induced in the laboratory and explain the results, and to explain how heart block might occur in the body.

11. To note the components of the microcirculatory system.

12. To identify an arteriole, venule, and capillaries in a frog's web, and to cite differences between relative size, rate of blood flow, and regulation of blood flow in these vessels.

13. To describe the effect of heat, cold, local irritation, and histamine on the rate of blood flow in the microcirculation, and to explain how these responses help maintain homeostasis.

MATERIALS

Frog
Dissecting pan and instruments
Petri dishes
Medicine dropper
Millimeter ruler
Frog Ringer's solutions (at room temperature, 5 C, and 32 C)
Kymograph recording apparatus and paper
Two heart levers
Signal marker
Stimulator and platinum electrodes
Thread, rubber bands, fine common pins
Frog board
Ring stand; right angle clamps (three)
Cotton balls
Freshly prepared solutions of the following in dropper bottles:
    5% atropine sulfate
    1% epinephrine
    5% potassium chloride (KCl)
    1% calcium chloride ($CaCl_2$)
    0.7% sodium chloride (NaCl)
    0.01% histamine
    0.01 N HCl

Investigation of heart and blood vessel function in human subjects is most interesting, but many areas obviously do not lend themselves to human experimentation. It would be tantamount to murder to inject a human subject with various drugs to observe their effects on heart activity, to expose the human heart to study the effect of heart block, or to demonstrate the intrinsic heart beat properties of the various regions of the heart by cutting the heart into small fragments. However, this type of investigation can be done on frogs or small laboratory animals and provides valuable data, since the physiologic mechanisms in these animals are similar if not identical to those in humans.

In this exercise, you will conduct the investigations just mentioned. In addition, you will observe the microcirculation in a frog's web and subject it to various chemical and thermal agents to demonstrate their influence on blood flow.

INTRINSIC PROPERTIES OF CARDIAC MUSCLE: AUTOMATICITY AND RHYTHMICITY

Cardiac muscle differs from skeletal muscle both functionally and in its fine structure. Skeletal muscle, whether within or removed from the body, must be electrically stimulated to contract. In contrast, heart muscle can and does depolarize spontaneously in the absence of external stimulation. This property, called **automaticity,** is believed to be due to "leaky" cell membranes that allow sodium ions to slowly enter the cells. This leakage causes muscle cells to slowly depolarize until the action potential threshold is reached and contraction ensues.

In addition, the spontaneous depolarization-repolarization events occur in a regular and continuous manner, a property referred to as **rhythmicity.** In the following experiment, you will observe these properties of cardiac muscle in an in vitro (outside, or removed from the body) situation. Work together in groups of three or four. (The instructor may choose to demonstrate this procedure if time or frogs are at a premium.)

1.   Obtain a frog, dissecting pan and instruments, two Petri dishes, frog Ringer's solution, and a medicine dropper.

2.   Doubly pith the frog, and quickly open the thoracic cavity and observe the heart rate in situ (at the site or within the body).

Record the heart rate _____ beats/min

3.   Dissect out the heart and gastrocnemius muscle of the calf and place the removed organs in separate Petri dishes containing frog Ringer's solution. (Note: Just in case you don't remember, the procedures for pithing the frog and the removal of the gastrocnemius muscle are provided on p. 140 in Exercise 16. The extreme care used in that procedure for the removal of the gastrocnemius muscle need not be exercised here.)

4.   Observe the activity of the two organs for a few seconds.

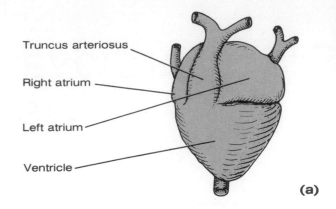

Truncus arteriosus

Right atrium

Left atrium

Ventricle

(a)

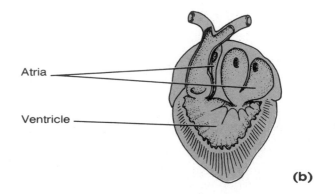

Atria

Ventricle

(b)

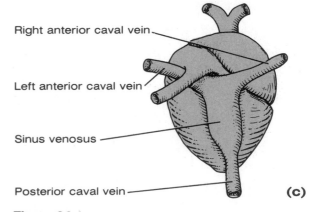

Right anterior caval vein

Left anterior caval vein

Sinus venosus

Posterior caval vein

(c)

Figure 34.1

Anatomy of the frog heart: (a) ventral view showing the single truncus arteriosus leaving the undivided ventricle; (b) longitudinal section showing the two atrial and single ventricular chambers; (c) dorsal view showing the sinus venosus (pacemaker).

Which is contracting? _____

At what rate? _____ beats/min

Is the contraction rhythmic? _____

5.   Sever the sinus venosus from the heart (Figure 34.1). The sinus venosus of the frog's heart corresponds to the SA node of the human heart. Does

the sinus venosus continue to beat? _____

If not, lightly touch it with a probe to stimulate it.

Record its rate of contraction _____

beats/min

6.   Sever the right atrium from the heart, then the left atrium. Does each atrium continue to beat?

_____ Rate _____ beats/min

Does the ventricle continue to beat? _____

Rate _____ beats/min

7.   Continue to fragment the ventricle to determine how small the ventricular fragments must be before the automaticity of ventricular muscle is abolished. Measure these fragments and record their approximate size.

_____ mm × _____ mm × _____ mm

Which portion of the heart exhibited the most

marked automaticity? _____

Which the least? _____

8.   Properly dispose of the frog and heart fragments before continuing with the next procedure.

RECORDING
FROG HEART ACTIVITY
UNDER VARIOUS
CONDITIONS

The heart's effectiveness as a pump is dependent on instrinsic (within the heart) and extrinsic (external to the heart) controls. In this experiment, you will investigate some of these factors.

The nodal system, in which the "pacemaker" imposes its depolarization rate on the rest of the heart, is one intrinsic factor that influences the heart's pumping action. If its impulses fail to reach the ventricles (as in heart block), the ventricles continue to beat but at their own inherent rate, which is much slower (see page 270). Although heart contraction is not dependent on nervous impulses, its rate can be modified by extrinsic impulses reaching it through the autonomic nerves. Additionally, cardiac activity is modified by various chemicals, hormones, ions, and metabolites, whose effects we will examine in the next experimental series.

The frog heart has two atria and a single, incompletely divided ventricle (see Figure 34.1).

The pacemaker is located in the sinus venosus, an enlarged region between the vena cavae and the right atrium. Many believe that the SA node of mammals evolved from the sinus venosus.

Record frog heart activity with kymograph apparatus. Work in groups of four two students handling the equipment setup and two preparing the frog for experimentation.

Kymograph Apparatus Setup

1.   Obtain a kymograph, recording paper, ring stand, frog board, three right-angle clamps, signal magnet, stimulator, platinum electrodes, and two heart levers.
2.   If not using an ink-writing kymograph apparatus, smoke the drum lightly; it is important that the recording be made with minimum pressure against the drum.
3.   According to the setup depicted in Figure 34.2 connect the electrodes to the two output terminals on the stimulator, and attach the signal magnet to the two binding posts on the back of the stimulator. Connect the signal magnet to the ring stand with a clamp, so it will record close to the bottom of the drum when recording is done. Attach the two heart levers to the stand in line with each other above the signal magnet.

Preparation of the Frog

1.   Obtain a frog, room temperature frog Ringer's solution, medicine dropper, dissecting instruments and pan, fine common pins, and some thread.
2.   Doubly pith the frog. (See p. 140 if you don't remember how to do this.)
3.   Make a longitudinal incision through the abdominal and thoracic walls with scissors, and then cut through the sternum to expose the heart.
4.   Grasp the pericardial sac with forceps, and cut it open so that the beating heart can be observed.

Is it an atrial-ventricular sequence? _____

5.   At the same time, locate the vagus nerve, which runs down the side of the neck and parallels the trachea and carotid artery. In good light, it appears to be striated. Slip an 18-inch length of thread under the vagus nerve so that it can later be lifted away from the surrounding tissues by the thread. Then place a saline-soaked cotton ball over the nerve to keep it moistened until you stimulate it later in the procedure.
6.   Flush the heart with the saline (Ringer's) solution. *From this point on the heart must be*

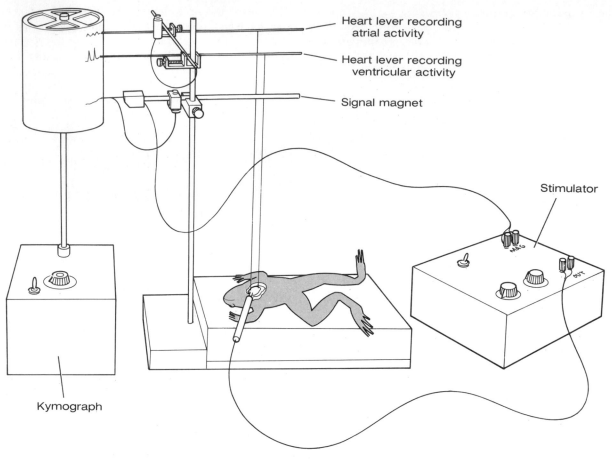

Heart lever recording
atrial activity

Heart lever recording
ventricular activity

Signal magnet

Stimulator

Kymograph

Figure 34.2

Apparatus setup for recording the activity of the frog heart.

kept continually moistened with room temperature Ringer's solution unless other solutions are being used for the experimentation.

7. Attach the frog to the frog board.
8. Bend a common pin to a 90-degree angle, and tie an 18- to 20-inch length of thread to its head. Force the pin through the apex of the heart (do not penetrate the ventricular chamber) until it is well secured in the angle of the pin. Prepare a second pin in the same manner, and insert it through the edge of one atrium.
9. Tie the thread from the apex to the end of the lower muscle lever directly over the heart, and tie the thread from the atrium to the upper muscle lever. Try to keep the threads well apart so that the activities of the two chambers will not interfere with one another. Do not pull the threads too tightly. The threads should be taut but should *not* stretch the heart.
10. Adjust the writing tips of the heart levers so that they are directly in line with one another and with the stylus of the signal marker. This will allow the time relationships between the contractions of the two chambers to be assessed easily. Leave enough room between the three writing tips to give each adequate

working distance. When properly set up, the heart levers should be in a horizontal position.

Recording Base Line Frog Heart Activity

1. Set the drum speed at 7, and turn on the kymograph briefly to ensure that the styluses are recording properly.
2. Set the signal magnet at 1/sec.
3. Record several normal heartbeats (12 to 15), and then adjust the paper speed so that the peaks of ventricular contractions are approximately 2 cm apart. (Peaks indicate systole; troughs indicate diastole.) Note the relative force of heart contractions while recording.
4. Count the number of ventricular contractions

per minute and record. _____ beats/min

5. Compute the A-V interval (period from the beginning of atrial contraction to the beginning of ventricular contraction.) _____ sec

How do the two tracings compare in time?

_____

_____

6. Mark the atrial and ventricular systoles and diastoles on the record. *Remember to keep the heart moistened with Ringer's solution.*

REFRACTORY PERIOD OF CARDIAC MUSCLE

When you investigated the physiology of skeletal muscle in Exercise 16, you saw that repeated rapid stimuli could cause the muscle to remain in the contracted state. In other words, the muscle was tetanized. This was possible because of the relatively short refractory period of skeletal muscle. In this experiment, you will investigate the refractory period of cardiac muscle and its response to stimulation. During the procedure, one student should keep the stimulating electrodes in constant contact with the frog heart ventricle.

1. Set the stimulator to deliver shocks of 20 V, and begin recording.

2. Deliver single shocks at the beginning of ventricular contraction, the peak of ventricular contraction, and then later and later in the cardiac cycle.

3. Observe the recording for **extrasystoles,** which are extra beats that show up riding on the ventricular contraction peak. Also note the **compensatory pause,** which allows the heart to get back on schedule after an extrasystole.

In which portion of the cardiac cycle was it possible to induce an extrasystole? _____

_____

4. Attempt to tetanize the heart by applying a tetanizing stimulation of 20 to 30 impulses per second. What were the results? _____

_____

_____

Considering the function of the heart, why is it important that heart muscle cannot be tetanized?

_____

_____

PHYSICAL AND CHEMICAL FACTORS MODIFYING HEART RATE

Now that you have observed normal frog heart activity, you will have an opportunity to investigate various modifying factors and their effects on heart activity. In each case, record a few normal heartbeats before introducing the modifying factor. After removing the agent, allow the heart to return to its normal rate before continuing with the testing. Indicate the point of introduction and removal of the modifying agent on each record.

Temperature

1. Obtain 5 C and 32 C frog Ringer's solutions and medicine droppers.

2. Increase the speed of the drum so that heart beats appear as spikes 4 to 5 mm apart.

3. Flood the heart with 5 C saline, and continue to record until the recording indicates a change in cardiac activity.

4. Stop recording, pipette off the cold Ringer's solution, and flood with room-temperature saline.

5. Start recording again to determine the resumption of the normal heart rate. When this has been achieved, flood the heart with 32 C Ringer's solution and again record until a change is noted.

6. Stop the recording, pipette off the saline, and flood the heart with room-temperature Ringer's solution once again.

7. Count the heart rate at the two temperatures and record.

_____ beats/min at 5 C

_____ beats/min at 32 C

What change occurred with the cold (5 C) Ringer's solution? _____

_____

What change occurred with the warm (32 C) Ringer's solution? _____

_____

Vagus Nerve Stimulation

The vagus nerve carries parasympathetic impulses to the heart, which modify heart activity.

1. Remove the cotton placed over the vagus nerve, lift the nerve away from the tissues, using previously tied thread, and place it on the platinum electrodes.

2. Stimulate the nerve at a rate of 50/sec for 0.5 msec at a voltage of 1 mV. Continue nerve stimulation until the heart stops momentarily and then begins to beat again (vagal escape). If no effect is observed, increase the stimulus intensity and try again.

3. Discontinue stimulation after you observe vagal escape, and flush the heart with saline until the normal heart rate resumes. What was the effect of vagal stimulation on heart rate? _____

_____

The phenomenon of vagal escape just observed demonstrates that many factors are involved in heart regulation and that any deleterious factor (in this case, excessive vagal stimulation) will be overcome if possible by other physiologic mechanisms.

Atropine Sulfate

Apply a few drops of atropine sulfate to the frog's heart and observe the recording. If no changes are observed within 2 minutes, apply a few more drops. When you observe a response, pipette off the excess atropine sulfate and flood the heart with Ringer's solution. What happens when the atropine sulfate is added? _____

_____

_____

Atropine is a drug that blocks the effect of the neurotransmitter acetylcholine, which is liberated by the parasympathetic nerve endings. Do your results accurately reflect this effect of atropine?

_____

Epinephrine

Flood the frog heart with epinephrine solution, and continue to record until a change in heart activity is noted.

What are the results? _____

_____

Which division of the autonomic nervous system does its effect imitate? _____

_____

Effect of Various Ions

To test the effect of various ions on the heart, apply the designated solution until you observe a change in heart rate or in strength of contraction; pipette off the solution, flush with Ringer's solution, and allow the heart to resume its normal rate before continuing. *Do not allow the heart to stop.* If the rate decreases dramatically, flood the heart with Ringer's solution.

Effect of Ca^{++} (use 1% $CaCl_2$) _____

_____

Effect of Na^+ (use 0.7% NaCl) _____

_____

Effect of K^+ (use 5% KCl) _____

_____

Potassium ion concentration is normally higher within the cells than in the extracellular fluid. Hyperkalemia tends to decrease the resting potential of the cell membranes, thus reducing the difference in potential between the inside and outside of the cell and decreasing the force of heart contraction. In some cases, the conduction rate of the heart is so depressed that **ectopic pacemakers** (pacemakers appearing erratically and at abnormal sites in the heart muscle) appear in the ventricles, and fibrillation may occur. Was there any evidence of premature beats in the recording of potassium ion effects? ____

Was arrhythmia produced with any of the ions tested? _____

If so, which? _____

Intrinsic Conduction System Disturbance (Heart Block)

1. Moisten a 10-inch length of thread and make a Stannius ligature (loop the thread around the heart at the junction of the atria and ventricle).

2. Decrease the drum speed to achieve intervals of approximately 2 cm between the ventricular contractions, and record a few normal heart beats.

3. Tighten the ligature in a stepwise manner while observing the atrial and ventricular contraction curves. As heart block occurs, the atria and ventricle will no longer show a 1:1 contraction ratio. Record a few beats each time you observe a different degree of heart block—a 2:1 atria to ventricle, 3:1, 4:1, and so on. As long as you can continue to count a whole number ratio between the two chamber types, the heart is in **partial heart block.** When you can no longer count a whole number ratio, the heart is in **total,** or **complete, heart block.**

4. When total heart block occurs, release the ligature to see if the normal A-V rhythm is re-

established. What is the result? _____

_____

5. Attach properly labeled recordings (or copies of the recordings), made during this procedure, to this sheet for future reference.

OBSERVATION OF THE MICROCIRCULATION IN A FROG'S WEB AND INVESTIGATION OF FACTORS AFFECTING LOCAL BLOOD FLOW

The thin web of a frog's foot provides an excellent opportunity to observe the flow of blood to, from, and within the capillary beds, where the real business of the circulatory system occurs. The collection of vessels involved in the exchange mechanism is referred to as the **microcirculation;** it consists of arterioles, venules, capillaries, and thoroughfare channels called metarterioles. The total cross-sectional area of the capillaries in the body is much greater than that of the veins and arteries combined. Thus, the flow through the capillary beds is quite slow. Capillary flow is also intermittent, since if all capillary beds were filled with blood at the same time, there would be no blood at all in the large vessels. The flow of blood into the capillary beds is regulated by the activity of muscular arterioles, which feed the beds, and by precapillary sphincters at the bed entrance. The amount of blood flowing into the true capillaries of the bed is regulated by nervous control (the vasomotor center of the medulla) and local chemical controls (local concentrations of carbon dioxide, histamine, pH). Thus a capillary bed may be flooded with blood or almost entirely bypassed (via the metarterioles), depending

on what is happening within the body or a particular body region at any one time. You will investigate some of the local chemical controls in the next group of experiments.

1. Obtain a frog, frog board (with a hole at one end), dissecting pins, frog Ringer's solution, 0.01 N HCl, 0.01% histamine solution, 1% epinephrine solution, and some paper towels.

2. Moisten several paper towels with room-temperature Ringer's solution, and wrap the frog's body securely with them. One hind leg should be left unsecured and extending beyond the paper cocoon.

3. Attach the frog to the frog board (or other supporting structure) with a large rubber band and then carefully spread (but do not stretch) the web of the exposed hindfoot over the hole in the support. Secure the toes to the board with dissecting pins.

4. Obtain a compound microscope, and observe the web under low power to find a capillary bed. Focus on the vessels in high power. Keep the web moistened with Ringer's solution as you work. If the circulation seems to stop during your observations, massage the hind leg of the frog gently to restore blood flow.

5. Observe the red blood cells of the frog. Note that unlike human RBCs, they are nucleated. Watch their movement through the smallest vessels—the capillaries. Do they move in single file or do they flow

through two or three cells abreast? _____

Are they flexible? _____ Explain. _____

_____

Can you see any white blood cells in the capillaries?

_____ If so, which types? _____

6. Note the relative speed of blood flow through the blood vessels. Differentiate between the arterioles, which feed the capillary bed, and the venules, which drain it. This may be tricky, because images are reversed in the microscope. Thus, the vessel that appears to feed into the capillary bed will actually be draining it. You can distinguish between the vessels, however, if you consider that the flow is more pulsating and turbulent in the arterioles and smoother and steadier in the venules. How does the rate of flow in the arterioles compare with that in

the venules? _____

In the capillaries? _____

What is the relative difference in the diameter of the

arterioles and capillaries? _____

Effect of Temperature on Blood Flow

To investigate the effect of temperature on blood flow, flood the web with cold saline two or three times to chill the entire area. Is a change in vessel

diameter noticeable? _____

Which vessels are affected? _____

How? _____

Blot the web gently with a paper towel, and then flood the web with warm saline. Note your observations.

_____

Effect of Inflammation on Blood Flow

Pipette 0.01 N HCl onto the web. Hydrochloric acid will act as an irritant and cause a localized inflammatory response. Is there an increase or decrease in the blood flow into the capillary bed

following the application of HCl? _____

_____

What purpose do these local changes serve during a localized inflammatory response? _____

_____

_____

Flush the web with room-temperature Ringer's solution and blot.

Effect of Histamine on Blood Flow

Histamine, which is released in large amounts during allergic responses, causes extensive vasodilation. Investigate this effect by adding a few drops of histamine solution to the frog web. What

are its effects? _____

_____

How does this response compare to that produced by HCl? _____

Blot the web and flood with warm saline as before. Now add a few drops of 1% epinephrine solution, and observe the web. What are its

effects on the blood vessels? _____

(Epinephrine is used clinically to reverse the vasodilation seen in severe allergic attacks (such as asthma) which are mediated by histamine and other vasoactive molecules.)

The Lymphatic System

EXERCISE 35

OBJECTIVES

1. To name the components of the lymphatic system.

2. To relate the function of the lymphatic system to that of the blood vascular system.

3. To describe the formation and composition of lymph and to describe how it is transported through the lymphatic vessels.

4. To describe the structure and function of lymph nodes.

MATERIALS

Large anatomic chart of the human lymphatic system
Prepared microscope slides of lymph nodes
Compound microscope
Dissection animal, tray, and instruments

DISTRIBUTION OF THE LYMPHATIC VESSELS AND LYMPH NODES

The lymphatic system consists of a network of successively larger lymph vessels and the lymph nodes. Its overall function is twofold. It returns tissue fluid (lymph) to the blood vessels. In addition, it protects the body by removing foreign material such as bacteria from the lymphatic stream and by producing agranular white blood cells.

As blood circulates through the body, the hydrostatic and osmotic pressures operating at the capillary beds result in an outward flow of fluid at the arterial end of the bed and its return at the venous end. However, not all of the lost fluid is returned to the bloodstream by this mechanism, and the fluid that lags behind in the tissue spaces must eventually return to the blood if the vascular system is to operate properly. If it does not, fluid accumulates in the tissues, producing a condition called edema.

The lymphatic system is a one-way system, since lymph flows only toward the heart. The microscopic, blind-ended lymph capillaries ramify through all the tissues of the body where they pick up the leaked fluid (primarily water and a small amount of dissolved proteins) and carry it through successively larger vessels (lymph venules to lymph veins) until the lymph finally returns to the venous system through one of the two large ducts in the thoracic region. The **right lymphatic duct** drains the lymph from the right upper extremity, head, and thorax; the large **thoracic duct** receives lymph from the rest of the body. In humans, both ducts empty the lymph into the venous circulation at the junction of the internal jugular vein and the subclavian vein on their own side of the body (Figure 35.1).

Like the venules and veins of the blood vascular system, the lymph venules and veins are thin-walled and the larger vessels are equipped with valves. Since the lymphatic system is a pumpless system, lymph transport depends on the milking action of the skeletal muscles and the pressure changes within the thorax during breathing.

As lymph is transported, it filters through oval or bean-shaped **lymph nodes,** which cluster along the lymphatic vessels of the body. There are hundreds of lymph nodes, but because they are usually embedded in connective tissue they are not ordinarily seen. Within the lymph nodes are **phagocytic cells (macrophages),** which destroy bacteria, cancer cells, and other foreign matter in the lymphatic stream,

309

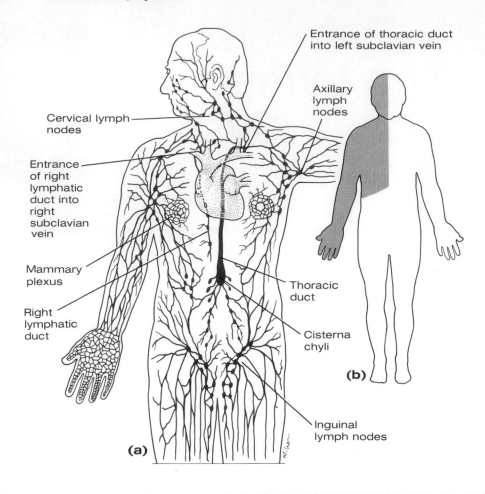

Cervical lymph nodes

Entrance of thoracic duct into left subclavian vein

Axillary lymph nodes

Entrance of right lymphatic duct into right subclavian vein

Mammary plexus

Right lymphatic duct

Thoracic duct

Cisterna chyli

(b)

Inguinal lymph nodes

(a)

Figure 35.1

Lymphatic system: (a) distribution of lymphatic vessels and lymph nodes; (b) dark area represents body area drained by the right lymphatic duct.

thus rendering many harmful substances or cells harmless before the lymph enters the bloodstream. Agranular white blood cells (see Exercise 29) originate in the lymph nodes, which release them to the lymph. Particularly large collections of lymph nodes are found in the inguinal, axillary, and cervical regions of the body. Although we are not usually aware of the filtering and protective nature of the lymph nodes, most of us have experienced "swollen glands" during an active infection. This swelling is a manifestation of the trapping function of the nodes.

The tonsils, thymus, and spleen are generally considered to be lymphoid organs, since they resemble the lymph nodes histologically, but they are not normally considered with a discussion of the lymphatic system.

Study the large anatomic chart to observe the general plan of the lymphatic system. Note the distribution of the lymphatic vessels and lymph nodes, and the location of the right lymphatic duct and thoracic duct. Also identify the **cisterna chyli,** the enlarged terminus of the thoracic duct that receives lymph from the digestive viscera.

MICROSCOPIC ANATOMY OF A LYMPH NODE

Obtain a prepared slide of a lymph node and a compound microscope. As you examine the slide, note the following anatomic features. The node is enclosed within a fibrous **capsule** from which connective tissue septa (**trabeculae**) extend inward to divide the node into several compartments. Very fine strands of reticular connective tissue issue from the trabeculae, forming the substrate of the gland within which the cells are found. In the outer region of the node, the **cortex,** some of the cells are arranged in globular masses, which are referred to as **germinal centers.** The germinal centers contain rapidly dividing B-lymphocytes. The rest of the cortical cells are primarily T-lymphocytes that circulate continuously between the lymph nodes and lymphatic stream. In the internal portion of the gland, the **medulla,** the cells are arranged in cordlike fashion. Most of the medullary cells are macrophages. Can you see cells that look like lymphocytes?

Lymph enters the node through a number of afferent vessels, circulates through sinuses within the node, and leaves the node through efferent vessels at the **hilus.** Since each node has fewer efferent vessels, the lymph flow stagnates somewhat within the node, allowing time for the phagocytic cells to remove debris from the lymph before it reenters the blood vascular system.

Draw a pie-shaped section of a lymph node, showing the detail of cells in a germinal center, sinusoids, and afferent and efferent vessels in the space provided here. Label all elements.

UNIT 12

THE RESPIRATORY SYSTEM

Anatomy of the Respiratory System

EXERCISE
36

OBJECTIVES

1. To define the following terms: *respiratory system, cellular respiration, external respiration* (or breathing).

2. To label the following structures on a diagram (or identify on a model or dissection) and describe the function of each:

pharynx	alveoli
epiglottis	bronchi
diaphragm	external nares
bronchioles	tongue
nasal conchae	larynx
nasal cavity	trachea
nasal septum	thyroid cartilage
clavicle	cricoid cartilage
lungs	paranasal sinuses
glottis	palate (soft and hard)
vocal cords	pleurae

3. To recognize pseudostratified columnar ciliated epithelium and lung tissue on prepared histologic slides, and describe the functions the observed structural modifications serve.

MATERIALS

Human torso model

Respiratory organ model and/or chart of the respiratory system

Larynx model (if available)

Sheep pluck (preserved or fresh from the slaughterhouse)

Dissection animals, trays, and instruments

Source of compressed air

2-foot length of laboratory rubber tubing

Alcohol swabs

Histologic slides of the following (if available): trachea (cross section), lung tissue, pathologic specimens such as section taken from lung tissues exhibiting bronchitis, pneumonia, emphysema, atelectasis, or other conditions

Compound and dissecting microscopes

Bronchogram (if available)

Suggested film: *Respiration in Man* (color, sound, 16 mm, 26 minutes, Encyclopaedia Britannica Films)

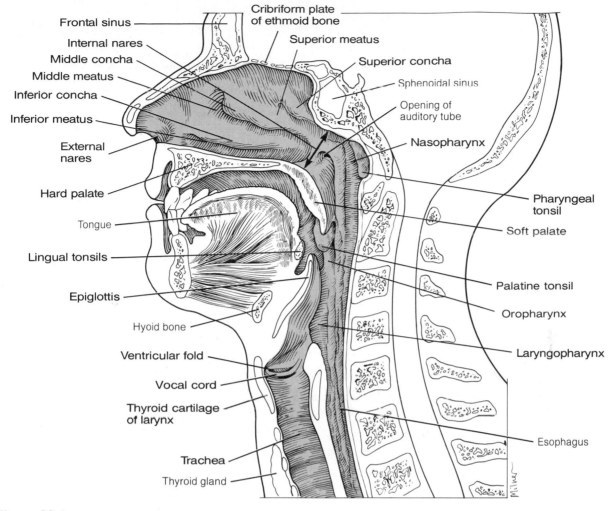

Figure 36.1

Structures of the upper respiratory tract (sagittal section).

Body cells require an abundant and continuous supply of oxygen to carry out their vital processes. As the cells use oxygen, they release carbon dioxide, a waste product that the body must eliminate. These oxygen-using cellular processes are collectively referred to as *internal,* or *cellular, respiration.*

The circulatory and respiratory systems are intimately involved in the acquisition and delivery of oxygen and the removal of carbon dioxide. The structures concerned with gas exchange between the blood and the external environment (*external respiration,* or breathing) are collectively referred to as the *respiratory system.* The transport of respiratory gases between the lungs and the cells is accomplished by the structures of the circulatory system, with blood as the transport medium. Should either system fail, the cells begin to die from oxygen starvation and the accumulation of carbon dioxide. If uncorrected, this situation soon causes death of the entire organism.

UPPER RESPIRATORY SYSTEM STRUCTURES

The upper respiratory system structures are shown in Figure 36.1 and described below. As you read through the descriptions, identify each structure by referring to this figure.

Air generally passes into the internal nose through the **external nares** (nostrils), into the paired internal **nasal cavities** (separated by the **nasal septum**), and then flows posteriorly over three pairs of lobelike structures, the **inferior, superior,** and **middle nasal conchae,** which warm the incoming air. As the air passes through the nasal cavity, it is also moistened and filtered by the mucosa. The air that flows directly beneath the upper part of the nasal cavity may chemically stimulate the olfactory receptors located in the mucosa of that region. The nasal

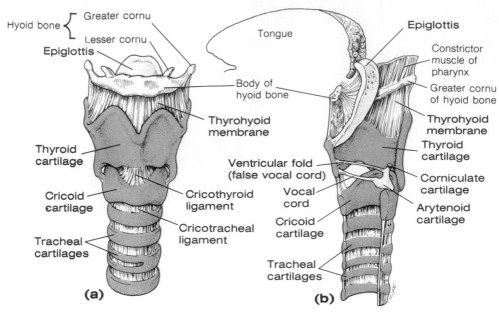

Figure 36.2

Structure of the larynx: (a) anterior view; (b) sagittal section.

cavity is surrounded by the **paranasal sinuses** in the frontal, sphenoid, ethmoid, and maxillary bones. These sinuses possibly act as resonance chambers in speech. Because of the close communication of the sinuses with the nasal passages and the continuity of the mucosa through both areas, nasal infections often invade the sinus areas, causing sinusitis that is difficult to treat. The nasal passages are separated from the oral cavity below by a partition composed anteriorly of the **hard palate** and posteriorly by the **soft palate.** The genetic defect called cleft palate (failure of the palatine bones and/or the palatine processes of the maxillary bones to fuse medially) causes difficulty in breathing and oral cavity functions such as mastication and speech.

Needless to say, in addition to entering the nasopharynx through the nasal cavities, air may also enter the body via the mouth and pass through the oral cavity into the **oropharynx** (throat) posteriorly, where the oral and nasal cavities are joined temporarily. The pharynx thus accommodates both ingested food and air. From the oropharynx, air enters the lower respiratory passageways by passing through the laryngopharynx and **larynx** (voice box) into the **trachea** below.

The larynx (Figure 36.2) consists of nine cartilages, the three most prominent being the large shield-shaped **thyroid cartilage,** whose anterior medial prominence is commonly referred to as the Adam's apple, the inferiorly located, ring-shaped **cricoid cartilage,** whose widest dimension faces posteriorly, and the flaplike **epiglottis,** located superior to the opening of the larynx. The epiglottis, sometimes referred to as the guardian of the airways, forms a lid over the larynx when we swallow. This closes off the respiratory passageways to incoming

food or drink, which is routed into the posterior esophagus, or food chute.

• Palpate your larynx by placing your hand on the anterior neck surface approximately halfway down its length. Swallow. Can you feel the cartilaginous larynx rising?

If anything other than air enters the larynx, a cough reflex attempts to expel the substance. Note that this reflex is operative only when a person is conscious; thus you should never try to feed or pour liquids down the throat of an unconscious person.

The mucous membrane of the larynx is thrown into two pairs of folds—the upper **false vocal folds** and the lower **true vocal folds,** or cords, which vibrate with expelled air for speech. The vocal cords are attached posterolaterally to the small triangular **arytenoid cartilages.** The slitlike passageway between the folds is called the **glottis.**

LOWER RESPIRATORY SYSTEM STRUCTURES

Air entering the trachea, or windpipe, from the larynx travels down its length (about 11.5 cm) to the level of the fifth thoracic vertebra, where the passageway divides into the right and left **primary bronchi,** which plunge into their respective lungs at an indented area called **hilus.** The right primary bronchus is more vertical, larger in diameter, and shorter than the left primary bronchus; as a result, foreign objects that enter the respiratory passageways are more likely to become lodged in it.

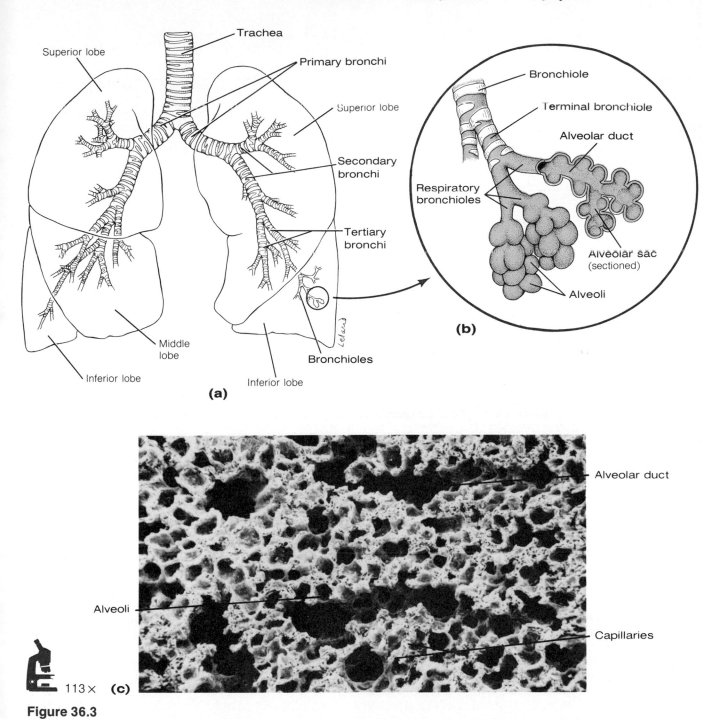

Trachea

Superior lobe

Primary bronchi

Superior lobe

Secondary bronchi

Tertiary bronchi

Middle lobe

Bronchioles

Inferior lobe

Inferior lobe

(a)

Bronchiole

Terminal bronchiole

Alveolar duct

Respiratory bronchioles

Alveolar sac (sectioned)

Alveoli

(b)

Alveolar duct

Alveoli

Capillaries

113× **(c)**

Figure 36.3

Structures of the lower respiratory tract. Inset (b) shows enlarged view of alveoli. Photomicrograph (c) shows the spongy nature of lung tissue.

The trachea is lined with ciliated mucus-secreting epithelium as are many of the other respiratory system passageways. The cilia beat in unison and propel mucus (produced by goblet cells) laden with dust particles, bacteria, and other debris away from the lungs and toward the throat, where it can be expectorated or swallowed. The walls of the trachea are reinforced with C-shaped cartilage rings, the incomplete portion being located posteriorly. These C-shaped cartilages serve a double function:

the incomplete parts allow the esophagus to expand anteriorly when a large food bolus is swallowed; the solid portions reinforce the trachea walls to maintain its passageway regardless of the pressure changes that occur during breathing (Figure 36.3).

The primary bronchi further divide into smaller and smaller branches (the secondary, tertiary, on down), finally becoming the **bronchioles,** whose terminal branches are termed **respiratory bronchioles.** All but the most minute branches contain

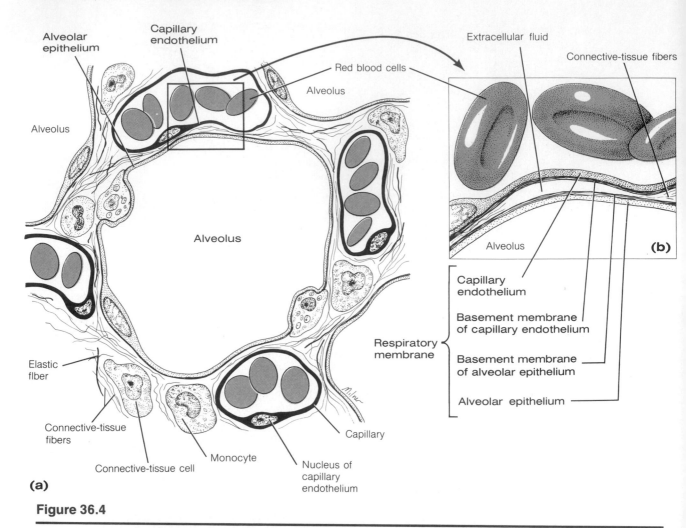

Figure 36.4

Diagrammatic view of the relationship between the alveoli and pulmonary capillaries involved in gas exchange: (a) one alveolus surrounded by capillaries; (b) enlargement of the respiratory membrane.

cartilaginous reinforcements in their walls, usually in the form of small plates of hyaline cartilage rather than cartilaginous rings. The respiratory bronchioles in turn subdivide into several **alveolar ducts,** which terminate in alveolar sacs that rather resemble clusters of grapes. The walls of the **alveolar sacs,** or **alveoli,** are composed of a single very thin layer of squamous epithelium overlying a thin basement membrane. The external surfaces of the alveoli are literally spider-webbed with a network of pulmonary capillaries (Figure 36.4). It is through these thin walls, the alveolar and capillary walls, that the gas exchanges occur by simple diffusion—the oxygen passing from the alveolar air into the capillary blood and the carbon dioxide leaving the capillary blood to enter the alveolar air. Thus, all other respiratory passageways simply serve as access or exit routes to and from these gas exchange chambers. Since the other passageways (trachea and bronchi) have no exchange function, they are termed anatomic dead space.

The continuous branching of the respiratory passageways in the lungs is often referred to as the respiratory tree. The comparison becomes much more meaningful if you observe a bronchogram. (Do so, if one is available for observation in the laboratory or refer to Figure 36.5.)

The paired lungs are separated from one another by the structures of the mediastinum (heart, trachea, bronchi, major blood vessels, esophagus). The substance of the lungs, other than the respiratory passageways that make up the bulk of their volume, is primarily elastic connective tissue that allows the lungs to recoil passively during expiration. Each lung is enclosed in a double-layered sac of serous membrane called the **pleura.** The outer layer, the **parietal pleura,** is attached to the thoracic walls and the **diaphragm;** the inner layer, covering the lung tissue, is the **visceral pleura.** The two pleural layers are separated by a space, which is more of a potential space than an actual one, since the production of a lubricating serous fluid by the pleural layers causes the parietal and visceral layers to adhere closely to one another, thus holding the lungs to the thoracic wall and allowing them to move easily against one another in a frictionless environment during the movements of breathing.

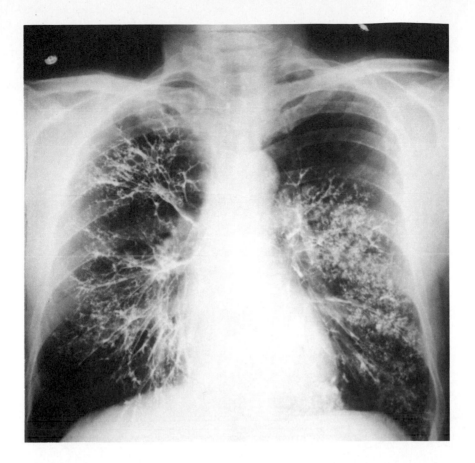

Figure 36.5

Bronchogram showing the extensive branching of the respiratory tree.

Before proceeding, be sure to locate all the respiratory structures described on the torso model, thoracic cavity structures model, or an anatomic chart.

SHEEP PLUCK DEMONSTRATION

A sheep pluck includes the larynx, trachea with attached lungs, the heart and pericardium, and portions of the major blood vessels found in the mediastinum (aorta, pulmonary artery and vein, vena cava). Obtain a sheep pluck and identify the lower respiratory system organs. Once you have completed your observations, insert a hose from an air compressor (vacuum pump) into the trachea and alternately allow air to flow in and out of the lungs. Notice how the lungs inflate. This observation is educational in a preserved pluck but is a spectacular sight in a fresh one. (If air compressors are not available, the same effect may be obtained by using a length of laboratory rubber tubing to blow into the trachea. Clean the end with an alcohol swab before inserting it into your mouth.)

DISSECTION OF FETAL PIG RESPIRATORY SYSTEM

1. Before beginning the dissection, examine the external nares and nasal cavity of the pig. The nostrils, or external nares, of the pig are small, relatively speaking, and located on the flat anterior surface of the **rostrum,** or snout. To view the interior of the nasal cavity, which is long and narrow in the pig, make a transverse cut through the snout posterior to the external nares. Examine the medial wall of the nasal passages, noting that the mucous membrane is thrown into convolutions in this area, since it overlies the turbinate bones, which project laterally from the nasal septum. What function does this mucosa perform for the respiratory system?

2. To examine the continuity between the oral pharynx and the nasal pharynx, place a probe deep into the nasal cavity so that it enters the nasopharynx behind. Leave the probe in place. Open the mouth of the pig, and observe the posterior aspect of the oral cavity. You should be able to see the tip of the probe hanging downward into the oral pharynx where the nasal (respiratory) and oral (digestive) passageways unite temporarily.

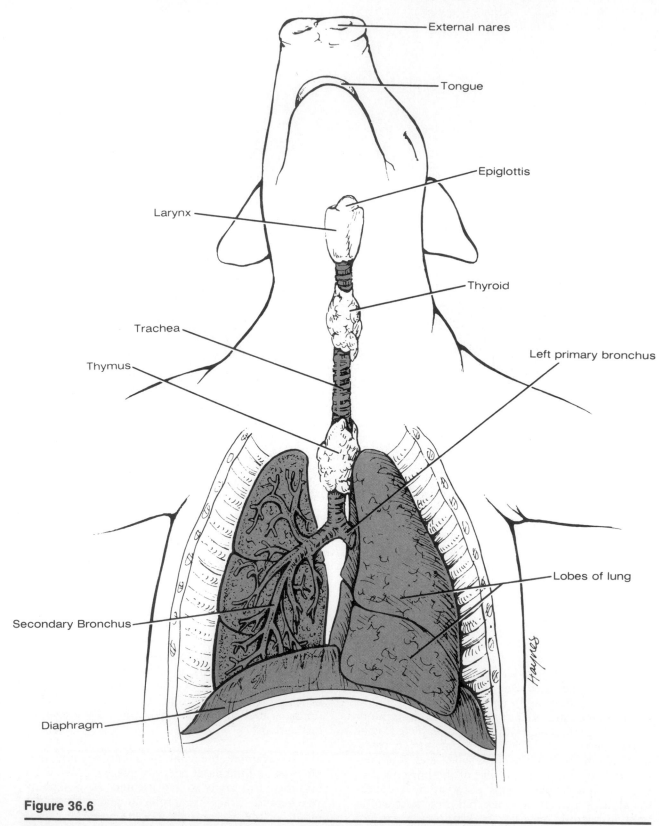

External nares

Tongue

Epiglottis

Larynx

Thyroid

Trachea

Left primary bronchus

Thymus

Lobes of lung

Secondary Bronchus

Diaphragm

Figure 36.6

Respiratory system of the pig. (Right lung partially dissected to show some R. bronchus branching.)

3. Secure the animal to the dissecting tray, and expose the respiratory structures by retracting the cut muscle and the rib cage. (Do not sever nerves and blood vessels located on either side of the trachea if these have not previously been studied.) If you have not opened the thoracic cavity previously, make a medial longitudinal incision through the neck muscles and thoracic musculature to view the thoracic organs.

4. Using Figure 36.6 as a guide, identify the following structures:

Trachea: determine by finger examination whether the cartilage rings are complete or incomplete posteriorly.

Thymus gland: a large gland overlying the heart

Thyroid gland: located on the trachea inferior to the larynx

Larynx: remove attached muscle from the larynx and pull it anteriorly for ease of examination. Identify the thyroid and cricoid cartilages and the flaplike epiglottis. Find the hyoid bone, located anterior to the larynx. Make a longitudinal incision through the ventral wall of the larynx and locate the true and false vocal cords on the inner wall.

Vagus nerve: a conspicuous white band, which lies alongside the trachea

5. To examine the contents of the thoracic cavity, follow the trachea as it bifurcates into the two primary bronchi, which plunge into the lungs. Note that in the pig, the trachea gives off an additional apical bronchus at the level of the fourth rib before it diverges into the right and left primary bronchi. The apical bronchus supplies the apical lobe of the right lung, the right primary bronchus supplies the cardiac, intermediate, and diaphragmatic lobes of the right lung, and the left primary bronchus supplies the left apical, cardiac, and diaphragmatic lobes of the left lung. Note the pericardial sac containing the heart located in the mediastinum (if it is still present). Examine the pleura, and note its exceptionally smooth texture. Locate the **diaphragm** and the **phrenic nerve** (a conspicuous white thread running along the pericardium to the diaphragm). The phrenic nerve controls the activity of the diaphragm in breathing. Lift one lung and find the esophagus beneath the parietal pleura. Follow it through the diaphragm to the stomach.

6. Make a longitudinal incision in the outer tissue of one lung lobe beginning at a primary bronchus. Attempt to follow a part of the respiratory tree from this point down into the smaller subdivisions. Carefully observe the cut lung tissue (under a dissection scope called a stereomicroscope if one is available), noting the richness of the vascular supply. As you work, observe that the fetal lung tissue feels firm and solid rather than spongy, the texture of the human lung. This textural difference reflects the fact that the fetal lungs are nonfunctional, that is, they have not yet been inflated. Cut off a section of the fetal lung and note how dense it is. Place the lung segment in a beaker of water and observe what happens. Does it float or sink? Explain the significance of your

observation. _____

_____

_____

Respiratory System Physiology

OBJECTIVES

1. To define the following (and be prepared to provide volume figures if applicable):

inspiration	expiratory reserve volume
expiration	expiratory end point
tidal volume	inspiratory reserve volume
vital capacity	minute respiratory volume

2. To explain the role of muscles and volume changes in the mechanical process of respiration.

3. To demonstrate proper usage of the spirometer.

4. To explain the relative importance of various mechanical and chemical factors in producing respiratory variations.

5. To describe bronchial and vesicular breathing sounds.

6. To explain the importance of the carbonic acid–bicarbonate buffer system in maintaining blood pH.

7. To describe the effect of temperature and gravity on ciliary action of the respiratory mucosa.

MATERIALS

Model lung (bell jar demonstrator)
Tape measure
Spirometer
Disposable mouthpieces
Nose clips
Alcohol swabs
Paper bag
Pneumograph and recording attachments
Recording apparatus (kymograph or physiograph)
Stethoscope
pH meter (standardized with buffer of pH 7)
Buffer solution (pH 7)
Concentrated HCl and NaOH
250-ml beakers
Glass stirring rod
Stethoscope
Frog
Dissecting pan (wax surface)
Dissecting instruments and fine dissecting pins
Frog Ringer's solution (room temperature and 15 C) in dropper bottles
Cork

MECHANICS OF RESPIRATION

External respiration, or breathing, consists of two phases: **inspiration,** during which air is taken into the lungs, and **expiration,** during which air passes out of the lungs. As the inspiratory muscles (external intercostals and diaphragm) contract during inspiration, the size of the thoracic cavity increases. The diaphragm moves from its relaxed dome shape to a flattened position, increasing the superoinferior volume, and the external intercostals lift the rib cage, increasing the anteroposterior and lateral dimensions (Figure 37.1). Since the lungs adhere to the thoracic walls like flypaper because of the cohesive character of the pleurae, the intrapulmonary (within the lungs) volume also increases, lowering the air pressure inside the lungs. The gases then expand to fill the available space creating a partial vacuum, which causes air to flow into the lungs—constituting the act of inspiration. During expiration, the inspiratory muscles relax. Their antagonists, the internal intercostal muscles and the natural tendency of the elastic lung tissue to recoil, act together to decrease the intrathoracic and intrapulmonary volume. As the gas molecules within the lungs are forced more closely together, the intrapulmonary pressure rises

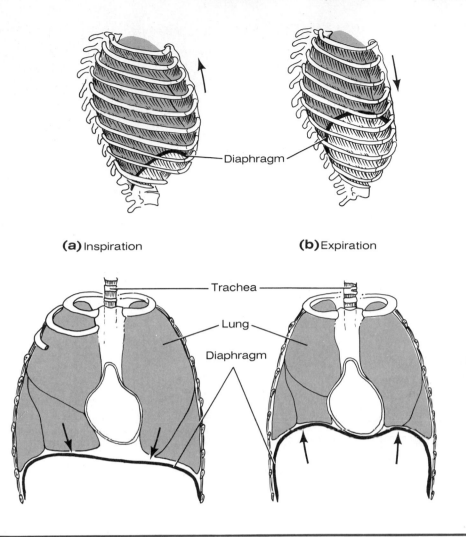

Figure 37.1

Rib cage and diaphragm positions during breathing: (a) at the end of a normal inspiration; chest expanded, diaphragm depressed; (b) at the end of a normal expiration; chest depressed, diaphragm elevated.

to a point higher than atmospheric pressure. This causes gases to flow from the lungs to equalize the pressure within and without the lungs—the act of expiration.

 Observe the model lung, which demonstrates the principles involved in the filling and emptying of the lungs. It is a simple apparatus with a bottle "thorax," a rubber membrane "diaphragm," and "balloon lungs."

1. Go to the demonstration area and work the model lung by moving the rubber diaphragm up and down, noting changes in balloon (lung) size as the volume of the thoracic cavity is alternately increased and decreased.

2. Check the appropriate columns in the chart on page RS 135 of the Lab Review section concerning these observations before proceeding.

3. After observing the operation of the model lung, conduct the following tests on your lab partner. Use the tape measure to determine chest circumference by placing the tape around the chest as high up under the armpits as possible. Record the measurements in inches in the appropriate space for each of the conditions below.

Quiet breathing:

inspiration _____ expiration _____

Forced breathing:

inspiration _____ expiration _____

Do the results coincide with what you expected on

the basis of what you have learned thus far? _____

_____

RESPIRATORY VOLUMES AND CAPACITIES— SPIROMETRY

A person's size, sex, age, and physical condition produce variations in respiratory volumes. Normal quiet breathing moves about 500 ml of air in and out of the lungs with each breath. As you have seen in the previous experiment, a person can usually forcibly inhale or exhale much more air than is normally exchanged in quiet breathing. The terms given to these measurable respiratory volumes are defined in the instructions for the experimental procedures below. These terms and their normal values should be memorized.

Respiratory capacities will be measured with an apparatus called a **spirometer.** The spirometer consists of a cylinder within a tank into which air can be added or removed. The outer tank contains water and has a tube running through it to carry air above the water level. The floating bottomless inner cylinder is inverted over the water-containing tank and connected to a volume indicator. The volume of air exhaled can be read off the indicator, which moves to indicate the changes in air volume within the apparatus (Figure 37.2).

 1. Examine a spirometer indicator scale *before beginning* to ensure proper reading of the scale. Each number from 1 to 6 indicates the number of liters, and each smaller marking between the liter marks is equivalent to 0.1 liter, or 100 ml.

2. Work in pairs, with one person recording data while the other conducts the volume determinations. Make sure the indicator is set at zero before beginning each trial. (If disposable cardboard mouthpieces are not available, sterilize the spirometer mouthpiece with an alcohol swab or a sterile cotton pad saturated with 79% alcohol.) Before beginning, practice without exhaling through the nose, or use the nose clips.

3. Conduct the test three times for each required measurement. Record the data here, and then find the average volume figure for that respiratory measurement. After you have completed the trials and computed the averages, enter the average values on the table prepared on the chalkboard for tabulation of class data,* and copy all averaged data on the chart provided in the laboratory review section.

*Note to the instructor: the format of class data tabulation can be similar to that shown here. However, it would be interesting to divide the class into smokers and nonsmokers and then compare the mean average VC and ERV for each group. Such a comparison might help to determine if smokers are handicapped in any way. It also might be a good opportunity for an informal discussion of the early warning signs of bronchitis and emphysema, which are primarily smoker's diseases.

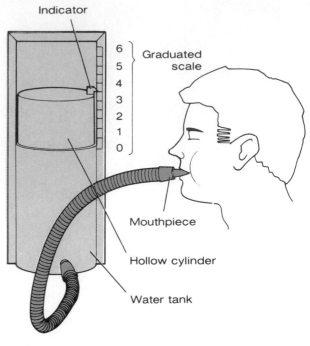

Figure 37.2

Operation of a spirometer. Breathing into the mouthpiece causes the inner cylinder to rise because of the increase in volume (and pressure) of air within the cylinder. Water provides an airtight seal.

4. Tidal volume measurement (TV). The volume of air inhaled and exhaled with each normal respiration is approximately 500 ml. To conduct the test, inhale a normal breath, and then exhale a normal breath of air into the spirometer mouthpiece. (Do not force the expiration!) Record the volume and repeat the test twice.

trial 1 _____ ml   trial 2 _____ ml

trial 3 _____ ml   average TV _____ ml

5. Inspiratory reserve volume measurement (IRV). The volume of air that can be forcibly inhaled following a normal inspiration is approximately 2800 ml.

Breathe normally two or three times; then make the deepest possible inspiration and exhale into the spirometer only to the expiratory end point. Do not force the expiration. (The expiratory end point is the position of the rib cage at the end of a normal expiration.) Record the volume and repeat twice. Subtract the TV average from each figure to find the IRV value, and then compute the average IRV. (Respiratory volume – TV average = IRV.)

trial 1 _____ ml   trial 2 _____ ml

trial 3 _____ ml   average IRV _____ ml

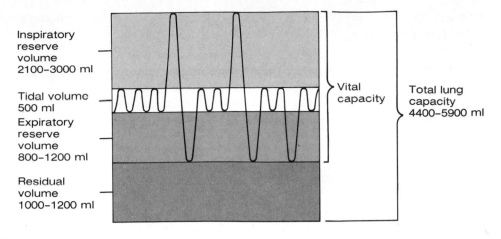

Figure 37.3

Idealized tracing of the various respiratory volumes.

6. Expiratory reserve volume measurement (ERV). The volume of air that can be forcibly exhaled after a normal expiration is approximately 1000 ml. The ERV is dramatically reduced in conditions in which the elasticity of the lungs is decreased by a chronic obstructive pulmonary disease (COPD) such as *emphysema*. Since energy must be used to *deflate* the lungs in such conditions, expiration is physically exhausting to individuals suffering from COPD. Inhale and exhale normally two or three times; then insert the spirometer mouthpiece, and exhale forcibly as much of the additional air as you can.

trial 1 _____ ml   trial 2 _____ ml

trial 3 _____ ml   average ERV _____ ml

7. Vital capacity measurement (VC). The total exchangeable air of the lungs (the sum of TV + IRV + ERV) is approximately 4500 ml. Breathe in and out normally two or three times, take the deepest possible inspiration, insert the mouthpiece, and exhale as forcibly as possible. Record and repeat two times.

trial 1 _____ ml   trial 2 _____ ml

trial 3 _____ ml   average VC _____ ml

8. Compute the minute respiratory volume using the following formula:

TV × respirations/min =
                        minute respiratory volume

minute respiratory volume = _____ ml/min

A respiratory volume that cannot be experimentally demonstrated here is the residual volume (RV), which is the amount of air remaining in the lungs even after a maximal expiratory effect. The presence of air (about 1200 ml) that cannot be voluntarily flushed from the lungs is important, since it allows gas exchange to go on continuously—even between respirations.

Figure 37.3 is an idealized tracing of the respiratory volumes described and tested in this exercise. Examine it carefully. Do your test results compare

closely to the values in the tracing? _____

Many disease states alter these volumes; however, the small variations you are likely to see during this lab are of little significance.

USE OF THE PNEUMOGRAPH TO DETERMINE FACTORS INFLUENCING RATE AND DEPTH OF RESPIRATION*

The neural centers that control respiratory rhythm are located in the medulla and pons. Oscillations between the two centers maintain a rate of 14 to 18 respirations per minute. On occasion, input from the stretch receptors in the lungs (via the vagus nerve to the medulla) modifies the respiratory rate, as in cases of overinflation or extreme deflation of the lungs. Death occurs when the medulla centers are completely suppressed, as from an overdose of sleeping pills or gross overindulgence in alcohol, and respiration ceases completely.

*Note to instructor: This exercise may also be done without using the recording apparatus by simply having the students count the respiratory rate visually.

Although these nervous system centers initiate the basic rhythm of breathing, there is no question that physical phenomena such as talking, yawning, coughing, exercise, and chemical factors such as changes in oxygen or carbon dioxide concentrations in the blood or fluctuations in blood pH can modify the rate and depth of respiration. Carbon dioxide changes in blood levels seem to act directly on the medulla control centers, whereas changes in oxygen concentrations are monitored by chemoreceptor regions in the aortic arch and carotid body, which in turn send input to the medulla. The experimental sequence in this section is designed to test the relative importance of various physical and chemical factors in the process of respiration.

The **pneumograph,** an apparatus that records variations in breathing patterns, is the best means of observing respiratory variations resulting from physical and chemical factors. The chest pneumograph is a coiled rubber hose that is attached around the thorax. As the subject breathes, chest movements produce pressure changes within the pneumograph that are transmitted to a recorder.

The instructor will demonstrate the method of setting up the pneumograph and discuss the interpretation of the results. Work in pairs so that one person can mark the record to identify the test for later interpretation. Ideally, the student being tested should not observe the recording process to prevent voluntary modification of the record.

1. Attach the pneumograph tubing firmly but not restrictively around the thoracic cage at the level of the sixth rib, leaving room for chest expansion during testing. If the subject is female, the tubing should be positioned above the breasts to prevent slippage during testing. Set the pneumograph speed at 1 or 2 and the time signal at 10-second intervals. Record quiet breathing for 1 minute with the subject in a sitting position. Record

breaths per minute. _____

2. Record the subject's breathing as he performs activities from the list below. Make sure the record is marked accurately to identify each test conducted. Record your results in the laboratory review-study section.

talking   swallowing water
yawning   coughing
laughing   lying down
standing   running in place
doing a math problem
(concentrating)

3. Without recording, have the subject breathe normally for 2 minutes, then inhale deeply and hold his breath for as long as he can. Time the breath-

holding interval: _____ sec. As the sub-

ject exhales turn on the recording apparatus and record the recovery period (time to return to normal breathing—usually slightly over 1 minute):

_____ sec. Did the subject have the urge

to inspire *or* expire during breath holding? _____

Without recording, repeat the above experiment, but exhale completely and forcefully after taking the deep breath. What was observed this time?

_____

_____

Explain the results. (Hint: the vagus nerve is the sensory nerve of the lungs and plays a role here.)

_____

_____

4. Without recording, have the subject hyperventilate (breathe deeply and forcefully at the rate of 1 breath/4 sec) for about 30 seconds.* Record after hyperventilation. Is the respiratory rate faster *or*

slower than during normal quiet breathing? _____

_____

5. Repeat the above test, but after hyperventilating, have the subject hold his breath as long as he can. Can the breath be held for a longer or shorter period

of time after hyperventilation? _____

6. Without recording, have the subject breathe into a paper bag for 3 minutes, then record breathing movements. (Caution: the subject's partner should watch the subject carefully during the bag-breathing exercise for any untoward reactions.) Is the breathing rate faster *or* slower than that recorded during

normal quiet breathing? _____

After hyperventilating? _____

7. Run in place for 2 minutes, then determine the

length of breath-holding. _____ sec

---

*A sensation of dizziness may develop. (As the carbon dioxide is washed out of the blood by overventilation, the blood pH increases, leading to a decrease in blood pressure and a decrease in cerebral circulation.) The subject may experience a lack of desire to breathe after forced breathing is stopped. (If the period of breathing cessation—**apnea**—is extended, cyanosis of the lips may occur.)

8. To prove that respiration has a marked effect on circulation, conduct the following test: Have your lab partner note the rate and relative force of your radial pulse before beginning.

rate _____ beats/min   relative force _____

Inspire forcibly. Immediately close your mouth and nose to retain the inhaled air, and then make a forceful and prolonged expiration. Your lab partner should observe and record the condition of the blood vessels of your neck and face, and again immediately palpate the radial pulse.

Observations _____

_____

radial pulse _____ beats/min relative force _____

Explain the changes observed. _____

_____

_____

Keep the pneumograph records to interpret results and hand them in if requested by the instructor. Observation of the test results should enable you to determine which chemical factor, carbon dioxide or oxygen, has the greatest effect on modifying the respiratory rate and depth.

RESPIRATORY SOUNDS

As air flows in and out of the respiratory tree, it produces two characteristic sounds that can be picked up with a stethoscope. The **bronchial sounds** are produced by air rushing through the large respiratory passageways (the trachea and the bronchi); the second sound type, **vesicular breathing sounds,** apparently results from air filling the alveolar sacs and resembles the sound of a rustling or muffled breeze.

1. Place the diaphragm of the stethoscope on the throat of the test subject just below the larynx. Listen for bronchial sounds on inspiration and expiration.
2. Move the stethoscope downward toward the bronchi until you can no longer hear sounds.
3. Place the stethoscope over the following chest areas and listen for vesicular sounds during respiration (heard primarily during inspiration).

   - At various intercostal spaces
   - Beneath the scapula
   - Under the clavicle

Diseased respiratory tissue, mucus, or pus can produce abnormal chest sounds such as rales (a rasping sound) and wheezing (a whistling sound).

ROLE OF THE RESPIRATORY SYSTEM IN ACID-BASE BALANCE OF BLOOD

As you have already learned, respiratory ventilation is necessary for the continuous oxygenation of the blood and the removal of carbon dioxide (a waste product of cellular respiration) from the blood. Blood pH must be relatively constant for the cells of the body to function optimally, and therefore the carbonic acid–bicarbonate buffer system of the blood is extremely important, since it helps stabilize arterial blood pH at 7.4 ± 0.02.

When carbon dioxide diffuses into the blood from the tissue cells, much of it enters the red blood cells, where it combines with water to form carbonic acid:

$$H_2O + CO_2 \xrightarrow[\text{Enzyme present in RBC}]{\text{Carbonic anhydrase}} H_2CO_3$$

Some carbonic acid formation also occurs in the plasma, but its formation is very slow because of the lack of the carbonic anhydrase enzyme. Shortly after it forms, carbonic acid dissociates to release bicarbonate (HCO_3^-) and hydrogen (H^+) ions. The hydrogen ions that remain in the cells are neutralized, or buffered, when they combine with the hemoglobin molecules. If they were not neutralized, the intracellular pH would become very acidic due to the accumulation of H^+ ions. The bicarbonate ions then diffuse out of the red blood cells into the plasma, where they become part of the carbonic acid–bicarbonate buffer system.

Acids (more precisely, H^+) released into the blood by the body cells tend to lower the pH of the blood and cause it to become too acidic. On the other hand, basic substances that enter the blood tend to cause the blood to become too alkaline and the pH to rise. Both of these tendencies are resisted in large part by the carbonic acid–bicarbonate buffer system. If H^+ concentration in the blood begins to increase, the H^+ ions combine with bicarbonate ions to form carbonic acid (a weak acid that does not tend to dissociate at physiologic pH) and is thus removed.

$$H^+ + HCO_3^- \longrightarrow H_2CO_3$$

Likewise, as blood H^+ concentration drops below what is desirable, H_2CO_3 dissociates to release the bicarbonate ions and H^+ ions to the blood, thus lowering the pH again.

$$H_2CO_3 \longrightarrow H^+ + HCO_3^-$$

In cases of excessively slow or shallow breathing (hypoventilation) or fast deep breathing (hyperventilation), the amount of carbonic acid in the blood can be greatly changed or modified—it may dramatically increase during hypoventilation and decrease substantially during hyperventilation. In both situations, the buffering ability of the blood may be inadequate, and acidosis or alkalosis will result. It is therefore important to maintain the normal rate and depth of breathing for proper control of blood pH.

To observe the ability of a buffer system to stabilize the pH of a solution, obtain five 250-ml beakers, and set up the following experimental samples:

Beaker 1
(150 ml distilled water)   pH _____

Beaker 2
(150 ml distilled water and
1 drop concentrated HCl)   pH _____

Beaker 3
(150 ml distilled water and
1 drop concentrated NaOH)   pH _____

Beaker 4
(150 ml standard buffer solution
[pH 7] and 1 drop concentrated
HCl)   pH _____

Beaker 5
(150 ml standard buffer solution
[pH 7] and 1 drop concentrated
NaOH)   pH _____

Using a pH meter standardized with a buffer solution of pH 7, determine the pH of the contents of each beaker and record above.

Add 3 more drops of concentrated HCl to beaker 4, stir, and record the pH _____

Add 3 more drops of concentrated NaOH to beaker 5, stir, and record the pH _____

How successful was the buffer solution in resisting pH changes when an acid (HCl) or a base (NaOH) was added? _____

_____

CILIARY ACTION OF THE RESPIRATORY MUCOSA

If time is limited, the instructor may demonstrate this procedure. Otherwise, students should work together in groups of four to six.

 1. Obtain a frog, dissecting instruments, pins and pan, room temperature frog Ringer's solution, bits of cork, and a millimeter ruler, and bring them to the laboratory bench.

2. Doubly pith the frog (see Exercise 16, page 140), and then cut the lower jaw away with scissors.

3. Use fine scissors to make a longitudinal cut through the ventral wall of the pharynx, and continue the incision until you have cut the length of the trachea.

4. Flush the pharynx and trachea with saline. Using fine dissecting pins, pin the opened pharynx and trachea to the dissecting pan.

5. Place a tiny piece of cork (size of the head of a common pin) about halfway down the trachea and observe its movement toward the upper end. Time how long it takes for the bit of cork to

move 0.5 cm and record. _____

6. Obtain some cold saline (15 C), and flush the trachea and pharynx with the cold saline two or three times. Again observe the length of time required for the cork to move 0.5 cm and record.

_____

7. Flush the tissues several times with room temperature saline, and then raise one end of the dissecting pan so that the pharyngeal region is higher than the trachea. Again watch the movement of the cork. Record the time required for

it to move 0.5 cm. _____

8. Dispose of the remains of the frog in the proper organic waste container, and clean the dissecting equipment.

UNIT 13

THE DIGESTIVE SYSTEM

Anatomy of the Digestive System

EXERCISE

38

OBJECTIVES

1. To state the overall function of the digestive system.

2. To identify the following structures on an appropriate diagram, torso model, or dissected pig and name their subdivisions if any:

oral cavity	esophagus
teeth	pancreas
tongue	mesentery
salivary glands	stomach
liver	small intestine
gall bladder	large intestine
oropharynx	anus
nasopharynx	appendix
laryngopharynx	peritoneum

3. To describe the general function of all the above organs or structures relative to digestive system activity.

4. To describe the histologic structure of the wall of the alimentary canal and/or label a cross-sectional diagram of the wall with the following terms: mucosa (epithelium, lamina propria, muscularis mucosa), submucosa, muscularis externa, and serosa, or adventitia.

5. To describe the composition of saliva.

6. To list the major enzymes or enzyme groups produced by each of the following organs: salivary glands, stomach, small intestine, pancreas.

7. To name human deciduous and permanent teeth and describe the anatomy of the generalized tooth.

8. To list and explain the specializations of structure of the stomach and small intestine that contribute to their functional roles.

9. To recognize the histologic structure of the following organs by microscopic inspection or by viewing an appropriate diagram or photomicrograph:

small intestine	pancreas	stomach
salivary glands	tooth	

MATERIALS

Dissectible torso model
Anatomic chart of the human digestive system

Dissection animal, trays, instruments, and bone cutters
Model of a villus and liver (if available)
Jaw model or human skull
Prepared microscope slides of the liver, pancreas,

longitudinal sections of the esophageal-cardiac junction and tooth and cross sections of the stomach, duodenum, and ileum
Hand lens
Compound microscope

The digestive system provides the body with the nutrients, water, and electrolytes essential for metabolic processes and health. The organs of this system are responsible for food ingestion, digestion, absorption, and the elimination of the undigested remains as feces.

The digestive system consists of a hollow tube extending from the mouth to the anus, into which various accessory organs or glands empty their secretions. Food material within this tube, the alimentary canal, is technically outside the body, since it has contact only with the cells lining the tract. For the ingested food to become available to the body cells, it must first be broken down physically (chewing, churning) and chemically (enzymatic hydrolysis) into its smaller diffusible molecules—a process called **digestion.** The digested end products can then pass through the epithelial cells lining the tract into the blood for distribution to the body cells—a process termed **absorption.** In one sense, the digestive tract can be viewed as a disassembly line, in which food is carried from one stage of its digestive processing to the next by muscular activity, and its nutrients are made available to the cells of the body en route.

GROSS ANATOMY OF THE HUMAN DIGESTIVE SYSTEM

The organs of the digestive system are traditionally separated into two major groups: the **alimentary canal,** or **gastrointestinal tract,** and the **accessory digestive organs.** The alimentary canal is approximately 30 feet long in the living person and consists of the mouth, pharynx, esophagus, stomach, small and large intestines, and anus. The accessory structures consist of the salivary glands, gall bladder, liver, and pancreas, which secrete their products into the alimentary canal.

The sequential pathway and fate of food as it passes through the digestive tract is described here. Identify each structure on Figure 38.1 and on the torso model.

Alimentary Canal

MOUTH OR ORAL CAVITY   Food enters the digestive tract through the mouth, or oral cavity. Within this mucous membrane–lined cavity are the gums, teeth, tongue, and openings of the ducts of the salivary glands. The **lips (labia)** protect

the opening of the chamber anteriorly, the cheeks form its lateral walls, and the palate, its roof. The anterior portion of the palate is referred to as the **hard palate,** since bone (the palatine processes of the maxillae and the palatine bones) underlies it. The posterior **soft palate** is a fibromuscular structure that is unsupported by bone. The **uvula,** a fingerlike projection of the soft palate, extends downward from its posterior margin. The soft palate closes off the oral cavity from the nasal and pharyngeal passages during swallowing. The floor of the oral cavity is occupied primarily by the muscular **tongue,** which is largely supported by the **mylohyoid muscle** and is attached to the hyoid bone, mandible, styloid processes, and pharynx. A membrane called the **frenulum** secures the inferior midline of the tongue to the floor of the mouth. The space between the lips and cheeks and the teeth is the **vestibule;** the area containing the teeth, which is posterior to the alveolar arches, is the **oral cavity** proper. (Figures 38.1 to 38.3 depict the structures of the oral cavity.)

On each side of the mouth at its posterior end are masses of lymphatic tissue, the **palatine tonsils.** Each lies in a concave area bounded anteriorly and posteriorly by membranes, the **glossopalatine arch** (anterior membrane) and the **pharyngopalatine arch** (posterior membrane). The tonsils, in common with other lymphoid tissues, are part of the body's defense system. Very often in young children, the palatine tonsils become inflamed and enlarge, partially blocking the entrance to the pharynx posteriorly and making swallowing difficult and painful. Another mass of lymphatic tissue, the **lingual tonsil,** covers the base of the tongue, posterior to the oral cavity proper.

Three pairs of salivary glands duct their section, saliva, into the oral cavity. One component of saliva, salivary amylase, begins the digestion of starchy foods within the oral cavity. (The salivary glands are discussed in more detail on page 335.)

As food enters the mouth, it is mixed with saliva and masticated (chewed). The cheeks and lips help hold the food between the teeth during mastication, and the highly mobile tongue manipulates the food for chewing and initiates swallowing. Thus the mechanical and chemical breakdown of food begins before the food has left the oral cavity. As noted in Exercise 26, the surface of the tongue is covered with papillae, many of which contain taste buds, the receptors for taste sensation. Thus, in addition to its manipulative function, the tongue provides for the enjoyment and appreciation of the food ingested.

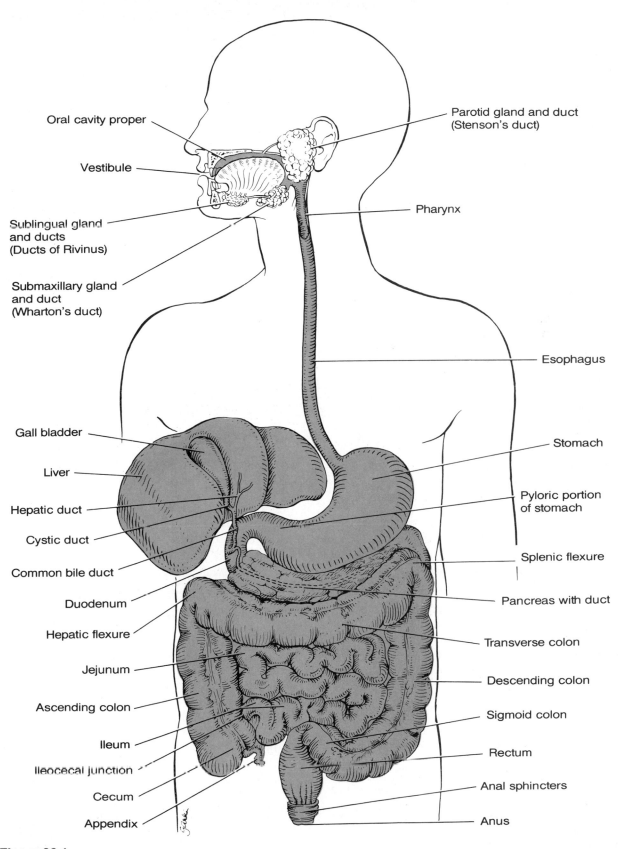

Figure 38.1

The human digestive system: alimentary tube and accessory organs. (Liver and gall bladder are reflected superiorly and to the right.)

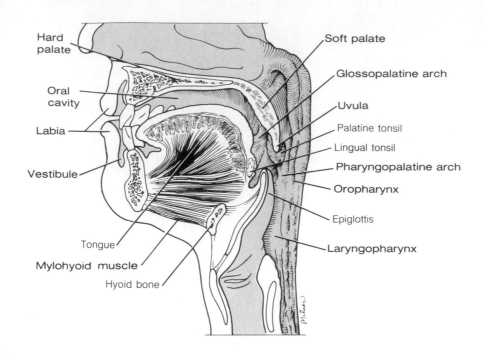

Figure 38.2

Sagittal view of the head showing oral, nasal, and pharyngeal cavities.

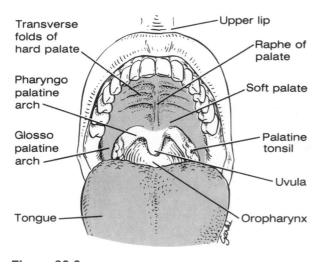

Figure 38.3

Anterior view of the oral cavity.

PHARYNX When the tongue initiates swallowing, the food passes posteriorly into the pharynx, a common passageway for food, fluid, and air. The pharynx is often subdivided anatomically into the **nasopharynx** (behind the nasal cavity), the **oropharynx** (behind the oral cavity extending from the soft palate to the epiglottis overlying the larynx), and the **laryngopharynx** (extending from the epiglottis to the base of the larynx), which is continuous with the esophagus.

The walls of the pharynx consist of two layers of skeletal muscles: an inner layer of longitudinal muscle (the levator muscles) and an outer layer of circular constrictor muscles, which initiate wavelike contractions that enable the pharynx to propel the food into the esophagus inferiorly. The levator muscles interdigitate with the contractors and reinforce them, thus adding flexibility to the walls of the pharynx.

ESOPHAGUS The esophagus, or gullet, extends from the pharynx through the diaphragm to the cardiac sphincter in the superior aspect of the stomach. It is approximately 10 inches long in humans and is essentially a food passageway that conducts food to the stomach in a wavelike peristaltic motion. The esophagus has no digestive or absorptive function. The walls of the esophagus at its superior end contain skeletal muscle, which is replaced by smooth muscle in the area nearing the stomach.

Because the tissue composition of the alimentary tube walls is modified in various areas along its length to serve specific functions, it is worthwhile to briefly describe a cross-sectional view of the general wall structure (Figure 38.4). The walls of the alimentary canal organs from the esophagus to the colon have four characteristic layers:

Mucosa: the innermost layer, which consists of an **epithelium,** a connective tissue basal layer, the **lamina propria,** and a thin smooth muscle layer, the **muscularis mucosa.**

Submucosa: deep to the mucosa and consisting primarily of connective tissue in which blood vessels, nerve endings, and lymphatic tissue are found.

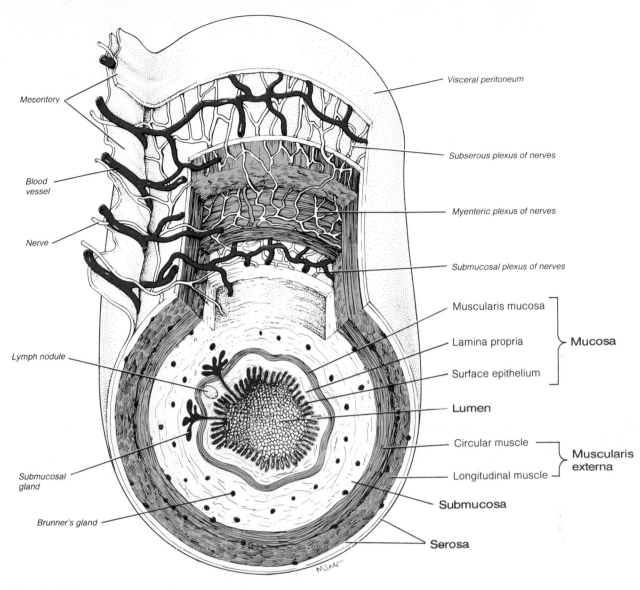

Figure 38.4

Structural pattern of the alimentary canal wall.

Muscularis externa: commonly composed of two layers of smooth muscle, the inner layer in circular orientation, and the outer layer oriented longitudinally.

Serosa: the outermost layer, consisting of a single layer of serous fluid–producing cells and commonly referred to as the **visceral peritoneum.** (In some areas, this outermost layer may be an adventitia composed of fibrous tissue rather than a serosa.)

STOMACH The stomach (Figure 38.5) is on the left side of the abdominal cavity and is hidden by the liver and diaphragm. Different regions of the saclike stomach are the **cardiac region** (the area surrounding the cardiac sphincter through which food enters the stomach from the esophagus), the **fundus** (the expanded portion of the stomach, superolateral to the cardiac region), the **body** (mid-portion of the stomach, inferior to the fundus), and the **pylorus** (the terminal part of the stomach, which is continuous with the small intestine through the pyloric valve).

The concave medial surface of the stomach is called the **lesser curvature,** and its convex lateral surface is called the **greater curvature.** The **lesser omentum,** a double layer of peritoneum, extends from the liver to attach to the lesser curvature of the stomach. The **greater omentum,** a saclike extension of the peritoneum, extends from the greater curvature of the stomach, reflects downward over the abdominal contents to cover them in an apronlike fashion, and then attaches posteriorly to the body wall. Figure 38.6 illustrates the omenta as well as the other peritoneal attachments of the abdominal organs.

The stomach is a temporary storage region for food as well as a site for the mechanical and chemi-

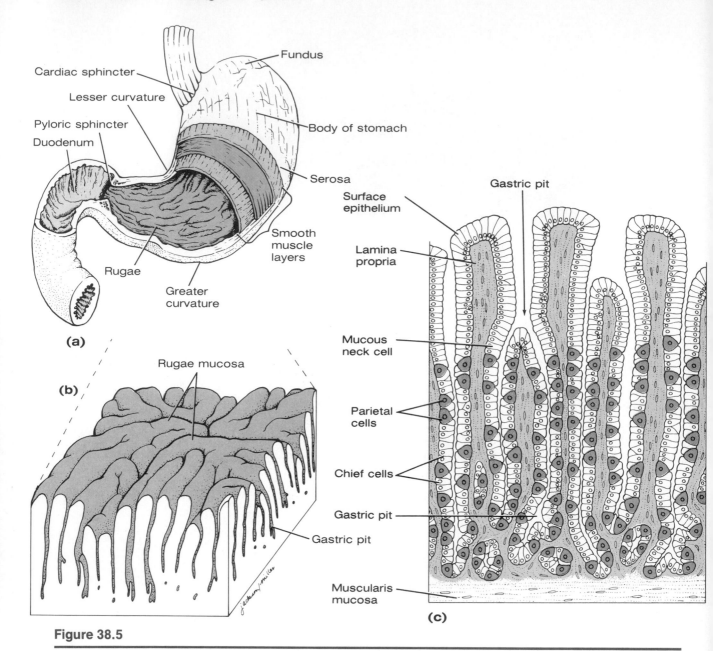

Cardiac sphincter

Lesser curvature

Pyloric sphincter

Duodenum

Rugae

Fundus

Body of stomach

Serosa

Smooth muscle layers

Greater curvature

(a)

(b)

Rugae mucosa

Gastric pit

Gastric pit

Surface epithelium

Lamina propria

Mucous neck cell

Parietal cells

Chief cells

Gastric pit

Muscularis mucosa

(c)

Figure 38.5

Anatomy of the stomach: (a) gross internal and external anatomy; (b) section of the stomach wall showing rugae and gastric pits; (c) enlarged view of gastric pits (longitudinal section).

cal breakdown of food. It contains a third obliquely oriented layer of smooth muscle in its muscularis externa that allows it to churn, mix, and pummel the food, physically reducing it to smaller fragments. Gastric glands of the mucosa secrete hydrochloric acid (HCl) and hydrolytic enzymes (primarily pepsinogen, a protein-digesting enzyme), which begin the enzymatic, or chemical, breakdown of protein foods. The mucosal glands also secrete a viscous mucus that prevents the stomach itself from being digested by the proteolytic enzymes. Most digestive activity occurs in the pyloric region of the stomach. After the food has been processed in the stomach, it resembles a creamy mass (chyme), which enters the small intestine through the pyloric sphincter.

SMALL INTESTINE The small intestine is a convoluted tube, extending from the pyloric sphincter to the ileocecal valve. The small intestine is suspended by a double layer of peritoneum, the fan-shaped **mesentery,** from the posterior abdominal wall (see Figure 38.6), and it lies, framed laterally and superiorly by the large intestine, in the abdominal cavity. The small intestine has three subdivisions: (1) The **duodenum** extends from the pyloric sphincter for about 1 foot and curves around the head of the pancreas; most of the duodenum lies in a retroperitoneal position. (2) The **jejunum,** continuous with the duodenum, extends for 3 to 4 feet. (3) The **ileum,** the terminal portion of the small intestine, is about 7 feet

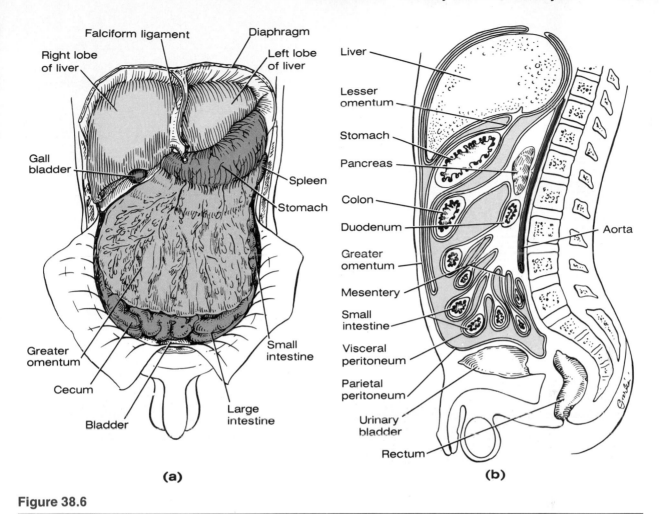

(a)   **(b)**

Figure 38.6

Peritoneal attachments of the abdominal organs: (a) anterior view, omentum in place; (b) sagittal view of a male torso.

long and joins the large intestine at the **ileocecal valve.** It is generally located inferiorly and to the right in the abdominal cavity.

Hydrolytic enzymes produced by the intestinal glands in the crypts of Lieberkühn and, more importantly, enzymes produced by the pancreas and ducted into the duodenum via the **pancreatic duct** complete the enzymatic digestion process in the small intestine. Bile (formed in the liver) also enters the duodenum via the **common bile duct** in the same area (see Figure 38.1). At the duodenum, the ducts join to form the **ampulla of Vater** and empty their products into the duodenal lumen through the **sphincter of Oddi.**

Nearly all nutrient absorption occurs in the small intestine, where three structural modifications that increase the absorptive area appear—the **microvilli, villi,** and **plicae circulares.** Microvilli are minute projections of the surface plasma membrane of the epithelial lining cells of the mucosa; the villi are the fingerlike projections of the mucosa layer that give it a velvety appearance and texture (Figure 38.7). The plicae circulares are deep folds of the mucosa and submucosa layers that extend partially or totally around the intestine. These structural modifications, which increase the surface

area, decrease in frequency and elaboration toward the end of the small intestine. Any residue remaining undigested and unabsorbed at the terminus of the small intestine enters the large intestine through the ileocecal valve. In contrast, the amount of lymphatic tissue in the submucosa of the small intestine (**Peyer's patches**) increases along the length of the small intestine and is very apparent in the ileum. This reflects the fact that the remaining undigested food residue contains large numbers of bacteria that must be prevented from entering the bloodstream.

LARGE INTESTINE The large intestine (see Figure 38.8) is about 5 feet long and extends from the ileocecal valve to the anus. It encircles the small intestine on three sides and consists of the following subdivisions: the **cecum, appendix, colon, rectum,** and **anal canal.** The colon is divided into several distinct regions. The **ascending colon** travels up the right side of the abdominal cavity and makes a right-angle turn (**hepatic flexure**) to cross the abdominal cavity as the **transverse colon.** It then turns (**splenic flexure**) and continues down the left side of the abdominal cavity as the **descending colon,** where it takes an S-shaped course as the **sigmoid**

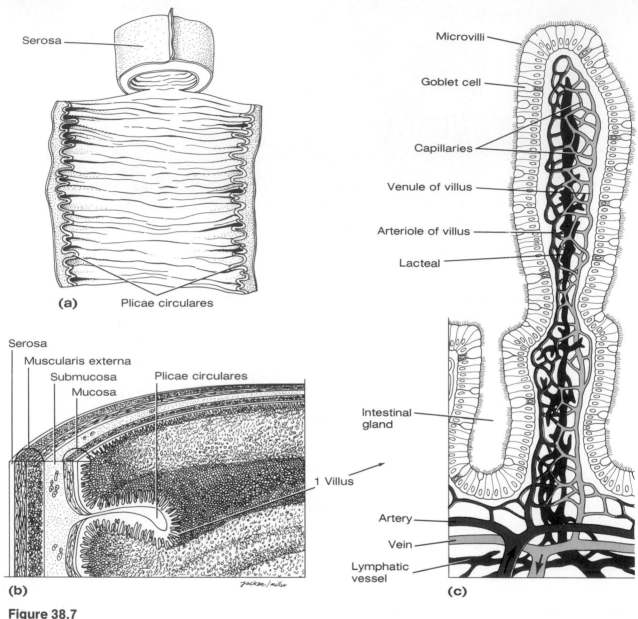

Figure 38.7

Structural modifications of the small intestine: (a) plicae circulares (circular folds) seen on the inner surface of the small intestine; (b) enlargement of one plica circulare to show villi; (c) detailed anatomy of a villus.

colon. The sigmoid colon, rectum, and the anal canal lie in the pelvis anterior to the sacrum and thus are not considered abdominal cavity structures. The anal canal terminates in the **anus,** the opening to the exterior of the body. The anus, which has an external sphincter of skeletal muscle (the voluntary sphincter) and an internal sphincter of smooth muscle (the involuntary sphincter), is normally closed except during defecation when the undigested remains of the food and bacteria are eliminated from the body as feces.

In the large intestine, the longitudinal muscle layer of the muscularis externa is reduced to three longitudinal muscle bands called the **taeniae coli.** Since these bands are shorter than the rest of the

wall of the large intestine, they cause the wall to pucker into small pocketlike sacs called **haustra.**

The blind tubelike appendix (approximately 3 inches long) is a trouble spot in the large intestine. Since it is generally twisted, it provides an ideal location for bacteria to accumulate and multiply. Inflammation of the appendix, or appendicitis, is the result.

The major functions of the large intestine include the following: it provides a site for the manufacture of some vitamins (B vitamins and vitamin K) by intestinal bacteria; the vitamins are then absorbed into the bloodstream. It also absorbs water from undigested food, thus conserving body water and forming the feces, or stool. Watery stools,

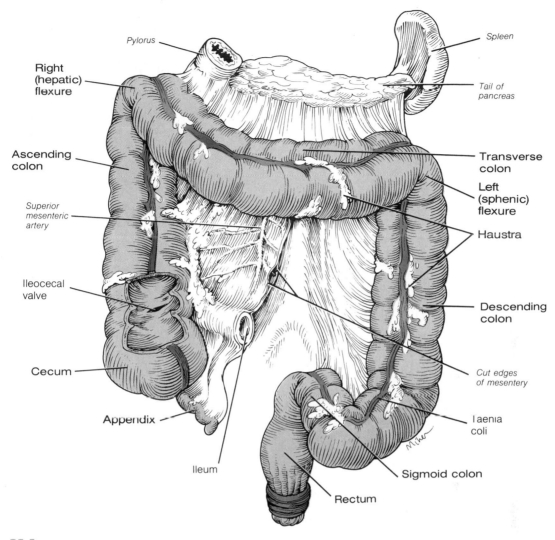

Figure 38.8

The large intestine (section of the cecum removed to show the ileocecal valve).

or diarrhea, result from any condition that rushes undigested food residue through the large intestine before it has had sufficient time to absorb the water (as in irritation of the colon by bacteria). Conversely, when food residue remains in the large intestine for extended periods (as with atonic colon or failure of the defecation reflex), excessive water is absorbed and the stool becomes hard and difficult to pass (constipation).

Accessory Digestive Organs

SALIVARY GLANDS   The three pairs of major salivary glands (see Figure 38.1) that empty their secretions into the oral cavity are the large **parotid glands,** located anteriorly to the ear and ducting into the mouth over the second upper molar through Stenson's duct; the **submaxillary** (or submandibular) **glands,** located inside the maxillary arch in the floor of the mouth and ducting under the tongue through Wharton's ducts; and the small **sublingual glands,** located most anteriorly in the floor of the mouth and

emptying under the tongue through several small ducts, the ducts of Rivinus. Food in the mouth and mechanical pressure (chewing rubber bands or wax) stimulates the salivary glands to secrete saliva. Saliva consists primarily of mucin (a viscous glycoprotein), which moistens the food and helps to bind it together into a mass called a bolus, and a clear serous fluid containing the enzyme, salivary amylase. Salivary amylase begins the digestion of starch (a large polysaccharide), breaking it down into disaccharides, or double sugars. The secretion of the parotid glands is mainly serous, whereas the submandibular and sublingual glands are mixed glands that produce mucin and serous components.

PANCREAS   The pancreas is a soft, triangular gland that extends horizontally across the posterior abdominal wall from the spleen to the duodenum (see Figure 38.1). Like the duodenum, it is a retroperitoneal organ (see Figure 38.6). It produces a whole spectrum of hydrolytic enzymes, which it secretes in an alkaline fluid into the duodenum through the pancreatic duct. The pancreatic juice, whose alkalinity reflects a high concentration of bicarbonate ion (HCO_3^-), neutralizes the acidic chyme as it enters the duodenum from the stomach, enabling the pancreatic and intestinal enzymes to operate at their optimal pH. (Optimal pH for digestive activity to occur in the stomach is very acidic and results from the presence of HCl; that for the small intestine is slightly alkaline.) The pancreas also has an endocrine function: it produces the hormones, insulin and glucagon (see Exercise 27, Anatomy and Basic Function of the Endocrine Glands).

LIVER AND GALL BLADDER The liver (see Figure 38.1), the largest gland in the body, is located under the diaphragm, more to the right than the left side of the body. As noted earlier, it hides the stomach from view in a superficial observation of stomach contents. The human liver has four lobes and is suspended from the diaphragm and anterior abdominal wall by the **falciform ligament.**

The liver is one of the body's most important organs, and it performs many metabolic roles. However, its digestive function is to produce bile, which leaves the liver through the **hepatic duct** and then enters the duodenum through the **common bile duct.** Bile has no enzymatic action but emulsifies (spreads thin or breaks up large particles into smaller ones) fats, thus creating a larger surface area for more efficient lipase activity. Without bile, very little fat digestion or absorption occurs.

When digestive activity is not occurring in the digestive tract, bile backs up the cystic duct and enters the **gall bladder,** which is a small, green sac on the inferior surface of the liver. It is stored there until needed for the digestive process. While in the gall bladder, bile is concentrated by the removal of water and some ions. When fat-rich food enters the duodenum, a hormonal stimulus causes the gall bladder to contract, releasing the stored bile and making it available to the duodenum.

TEETH By the age of 21, two sets of teeth have developed (Figure 38.9). The initial set, called the **deciduous,** or **milk teeth,** normally appear between the ages of 6 months and 2½ years. The first of these to erupt are the lower central incisors, an event which is usually applauded by the child's parents. The child begins to shed the deciduous teeth around the age of 6. The second set of teeth, the **permanent teeth,** gradually replace them. As the deeper permanent teeth progressively enlarge and develop, the roots of the deciduous teeth are resorbed, leading to their final shedding, and during the sixth to twelfth years, the child has mixed dentition—both permanent and deciduous teeth. Generally, by the age of 12, all of the deciduous teeth have been shed, or exfoliated.

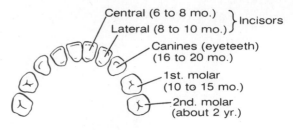

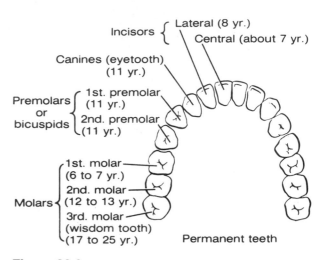

Figure 38.9

Human deciduous and permanent teeth. (Approximate time of teeth eruption shown in parentheses.)

The teeth are classified as **incisors, canines** (eye teeth), **premolars** (bicuspids), and **molars,** and dentition is described by means of a **dental formula,** which designates the numbers, types, and position of the teeth in one side of the jaw. (Since tooth arrangement is bilaterally symmetrical, it is only necessary to designate one side of the jaw.) The dental formula for the deciduous teeth from the medial aspect of the jaw and proceeding posteriorly is as follows:

$$\frac{\text{Upper teeth: 2 incisors, 1 canine, 0 premolars, 2 molars}}{\text{Lower teeth: 2 incisors, 1 canine, 0 premolars, 2 molars}} \times 2$$

This formula is generally abbreviated to read as follows:

$$\frac{2,1,0,2}{2,1,0,2} \times 2 = 20 \text{ (number of deciduous teeth)}$$

The 32 permanent teeth are then described by the following dental formula:

$$\frac{2,1,2,3}{2,1,2,3} \times 2 = 32 \text{ (number of permanent teeth)}$$

Although 32 is designated as the normal number of permanent teeth, not everyone develops a full complement. In many people, the No. 3 molars, commonly called the wisdom teeth, never erupt.

Teeth names reflect differences in relative structure and function. The incisors are chisel-shaped and

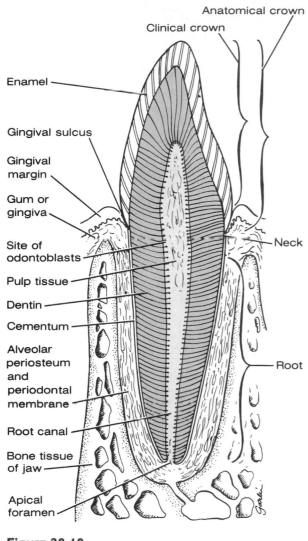

Anatomical crown

Clinical crown

Enamel

Gingival sulcus

Gingival margin

Gum or gingiva

Site of odontoblasts

Pulp tissue

Dentin

Cementum

Alveolar periosteum and periodontal membrane

Root canal

Bone tissue of jaw

Apical foramen

Neck

Root

Figure 38.10

Longitudinal section of human canine tooth.

lowing basic anatomic plan (Figure 38.10): the crown is the superior portion of the tooth; the portion of the crown visible above the **gum,** or **gingiva,** is referred to as the **clinical crown.** The entire area covered by **enamel** is called the **anatomic crown.** The crevice between the end of the anatomic crown and the upper margin of the gingiva is referred to as the **gingival sulcus** and its apical border as the **gingival margin.** The enamel is the hardest substance in the body and is fairly brittle. It consists of 95% to 97% inorganic calcium salts (chiefly $CaPO_4$) and thus is heavily mineralized. That portion of the tooth embedded in the alveolar portion of the jaw is the root, and the root and crown are connected by a slight constriction, the **neck,** or **cervix.** The outermost surface of the root is covered by **cementum,** which is similar to bone in composition and less brittle than enamel. The cementum attaches the tooth to the **periodontal membrane,** a fibrous membrane that holds the tooth in the alveolar socket and exerts a cushioning effect. **Dentin,** which comprises the bulk of the tooth, consists of bonelike material and is medial to the enamel and cementum. The **pulp cavity** occupies the central portion of the tooth. Pulp, connective tissue liberally supplied with blood vessels, nerves, and lymphatics, occupies this cavity and provides the source of nutrition for the tooth tissues. Specialized cells, **odontoblasts,** reside in the outer margins of the pulp cavity and produce the dentin. Since the pulp contains the nerve supply of the tooth, it also provides tooth sensation. As the pulp cavity extends into distal portions of the root is becomes the **root canal.** An opening at the root apex, the **apical foramen** provides a route of entry into the tooth for the blood vessels, nerves, and other structures from the tissues beneath.

exert a shearing action used in biting. The canines are cone-shaped or fanglike, the latter description being much more applicable to the canines of animals whose teeth are used for the tearing of food. Both the incisors and the canines have single roots. The premolars, or bicuspids, have two cusps (grinding surfaces) and typically two roots. The molars have relatively flat, broad superior surfaces specialized for the fine grinding of food and typically have two or three roots but may have more.

 Identify the four types of teeth (incisors, canines, premolars, and molars) on the jaw model or human skull.

A tooth is commonly considered to consist of two major regions, the **crown** and the **root.** A longitudinal section made through a tooth shows the fol-

DISSECTION OF THE FETAL PIG DIGESTIVE SYSTEM

 1. Obtain your pig, and secure it to the dissecting tray, lateral surface down, for the initial portion of this dissection procedure. Obtain all necessary dissecting instruments. If you have already completed the dissection of the circulatory and respiratory systems, the abdominal cavity is already exposed, and many of the digestive system structures have been previously identified. However, duplication of effort generally provides a good learning experience, so you will trace and identify all of the digestive system structures in this exercise.

2. To expose and identify the salivary glands, cut through the skin in a diagonal plane from an initial incision below the ear to the mouth. Continue the cut posteriorly to the anterior aspect of the forelimb.

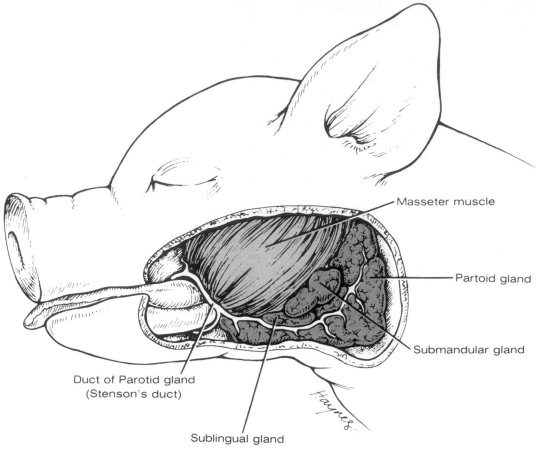

Figure 38.11

Diagrammatic view of the salivary glands of the fetal pig.

This will produce a triangular flap of skin that can be folded back to identify the salivary glands. Once the skin has been reflected back, carefully remove the underlying connective tissue and observe the tissues to differentiate between muscle tissue, appearing mostly as small parallel bundles of fibers in this region, and the glandular tissue, which is quite different in texture (bunches of nodular-looking tissue). Many lymph nodes are present in this area and should be removed if they obscure the salivary glands (Figure 38.11). Locate the large thin triangular-shaped parotid gland on the cheek, just inferior to the ear. In the fetal pig, this gland may be poorly developed and somewhat diffuse. Follow its duct (Stenson's duct) over the surface of the masseter muscle to the angle of the mouth. The small compact submandibular gland is posterior to the parotid gland and partially underlies it and is easily mistaken for a lymph node. The sublingual gland is small and located at the base of the tongue anterior

to the submandibular gland. The ducts of these two salivary glands run deep and parallel to each other and empty on the side of the frenulum of the tongue. These need not be identified in the pig.

3. To expose and identify the structures of the oral cavity, cut through the mandible with bone cutters just anterior to the angle to free the lower jaw from the maxilla. Observe the teeth of the pig. The dental formula for the fetal pig (deciduous teeth) is as follows:

$$\frac{3,1,4,0}{3,1,4,0} \times 2 = 32$$

Notice that many of the fetal teeth are incompletely developed or may not have yet emerged through the gums. If you have a very young fetus with no teeth in evidence, cut into the gum tissue to observe whether the deciduous teeth are present. Identify the hard and soft palates, and use a probe to trace the

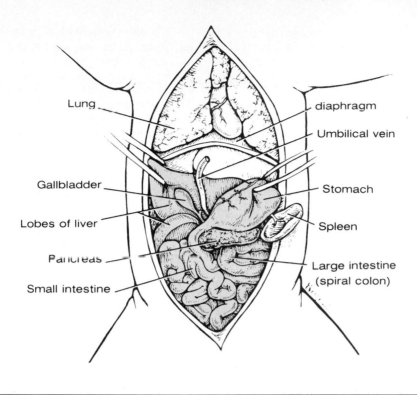

Figure 38.12

Digestive organs of the fetal pig (greater omentum removed; liver reflected superiorly.)

hard palate to its posterior limits. Does the pig have a uvula?

Identify the oropharynx at the rear of the oral cavity and the nasopharynx, the continuation of the nasal cavities, superior to it. Identify the tongue and rub your finger across its surface to feel the papillae. As in humans, the tongue plays a role in the manipulation of food in the mouth, and its papillae house the taste buds.

Locate the **frenulum** attaching the tongue to the floor of the mouth. Trace the tongue posteriorly until you locate the **epiglottis,** which appears as a small white tissue tab and that closes the respiratory passageway when swallowing occurs. Also identify the esophagus posterior to the epiglottis.

4.   Locate the abdominal alimentary tube structures. If you have not opened the abdominal cavity previously, make a midline incision from the rib cage to the umbilical cord and then make two lateral cuts to encircle the umbilicus, and continue these cuts, staying lateral to the body midline, posteriorly until

the pubic bone is reached. Make four lateral cuts, two parallel to the rib cage and two at the inferior margin of the abdominal cavity so that the abdominal wall can be reflected back for examination of the abdominal contents. Also cut the umbilical vein at the lower margin of the liver (Figure 38.12). Observe the shiny membrane lining the inner surface of the abdominal wall, which is the parietal peritoneum. Identify the large reddish brown liver just beneath the diaphragm. The pig's liver has four main lobes (right lateral and right central and left lateral and left central), plus a small posterior caudate lobe. How does this compare to the number of liver lobes seen in the human liver?

Lift the liver and examine its inferior surface to locate the gall bladder embedded in its ventral surface. In preserved specimens, the gall bladder is often shrunken and nearly colorless.

Identify the falciform ligament, a delicate layer of mesentery separating the main lobes of the liver and attaching the liver superiorly to the abdominal wall. Notice that this structure surrounds the umbilical vein of the fetus. (Later in life, the

the common bile duct and trace its course superiorly to the point where it diverges into the cystic duct (gall bladder duct) and the hepatic duct (duct from the liver). Note that the duodenum assumes a looped position.

Lift the small intestine to note the manner in which it is attached to the posterior body wall by the mesentery. Observe the mesentery closely. What types of structures do you see in this double

peritoneal fold? _____

_____

Other than providing support for the intestine, what

other functions does the mesentery have? _____

_____

Trace the course of the small intestine from its proximal, or duodenal, end to its distal, or ileal, end. Can you see any obvious differences in the external anatomy of the small intestine from one end to the

other? _____

With a scalpel, slice open the distal portion of the ileum and flush out the inner surface with water. Feel the inner surface with your finger tip. How

does it feel? _____

_____

Use a hand lens to see if you can see any villi and to locate the areas of lymphatic tissue called Peyer's patches, which appear as scattered white patches on the inner intestinal surface.

Return to the duodenal end of the small intestine. Make an incision into the duodenum. As before, flush the surface with water, and feel the inner surface. Does it feel any different than the ileal

mucosa? _____

If so, describe the difference. _____

_____

_____

Use the hand lens to observe the villi. What differences do you see between the villi in the two

areas of the small intestine? _____

Make an incision into the junction between the ileum and cecum to locate the ileocecal valve. Observe the cecum (lymph nodes may have to be removed from this area to observe it clearly). Does

the cat have an appendix? _____

Identify the ascending, transverse, and descending portions of the colon and the **mesacolon,** a membrane that supports the colon in the cat. (Humans have no mesacolon, since the large intestine adheres tightly to the posterior body wall.) Trace the descending colon to the rectum, which penetrates the body wall, and identify the anus.

Identify the two portions of the peritoneum, the parietal peritoneum lining the abdominal wall (identified previously) and the visceral peritoneum, which is the outermost layer of the wall of the abdominal organs (serosa).

5. Prepare your cat for storage by wrapping it in paper towels wet with embalming fluid, return it to the plastic bag, and attach your name label. Wash the dissecting tray and instruments before continuing or leaving the laboratory.

MICROSCOPIC ANATOMY OF SELECTED DIGESTIVE SYSTEM AND ACCESSORY ORGANS

 Obtain a microscope and the following slides in preparation for the histologic study: salivary glands (submandibular or sublingual), pancreas, liver, cross sections of the duodenum, ileum, liver, and the stomach, and longitudinal sections of a tooth and the cardioesophageal junction.

Salivary Glands

Examine the glandular tissue under low power and then high power to become familiar with the appearance of a glandular tissue. Note the clustered arrangement of the cells around their ducts. The cells are basically triangular with their pointed ends facing the duct orifice. If possible, differentiate between the serous cells, which produce the clear enzyme-containing fluid and have granules in their cytoplasm, and the mucus-producing cells, which look hollow or have a clear cytoplasm. In many cases the serous crescents are distal to the duct and the mucus-producing cells. Draw a small portion of the salivary gland tissue here and label appropriately.

Pancreas

Observe the pancreas tissue under low power and then high power to distinguish between the lighter-staining, endocrine-producing clusters of cells (islets of Langerhans) and the deeper-staining parenchyma cells (acinar cells), which produce the hydrolytic enzymes and form the major portion of the pancreatic tissue. Note the arrangement of the exocrine parenchyma cells around their central ducts. Draw a small portion of the pancreatic tissue in the space provided. Include both exocrine and endocrine tissues. Label appropriately.

Small Intestine

DUODENUM Observe the tissue under low power to identify the four basic layers of the wall of the intestine (mucosa, submucosa, muscularis externa, and serosa). Identify the scattered Brunner's glands (mucus-producing glands) in the submucosa. What type of epithelium do you see

here? _____

Note the large leaflike villi, which increase the surface area for absorption. Note also the crypts of Lieberkühn, invaginated areas of the mucosa where the cells are involved in the production of intestinal enzymes. Sketch and label a small section of the duodenal wall, showing all layers and villi.

ILEUM The structure of the ileum is similar to that of the duodenum except that the villi tend to be less elaborate (most of the absorption has occurred by the time the ileum is reached). Observe the villi, and identify the four layers of the wall and the large Peyer's patches.

What tissue comprises Peyer's patches? _____

_____

Stomach

Examine the tissue under low power to locate the muscularis externa; then move to high power to closely examine this layer. How many smooth

muscle layers are visible? _____ How does this correlate with the churning movements per-

formed by the stomach? _____

_____

_____

_____

Identify the gastric glands and the gastric pits. If the section is taken from the stomach fundus and appropriately stained, you can identify the blue-staining **chief cells,** which produce pepsinogen, and the red-staining **parietal cells,** which secrete HCl, in the gastric glands. Draw a small section of the stomach wall and label it appropriately.

Cardiac-Esophageal Junction

Examine the slide under low power, and scan it to localize the junction between the end of the esophagus and the beginning of the stomach, the cardiac-esophageal junction. How does the epithelium of the esophagus differ from that of

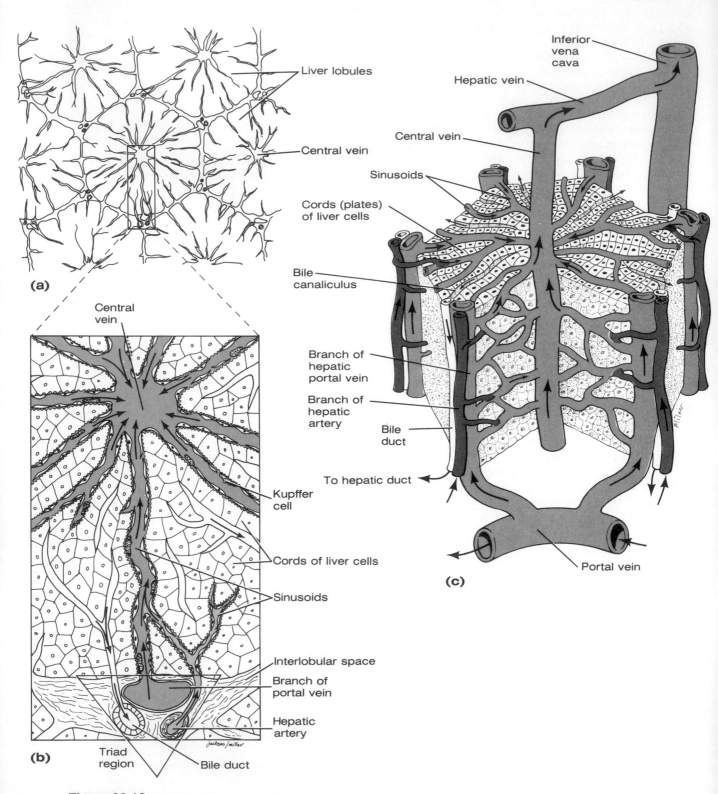

(a)

Liver lobules

Central vein

(b)

Central vein

Kupffer cell

Cords of liver cells

Sinusoids

Interlobular space

Branch of portal vein

Hepatic artery

Triad region

Bile duct

(c)

Inferior vena cava

Hepatic vein

Central vein

Sinusoids

Cords (plates) of liver cells

Bile canaliculus

Branch of hepatic portal vein

Branch of hepatic artery

Bile duct

To hepatic duct

Portal vein

Figure 38.13

Microscopic anatomy of the liver (diagrammatic view): (a) several liver lobules (cross section); (b) enlarged view of a portion of one liver lobule (cross section); (c) portion of one liver lobule (longitudinal section). Arrows show direction of bile and blood flow.

the stomach? _____

_____

Why do you suppose this is so? _____

_____

The Liver

The liver (Figure 38.13) is very important in the initial processing of the nutrient-rich blood draining the digestive organs. It is composed of structural and functional units called **lobules.** Each lobule is a basically cylindrical structure consisting of cordlike arrangements of parenchyma cells, which radiate outward from a central vein running upward in the longitudinal axis of the lobule. At each of the six corners of the lobule is a **triad** region, so named because three basic structures are always present: a branch of the hepatic artery (the functional blood supply of the liver), a branch of the hepatic portal vein (carrying nutrient-rich blood from the digestive viscera), and a bile duct. Between the liver parenchyma cells are blood-filled spaces, or **sinusoids,** through which blood from the hepatic portal vein and hepatic artery percolates past the parenchyma cells. Special phagocytic cells, **von Kupffer cells,** line the sinusoids and remove debris such as bacteria from the blood as it flows past, while the parenchyma cells pick up oxygen and nutrients. Much of the glucose transported to the liver from the digestive system is stored as glycogen in the liver for later use, and amino acids are taken from the blood by the liver cells and utilized to make plasma proteins. The sinusoids empty into the central vein, and the blood ultimately drains from the liver into the hepatic vein.

Bile is also continuously being made by the parenchyma cells. It flows through tiny canals, the **bile canaliculi,** which run between adjacent parenchyma cells toward the bile duct branches in the triad regions, where the bile eventually leaves the liver. Note that the direction of blood and bile flow in the liver lobule is exactly opposite.

Examine a slide of liver tissue and identify as many of the structural features in Figure 38.13 as possible. (Also examine a three-dimensional model of the liver if this is available). Draw your observations below.

Tooth

Observe a slide of a longitudinal section of a tooth and compare your observations with the structures detailed in Figure 38.10. Identify as many of these structures as possible.

Mechanisms of Food Propulsion and the Physiology of Smooth Muscle

EXERCISE 39

OBJECTIVES

1. To explain why swallowing is both a voluntary and a reflex activity.

2. To discuss the role of the tongue and larynx in swallowing.

3. To state the function of the gastroesophageal sphincter.

4. To describe the following digestive system movements: pendular, segmental, peristaltic, and mass.

5. To define the following terms as applied to the physiologic activity of unitary smooth muscle: *automaticity* and *plasticity*.

6. To discuss the effect of the autonomic nervous system, local ionic conditions, and temperature on the activity of visceral smooth muscle.

7. To differentiate clearly between skeletal and smooth (unitary) muscle relative to factors that stimulate them to contract.

MATERIALS

Supply area 1:
Water pitcher
Paper cups
Stethoscope

Supply area 2:
Rats or guinea pigs that have fasted for 24 hours. (The rats should be force-fed cream and meat extract 2 hours before laboratory experiment.)
Large bell jar
Ether
Absorbent cotton
Dissecting instruments and pan
Hand lens
Physiologic saline

Supply area 3:
Materials in supply area 2, plus physiograph or kymograph recording apparatus (as employed in Exercise 16) with muscle transducers or isotonic muscle levers for recording muscle activity.
L-shaped glass tubing
Rubber hose length
Tyrode's solution
Water bath set at 37 C
250-ml beakers
Laboratory thermometer (°C)
Metric ruler
Alcohol burner or Bunsen burner
Medicine dropper
Thread
0.5-ml pipettes or 1-cc prepackaged syringes with needles
Ice
Epinephrine 1:10,000 and acetylcholine 1:10,000 solutions
Solutions in dropper bottles: 7.5% KCl, 2% NaOH, 20% $CaCl_2$, and 2% HCl

FOOD PROPULSION MECHANISMS

Deglutition (Swallowing)

Swallowing, or deglutition, occurs in three phases: buccal (mouth), pharyngeal, and esophageal. The initial phase—the buccal—is voluntarily controlled. Once begun, the process continues involuntarily in the pharynx and esophagus, through peristalsis, resulting in the delivery of the swallowed contents to the stomach.

Obtain a pitcher of water, a stethoscope, and a paper cup before making the following observations:

1. While swallowing a mouthful of water, consciously note the movement of your tongue during the process. Record your

observations. _____

_____

_____

2. Repeat the swallowing process while your laboratory partner watches the externally obvious movements of your larynx. (This movement is more obvious in a male, who has a larger Adam's apple.) Record your observations. _____

_____

_____

What do these movements accomplish? _____

_____

_____

3. Have your partner place the diaphragm of the stethoscope over your abdominal wall approximately 1 inch below the xiphoid process and slightly to the left to listen for sounds as you again take two or three swallows of water. There should be two audible sounds—one when the water splashes against the closed gastroesophageal sphincter and the second when the peristaltic wave of the esophagus arrives at the sphincter and the sphincter opens, allowing the water to gurgle into the stomach. Determine the time interval between these two sounds as accurately as possible and record it below.

Interval between arrival of water at the sphincter

and opening of the sphincter _____ sec

This interval gives a fair indication of the time it takes for the peristaltic wave to travel down the 10 inch long esophagus. (Actually the time interval is slightly less than it seems, since pressure causes the sphincter to relax before the peristaltic wave reaches it.)

OBSERVATION OF DIGESTIVE SYSTEM MOVEMENTS IN A SMALL ANIMAL

Visceral smooth muscle (unitary smooth muscle) composes the muscularis externa portion of the walls of organs of the digestive tract. Smooth muscle activity produces various types of movement in the digestive viscera, which promote digestion, absorption, and food movement. You will observe the three major types of movement: **pendular, segmental,** and **peristaltic.**

Pendular movements, which cause the intestine to sway from side to side, do not have a role in moving the chyme along the tract.

Segmental movements cause local constrictions of the intestine and occur rhythmically. Like the pendular movements, they are relatively ineffective in moving substances along the tract. The segmental movements primarily serve to mix chyme with the digestive juices and to increase the absorption rate by continually moving different portions of the chyme mass over adjacent regions of the intestinal wall.

Peristaltic movements, the major means of propelling food through the digestive viscera, are waves of contraction that move along the long axis of the intestine, followed by waves of relaxation (Figure 39.1). Unlike segmental movements, on which they are superimposed, peristaltic waves are arrhythmic.

Both peristaltic and segmental movements can occur in the absence of extrinsic nerve stimulation; however, if all nerve supply is cut off, peristaltic movements cease, whereas segmental movements continue. This fact indicates that peristalsis is regulated by the intrinsic plexus (myenteric plexus) and that segmentation is initiated within smooth muscle itself. Local stimuli such as pressure, acid, hypertonicity or hypotonicity of the chyme, as well as autonomic nervous system stimulation strongly influence small intestine motility.

Two other movements deserve mention. **Mass movements,** which occur principally in the large intestine, are long slow-moving waves (mass peristalsis) that traverse large areas three to four times daily to move the contents of the large bowel toward the rectum. A **villi-promoted pumping action** occurs when the fibers of the muscularis mucosa attached to the villi produce a ripplelike motion. You will not observe these movements during this experiment.

Your laboratory instructor may demonstrate intestinal movements on an animal prepared for observation, or four- or five-member student groups may sacrifice, dissect, and observe the animals.

1. Anesthetize the rat or guinea pig by placing it under a large bell jar with ether-soaked absorbent cotton sponges. (Do this under a ventilating hood with the fan on to remove the ether fumes from the laboratory.) When the excitement phase (indicated by jerking, uncontrolled movements) is over and the animal no longer responds to prodding with a probe, you can begin the dissection to expose the small intestine.

2. Place the animal on a dissecting tray and quickly make a median incision through the skin and muscle of the abdomen from the sternum to the pubic symphysis. Make four lateral cuts (two superiorly and two inferiorly) from the midline incision so that the abdominal musculature can be reflected back to reveal the abdominal contents. Lift the greater omentum and carefully reflect it back. Quickly identify the stomach (under the left aspect of the liver) and the small and large intestines.

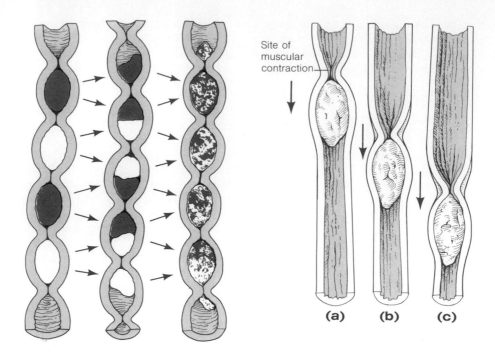

Figure 39.1

Segmental and peristaltic movements of the digestive tract. (a) Segmentation; single segments of intestine alternately contract and relax. As inactive segments exist between active segments, the food is mixed but not moved along the tract. (b) Peristalsis; superimposed on segmentation. Neighboring segments of the intestine alternately contract and relax, moving food along the tract.

3. Observe the intestine to identify the segmental, peristaltic, and pendular movements described above. Compare the frequency and vigor of movements or peristalsis in the stomach, small intestine, and large intestine.

stomach _____

_____

small intestine _____

_____

large intestine _____

_____

4. Immerse some gauze in warm saline and place it on the small intestine. After approximately 1 minute, remove the gauze and again observe the frequency of peristalsis. What changes have occurred?

_____

_____

_____

_____

5. After all members of the group have made their observations, prepare segments of the small intestine (as noted in step 3 of the following experiment) for the study of smooth muscle physiology. If you do not perform the next experiment, sacrifice the animal by a sharp blow to the head and properly dispose of it.

PHYSIOLOGY OF SMOOTH MUSCLE

Visceral smooth muscle (unitary smooth muscle), which provides the propulsive force for moving foods along the digestive tract, has highly specialized characteristics well suited to the work it performs in the walls of the digestive organs. The cells lie closely apposed in layers separated by minimal amounts of connective tissue, the contact so intimate that the cells act as a functional syncytium.

With few exceptions, smooth muscle, whose contraction rate is relatively slow, exhibits **automaticity:** it contracts in the absence of nervous system stimulation, and very often in a rhythmic manner.*

---

*Multiunit smooth muscle found in the iris and ciliary muscles of the eye and in the arterioles has individual muscle fiber innervation and is not characterized by automaticity.

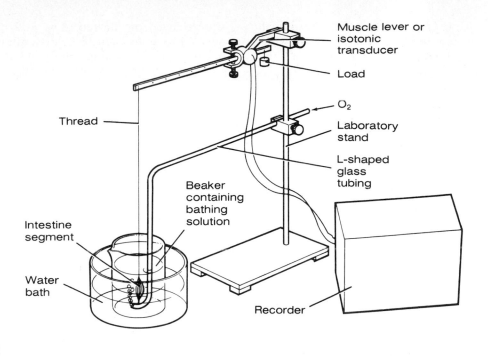

Figure 39.2

Apparatus setup for investigation of the effects of environmental factors on smooth muscle activity.

In addition, the visceral muscle cells are extremely sensitive to local environmental factors such as changes in pH and ionic concentrations. The role of the autonomic nervous system in visceral muscle function is to regulate the rate and relative strength of contraction rather than to initiate it.

Visceral muscle also exhibits **plasticity:** it can be stretched without responding with increased contractile strength. This is undeniably an important function, since it permits the internal organs to enlarge as their internal contents increase without initiating vigorous expulsive action upon stretch stimulation. There are, however, limits to such plasticity and when a hollow organ is extremely overfilled, expulsive efforts eventually occur.

You will observe many of these properties of smooth muscle in the following experiment. If students perform the experiment, one or two should dissect the rat (if it has not been dissected previously), and prepare the intestinal segments as specified in steps 1 through 3. The remaining two or three students should set up the recording apparatus (step 4), and complete the apparatus preparation (Figure 39.2).

Preparation of Intestinal Muscle and Base-Line Tracing

1.   Prepare two 250-ml beakers by pouring 150 ml of room-temperature Tyrode's solution into each.

2.   Following the directions given on p. 345, expose the small intestine.

3.   Cut two 1-inch (2.5-cm) segments from the small intestine and place them in one beaker containing Tyrode's solution. Cut the remainder of the intestine free from its large intestinal attachment and place it into the second beaker. Place this beaker into the water bath set at 37 C. Cut off additional sections as needed to complete the experimental set.

4.   Set up the physiograph/kymograph apparatus for recording as instructed in Exercise 16, Figure 16.1 or 16.2, except for the attachment of the muscle to the muscle lever or transducer. Complete apparatus preparation as shown in Figure 39.2.

5.   Swish the intestinal segment through the Tyrode's solution with forceps to flush out the inner walls and then tie an 18- to 20-inch length of thread to each end of the intestinal segment. Attach one thread end to the short portion of the L-shaped glass tubing, and attach the other thread to the transducer or muscle lever according to the arrangement shown in Figure 39.2. Adjust the tension on the thread to take up any slack, and place the thermometer in the beaker.

6.   Attach one end of the rubber tubing to the glass tubing and the other end to an air supply outlet. Turn on the air valve to provide constant aeration at the rate of 5 to 10 bubbles/sec.

7.   After a few seconds, the smooth muscle will begin to contract. When this occurs, start the recording apparatus and set it to record on a

slow-moving paper or drum. Note the initial temperature of the beaker solution, record the contractions for about 2 minutes, and then shut off the recorder.

Record the number of contractions/min _____

Describe the relative height and rhythm of the contractions.

relative height of contraction _____ mm

rhythm of contraction _____

Influence of Stretch on Smooth Muscle

1. Attach a 10-g weight on the transducer lever (or muscle lever) and then record contractile activity for 2 minutes.

2. Increase the weights by 10-g units, recording muscle activity for 2 minutes after each weight is added. Mark the record carefully to indicate each weight increase.

3. Describe the effect of increased load (stretch)

on smooth muscle. _____

_____

_____

How does this correlate with what you observed relative to the stretching of skeletal muscle in

previous experiments? _____

_____

_____

Effect of Environmental Factors on Smooth Muscle Contraction
TEMPERATURE

1. Put ice in the beaker to lower the temperature of the solution to 15 C. When this temperature has been reached, record contractile activity for 2 minutes. Stop the recording, and remove the ice. Label the recording appropriately (15 C).

2. Light the alcohol lamp or Bunsen burner under the beaker to warm the solution. (Keep the Bunsen burner flame on low.) Start the recording apparatus after noting the initial temperature, and record muscle activity continuously until the temperature reaches 38 C. (Do not allow the temperature to go over 40 C.) Note the tempera-

ture achieved on the recording at 2-minute intervals. When it reaches 38 C, discontinue recording and shut off the heating source. What changes did you note in the following as the temperature increased?

contraction amplitude (strength) _____

_____

contraction frequency _____

_____

contraction regularity _____

_____

At what temperatures were the contractions

strongest? _____ C weakest? _____ C

Explain. _____

_____

_____

CHANGES IN HYDROGEN ION, Ca^{++}, AND K^+ CONCENTRATIONS   Run each test according to the directions accompanying each. After adding the appropriate chemical, record muscle activity until a definite effect has been demonstrated; then stop the recording apparatus, mark the record appropriately, and replace the bathing solution with fresh 37 C Tyrode's solution (except in test 2). Obtain the necessary test solutions (NaOH, HCl, KCl, and $CaCl_2$) before beginning.

Test 1: To establish a control, record the contractile activity for 1 minute without chemical supplementation.

contractions/min _____

Test 2: Add a few drops of NaOH. Record activity.

Time for demonstrable change in activity _____

What change in activity do you see? _____

_____

Test 3: Add a few drops of HCl. Record activity. Time for demonstrable change in activity

_____

What change in activity do you see? _____

_____

Test 4: Add a few drops of KCl. Record activity. Time for demonstrable change in activity

_____

What change in activity do you see? _____

_____

Test 5: Add a few drops of $CaCl_2$. Record activity time for demonstrable change in activity

_____ What change in activity do you

see? _____

_____

How did a change in pH affect smooth muscle

activity, that is, addition of HCl or NaOH? _____

_____

_____

Explain the effects of increase of KCl and $CaCl_2$

on muscle activity. _____

_____

_____

NEURAL AND HUMORAL CONTROL OF SMOOTH MUSCLE

In this experiment, you will test the effect of two drugs (neurotransmitters) on the activity of the smooth muscle of the gut. Obtain dropper bottles of epinephrine and acetylcholine, 0.5-ml pipettes or 1-cc syringes with needles.

1. Replace the solution in the beaker with fresh Tyrode's solution (37 C). Record the temperature, and then make a tracing of contractile activity for 2 minutes.

2. Without stopping the recorder, make a mark on the recording and quickly add 0.5 ml of epinephrine with the pipette or syringe, and continue recording for 5 minutes. Stop the recording, and mark appropriately. What was the effect of epinephrine on the activity of smooth muscle?

_____

_____

3. Repeat steps 1 and 2 _except_ use 0.5 ml of acetylcholine as the additive. What was the effect of acetylcholine on the activity of smooth muscle?

_____

The drugs just tested are similar, if not identical, to the neurotransmitters of the autonomic nervous system. Which corresponds to that released by the

sympathetic nervous system? _____

Is the activity of the smooth muscle observed under the influence of this drug what you would expect if the sympathetic nervous system was stimulating the

muscle? _____

Which drug corresponds to the neurotransmitter released by the parasympathetic nervous system?

_____

Which of these subdivisions of the autonomic ner-

vous system is the "fight or flight" system? _____

_____

How do the effects illustrated here aid in this func-

tion? _____

_____

Chemical Breakdown of Foodstuffs: Enzymatic Action

EXERCISE 40

OBJECTIVES

1. To summarize the digestive system enzymes involved in the digestion of proteins, fats, and carbohydrates, and to state their site of origin and the environmental conditions promoting their optimal function.

2. To name the end products of digestion of proteins, fats, and carbohydrates.

3. To perform the appropriate chemical tests to determine if digestion of a particular foodstuff has occurred.

4. To cite the function(s) of bile in the digestive process.

5. To discuss the role of temperature and pH in the regulation of enzyme activity.

6. To define *enzyme, catalyst, control, substrate,* and *hydrolase.*

MATERIALS

Supply area 1:
50-ml or 100-ml graduated cylinders
Test tubes and rack
Paraffin
Dropper bottles with Lugol's IKI (Lugol's iodine) and Benedict's solution

0.1% starch solution
1.0% maltose solution
1 N NaOH and 1 N HCl in dropper bottles
Medicine dropper
Glass stirring rod
Wide-range pH paper
Wax markers
Hot plate and 500-ml beakers
Water bath at 37 C
Ice bath

Supply area 2:
5% pepsin solution
Fibrin or finely minced egg white; 1% albumin solution or egg white
0.8% HCl, 0.5% NaOH, 10% NaOH, 1% $CuSO_4$
Evaporating dishes; test tubes and test tube rack
10-ml graduated cylinder; glass stirring rod
Freshly prepared 0.1% Ninhydrin solution
0.1% alcoholic solution of glycine or alanine

Supply area 3:
1% pancreatin solution in 0.2% Na_2CO_3
Litmus cream (fresh cream to which powdered litmus has been added to achieve a blue color)
0.1 N HCl; glass stirring rod
Bile salts (sodium taurocholate)
Test tubes and test tube rack
10-ml graduated cylinder
Vegetable oil in dropper bottle

Since nutrients can only be absorbed when broken down to their monomer forms, food digestion is a prerequisite to food absorption. You have already studied mechanisms of passive and active absorption in Exercise 5. Before proceeding, review that material on pages 30 to 35.

Enzymes are large protein molecules produced by body cells. They are biologic catalysts, which increase the rate of a chemical reaction without themselves becoming part of the product. The digestive enzymes are hydrolytic enzymes, or hydro-lases, which break down organic food molecules by adding water to the molecular bonds, thus cleaving the bonds between the subunits, or monomers.

The various hydrolytic enzymes are highly specific in their action. Each enzyme hydrolyzes only one or a small group of substrate molecules, and very specific environmental conditions are necessary for it to function optimally. Since digestive enzymes act functionally outside the body cells in the digestive tract, their hydrolytic activity can also be studied in a test tube. Such an in vitro study provides a convenient laboratory environment for the variation of environmental factors to investigate the effect of such variations on enzymatic activity.

Figure 40.1 is a flow sheet of the progressive digestion of proteins, fats, and carbohydrates, which indicates specific enzymes involved and their site of formation. Acquaint yourself with this flow sheet before beginning this experiment, and refer to it during the laboratory as necessary. Work in groups of four, with each group taking responsibility for setting up and conducting one of the following experiments. Each group should then communicate its results to the rest of the class by recording the results of their experiment in a chart on the chalkboard. Additionally, all members of the class should observe the controls as well as the positive and negative examples of all experimental results. All members of the class should be able to explain the tests used and the results observed and anticipated for each experiment.

STARCH DIGESTION BY SALIVARY AMYLASE

 From supply area 1, obtain a test tube rack, 10 test tubes, paraffin, Lugol's solution, Benedict's solution, wax marking pencils, and a beaker of distilled water. Two students should prepare the controls (steps 1 to 4) while the other two collect saliva and prepare the experimental samples (steps 5 to 7). Since in this experiment you will investigate the hydrolysis of starch to maltose by **salivary amylase** (the enzyme produced by the salivary glands and secreted into the mouth), it is important to be able to identify the presence of these substances to determine to what extent the enzymatic activity has occurred. Thus controls must be prepared to provide a known standard against which comparisons can be made. (Starch decreases and sugar increases as digestion occurs.)

1. Mark test tube 1 with the wax marker. Place 1 ml of starch solution in the test tube and add two drops of IKI solution. The presence of a blue-black color when IKI is added indicates the presence of starch and is referred to as a **positive starch test.** (As the starch is progressively hydrolyzed, the color with IKI changes from blue-black to blue-red to faint red and then finally disappears when all the starch has been digested.)

2. To obtain a negative starch test, place 1 ml of distilled water in test tube 2, and add two drops of IKI. The unchanged color of the solution indicates the absence of starch. What color did you obtain with water and IKI? _____

3. To obtain a **positive sugar test,** place 1 ml of maltose solution in test tube 3, and add five drops of Benedict's solution. Mix well. Place the test tube in a water bath (a beaker of water on a hot plate) and boil for 5 minutes. The presence of a bright yellow to deep red precipitate (cuprous oxide) indicates a positive test for maltose, sucrose, or any other reducing sugar. (A change to green is also considered a positive test for sugar but indicates the presence of a smaller amount.)

4. To obtain a negative sugar test, place 1 ml of distilled water in test tube 4, add five drops of Benedict's solution, and boil for 5 minutes. If the solution's color remains unchanged from the blue of Benedict's solution, the absence of sugar is indicated.

5. To collect saliva, rinse your mouth out well two or three times with distilled water. Soften a piece of paraffin in your mouth, and then chew it to stimulate the secretion of saliva. Collect the saliva by expectorating into a clean beaker until you have collected between 20 and 25 ml of saliva. Measure the saliva in a small graduated cylinder and add an equal amount of distilled water. Stir well and determine the pH of the mixture. You now have a 50% solution of saliva.

6. Mark six test tubes with the numbers 5 to 10 and prepare them for incubation as described below. Note the time that incubation begins in each case, and continually observe the tubes for color changes once the incubation period has started.

Tube	Additives	Incubation condition
5	4 drops IKI, 5 ml starch, 5 ml saliva	Room temperature
6	4 drops IKI, 5 ml starch, 5 ml saliva	37 C
7	4 drops IKI, 5 ml starch, 5 ml saliva	0 C
8	4 drops IKI, 5 ml starch, 5 ml saliva; boil 4 minutes before incubating	37 C
9	4 drops IKI, 5 ml starch, 5 ml saliva; add 1 N HCl until a pH of 3 is achieved (stir after each addition); incubate	37 C
10	4 drops IKI, 5 ml starch, 5 ml saliva; add 1 N NaOH until a pH of 9 is achieved (stir after each addition); incubate	37 C

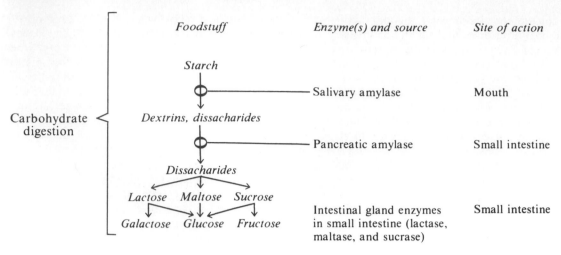

Absorption: The monosaccharides (glucose, galactose, and fructose) are absorbed into the capillary blood in the villi and transported to the liver via the hepatic portal v.

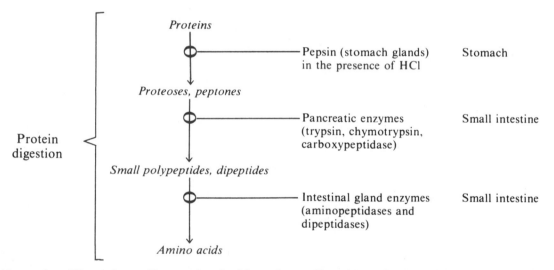

Absorption: The amino acids are absorbed into the capillary blood in the villi and transported to the liver via the hepatic portal v.

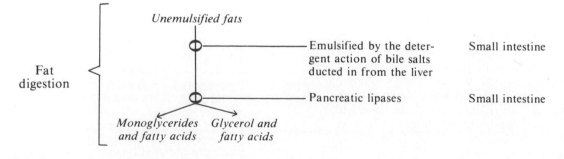

Absorption: Absorbed primarily into the lacteals of the villi (glycerol and short-chain fatty acids are absorbed into the capillary blood in the villi). Transported to the liver via the systematic circulation (hepatic artery), which receives the lymphatic flow from the thoracic duct or via the hepatic portal v.

Figure 40.1

Flow sheet of digestion and absorption of foodstuffs.

7. As the tubes become colorless (negative starch test), note the time and discontinue their incubation. Perform the *Benedict's test* on each sample. (Add 1 ml of the incubated solution to a test tube, and then add five drops of Benedict's solution to the tube. Boil for 5 minutes.) After 1½ hours of incubation, discontinue the incubation of any remaining tubes and conduct the Benedict's test on their contents. Record the results below on Chart 1 and on the chalkboard.

PEPSIN DIGESTION OF PROTEIN

 Obtain a test tube rack, seven test tubes, wax marking pencils, wide range pH paper, plus one of each of the materials/supplies listed under *Supply area 2*. Two students should prepare the control samples according to the directions in steps 1 and 2 while the other members of the group set up the experimental samples (steps 3 and 4). **Pepsin,** produced by the chief cells of the stomach glands, hydrolyzes proteins to small fragments (proteoses, peptones, and polypeptides). To identify the extent of protein digestion, prepare controls as standards for comparison. (Protein decreases, and breakdown products such as proteoses and amino acids increase as digestion proceeds.)

1. To prepare the first control sample, perform a *biuret test* to indicate the presence of protein. Add 3 ml of 1% albumin solution or egg white to test tube 1, and then add 3 ml of 10% NaOH and two or three drops of 1% $CuSO_4$. Mix well with a glass stirring rod. If the mixture turns violet, protein is present. If it does not, continue to add counted drops of $CuSO_4$ to the sample until a violet color appears. *Note the total number of drops used,* and add the same volume of $CuSO_4$ when conducting the biuret test on the experimental samples. A negative biuret test (no color change) indicates complete protein digestion.

2. To prepare the second control sample, perform a *Ninhydrin test* to indicate the presence of amino acids. Add 5 ml of 0.1% alcoholic solution of alanine or glycine to test tube 2, and then add 0.5 ml of 0.1% Ninhydrin solution, and heat.

Record the resulting color. _____

This color indicates the presence of amino acids (breakdown products of protein digestion).

3. To prepare the experimental samples, mark five test tubes with numbers 3 to 7. Preparation and incubation conditions for the experimental samples are listed below. Determine the pH of each sample *before* adding a pea-sized amount of fibrin or finely minced egg white, and record the resultant pH.

pH	Sample no.	Conditions	Temp. (C)
	3	5 ml 5% pepsin; 5 ml 0.8% HCl; egg white or fibrin; incubate	37
	4	5 ml 5% pepsin; 5 ml distilled H_2O; egg white or fibrin; incubate	37
	5	5 ml 5% pepsin; 5 ml 0.5% NaOH; egg white or fibrin; incubate	37
	6	5 ml 5% pepsin (boiled); 5 ml 0.8% HCl; egg white or fibrin; incubate	37
	7	5 ml distilled H_2O; 5 ml 0.8% HCl; egg white or fibrin; incubate	37

Chart 1   Salivary Amylase Digestion of Starch

Tube no.	Contents/conditions	Time of initiation	Time for negative IKI test	Benedict's test +	−
5					
6					
7					
8					
9					
10					

Continue incubation for 2 hours. Shake the tubes occasionally and determine if there is any evidence of digestive activity. The fibrin will swell initially (this is not evidence of hydrolysis) and then it or the egg white will become increasingly more transparent and decrease in mass as digestion occurs.

4. Perform the biuret test and the Ninhydrin test on two separate samples from each test tube after the 2-hour incubation period. Record the results on chart 2, p. 355 and on the chalkboard. (A− indicates a negative test; a + indicates a test that is slightly positive; a + + indicates a strongly positive test.)

PANCREATIC LIPASE DIGESTION OF FATS AND THE ACTION OF BILE SALTS

Pancreatin describes the enzymatic product of the pancreas, which includes protein, carbohydrate, and fat-digesting enzymes. It is used here to investigate the properties of pancreatic lipase, which hydrolyzes fats to their component fatty acids and glycerol.

1. Obtain a test tube rack, six test tubes, and one sample each of the substances and supplies listed under *Supply area 3*. One student should prepare the control (step 2), another should set up the demonstration of the action of bile on fats (step 3), while the other two group members prepare the experimental samples (step 4).

2. To prepare the control, add 5 ml of litmus cream to test tube 1. Then add 0.1 N HCl drop by drop (stirring after each addition with a glass stirring rod) until the cream turns pink. This change in color indicates that the test tube contains an acidic product and will identify those tubes in which fat hydrolysis has occurred. (Although fats are not normally considered to be acidic, their hydrolysis products, the fatty acids, are organic acids.)

3. Although bile, an excretory product of the liver, is not an enzyme, it is important to fat digestion because of its emulsifying action (the breakdown of larger particles into smaller ones) on fats. Emulsified fats provide a larger surface area for enzymatic activity. To demonstrate the action of bile on fats, prepare two test tubes, and label them A and B. To tube A, add 5 ml H_2O and five drops of vegetable oil. To tube B, add 5 ml of H_2O, five drops of vegetable oil, and a pinch of bile salts. Shake each tube vigorously and allow the tubes to stand in a test tube rack at room temperature for 10 to 15 minutes. Observe both tubes. If emulsification has not occurred the oil will be floating on the surface of the water. If emulsification has occurred, the fat droplets will be suspended throughout the water, forming an emulsion. In which tube has emulsification occurred? _____

4. Prepare the experimental samples as indicated in chart 3, p. 355. Then shake each test tube well and incubate each in a 37 C water bath until a color change (blue to red) becomes apparent. Note the time incubation begins, the time it ends, and any changes in color and odor of the samples. After 2 hours, discontinue incubation of any samples remaining, regardless of whether a color change has occurred or not. Record results on the chalkboard and in the chart. (If there was no color change, write N.C.)

Chart 2   Pepsin Digestion of Protein

Tube no.	Transparency +	Transparency −	Biuret test −	Biuret test +	Biuret test ++	Ninhydrin test −	Ninhydrin test +	Ninhydrin test ++
3								
4								
5								
6								
7								

Chart 3   Pancreatic Lipase Digestion of Fats

Tube No.	Contents	Incubation Began	Ended	Change in color	Change in odor
2	5 ml litmus cream; 5 ml pancreatin				
3	5 ml litmus cream; 5 ml pancreatin; pinch bile salts				
4	5 ml litmus cream; 5 ml distilled H_2O; pinch bile salts				

UNIT 14

THE URINARY SYSTEM

Anatomy of the Urinary System

EXERCISE
41

OBJECTIVES

1. To describe the overall function of the urinary system.

2. To identify the following organs on an appropriate diagram, torso model, or dissection specimen and/or to describe their anatomic structure:

kidneys	urinary bladder
urethra	ureters

3. To describe the general functions of each of the organs listed in Objective 2.

4. To define *micturition*.

5. To explain the pertinent differences in the control of the two bladder sphincters (internal and external).

6. To compare the course and length of the urethra in the male with that in the female.

7. To identify the following regions of the dissected kidney (longitudinal section): hilus, cortex, medulla, medullary pyramids, major and minor calyces, pelvis, renal columns, and capsule layers.

8. To trace the blood supply of the kidney from the renal artery to the renal vein.

9. To define the nephron as the physiologic unit of the kidney and describe its anatomy: Bowman's capsule, proximal convoluted tubule, loop of Henle, distal convoluted tubule, plus the associated capillary networks (glomerulus and peritubular capillaries).

10. To define glomerular filtration, tubular reabsorption, and tubular secretion, and note the nephron areas involved in these processes.

11. To recognize microscopic or diagrammatic views of the histologic structure of the kidney and bladder.

MATERIALS

Human dissectible torso model and/or anatomic
   chart of the human urinary system
3-dimensional model of the cut kidney and of the
   nephron (if available)

Dissection animal, tray, and instruments
Pig or sheep kidney, doubly or triply injected
Prepared histologic slides of a longitudinal section
   of kidney and cross sections of the bladder
Compound microscope

Metabolism of nutrients by the body produces various wastes (carbon dioxide, nitrogenous wastes, ammonia, and so on) that must be eliminated from the body if normal function is to continue. Although excretory processes involve several organ systems (the lungs excrete carbon dioxide and the skin glands excrete salts and water), it is the urinary system that is primarily concerned with the removal of nitrogenous wastes from the body. In addition to this purely excretory function, the kidney maintains the electrolyte, acid-base, and fluid balance of the blood and is thus a major, if not *the* major, homeostatic organ of the body.

To perform its functions, the kidney acts much like a blood filter. It allows toxins, metabolic wastes, and excess ions to leave the body in the urine, while simultaneously retaining needed substances and returning them to the blood. Malfunction of the urinary system, particularly of the kidneys, leads to a failure in homeostasis, which results, unless corrected, in death.

GROSS ANATOMY OF THE HUMAN URINARY SYSTEM

The urinary system (Figure 41.1) consists of the paired kidneys and ureters and the single urinary bladder and urethra. The kidneys perform the functions described above and manufacture urine in the process. The remaining organs of the system provide temporary storage reservoirs for urine or transport urine from one body region to another.

 Examine the human torso model, a large anatomic chart, or a three-dimensional model of the urinary system to locate and study the anatomy and relationships of the urinary organs.

1.   Locate the paired **kidneys** on the dorsal body wall in the superior lumbar region. Note that they are not positioned at exactly the same level: the right kidney is slightly lower than the left kidney. Why do

you suppose this is so? _____

_____

In the living person, fat deposits hold the kidneys in place in a retroperitoneal position. When the fatty material is reduced or deficient in amount (in cases

of rapid weight loss or in very thin individuals), the kidneys are less securely anchored to the body wall and may drop to a lower or more inferior position in the abdominal cavity. This phenomenon is called **ptosis.**

2.   Observe the two **renal arteries** as they diverge from the descending aorta and plunge into the indented medial region **(hilus)** of each kidney. Note also the two **renal veins,** which drain the kidneys (circulatory drainage) and the two **ureters,** which drain the urine from the kidneys and conduct it to the bladder by peristalsis for temporary storage.

3.   Locate the urinary **bladder,** and observe the point of entry of the two ureters into this organ. Also locate the single **urethra,** which drains the bladder. The triangular region of the bladder, which is delineated by these three openings (two ureter and urethral orifices), is referred to as the **trigone** (Figure 41.2). Although the formation of urine by the kidney is a continuous process, urine is usually removed from the body when voiding is convenient. In the meantime the bladder provides temporary storage for urine.

Voiding, or **micturition,** is the process in which urine empties from the bladder. Two sphincter muscles or valves, the **internal sphincter** (more superiorly located) and the **external sphincter** (more inferiorly located) control the emptying of urine from the bladder. Ordinarily, the bladder continues to collect urine until about 300 ml have accumulated, at which time the stretching of the bladder wall activates stretch receptors. Impulses transmitted to the central nervous system in turn produce reflex contractions of the bladder wall through parasympathetic nervous system pathways. As the contractions increase in force and frequency, the stored urine is forced past the internal sphincter, which is a smooth-muscle involuntary sphincter, into the upper part of the urethra. It is then that a person feels the urge to void. The lower external sphincter consists of skeletal muscle and is voluntarily controlled. If it is not convenient to void, the opening of this sphincter can be inhibited. Conversely, if the time is convenient, the sphincter may be relaxed and the stored urine flushed from the body. If voiding is inhibited, the reflex contractions of the bladder cease temporarily and urine continues to accumulate in the bladder. After 200 to 300 ml more have been collected, the micturition reflex will again be initiated. Lack of voluntary control over the ex-

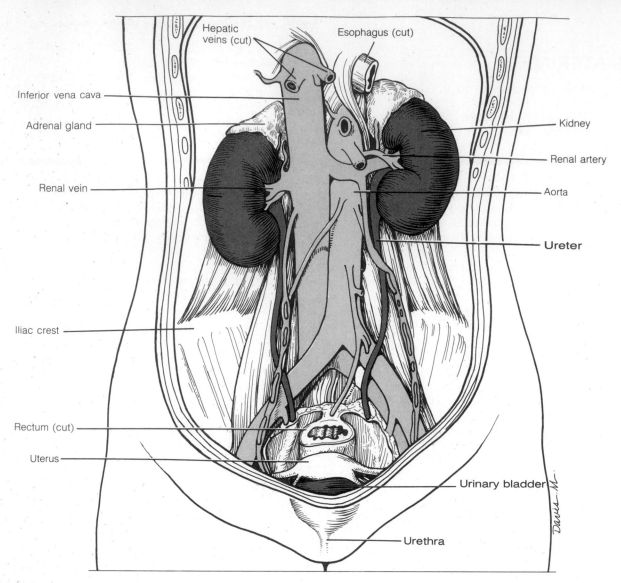

Figure 41.1

Anterior view of the urinary organs of a female. (Most unrelated abdominal organs have been removed.)

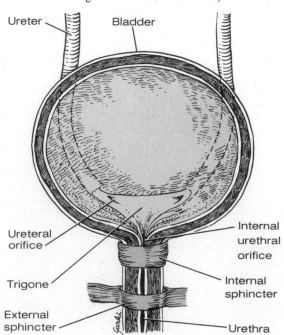

Figure 41.2

Detailed structure of the bladder.

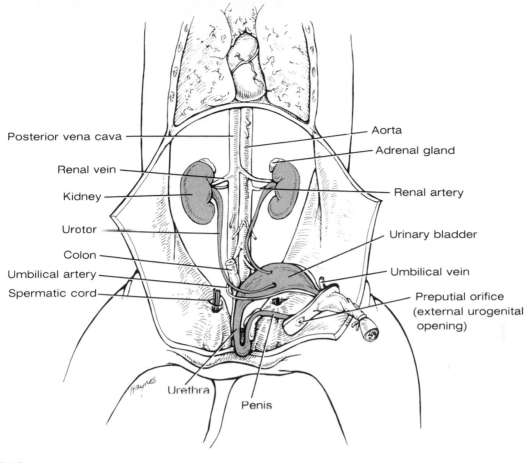

Posterior vena cava

Aorta

Adrenal gland

Renal vein

Renal artery

Kidney

Ureter

Urinary bladder

Colon

Umbilical artery

Umbilical vein

Spermatic cord

Preputial orifice
(external urogenital
opening)

Urethra

Penis

Figure 41.3

Urinary system of the male fetal pig (reproductive structures also indicated).

ternal sphincter is referred to as **incontinence.** Incontinence is a normal phenomenon in children 2 years old or younger, as they have not yet gained control over the voluntary sphincter. Past this age, incontinence is generally a result of emotional problems, bladder irritability, or some other pathologic condition of the urinary tract.

4.   Follow the course of the urethra to the body exterior. In the male, it is approximately 8 inches (20 cm) long and opens at the tip of the penis after traveling the length of the penis. The urethra of the male has a dual function: it is a urine conduit to the body exterior, and it provides a passageway for the ejaculation of sperm. Thus, in the male, the urethra is part of both the urinary and reproductive systems. In the female, the urethra is very short, approximately 1½ inches (4 cm) long, and it travels downward and slightly forward from the bladder to the external urethral opening or orifice. There are no common urinary-reproductive pathways in the female, and the female urethra serves only to transport urine to the body exterior.

DISSECTION OF THE FETAL PIG URINARY SYSTEM

The structures of the reproductive and urinary systems are often considered together as the urogenital system, since they have common embryologic origins. However, the emphasis in this dissection is on identifying the structures of the urinary tract (Figures 41.3 and 41.4) with only a few references to contiguous reproductive structures. (Exercise 44 is a study of the anatomy of the reproductive system.)

 1.   Obtain your dissection specimen, and pin or tie its limbs to the dissection tray. Reflect the abdominal viscera (small intestines) to locate the kidneys high on the dorsal body wall. Note that the kidneys in the pig, as in the human, are retroperitoneal (behind the peritoneum).

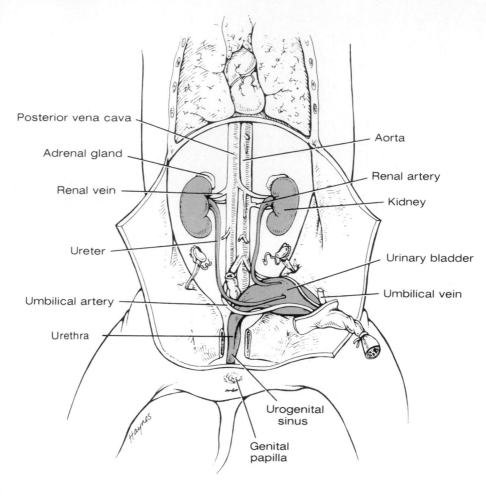

Figure 41.4

Urinary system of the female fetal pig (reproductive structures also indicated).

2.   Carefully remove the peritoneum, and clear away the bed of fat that invests the kidneys. Locate the adrenal (suprarenal) glands, which appear as bandlike pale orange glands lying on the anteromedial surface of each kidney.

3.   Identify the renal artery (red latex–injected), the renal vein (blue latex–injected), and the ureter at the hilus region of the kidney.

4.   Trace the ureters posteriorly along the dorsal body wall to where they turn ventrally to enter the **allantoic bladder,** or fetal urinary bladder, a collapsed elongated sac lying between the umbilical arteries. Trace the posterior portion of the bladder to the point where it narrows to become the urethra, which enters the pelvic cavity. Backtrack to the point where

the ureters enter the allantoic bladder, and then trace the bladder into the umbilical cord, where it continues as the **allantoic stalk.** After birth, when the allantois becomes nonfunctional, the allantoic bladder becomes the urinary bladder.

5.   Cut through the bladder wall and examine the region of the urethral exit to see if you can discern any evidence of the internal sphincter.

6.   Using a probe, trace the urethra as it exits from the bladder to its terminus in the urogenital sinus which also opens into the vagina in the female pig and into the penis of the male. (In the human female, the urethra does not empty into the vagina but has a separate external opening located superior to the vaginal orifice.) Identify the external urogenital

opening in the male pig located just below the umbilicus. Do not expose the urethra along its entire length at this time, because you may damage the reproductive structures, which will be studied later.

7.   Before cleaning up the dissection materials, observe a pig of the opposite sex. Also, if the larger pig or sheep kidneys are not available for the observations of gross internal anatomy described below, remove one of the fetal pig kidneys to make the observations.

GROSS INTERNAL ANATOMY OF THE PIG OR SHEEP KIDNEY

1.   Obtain a preserved sheep or pig kidney, dissecting pan, and instruments. Observe the kidney to identify the **renal capsule,** a smooth transparent membrane that adheres tightly to the kidney tissue.

2.   Find the ureter, renal vein, and renal artery at the hilus (indented) region. The renal vein has the thinnest wall and will be collapsed. The ureter is the largest of these structures and has the thickest wall.

3.   Make a cut through the longitudinal axis (frontal section) of the kidney and locate the anatomic areas described below and depicted in Figure 41.5.

Kidney cortex: the outer kidney region, which is lighter in color. (If the kidney is double-injected with latex, you will see a predominance of red and blue latex specks in this region indicative of the rich vascular supply.)
Medullary region: deep to the cortex; a darker, reddish-brown color. The medulla is segregated into triangular regions that have a striped, or striated, appearance—the **medullary pyramids.** The base of each pyramid faces toward the cortex; its **apex,** or **papilla,** points to the innermost kidney region.
Renal columns: Areas of tissue more like the cortex in appearance, which segregate and dip downward between the pyramids.
Renal pelvis: Medial to the hilus; a relatively flat, basinlike cavity that is continuous with the **ureter,** which exits from the hilus region. Finger-like extensions of the pelvis should be visible. The larger, or primary, extensions are called the **major calyces;** subdivisions of the major calyces are the **minor calyces.** Note that the minor calyces terminate in cuplike areas that enclose the apexes of the medullary pyramids and collect urine draining from the pyramidal tips into the pelvis.

4.   If the preserved kidney is doubly or triply injected, follow the renal blood supply from the renal artery to the **glomeruli.** The glomeruli appear as little red and blue specks in the cortex region. (See Figures 41.5 and 41.6 and the discussion below.)

Since the kidneys have a major responsibility for removing wastes from the blood and constantly monitoring its fluid and electrolyte balance, it is not surprising that they have a rich vascular supply. Approximately a fourth of the total blood flow of the body passes through the kidneys each minute, entering the kidneys via the **renal artery.** Having entered the pelvis, the renal artery breaks up into several branches called **interlobar arteries,** which ascend toward the cortex in the renal column areas. At the top of the medullary region, these arteries give off arching branches, the **arcuate arteries,** which curve over the bases of the medullary pyramids. Small **interlobular arteries** branch off the arcuate arteries and ascend into the cortex, giving off the individual **afferent arterioles,** which provide the capillary networks (**glomeruli** and **peritubular capillary beds**) supplying the nephrons, or functional units, of the kidney. Blood draining from the nephron capillary networks in the cortex enters the **interlobular veins** and then drains through the **arcuate veins** and the **interlobar veins** to finally enter the **renal vein** in the pelvis region.

MICROSCOPIC ANATOMY OF THE KIDNEY AND BLADDER

Obtain prepared slides of kidney and bladder tissue, and a compound microscope.

Kidney

Each kidney contains approximately one million nephrons, which are the anatomic units responsible for the filtration, reabsorption, and secretion activities of the kidney. Figure 41.6 depicts the detailed structure and the relative positioning of the nephrons in the kidney.

Each nephron consists of two major structures: the **glomerulus** (a capillary knot) and the **renal tubule.** During embryologic development, each renal tubule begins as a blind-ended tubule that gradually encloses an adjacent capillary cluster, or glomerulus. The enlarged end of the tubule encasing the glomerulus is **Bowman's capsule,** and its inner, or visceral, wall consists of highly specialized cells called **podocytes.** Podocytes have long branching processes that interdigitate with those of other podocytes and cling to the endothelial wall of the glomerular capillaries, thus forming a very porous epithelial membrane surrounding the glomerulus. The glomerular-Bowman's capsule complex is sometimes called the **renal corpuscle.**

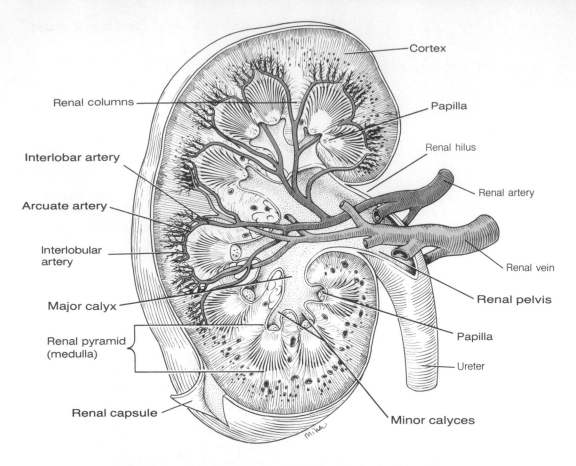

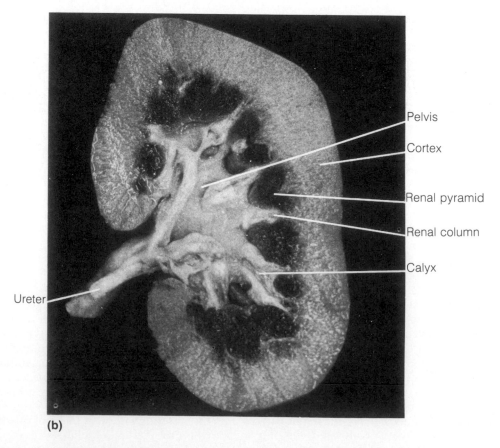

(b)

Figure 41.5

Frontal section of a kidney: (a) diagrammatic view, showing larger arteries supplying the kidney tissue; (b) photograph of a pig kidney. (Photo courtesy of Ann Allworth.)

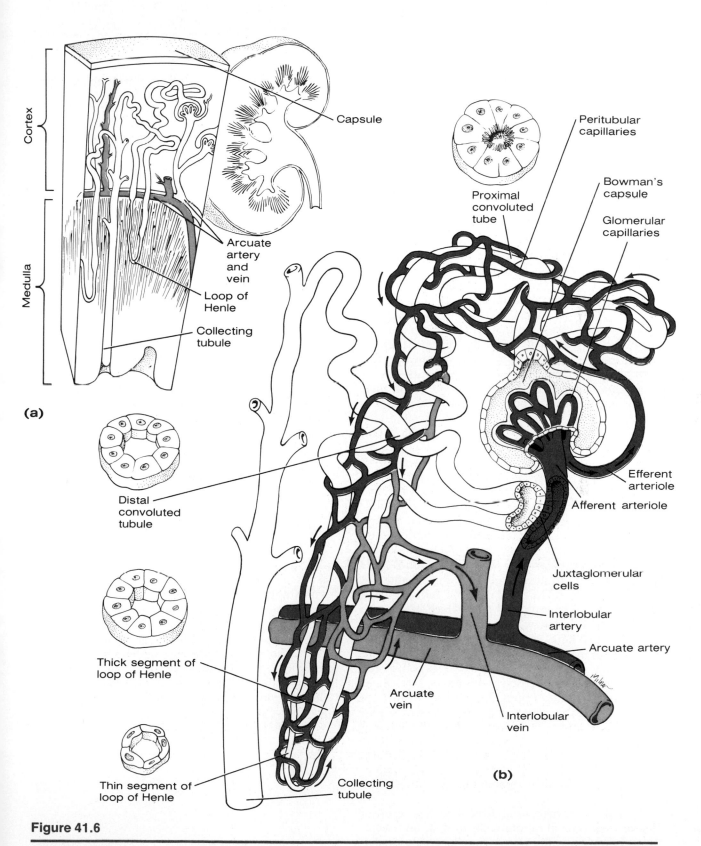

Figure 41.6

Structure of a nephron: (a) wedge-shaped section of kidney tissue, indicating the position of the nephrons in the kidney; (b) detailed nephron anatomy and associated blood supply.

The rest of the tubule is approximately 2 inches long. As it emerges from Bowman's capsule, it becomes highly coiled and convoluted, drops down into a long hairpin loop, and then again becomes highly coiled and twisted before entering a collecting duct, or tubule. The anatomic areas of the tubule are, in order from Bowman's capsule: the **proximal convoluted tubule, loop of Henle** (descending and ascending limbs), and the **distal convoluted tubule.** The wall of the renal tubule is composed almost entirely of cuboidal epithelial cells, with the exception of the descending limb of the loop of Henle, which is simple squamous epithelium. The lumen surfaces of the cuboidal cells in the proximal convoluted tubule are densely covered with microvilli (a cellular modification that greatly increases the surface area exposed to the lumen contents, or filtrate). Microvilli also occur on cells of the ascending limb of the loop of Henle and the distal convoluted tubule but in much reduced numbers.

The renal corpuscle, proximal convoluted tubule, distal convoluted tubule, and much of the loop of Henle are located in the cortical region, although portions of the loop of Henle dip into the medullary region. The **collecting tubules,** each of which receives urine from many nephrons, run downward through the medullary pyramids, giving them their striped appearance, to empty the final urinary product into the calyces and pelvis of the kidney.

The function of the nephron depends on several unique features of the renal circulation. The capillary vascular supply consists of two distinct capillary beds, the **glomerulus** and the **peritubular capillary bed.** Vessels leading to and from the first capillary bed, the glomerulus, are both arterioles: the **afferent arteriole** feeds the bed while the **efferent arteriole** drains it. The efferent arteriole then breaks up into the second capillary network (the peritubular capillaries), which surrounds the proximal and distal convoluted tubules and the loop of Henle. (The peritubular capillaries then drain into an interlobular vein, which leaves the cortex.) The glomerular capillary bed has no parallel elsewhere in the body. It is a high-pressure bed along its entire length. This high pressure is a result of two major factors: the bed is fed and drained by arterioles (arterioles are high-resistance vessels as opposed to venules, which are low-resistance vessels), and the afferent feeder arteriole is larger in diameter than the efferent arteriole that drains the bed. The high hydrostatic pressure created by these two anatomic features forces out fluid and blood components smaller than proteins from the glomerulus into Bowman's capsule.

As noted earlier, urine formation is a result of three processes: filtration, reabsorption, and secretion. Filtration is the role of the glomerulus and is largely a passive process in which a portion of the blood passes from the glomerular bed into Bowman's capsule. This filtrate then enters the proximal convoluted tubule where tubular reabsorption and secretion begin. During tubular reabsorption, many of the filtrate components move through the tubule cells and return to the blood in the peritubular capillaries. Some of this reabsorption is passive such as that of water, which passes by osmosis, but the reabsorption of most substances depends on active transport processes and is highly selective. Which substances are reabsorbed at any time depends on the composition of the blood and needs of the body at that time. Substances that are almost entirely reabsorbed from the filtrate include water, glucose, and amino acids. Various ions are selectively reabsorbed or allowed to go out in the urine according to what is required to maintain appropriate blood pH and electrolyte composition. Waste products (urea, creatinine, uric acid, and drug metabolites) are reabsorbed to a much lesser degree or not at all. Much, if not most, of the tubular reabsorption occurs in the proximal convoluted tubule, with small amounts occurring in other tubular areas, primarily the distal convoluted tubule.

Tubular secretion is essentially the reverse process. Substances such as hydrogen and potassium ions and ammonia move either from the blood of the peritubular capillaries through the tubular cells or from the tubular cells into the filtrate to be disposed of in the urine. This process is particularly important for the disposal of selected metabolites not already in the filtrate or as an adjunct method for controlling blood pH.

 Observe a model of the nephron before continuing on with the microscope study of the kidney.

Hold the longitudinal section of the kidney up to the light to identify cortical and medullary areas. Scan the slide under low power. Move the slide so that you can see the cortical area. Identify a glomerulus, which appears as a ball of tightly packed material containing many small nuclei. It is usually delineated by a vacant-appearing region that surrounds it. Note that the renal tubules are cut at various angles. Also try to differentiate between the thin-walled loop of Henle portion of the tubules and the cuboidal epithelium of the proximal convoluted tubule, which has microvilli. Draw a glomerulus and a small portion of the surrounding tissue below and label the glomerulus.

Bladder

Scan the bladder tissue. Note the heavy muscular wall (detrusor muscle), which consists of three irregularly arranged muscular layers. The innermost and outermost muscle layers are arranged longitudinally; the middle layer is arranged circularly. Attempt to differentiate the three layers.

Observe the mucosa with its highly specialized transitional epithelium. The plump, transitional epithelial cells have the ability to slide over one another, thus decreasing the thickness of the mucosa layer as the bladder fills and stretches to accommodate the increased urine volume. Depending on the degree of stretching of the bladder, the mucosa may be three to eight cell layers thick. Draw a small section of the bladder wall, and label all regions or tissue areas.

Urinalysis

OBJECTIVES

1. To list the physical characteristics of urine, and to indicate the normal pH and specific gravity ranges.

2. To list substances that are normal urinary constituents.

3. To conduct various urinalysis tests and procedures and use them to determine the substances present in a urine specimen.

4. To define the following urinary conditions:

calculi	*hematuria*
glycosuria	*hemoglobinuria*
albuminuria	*pyuria*
ketonuria	*casts*

5. To explain the implications and possible causes of conditions listed in Objective 4.

MATERIALS

Urine samples (collected in clean bottles by the students before the laboratory begins)

Pathologic urine specimens obtained from a health facility by the instructor and numbered

Urinometer
Hot plate
1000-ml beakers
Wide-range pH paper
Dip sticks (Clinistix, Ketostix, Albustix, Hemastix)
Ictotest reagent
Test reagents for sulfates: 10% barium chloride solution, dilute HCl (hydrochloric acid)
Test reagent for glucose (Benedict's reagent)
Test reagent for phosphates (dilute nitric acid, dilute ammonium molybdate)
Test reagent for proteins and urea (concentrated nitric acid in dropper bottle)
Test reagent for Cl: 3.0% silver nitrate solution ($AgNO_3$)
Test tubes, test tube rack, and test tube holders
10-cc graduated cylinders
Glass stirring rods
Medicine droppers
Wax marking pencils
Microscope slides
Cover slips
Compound microscope
Centrifuge and centrifuge tubes
"Sedi-stain"

Blood composition depends on three major factors: dietary intake, cellular metabolism, and urinary output. In 24 hours, the kidneys' two million nephrons filter approximately 150 to 180 liters of blood plasma through their glomeruli into the tubules, where it is selectively processed by tubular reabsorption and secretion. In the same period, urinary output, which contains by-products of metabolism and excess ions, is 1.0 to 1.8 liters. In healthy individuals, the kidneys can maintain blood constancy despite wide variations in diet and metabolic activity. Urine pH varies from 4.5 to 8.0, which reflects the ability of the renal tubules to excrete basic or acidic ions to maintain the blood plasma pH in the physiologic range of 7.35 to 7.45. With certain pathologic conditions, urine composition often changes dramatically.

CHARACTERISTICS OF URINE

Freshly voided urine is generally clear and pale yellow to amber in color. This normal yellow color is due to urochrome, a pigment metabolite arising from the body's destruction of hemoglobin. As a rule, color variations from pale yellow to deeper amber indicate the relative concentration of solutes to water in the urine. The greater the solute concentration, the deeper the color. Abnormal urinary color may be due to certain foods, such as beets, various drugs, bile, or blood.

The odor of freshly voided urine is characteristic and slightly aromatic, but bacterial action gives it an ammonia-like odor when left standing. Some drugs, vegetables (such as asparagus), and various disease processes (such as diabetes mellitus) change the characteristic odor of urine.

The pH of urine ranges from 4.5 to 8.0, but its average value, 6.0, is slightly acidic. Diet may markedly influence the pH of the urine. For example, a diet high in protein (meat, eggs, cheese) and whole wheat products increases the acidity of the urine. Such foods are called acid ash foods. On the other hand, a vegetarian diet (alkaline ash diet) increases the alkalinity of the urine. A bacterial infection of the urinary tract may also result in urine with a high pH.

Specific gravity is the relative weight of a specific volume of liquid compared with an equal volume of distilled water. The specific gravity of distilled water is 1.0, since 1 ml weighs 1 g. Since urine contains dissolved solutes, it weighs more than water, and its customary specific gravity ranges from 1.001 to 1.030. Urine with a specific gravity of 1.001 contains few solutes and is considered very dilute. Dilute urine commonly results when a person drinks excessive amounts of water, uses diuretics, or suffers from diabetes insipidus or chronic renal failure. Conditions that produce urine with a high specific gravity include limited fluid intake, fever, and a kidney inflammation called pyelonephritis. If the urine becomes excessively concentrated, some of the substances normally held in solution begin to precipitate or crystalize, forming **kidney stones,** or **renal calculi.**

Normal constituents of urine, in order of decreasing concentration, include water, urea,* sodium,† phosphorus, potassium, sulfur, creatinine,* ammonia, and uric acid.* Much smaller, but highly variable amounts of calcium, magnesium, and bicarbonate ions are also found in the urine. Abnormally high concentrations of any of these urinary constituents may indicate a pathologic condition.

ABNORMAL URINARY CONSTITUENTS

Abnormal urinary constituents are substances not normally present in the urine when the body is operating properly.

---

*Urea, uric acid, and creatinine are the most important nitrogenous wastes found in urine. Urea is an end product of protein breakdown; uric acid is a metabolite of purine breakdown; and creatinine is associated with muscle metabolism.

† Sodium ions appear in relatively high concentration in the urine because of reduced urine volume, not because large amounts are being secreted. Sodium is the major positive ion in the plasma; most of it is actively reabsorbed under normal circumstances.

Glucose

The presence of glucose in the urine, a condition called **glycosuria,** is indicative of abnormally high blood sugar levels. Normally blood sugar levels are maintained between 80 and 120 mg/100 ml of blood. At this level all glucose in the filtrate is reabsorbed by the tubular cells and returned to the blood. Glycosuria may result from carbohydrate intake so excessive that normal physiologic and hormonal mechanisms cannot clear it from the blood quickly enough. In such cases of glycosuria, the active transport reabsorption mechanisms of the tubules for glucose are exceeded, but only temporarily.

Pathologic glycosuria occurs in conditions such as uncontrolled diabetes mellitus, in which the body cells are unable to absorb glucose from the blood because the pancreatic islet cells produce inadequate amounts of the hormone insulin. Under such circumstances, the body cells increase their metabolism of fats, and the excess and unusable glucose spills out in the urine.

Albumin

Albuminuria, or the presence of albumin in the urine, is an abnormal finding. Albumin is the single most abundant blood protein and is very important in maintaining the osmotic pressure of the blood. Albumin, like other blood proteins, normally is too large to pass through the glomerular filtration membrane; thus albuminuria is generally indicative of an abnormally increased permeability of the glomerular membrane. Certain nonpathologic conditions, such as excessive exertion, pregnancy, or excessive protein intake, can temporarily increase the membrane permeability, leading to **physiologic albuminuria.** Pathologic conditions leading to the presence of albumin in the urine include events that damage the glomerular membrane, such as kidney trauma due to blows, the ingestion of heavy metals, bacterial toxins, glomerulonephritis, and hypertension.

Ketone Bodies

Ketone bodies (acetoacetic acid, betahydroxybutyric acid, and acetone) normally appear in the urine in very small amounts. **Ketonuria** or acetonuria, the presence of these intermediate products of fat metabolism in excessive amounts, generally indicates that abnormal metabolic processes are occurring. The result may be acidosis and its complications. Ketonuria is an expected finding during starvation, when inadequate food intake forces the body to use its fat stores. Ketonuria coupled with a finding of glycosuria is generally diagnostic for diabetes mellitus.

Red Blood Cells

Hematuria, the appearance of red blood cells, or erythrocytes, in the urine, almost always indicates pathology of the urinary tract, since erythrocytes are too large to pass through the glomerular pores. Possible causes include irritation of the urinary tract organs by calculi (kidney stones), which produces frank bleeding, infections, or physical trauma to the urinary organs.

Hemoglobin

Hemoglobinuria, the presence of hemoglobin in the urine, is a result of the fragmentation, or hemolysis, of red blood cells and the liberation of the hemoglobin into the plasma with its subsequent appearance in the kidney filtrate. Hemoglobinuria may be indicative of various pathologic conditions, including hemolytic anemias, transfusion reactions, burns, or renal disease.

Bile Pigments

The appearance of bilirubin or biliverdin (bile pigments) in the urine is an abnormal finding, and most often reflects liver pathology such as hepatitis or cirrhosis.

White Blood Cells

Pyuria is the presence of white blood cells or other pus constituents in the urine. It indicates an inflammatory process in the urinary tract.

Casts

Any complete discussion of the varieties and implications of casts is beyond the scope of this exercise; however, because they always represent a pathologic condition of the kidney or urinary tract, they should be mentioned. Casts are hardened cell fragments, usually cylindrical, which are flushed out of the urinary tract. White blood cell casts are a common finding with pyelonephritis, red blood cell casts are commonly seen with glomerulonephritis, and fatty casts are indicative of severe renal damage.

ANALYSIS OF URINE SAMPLES

 In this part of the exercise, you will use various types of prepared dip sticks and perform chemical tests to determine the characteristics of normal urine as well as to identify abnormal urinary components.* You will investigate a student-collected specimen and an unknown specimen of urine. Make the following determinations on both samples and record your results by circling the appropriate item or description or by adding data to complete Table

* Many of the chemical tests used here are no longer used in most large clinical agencies that rely almost entirely on dipsticks or computerized urinalyses. However, the tests are not difficult and provide comparison results.

42.1. If you have more than one unknown sample, accurately identify each sample used by number.

DETERMINATION OF THE PHYSICAL CHARACTERISTICS OF URINE

1. Determine the color, transparency, and odor of your sample and one of the numbered pathologic samples, and circle the appropriate descriptions in Table 42.1.
2. Obtain a roll of wide-range pH paper to determine the pH of each sample. Use a fresh piece of paper for each test, and dip the strip into the urine to be tested two or three times before comparing the color obtained with the chart on the dispenser. Record your results.
3. To determine specific gravity, obtain a urinometer cylinder and float. Mix the urine well, and fill the urinometer cylinder about two-thirds full with urine. Examine the urinometer float to determine how to read its markings. (In most cases, the scale has numbered lines separated by a series of unnumbered lines. The numbered lines give the reading for the first two decimal places. You must determine the third decimal place by reading the lower edge of the meniscus—the curved surface representing the urine-air junction—on the stem of the float.)
4. Carefully lower the urinometer float into the urine. Make sure it is floating freely before attempting to take the reading. Record the specific gravity of both samples on the table. *Do not dispose of this urine if the samples that you have are less than 200 ml in volume* because you will need to make several more determinations.

DETERMINATION OF INORGANIC CONSTITUENTS IN URINE

Sulfates (Sulfur)

Add 5 ml of urine to a test tube, and then add a few drops of dilute hydrochloric acid and 2 ml of 10% barium chloride solution. The appearance of a white precipitate (barium sulfate) indicates the presence of sulfates in the sample. Clean the test tubes well after use. Record your results. Are

sulfates a *normal* constituent of urine? _____

Phosphates (Phosphorus)

Add 5 ml of urine to a test tube, and then add three or four drops of dilute nitric acid and 3 ml of ammonium molybdate. Mix well with a glass stirring rod, and then heat gently in a hot water

TABLE 42.1 Urinalysis Results*

Observation or test	Normal values	Student urine specimen	Unknown specimen (#)
Physical characteristics			
Color	Pale yellow	Yellow: pale medium dark other _____	Yellow: pale medium dark other _____
Transparency	Transparent	Clear slightly cloudy cloudy	Clear slightly cloudy cloudy
Odor	Characteristic	Describe _____	Describe _____
pH	4.5–8.0	_____	_____
Specific gravity	1.001 1.030	_____	_____
Inorganic components			
Sulfates	Present	Present Absent	Present Absent
Phosphates	Present	Present Absent	Present Absent
Chlorides	Present	Present Absent	Present Absent
Organic components			
Urea	Present	Present Absent	Present Absent
Glucose			
Clinistix	Negative	Record results: _____	Record results: _____
Benedict's test	Negative—color unchanged	Negative—color unchanged 1+—yel-green (0.5 g/100 ml) 2+—olive green (1 g/100 ml) 3+—yel-orange (1.5 g/100 ml) 4+—brick red (2+ g/100 ml)	Negative—color unchanged 1+—yel-green (0.5 g/100 ml) 2+—olive green (1 g/100 ml) 3+—yel-orange (1.5 g/100 ml) 4+—brick red (2+ g/100 ml)
Albumin			
Albustix	Negative	Record results: _____	Record results: _____
Heller's Test	Negative—no turbidity at urine/acid interface	Negative—no turbidity at interface Protein trace—slight granular turbidity +1 — moderate turbidity +2 — flocculent, thick +3 — curdy cloud +4 — solid mass	Negative—no turbidity at interface Protein trace—slight granular turbidity +1 — moderate turbidity +2 — flocculent, thick +3 — curdy cloud +4 — solid mass
Hemoglobin	Negative	Record results: _____	Record results: _____
Bilirubin (Ictotest)	Negative (no color change)	Negative Positive (purple)	Negative Positive (purple)

*In recording urinalysis data, circle the appropriate description if provided; otherwise record the results you observed.

bath. The formation of a yellow precipitate indicates the presence of phosphates in the sample.

Chlorides

Place 5 ml of urine in a test tube, and add several drops of silver nitrate ($AgNO_3$). The appearance of a white precipitate (silver chloride) is a positive test for chlorides. Record your results.

DETERMINATION OF ORGANIC CONSTITUENTS IN URINE

Urea

Put two drops of urine on a clean microscope slide and _carefully_ add two drops of concentrated nitric acid to the urine. Slowly warm the mixture on a hot plate until it begins to dry at the edges, but do not allow it to boil or to evaporate to dryness. When the slide has cooled, examine the edges of the preparation under low power to identify the rhombic or hexagonal crystals of urea nitrate, which form when urea and nitric acid react chemically. Keep the light low for best contrast. Record your results.

Glucose

1. Obtain a vial of Clinistix and conduct the dip stick test according to the instructions on the vial. Record the results on Table 42.1.
2. Add 5 ml of Benedict's solution to each of two test tubes, and then add eight drops of the urine to be tested to each test tube. Label appropriately. Place the tubes in a water bath, and bring to a boil. Boil for 5 minutes, and then allow the tubes to cool slightly. Observe the final color of the solutions and circle the appropriate observation(s) in the table.

Albumin

1. Obtain the Albustix dip sticks, and conduct the determinations as indicated on the vial. Record your results.
2. Heller's test for the presence of albumin. Carefully pipette or pour 4 ml of concentrated nitric acid into a test tube. Then incline the test tube and slowly allow 4 ml of urine to run down the side of the test tube to form a distinct layer overlying the acid. Observe the interface between the urine and the acid to see if a white precipitate has formed (which indicates the presence of protein in the

urine). Record your results by circling the most appropriate description on Table 42.1.

Hemoglobin

Test your urine samples for the presence of hemoglobin by using a Hemastix dip stick according to the directions on the vial. (Usually a short drying period is required before making the reading, so read the directions carefully.) Record your results.

Bilirubin

Using a medicine dropper, place one drop of urine in the center of one of the special test mats provided with the Ictotest reagent tablets. Place one of the reagent tablets over the drop of urine, and then add two drops of water directly to the tablet. If the mixture turns purple when you add water, bilirubin is present. Record your results.

MICROSCOPIC ANALYSIS OF URINE SEDIMENT (OPTIONAL)

 If a centrifuge is available, a microscopic analysis of the urine sediment can be done. The instructor will provide directions for the specific centrifuge to be used. After centrifugation, pour off the supernatant (the fluid that appears above the sediments at the bottom of the tube), and place a drop of the sediment on a clean glass slide. Add a drop of "Sedi-stain," and cover the specimen with a coverslip.

Using the lowest light source possible, examine the slide under low power to determine if any of the sediments illustrated in Figures 42.1 and 42.2 (and described here) can be visualized.

Unorganized sediments (Figure 42.1): chemical substances that form crystals or precipitate from solution; for example, calcium oxalates and carbonates, uric acid, ammonium ureates, and cholesterol. Also, if an individual has been taking antibiotics or certain drugs such as sulfa drugs these may often be detected in the urine in crystalline form. Normal urine contains very small amounts of crystals, but conditions such as urinary retention or urinary tract infection may result in the appearance of much larger amounts (and their possible consolidation into calculi). High power examination may be necessary to view the various crystals that tend to be much more minute than the organized (cellular) sediments.

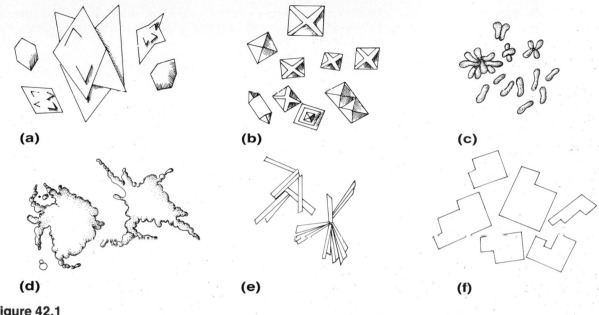

Figure 42.1

Examples of unorganized sediments: (a) uric acid crystals; (b) calcium oxalate crystals; (c) calcium carbonate crystals; (d) ammonium ureate crystals; (e) calcium phosphate crystals; (f) cholesterol crystals.

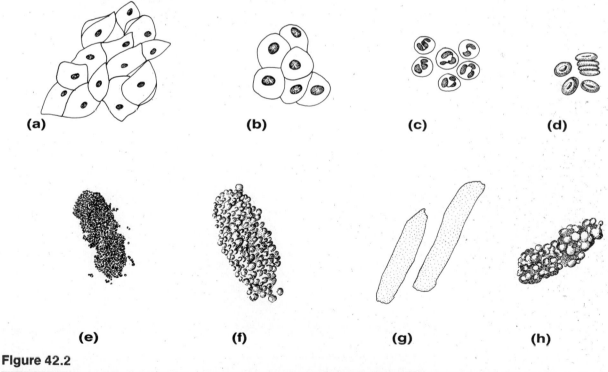

Figure 42.2

Examples of organized sediments: (a) squamous epithelial cells; (b) transitional epithelial cells; (c) white blood cells (pus); (d) red blood cells; (e) granular casts; (f) red blood cell casts; (g) hyaline casts; (h) fatty casts.

Organized sediments (Figure 42.2): include epithelial cells (rarely of any pathologic significance), pus cells (white blood cells), red blood cells, and casts. Urine is normally negative for organized sediments, and the presence of the last three categories mentioned (other than trace amounts) always indicates kidney pathology (infectious or noninfectious in nature).

Draw a few of your observations below and attempt to identify them by comparing with Figures 42.1 and 42.2.

Kidney Regulation of Fluid and Electrolyte Balance

OBJECTIVES

1. To explain the role of antidiuretic hormone in the water balance of the body and to state what factors elicit its release.

2. To explain the role of aldosterone in sodium and potassium balance in the body, and to state what factors elicit its release.

3. To explain the site of formation and the role of renin.

4. To examine the results of experiments conducted to determine and explain how the body counteracts fluid and electrolyte imbalance when various solutions are ingested.

5. To define *polyuria, dysuria, anuria, oliguria, and diuresis.*

MATERIALS

Distilled water
0.9% and 2.0% saline (NaCl) solutions, 0.2% $NaHCO_3$
Wide-range pH paper
Urinometer
Wax marking pencils
Colored pencils
20% potassium chromate in dropper bottle
2.9% $AgNO_3$ in dropper bottle ($AgNO_3$ should be made fresh daily)
500-ml and 1000-ml beakers
100-ml and 500-ml graduate cylinders (meticulously cleaned)
Test tubes and rack
Test tube holders

REGULATORY MECHANISMS

A major role of the kidney is to keep the solute concentration, or osmolarity, of the body fluids at a relatively constant level (approximately 300 milliosmols/liter), thus maintaining adequate body hydration and avoiding both dehydration and water intoxication. The reabsorption of water and electrolytes by the kidney is regulated by hormones, particularly antidiuretic hormone (ADH),* which is released by the posterior pituitary, and aldosterone, which is produced by the adrenal gland cortex.

ADH enters the blood when specialized neurons **(osmoreceptors)** in the hypothalamus are stimulated by an increase in the osmotic pressure of the blood (that is, increased solute concentration and decreased water concentration) as might occur in dehydration. In the kidney, ADH increases resorption

of water primarily by acting on the distal convoluted tubules and the collecting tubules to increase their permeability to water. The net result of this mechanism is that the water content of the blood increases, osmotic pressure decreases, and a small volume of highly concentrated or hypertonic urine is excreted. ADH is released continually except when inhibited by a decrease in blood osmotic pressure. In the absence of ADH release—for example in the hormonal imbalance known as diabetes insipidus—large volumes (up to 25 liters) of very dilute urine are lost daily.

The release of aldosterone from the adrenal cortex may be initiated by a decrease in blood volume or blood pressure, an increase in K^+ in the plasma, or a decrease in blood sodium. Any or all of these factors cause specialized cells in the kidney (juxtoglomerular cells) to release the enzyme called **renin.** Once released, renin catalyzes a series of reactions leading to the formation of a molecule called angiotensin II in the blood. Angiotensin II has two major functions: It acts directly as a vasoconstrictor, producing an increase in systemic blood pressure and greater efficiency in kidney func-

---

* The term antidiuretic is derived from **diuresis,** the flow of urine from the kidney. ADH depresses urine formation.

tion, and it stimulates the release of aldosterone from the adrenal cortex.

Sodium is the most important plasma ion that has to be regulated; normally it is reabsorbed actively along the entire length of the nephron tubules. In the presence of aldosterone, the renal reabsorption of sodium increases even more, especially in the ascending loop of Henle and the distal convoluted tubule. This mechanism helps to prevent a deficit of sodium in the blood, which would result in the dilution of body fluids and a loss in plasma volume to the body tissues, leading to edema and possible circulatory collapse. For each sodium ion that is reabsorbed, a potassium ion is secreted into the filtrate. Thus as the Na^+ concentration in the blood increases, the K^+ concentration decreases, bringing these two ions back into their normal electrolyte balance in the blood.

The volume and specific gravity of urinary output are important indications of a person's state of hydration. In individuals with excessively low fluid intake, the urine is much reduced in volume and its specific gravity is high. Conversely, if the fluid intake is excessive, a large volume of low-specific-gravity urine is excreted.

EFFECT OF FLUID AND SALT LOAD ON URINARY OUTPUT AND CHARACTERISTICS

As you investigate the alterations in volume and urine composition the kidneys make under conditions of known fluid and salt load, you should work in groups of four, each student being responsible for conducting one of the four procedures and reporting the results to the other members of the group. Record all data on the data sheet provided. When you have completed the experiments, graph the results. All group members should be prepared to report and interpret their results in a class discussion.

 1. Each student should void into a clean 500-ml beaker before beginning the experiment. Save this urine, label it with your name, the word "control," and the time of collection. You will need this sample for tests to be performed in the time intervals between experimental sample collections and testings.

2. Consult the list that follows, which indicates the experimental conditions for each group member. Each student is to drink the designated amount of one of the solutions in as short a time as is comfortable. (Do not attempt to drink the solution too rapidly, as this may cause emesis or vomiting.) After drinking the solution, record the time, and then void at 30 minute intervals for a period

of 2½ hours after obtaining the control sample. Collect each voiding in a premarked beaker (labeled with your name, time of collection, solution drunk) and measure the volume. If you are unable to void at any predetermined collection time, retain the urine until the next half-hour collection interval.

Student no.	Solution and volume to be drunk
1	700–1000 ml distilled water
2	700–1000 ml 0.9% saline solution
3	150–250 ml 2.0% saline solution*
4	500 ml 0.2% sodium bicarbonate solution

3. To determine the results, make the following observations or tests on each sample collected and on the control sample. Record the results in the appropriate column of the data sheet.

Transparency: record as clear or cloudy.
Color: record as pale, medium, or deep yellow.
Volume: measure the volume of the urine sample collected and calculate the rate of urine formation. Divide the total volume collected in the 30-minute interval by 30 to determine milliliters per minute excreted.
pH: use wide-range pH paper to determine the pH.
Specific gravity: fill a urinometer cylinder two-thirds full with urine and gently place the urinometer float in the urine. Read the specific gravity at the meniscus as described on page 368.
Chloride estimation: using a medicine dropper, put 10 drops of urine into a test tube, add 1 drop of potassium chromate, and then add $AgNO_3$, 1 drop at a time. Count the number of drops necessary to change the solution from bright yellow to brown. Each drop of the 2.9% $AgNO_3$ solution required to bring about this color change corresponds to 1 g/liter of NaCl in the urine (or 1000 mg/1000 ml). Calculate the total number of milligrams of NaCl per milliliter in the sample and enter the information on the data sheet.

4. When all the information has been collected, complete your data sheet with the information from other students in the group, and make a graphic

---

*The 2.0% saline solution is difficult to ingest. If any nausea, vomiting, or abnormal cramping occurs, discontinue the experiment.

analysis of the group results. Make four separate plots as indicated below for each set of test results on the labeled graphic sheets provided. Indicate the appropriate intervals on the vertical axis.

- specific gravity vs. time
- volume plot: ml/min vs. time
- mg NaCl/ml vs. time
- pH vs. time

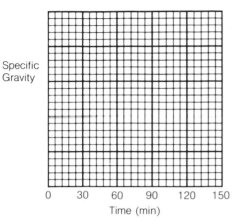

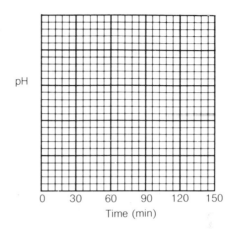

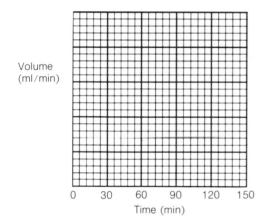

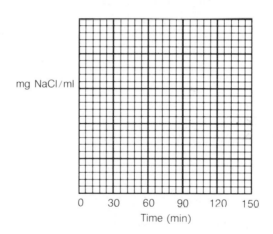

Student no. and solution ingested	Time (min)	Color	Transparency	pH	sp. gr.	Vol. (ml/min)	NaCl (mg/ml)
1 700–1000 ml distilled water	control						
	30						
	60						
	90						
	120						
	150						

(*continued*)

Student no. and solution ingested	Time (min)	Color	Transparency	pH	sp. gr.	Vol. (ml/min)	NaCl (mg/ml)
2 700—1000 ml 0.9% saline	control						
	30						
	60						
	90						
	120						
	150						
3 150—250 ml 2.0% saline	control						
	30						
	60						
	90						
	120						
	150						
4 500 ml 0.2% sodium bicarbonate	control						
	30						
	60						
	90						
	120						
	150						

UNIT 15

THE REPRODUCTIVE SYSTEM

Anatomy of the Reproductive System

EXERCISE
44

OBJECTIVES

1. To discuss the general function of the reproductive system.

2. To identify the following structures of the male reproductive system when provided with an appropriate model or diagram, and to discuss the general function of each.

testis	ejaculatory duct
ductus deferens	Cowper's glands
seminal vesicles	penis
urethra and its divisions	scrotum
epididymis	prostate

3. To define *semen,* discuss its composition, and name the glands involved in its production.

4. To trace the pathway followed by a sperm from its site of formation to the external environment.

5. To describe the structure of the penis and relate its structure to its erectile function.

6. To name the exocrine and endocrine products of the testes, indicating the cell types or structures responsible for the production of each.

7. To describe the microscopic structure of the epididymis, and to relate structure to function.

8. To define *ejaculation, erection,* and *gonad.*

9. To identify the following structures of the female reproductive system when provided with an appropriate model or diagram, and to discuss the general function of each.

ovary	uterine tube
uterus	vagina

    greater vestibular (Bartholin's) glands
    supporting structures: broad, round, and ovarian ligaments
    external genitalia: mons pubis, labia minora and majora, clitoris

10. To discuss the function of the fimbriae and ciliated epithelium of the uterine (fallopian) tubes.

11. To identify homologous structures of the male and female systems.

12. To discuss the microscopic structure of the ovary and be prepared to identify the following ovarian structures: primary follicle, graafian follicle, and corpus luteum, and to state the hormonal products of the last two structures.

13. To define *endometrium, myometrium,* and *ovulation.*

14. To identify the fundus, body, and cervical regions of the uterus.

15. To identify the major reproductive structures of the male and female dissection animal.

16. To recognize and discuss pertinent differences between the reproductive structures of the human and the dissection animal.

MATERIALS

Models or large laboratory charts of the male and female reproductive tracts

Prepared microscope slides of cross sections of the penis, epididymis, ovary, and uterus (showing endometrium)

Dissection animal, tray, and instruments

Bone clippers

Related film: *Human Body: Reproductive System* (color, sound, 16 mm, Coronet Films)

Most simply stated, the biologic function of the reproductive system is to perpetuate the species. Thus the reproductive system is unique, since the other organ systems of the body function primarily to sustain the individual.

The essential organs of reproduction are those that produce the germ cells—the testes and the ovaries. The reproductive role of the male is to manufacture sperm and to deliver them to the female reproductive tract. The female, in turn, produces eggs. If the time is suitable, the combination of sperm and egg produces a fertilized egg, which is the first cell of a new individual. Once fertilization has occurred, the female uterus provides a nurturing, protective environment in which the embryo, later called the fetus, develops until birth.

Although the drive to reproduce is strong in all animals, in humans this drive is also intricately related to nonbiologic factors. Emotions and social considerations often enhance or thwart its expression.

GROSS ANATOMY OF THE HUMAN MALE REPRODUCTIVE SYSTEM

The primary reproductive organs of the male are the **testes,** the male gonads, which have both an exocrine (sperm production) and an endocrine (testosterone production) function. All other reproductive structures are conduits or sources of secretions, which aid in the safe delivery of the sperm to the body exterior or female reproductive tract.

As the following organs and structures are described, locate them on Figure 44.1, and then identify them on a three-dimensional model of the male reproductive system or on a large laboratory chart.

The paired oval testes lie in the **scrotal sac** outside the abdominopelvic cavity. The temperature

(94 to 95 F, or approximately 30 C), slightly lower than body temperature, that exists here is required to produce viable sperm.

The accessory structures forming the duct system are the epididymis, the ductus deferens, the ejaculatory duct, and the urethra. The **epididymis** is an elongated structure running up the side of the testis and capping its superior aspect. The epididymis forms the first portion of the duct system and provides a site for immature sperm that enter it from the testis to complete their maturation process. The **ductus deferens** (sperm duct) arches upward from the epididymis through the inguinal canal into the pelvic cavity and courses over the superior aspect of the bladder. In life, the ductus deferens (also called the *vas deferens*) is enclosed along with blood vessels and nerves in a connective tissue sheath called the **spermatic cord.** The terminus of the ductus deferens enlarges to form the region called the **ampulla,** which empties into the **ejaculatory duct.** Contraction of the ejaculatory duct propels the sperm through the prostate gland to the **prostatic urethra,** which in turn empties into the **membranous urethra** and then into the **penile urethra,** which runs through the length of the penis to the body exterior.

The accessory glands include the prostate gland, the paired seminal vesicles, and Cowper's glands. These glands produce **semen,** the liquid medium in which sperm leave the body. The **seminal vesicles,** which produce about 60% of the semen, are close to the terminus of the ductus deferens. They produce a viscous secretion containing fructose (a simple sugar) and other substances that nourish the sperm passing through the tract. The duct of each seminal vesicle merges with a ductus deferens to form an ejaculatory duct; thus sperm and seminal fluid enter the urethra together. The **prostate** encircles the urethra just inferior to the bladder. It secretes a milky alkaline fluid into the urethra, which is believed to have a role in activating the sperm. Prostate gland secretion accounts for about 20% of the volume of semen. Hypertrophy of the prostate gland, common-

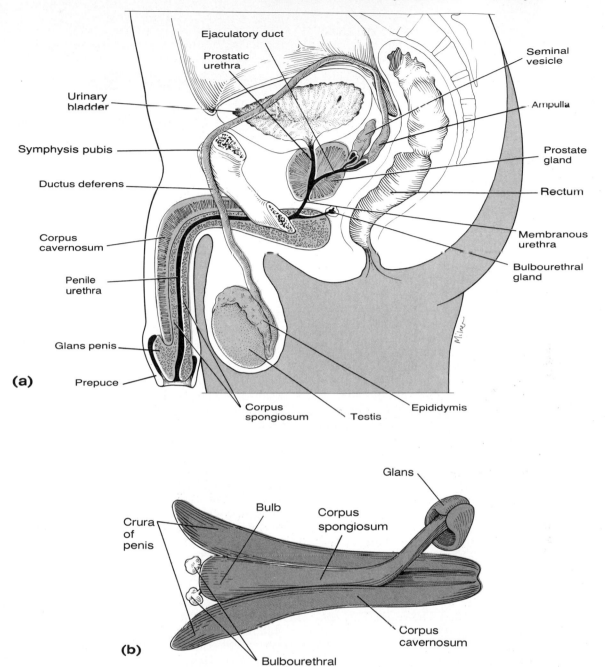

Ejaculatory duct

Prostatic urethra

Urinary bladder

Symphysis pubis

Ductus deferens

Corpus cavernosum

Penile urethra

Glans penis

(a)

Prepuce

Corpus spongiosum

Testis

Epididymis

Seminal vesicle

Ampulla

Prostate gland

Rectum

Membranous urethra

Bulbourethral gland

Glans

Crura of penis

Bulb

Corpus spongiosum

(b)

Bulbourethral glands

Corpus cavernosum

Figure 44.1

Reproductive system of the human male: (a) medial sagittal section; (b) inferior intact view of the penis.

ly seen in old age, constricts the urethra, a troublesome condition that makes urination difficult.

The **bulbourethral glands (Cowper's glands)** are tiny, pear-shaped glands inferior to the prostate. They produce a clear, viscous alkaline solution that drains into the membranous urethra. This secretion is believed to wash the residual urine out of the urethra when ejaculation occurs. The relative alkalinity of all of these glandular secretions may also have a role in buffering the sperm against the acidity of the female reproductive tract.

The **penis,** part of the external genitalia of the male along with the scrotal sac, is the copulatory organ of the male and is designed to deliver sperm into the female reproductive tract. It consists of a shaft, which terminates in an enlarged tip, the **glans.** The skin covering the penis is loosely applied, and it reflects downward to form a circular fold of skin, the **prepuce,** or **foreskin,** around the proximal end of the glans. Internally, the penis consists primarily of three elongated cylinders of erectile tissue, which become engorged with blood during sexual excitement. This causes the penis to become rigid and enlarged so that it may more adequately serve as a penetrating device. This event is called **erection.** The paired dorsal cylinders are the **corpora cavernosa.** The single ventral **corpus spongiosum** surrounds the penile urethra.

GROSS ANATOMY OF THE HUMAN FEMALE REPRODUCTIVE SYSTEM

The **ovaries** (female gonads) are the primary reproductive organs of the female. Like the testes of the male, the ovaries produce both an exocrine product (the eggs, or ova) and endocrine products (estrogens and progesterone). The other accessory structures of the female reproductive system transport, house, nurture, or otherwise serve the needs of the reproductive cells and/or the developing fetus.

The reproductive structures of the female are generally considered in terms of internal organs and external organs, or external genitalia. As you read the descriptions of these structures, locate them on Figures 44.2 and 44.3 and then on the female reproductive system model or large laboratory chart.

The external genitalia (**vulva**) consist of the mons pubis, labia majora and minora, the clitoris, urethral and vaginal orifices, the hymen, and the greater vestibular glands. The **mons pubis** is a rounded fatty eminence overlying the pubic symphysis. Running inferiorly and posteriorly from the mons pubis are two elongated, pigmented, hair-covered skin folds, the **labia majora,** which enclose two smaller hair-free folds, the **labia minora.** The labia majora are homologous to the scrotum of the male. The labia minora, in turn, enclose a region called the **vestibule,** which contains many structures: the clitoris, most anteriorly, followed by the urethral orifice and the vaginal orifice. The region between the posterior terminus of the labia minora folds and the anus is called the **perineum.**

The **clitoris** is a small protruding structure, homologous to the male penis, and like its counterpart is composed of highly sensitive, erectile tissue. It is hooded by skin folds of the anterior labia minora, referred to as the **prepuce of the clitoris.** The urethral orifice is the outlet for the urinary system and has no reproductive function in the female. The vaginal opening is partially closed by a thin fold of mucous membrane called the **hymen** and is flanked by the mucus-secreting **greater vestibular (Bartholin's) glands,** which lubricate the distal end of the vagina during coitus. (These glands are not depicted in the illustrations.)

The internal female organs include the vagina, uterus, uterine, or fallopian, tubes, ovaries, and the ligaments and supporting structures that suspend these organs in the pelvic cavity. The **vagina** extends for approximately 6 inches (15 cm) from the vestibule to the uterus superiorly. It serves as the copulatory

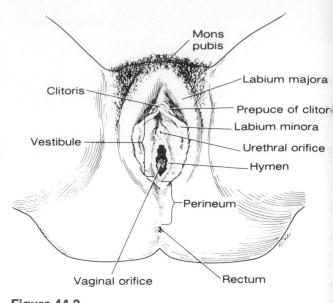

Figure 44.2

External genitalia of the human female.

organ and the birth canal and permits passage of the menstrual flow. The pear-shaped **uterus,** situated between the bladder and the rectum, is a highly muscular organ with its narrow end, the **cervix,** directed inferiorly. The major portion of the uterus is referred to as the **body,** and its superior rounded region above the entrance of the uterine tubes is called the **fundus.** The fertilized egg is implanted in the uterus, which houses the embryo or fetus during its development. In some cases, the fertilized egg may be implanted in a uterine tube or even on the abdominal viscera, creating an **ectopic pregnancy.** Such implantations are usually unsuccessful and may even endanger the mother's life, since the uterine tubes cannot accommodate the increasing size of the fetus. The thick mucosal lining of the uterus, the **endometrium,** sloughs off periodically (usually about every 28 days) in response to cyclic changes in the levels of ovarian hormones in the woman's blood. This sloughing-off process, which is accompanied by bleeding, is referred to as the **menstrual flow,** or **menses.** The **uterine,** or **fallopian, tubes** enter the superior region of the uterus and extend laterally for about 4 inches (10 cm) toward the **ovaries** in the peritoneal cavity. The distal ends of the tubes are funnel-shaped and have fingerlike projections called **fimbriae.** Unlike the male duct system, there is no actual contact between the female gonad and the initial part of the female duct system—the uterine tube. (Because of this open passageway between the female reproductive organs and the peritoneal cavity, reproductive system infections, such as gonorrhea, can spread to cause widespread inflammations of the pelvic viscera, a condition called *pelvic inflammatory disease* or PID.) The flattened almond-shaped ovaries lie adjacent to the uterine tubes but are not connected to them. As an egg is expelled from

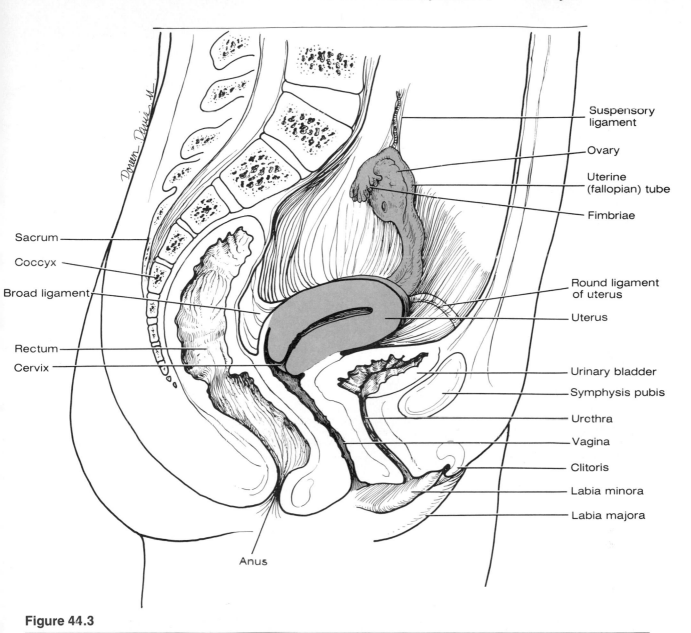

Figure 44.3

Sagittal section of the human female reproductive system.

the ovary, an event called **ovulation,** it enters the pelvic cavity. The waving fimbriae of the uterine tubes create fluid currents that (hopefully) waft the egg into the lumen of the uterine tube, where it begins its passage to the uterus, propelled by the cilia of the tubule walls. The usual and most desirable site of fertilization is the uterine tube, since the journey to the uterus takes about 5 days and an egg is usually nonviable after 24 to 48 hours. Thus, sperm must swim upward through the vagina and uterus and into the uterine tubes to reach the egg. This must be an arduous journey, since they must swim against the downward current created by ciliary action—rather like swimming against the tide!

The internal female organs are for the most part retroperitoneal, except the ovaries, which are covered by only a thin layer of epithelium. They are supported and suspended somewhat freely by ligamentous folds of peritoneum. The peritoneum takes an undulating course: from the pelvic cavity floor it moves superiorly over the top of the bladder, reflects over the anterior and posterior surfaces of the uterus, and then over the rectum, and up the posterior body wall. The fold that encloses the uterine tubes and uterus and secures them to the lateral body walls is referred to as the **broad ligament.** The portion of the broad ligament specifically anchoring the uterus is called the **mesometrium** and that anchoring the uterine tubes, the **mesosalpinx.** The **round ligaments,** fibrous cords that run from the uterus to the labia majora, also aid in attachment of the uterus to the body wall. The ovaries are supported posteriorly by a small fold of peritoneum, the **mesovarium** portion of the broad ligament, medially by the **ovarian ligament** (a thickening

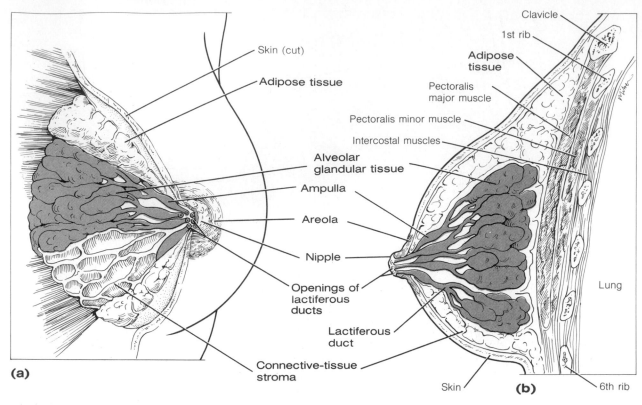

Figure 44.4

Female mammary gland: (a) anterior view; (b) sagittal section.

of the broad ligament that extends from the uterus to the ovary), and laterally by the **suspensory ligaments.**

The mammary glands or breasts exist, of course, in both sexes, but they have a reproductive function only in females. Since the function of the mammary glands is to produce milk to nourish the newborn infant, their importance is more closely associated with events that occur when reproduction has already been accomplished. Periodic stimulation by the female sex hormones, especially estrogens, increase the size of the female mammary glands at puberty. During this period, the duct system becomes more elaborate, and fat is deposited—fat deposition being the more important contributor to increased breast size.

The rounded, skin-covered mammary glands lie anterior to the pectoral muscles of the thorax, attached to them by connective tissue. Slightly below the center of each breast is a pigmented area, the **areola,** which surrounds a centrally protruding **nipple** (Figure 44.4).

Internally each mammary gland consists of 15 to 20 **lobes** separated by connective tissue and adipose, or fatty, tissue, which radiate around the nipple. Within each lobe are smaller chambers called **lobules,** containing the clusters of **alveolar glands** that produce the milk during lactation. The alveolar glands of each lobule pass the milk into a num-

ber of ducts, which join to form an expanded storage chamber, the **ampulla** as they approach the nipple. Ampullae of each lobe are drained by the **lactiferous ducts,** which terminate in the nipple.

MICROSCOPIC ANATOMY OF SELECTED REPRODUCTIVE ORGANS

Testis

Each testis is covered by a dense connective tissue capsule called the **tunica albuginea** (literally, white tunic). Extensions of this sheath enter the testis, dividing it into a number of lobes, each of which houses one to four highly coiled **seminiferous tubules,** the sperm-forming factories (Figure 44.5). The seminiferous tubules of each lobe converge to empty the sperm into another tubular region, the **rete testis,** at the mediastinum of the testis. Sperm traveling through the rete testis then enter the epididymis, located on the exterior aspect of the testis, as previously described. Lying between the seminiferous tubules and softly padded with connective tissue are the **interstitial cells,** which produce testosterone, the hormonal product of the testis. You will conduct a microscopic study of testes tissue in Exercise 45.

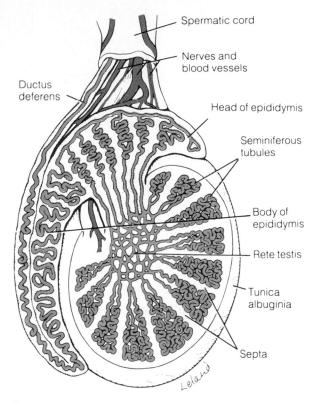

Spermatic cord

Nerves and
blood vessels

Ductus
deferens

Head of epididymis

Seminiferous
tubules

Body of
epididymis

Rete testis

Tunica
albuginia

Septa

Figure 44.5

Structure of the human testis: sagittal section, showing
internal structure; epididymis and ductus deferens also
shown.

EPIDIDYMIS   Obtain a cross section of the epididymis. Examine the composition of the tubule
wall carefully. Identify the smooth muscle layer
and the pseudostratified columnar epithelium.
Look for sperm in the lumen. What do you think
the function of the smooth muscle is?

_____

_____

Draw what you observe below and label appropriately.

Penis

Obtain a cross section of the penis. Scan the tissue under low power to identify the urethra and
the cavernous bodies. Observe the lumen of the
urethra carefully. What type of epithelium do you

see? _____

Explain the function of this type of epithelium.

_____

_____

Draw a diagram of the cross section of the penis below. Label the corpus cavernosum, corpus
spongiosum, and urethra.

Wall of the Uterus*

Obtain a cross-sectional view of the uterine wall.
Identify, sketch, and label the three layers of the
uterine wall (the endometrium, myometrium, and
serosa). As you study the slide, notice that the
bundles of smooth muscle are oriented in several
different directions. What is the function of the
myometrium (smooth muscle layer) during the
birth process? _____

_____

*A microscopic study of the ovary is described in Exercise 45,
p. 392. If that exercise is not to be conducted in its entirety, the
instructor might want to include the ovary study in the scope of
this exercise.

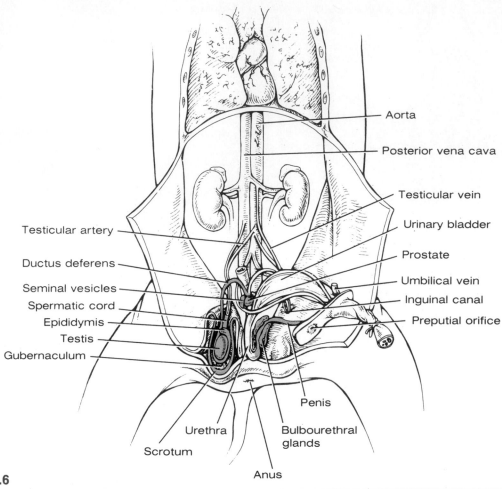

Figure 44.6

Reproductive system of the male fetal pig.

DISSECTION OF THE FETAL PIG REPRODUCTIVE SYSTEM

Obtain your pig, a dissection tray, and the necessary dissection instruments. After you have completed the study of the reproductive structures of your specimen, observe a pig of the opposite sex. (The following instructions assume that the abdominal cavity has been opened in previous dissection exercises.)

Male Reproductive System

Refer to Figure 44.6 as you identify the structures.

 1. Identify the **preputial orifice** on the ventral body surface just posterior to the umbilical cord; this is the external urethral orifice. Carefully cut through the skin of the orifice and identify the distal end of the penis, which lies within a fold of skin (the prepuce). Reflect the cut midventral strip of the body wall to identify the thin cordlike penis shaft, which lies immediately posterior to the urinary bladder. Cross section the penis to observe the relative positioning of the cavernous bodies.

2. Follow the penis posteriorly to the point where the urethra enters it. As in the human male, the male pig urethra serves as both a urine and sperm duct. Also identify the paired elongated Bulbourethral glands, which flank the penile-urethral junction laterally.

3. Continue to follow the urethra superiorly until you can see the point where the right and left ductus deferens (each encapsulated in a spermatic cord) loop over the ureters to enter the dorsal aspect of the urethra. (This positioning of the spermatic cord is due to the fact that initially the testis is in the same relative position as the ovary is in the female. In the fetus, the testis descends laterally and ventrally to the ureters to enter the scrotal sac.)

4. Identify the small paired seminal vesicles on the dorsal aspect of the urethra adjacent to the points of ductus deferens entry. Also, attempt to find the prostate gland at the junction of the bladder and

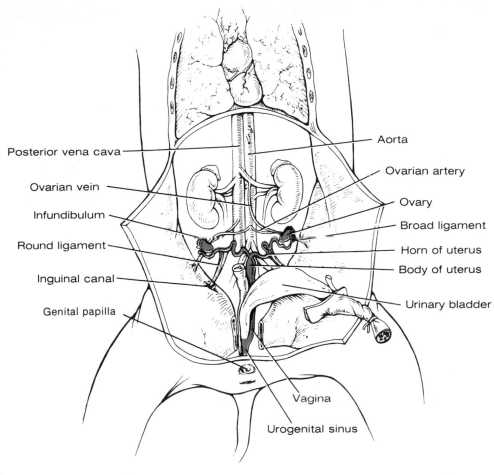

Figure 44.7

Reproductive system of the female fetal pig.

urethra by carefully dissecting the tissue between the seminal vesicles. In the adult pig, the prostate gland can be found in this position, but it is generally difficult to identify in the fetal pig.

5.   Trace the spermatic cord posteriorly through the inguinal canal into the scrotum. Identify the scrotal sac exteriorly on the ventral body surface anterior to the anus. Make a shallow incision into the scrotum to expose the testis lying in a membranous sac (inguinal sac). (It may be necessary to flush the interior of the scrotum with tap water to remove coagulated blood, which tends to accumulate there.) Gently grasp the inguinal sac and pull it up and out of the scrotum. As you do so, notice that the spermatic cord moves more posteriorly into the inguinal canal.

6.   Carefully cut open the inguinal sac and identify the following structures:

Testis: the male gonads
Epididymis: a fine convoluted tube running along the posterior, lateral, and anterior surfaces

of the testis; conveys sperm from the testis to the ductus deferens
Gubernaculum: a tough cord extending from the posterior end of the epididymis to the inguinal pouch and scrotum; its attachments are established early in development, and its failure to grow as rapidly as other body structures in the area results in its "pulling" the testis into the scrotum, thus causing its descent.

7.   Make a longitudinal cut through the testis and epididymis. Can you see the tubular nature of the epididymis and rete testis portion of the testis with the naked eye?

Female Reproductive System

Refer to Figure 44.7 as you identify the structures described below:

1.   Unlike the pear-shaped simplex, or one-part, uterus of the human, the uterus of the pig is Y-shaped (bipartite or bicornate) and consists of a

uterine body from which two **uterine horns** (cornua) diverge. Such an enlarged uterus enables the animal to produce litters. Examine the abdominal cavity, identify the bladder, and the body of the uterus lying just dorsal to it.

2. Follow one of the uterine horns as it travels superiorly in the body cavity, noting the thin mesentery (the broad ligament), which helps to anchor it and the other reproductive structures to the body wall. Approximately halfway up the length of the uterine horn, it should be possible to identify the more important round ligament, a cord of connective tissue extending laterally and posteriorly from the uterine horn to the region of the body wall that corresponds to the inguinal region of the male.

3. Identify the uterine tube and ovary at the distal end of the uterine horn just caudal to the kidney. Observe how the funnel-shaped end of the uterine tube curves around the ovary. As in the human, the distal end of the tube is fimbriated, or fringed, and the tube is lined with ciliated epithelium. The uterine tubes of the pig are tiny and relatively much shorter than in the human. Identify the ovarian ligament, a short thick cord that extends from the uterus to the ovary and anchors the ovary to the body wall. Also observe the ovarian artery and vein, passing through the mesentery to the ovary and uterine structures.

4. Return to the body of the uterus and follow it caudad to the bony pelvis. Use bone clippers to cut through the median line of the pelvis (pubic symphysis), cutting carefully so that the underlying urethra is not damaged. Expose the pelvic region by pressing the thighs dorsally. Follow the uterine body caudally to the vagina and note the point where the urethra draining the bladder and the vagina enter a common chamber, the **urogenital sinus.** How does this anatomic arrangement compare to that seen in the human female?

5. Observe the vulva of the pig, which is similar to the human vulva. Identify the external urogenital orifice, which is hooded by the **genital papilla.**

6. Insert one blade of a scissors into the lateral aspect of the urogenital orifice and cut anteriorly along the length of the urogenital sinus and vagina. Reflect the cut edges and identify the following structures:

Clitoris: a small elevation on the ventral surface of the sinus, just anterior to the genital papilla
Urethral opening into the urogenital sinus
Cervix: the muscular ring separating the vagina from the uterus. Notice that the vagina is comparatively much shorter in the pig than in the human (its limits are demarcated by the cervix anteriorly and the urogenital sinus posteriorly).

7. When you have completed your observations of the reproductive system of the male and female pigs, clean your dissecting instruments and tray, and properly wrap the pig for storage.

Physiology of Reproduction: Gametogenesis and the Female Cycles

OBJECTIVES

1. To define *meiosis, gametogenesis, oogenesis, spermatogenesis, spermiogenesis, synapsis, haploid, diploid,* and *menses,* and to state the similarities and differences between spermatogenesis and oogenesis.

2. To relate the stages of spermatogenesis to the cross-sectional structure of the seminiferous tubule.

3. To discuss the microscopic structure of the ovary and to be prepared to identify primary and graafian follicles and corpus luteum, and to state the hormonal products of the last two structures.

4. To relate the stages of oogenesis to follicular development in the ovary.

5. To cite similarities and differences between mitosis and meiosis.

6. To describe the anatomic structure of the sperm and relate it to function.

7. To discuss the phases and control of the menstrual cycle.

8. To describe the feedback relationship between anterior pituitary gonadotropins and ovarian hormones.

9. To describe the effect of FSH and LH (ICSH) on testicular function.

MATERIALS

Prepared microscope slides of testis, ovary, human sperm, and uterine endometrium (showing menses, proliferative, and secretory stages)

Three-dimensional models illustrating meiosis, spermatogenesis, and oogenesis

Compound microscope

Demonstration Area: microscopes set up to demonstrate the following stages of oogenesis in *Ascaris megalocephala:*

1. Primary oocyte with fertilization membrane, sperm nucleus, and aligned tetrads apparent
2. Formation of the first polar body
3. Secondary oocyte with dyads aligned
4. Second polar body formation and ovum
5. Fusion of the male and female pronuclei to form the fertilized egg

MEIOSIS

Every human being, so far, has developed from the union of gametes. The gametes, produced only in the testis or ovary, are unique cells, since they have only half the normal chromosome number (designated as **N,** or the **haploid complement**) seen in all other body cells. In humans, gametes have 23 chromosomes instead of the 46 in other tissue cells. Theoretically, every gamete has a full set of genetic instructions, a conclusion borne out by the observation that some lower animals can develop from an egg that is artificially stimulated, as by a pinprick, rather than by sperm entry. (This is less true of the sperm, since it has

an incomplete sex chromosome.) The reduction of chromosome number by half is important to maintain the characteristic chromosomal number of the species generation after generation; otherwise there would be a doubling of chromosome number with each succeeding generation and the cells would become so chock-full of genetic material there would be little room for anything else.

Egg and sperm chromosomes that carry genes for the same traits are called **homologous chromosomes,** and when the sperm and egg fuse to form the fertilized egg, it is said to contain 23 pairs of homologous chromosomes, or the **diploid (2N)** chromosome number of 46. The fertilized egg, once formed, then divides by mitosis to produce the cells

387

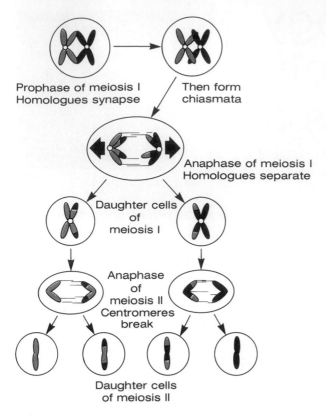

Figure 45.1

Events of meiosis. One pair of homologous chromosomes (male homologue is dark; female homologue is red).

needed to construct the multicellular human body. All cells of the developing human body have a chromosome content exactly identical in quality and quantity to that of the fertilized egg; this is assured by the nuclear division process called **mitosis.** (Mitosis was considered in depth in Exercise 4. You may want to review it at this time.)

To produce the reduced (haploid) chromosomal number, a specialized type of nuclear division occurs in the ovaries and testes. Before meiosis begins, the chromosomes are replicated in the mother cell or stem cell just as they are before mitosis. As a result, the mother cell briefly has double the normal diploid genetic complement. The stem cell then undergoes two consecutive nuclear divisions, termed meiosis I and II, or the first and second maturation divisions, without replicating the chromosomes before the second division. The result is that four haploid daughter cells are produced, rather than the two diploid daughter cells produced by mitotic division.

The entire process of meiosis is quite complex and is dealt with here only to the extent necessary to reveal important points of difference between this type of nuclear division and mitosis. Essentially, each meiotic division involves the same phases and events seen in mitosis (prophase, metaphase, anaphase, and telophase), but in the first maturation division (meiosis I) an event not seen in mitosis

occurs as the mother cell goes into prophase. The homologous chromosomes, each now a duplicated structure, begin to pair so that they are closely aligned along their entire length. This pairing is called **synapsis.** As a result, 23 **tetrads** (groupings of four chromosome strands) form, become attached to the spindle fibers, and begin to align themselves on the spindle equator. While in synapsis, two of the four strands (one from each homologue) in each tetrad wrap and coil around each other, forming many points of **crossover,** or **chiasmata.** (Perhaps this could be called the conjugal bed of the cell!) When anaphase of meiosis I begins, the homologues separate from one another, breaking and exchanging parts at points of crossover and move apart toward opposite poles of the cell. The centromeres holding the duplicated chromosome strands together do not break at this point (Figure 45.1).

During the second maturation, events parallel those in mitosis, except that the daughter cells do not replicate their chromosomes before this division and each daughter cell has only some of the homologous chromosomes rather than a complete set. The crossover events and the way in which the homologues align on the spindle equator during the first maturation division introduce an immense variability in the resulting gametes, which explains why we are all unique.

 Obtain a model depicting the events of meiosis, and follow the sequence of events during the first and second maturation divisions. Identify prophase, metaphase, anaphase, and telophase in each; the tetrads and chiasmata during the first maturation division; and the dyads in the second maturation division. Note in which ways the daughter cells resulting from meiosis I differ from the mother cell and how the gametes differ from both cell populations. (Use the key on the model, your textbook, or an appropriate reference as necessary to aid you in these observations.)*

SPERMATOGENESIS

Human sperm production begins at puberty and continues without interruption throughout life. The average male ejaculation contains more than a half-billion sperm. Since only one sperm fertilizes an ovum, it seems that nature has tried to assure that the perpetuation of the species will not be endangered for lack of sperm.

As explained in Exercise 44, spermatogenesis occurs in the seminiferous tubules of the testes. The primitive stem cells, or **spermatogonia,** found at the tubule periphery, undergo extensive mitotic activity to build up and retain the stem cell

*The instructor may wish you to observe meiosis in *Ascaris* at this point to provide cellular material for comparison (page 394).

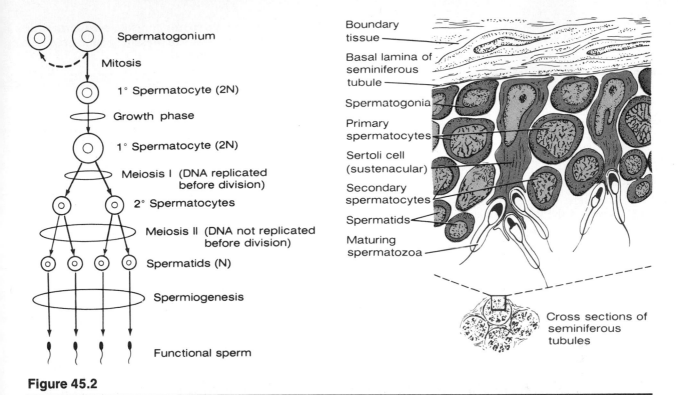

Figure 45.2

Spermatogenesis: left, flow sheet of meiotic events; right, diagrammatic view of seminiferous tubule. Redrawn with permission from C. R. Leeson and T. S. Leeson, *Histology*, 4th ed. (Philadelphia: W.B. Saunders, 1981).

line. Before puberty, all divisions produce more spermatogonia; at puberty, however, under the influence of FSH secreted by the anterior pituitary gland, each mitotic division of a spermatogonium results in the production of one spermatogonium and another cell, now called a **primary spermatocyte,** which is destined to undergo meiosis. As meiosis occurs, the dividing cells approach the lumen of the tubule. Thus the progression of meiotic events can be followed from the tubule periphery to the lumen. It is important to recognize that the **spermatids,** which are actually the product of meiosis (since they are haploid cells), are not functional gametes; they are nonmotile cells and have too much excess baggage to function well in a reproductive capacity. Another process, called **spermiogenesis,** which follows spermatogenesis, strips away the extraneous cytoplasm from the spermatid and converts it to a motile streamlined sperm.

 1. Obtain a slide of the testis and a microscope. Examine the slide under low power to identify the cross-sectional views of the cut seminiferous tubules. Then rotate the h.p. lens into position and observe the wall of one of the cut tubules. Refer to Figure 45.2 as you work to make the following identifications.

2. Scrutinize the cells at the periphery of the tubule. The cells in this area are the spermatogonia, which undergo frequent mitoses to increase their

number and maintain their population. About half of the spermatogonia differentiate to form primary spermatocytes. These cells begin meiosis, which leads to the formation of spermatid cells having half the usual genetic composition. (In humans, the spermatids have 23 chromosomes instead of the usual 46 found in other body cells.)

3. Observe the cells in the middle of the tubule wall. Do you see a large number of cells (spermatocytes) that are obviously undergoing a nuclear division

process? _____ Look for the chromosomes, visible only during nuclear division, that have the appearance of coiled springs. Attempt to differentiate between the larger primary spermatocytes and the somewhat smaller secondary sper-

matocytes. Can you see the tetrads? _____

Evidence of crossover? _____ Where

would you expect to see the tetrads? Closer to the

spermatogonia or to the lumen? _____

_____

4. Examine the cells at the tubule lumen. Identify the small round-nucleated spermatids, many of which may appear lopsided and as though they are starting to lose their cytoplasm. See if you can find

a spermatid embedded in an elongated cell type, a **Sertoli cell,** which extends from the periphery of the tubule. The Sertoli cells nourish the spermatids as they begin their transformation into sperm. Also in the lumen area, locate sperm, which can be identified by their tails. The sperm develop directly from the spermatids by the loss of extraneous cytoplasm and the development of a propulsive tail.

5. Identify the testosterone-producing interstitial cells lying external to and between the seminiferous tubules.

In the next stage of sperm development, spermiogenesis, all the superficial cytoplasm is sloughed off, and the remaining cell organelles are compacted into the three regions of the mature sperm. At the risk of oversimplifying, these anatomic regions are the head, the midpiece, and the tail, which correspond roughly to the activating and genetic region, the metabolic region, and the locomotor region. The mature sperm is a streamlined cell equipped with an organ of locomotion and a high rate of metabolism that enables it to move long distances in jig time to get to the egg. It is a prime example of the correlation of form and function.

The sperm head contains the DNA, or genetic material, of the chromosomes. Essentially it is the nucleus of the spermatid. Anterior to the nucleus is the **acrosome,** which contains enzymes involved in sperm penetration of the egg.

In the midpiece of the sperm is a centriole from which arise the filaments that structure the sperm tail. Wrapped tightly around the centriole are mitochondria, which apparently provide the ATP needed for the contractile activity of the tail filaments.

The tail is composed of filaments that arise from the centriole and are constructed of contractile proteins, much like those in muscle. The filaments propel the sperm when powered by ATP.

6. Obtain a prepared slide of human sperm and view it under high power or oil immersion. Identify the head, acrosome, and tail regions. Draw and appropriately label two or three sperm in the space below.

Very often deformed sperm, for example sperm with multiple heads, or tails, are present in such preparations. Did you observe any? _____ If so,

describe them. _____

_____

7. Examine the model of spermatogenesis to identify the spermatogonia, the primary and secondary spermatocytes, the spermatids, and the functional sperm.

OOGENESIS AND THE OVARIAN CYCLE

The gonadotropic hormones produced by the anterior pituitary influence the development of ova in the ovaries and their cyclic production of female sex hormones. Within an ovary, each immature ovum develops encased in layers of specialized cells called *follicle cells;* the entire structure is called a **follicle.**

The process of oogenesis, or female gamete formation, which occurs in the ovary, is very similar to spermatogenesis occurring in the testis, but there are some important differences. The process, schematically outlined in Figure 45.3, begins with the primitive stem cells called **oogonia,** located in the germinal epithelium of the developing female fetus. During fetal development, the oogonia undergo mitosis thousands of times until their number reaches approximately 400,000. They then become encapsulated by follicle cells and form the **primary follicles** of the ovary. By the time the female child is born, most of her oogonia have increased in size and have become **primary oocytes,** which are in the prophase stage of meiosis I. Thus at birth, the total potential for producing germ cells in the female is already determined since the primitive stem-cell line no longer exists or will exist for only a brief period after birth.

After birth, the primary oocytes become quiescent until puberty when, under the influence of FSH, one or sometimes more of the follicles begin to undergo maturation approximately every 28 to 30 days. Since the reproductive life of the female is about 38 years (from ages 12 to 50) and one follicle per month matures, only 400 to 500 eggs out of a potential of 400,000 are ever released. Again nature has provided a wide margin of safety.

As a follicle grows, it begins to produce estrogen, and the primary oocyte completes its first maturation division, producing two daughter cells that are very disproportionate in size. One of these is the **secondary oocyte,** which contains nearly all of the cytoplasm in the primary oocyte; the other is the tiny **first polar body.** The first polar body then completes the second maturation division, producing two more polar

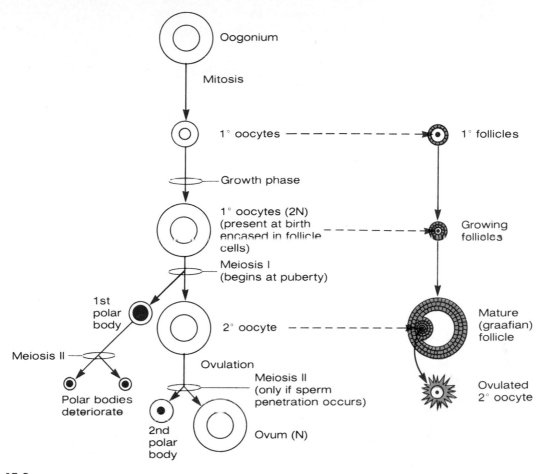

Figure 45.3

Oogenesis: left, flow sheet of meiotic events; right, correlation with follicular development and ovulation in the ovary.

bodies, which are haploid. These eventually disintegrate for lack of sustaining cytoplasm.

As the follicle containing the secondary oocyte continues to enlarge and blood levels of estrogen rise, estrogen stimulates the anterior pituitary to release luteinizing hormone (LH) and perhaps lactogenic hormone (LTH). At the same time, it inhibits the anterior pituitary production of FSH. Page 393 depicts these relationships. Approximately in the middle of the 28-day cycle, the follicle reaches the mature **graafian follicle** stage. Ovulation occurs in response to a sudden burst of LH released by the anterior pituitary. The secondary oocyte is extruded with its corona radiata intact and begins its journey down the uterine tube to the uterus. If penetrated en route by a sperm, the secondary oocyte will undergo meiosis II, producing one large **ovum** and a tiny polar body. When the second maturation division is complete, the chromosomes of the egg and sperm combine to form the diploid nucleus of the fertilized egg. If sperm penetration does not occur, the secondary oocyte simply disintegrates without ever producing the female gamete in human females.

Thus in the female, meiosis produces only one functional gamete in contrast to the four produced

in the male. Another major difference is in the relative size and structure of the functional gametes. The sperm are tiny and are equipped with tails for locomotion. They have virtually no nutrient-containing cytoplasm and the nutrients contained in semen are thus essential to their survival. In contrast, the egg is a relatively large nonmotile cell, well stocked with cytoplasmic reserves that nourish the developing embryo until implantation can be accomplished.

Once the secondary oocyte has been extruded from the ovary, LH transforms the ruptured follicle into the **corpus luteum,** which begins producing progesterone. Like estrogen, progesterone inhibits FSH release by the anterior pituitary. As FSH declines, its stimulatory effect on follicular production of estrogen ends, and estrogen blood levels begin to decline. Since increased estrogen levels triggered LH release by the anterior pituitary, lower estrogen levels result in the decline of LH in the blood. Since corpus luteum secretory function is maintained by high blood levels of LH, as LH blood levels begin to decline toward the end of the 28-day cycle, progesterone production ceases, the corpus luteum begins to degenerate and is replaced by scar tissue.

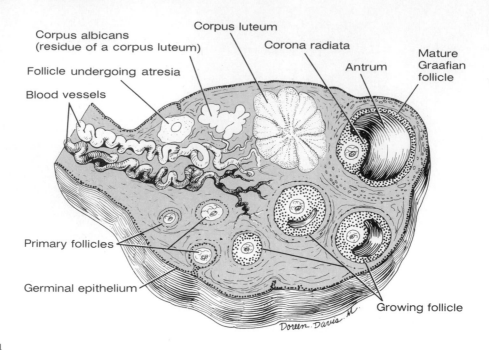

Figure 45.4

Diagrammatic view of a human ovary.

Since many different stages of ovarian development exist within the ovary at any one time, a single microscopic preparation will contain follicles at many different identifiable stages of development. Obtain a cross section of ovary tissue, and identify the following structures. Refer to Figure 45.4 as you work.

Germinal epithelium: outermost layer of the ovary.

Primary follicle: one layer of follicle cells surrounding the larger central developing ovum.

Growing follicles: follicles consisting of two or more layers of follicle cells surrounding the central developing ovum.

Graafian or **mature follicle:** at this stage of development, the follicle has formed a central cavity (**antrum**) containing fluid produced by the follicle cells. The developing ovum is pushed to one side of the follicle and is surrounded by a capsule of several layers of follicle cells called the **corona radiata** (radiating crown). When the immature ovum is released, it enters the uterine tubes with its corona radiata intact. The connective tissue stroma (background tissue) adjacent to the mature follicle forms a capsule that encloses the follicle and is called the **theca.**

Corpus luteum: a solid glandular structure or a structure containing a scalloped lumen.

Examine the model of oogenesis and compare it with the spermatogenesis model. Note differences in the size and structure of the functional gametes.

THE MENSTRUAL CYCLE

The menstrual cycle, sometimes referred to as the uterine cycle, is hormonally controlled by the ovarian production of estrogen and progesterone. It is normally divided into three stages: menstrual, proliferative, and secretory. The stages shown in the bottom portion of Figure 45.5 are described as follows:

Menstrual stage: approximately days 1 to 5; sloughing off of the thick endometrial lining of the uterus, accompanied by bleeding.

Proliferative stage: approximately days 6 to 14. Under the influence of estrogens produced by the growing follicle of the ovary, the endometrium is repaired, glands and blood vessels proliferate, and the endometrium thickens. Ovulation occurs at the end of this stage.

Secretory stage: approximately days 15 to 28. Under the influence of progesterone produced by the corpus luteum, the vascular supply to the endometrium increases still more, and the glands increase in size and begin to secrete nutrient substances, which sustain a developing embryo, if present, until implantation can occur. If fertilization has occurred, the embryo will produce a hormone much like LH, which will maintain the function of the corpus luteum. Otherwise, as the corpus luteum begins to deteriorate, lack of ovarian hormones in the blood causes blood vessels supplying the

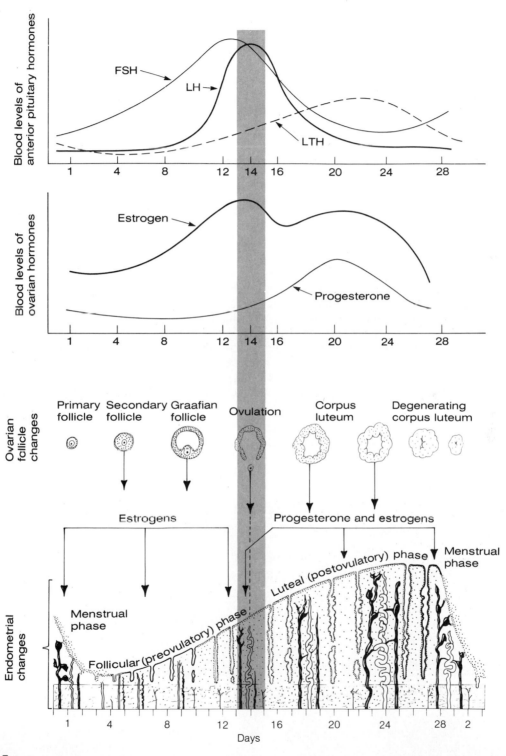

Figure 45.5

Hormonal interactions of the female cycles. Relative levels of anterior pituitary hormones correlated with follicular and hormonal changes of the ovary. Menstrual cycle also depicted.

endometrium to kink and become spastic, settting the stage for menstruation to begin by the 28th day.

Although the foregoing explanation assumes a classic 28-day cycle, the length of the menstrual cycle is highly variable, sometimes as short as 21 days or as long as 38. Only one interval is relatively constant in all females: the time from ovulation to the onset of menstruation is almost always 14 to 15 days.

Obtain slides showing the menstrual, secretory, and proliferative phases of the uterine endometrium. Observe each carefully, comparing their relative thicknesses and vascularity.

DEMONSTRATION OF OOGENESIS IN ASCARIS (OPTIONAL)

Oogenesis in mammalian eggs is difficult to demonstrate in a laboratory situation, but the process of oogenesis (and mechanics of meiosis) may be studied rather easily in the transparent eggs of *Ascaris megalocephala,* an invertebrate roundworm parasite found in the intestine of mammals. Such a study is significant, since the process in *Ascaris* is much* the same as in humans and, since its diploid chromosome number is 4, the chromosomes are easily counted.

Go to the demonstration area where the slides are set up and make the following observations:

---

*In *Ascaris,* meiosis does not begin until the sperm has penetrated the primary oocyte, whereas in humans, meiosis I occurs before sperm penetration.

1.   Scan the first demonstration slide to identify the primary oocyte. It will have what appears to be a relatively thick cell membrane; this is the fertilization membrane that the oocyte produces after sperm penetration. Find and study a primary oocyte that is undergoing the first maturation division. Look for a barrel-shaped spindle with two tetrads (two groups of four beadlike chromosomes) on it. Most often the spindle is located at the periphery of the cell. (The sperm nucleus may or may not be seen, depending on how the cell was cut.)

2.   Observe slide 2. Locate a cell in which half of each tetrad (a dyad) is being extruded from the cell surface into the first polar body. Both the polar body and the larger cell (now the secondary oocyte) are diploid.

3.   On slide 3 attempt to locate a secondary oocyte in the second maturation division. In this view, two dyads (two groups of two beadlike chromosomes) will be seen on the spindle.

4.   Locate a cell in which the second polar body is being formed on slide 4. In this case, both it and the ovum will now contain two chromosomes, the haploid number for *Ascaris.*

5.   On the fifth slide, identify a fertilized egg or a cell in which the sperm and ovum nuclei (actually pronuclei) are fusing to form a single nucleus containing four chromosomes.

Survey of Embryonic Development

EXERCISE 46

OBJECTIVES

1. To define *fertilization* and *zygote.*

2. To define *cleavage* and discuss its function.

3. To define and discuss the *gastrulation* process.

4. To name the three primary germ layers and discuss the importance of each.

5. To differentiate between the blastula and gastrula forms of the sea urchin and human when provided with appropriate models or diagrams.

6. To identify blastocyst and/or chorionic vesicle.

7. To identify the following structures of a human chorionic vesicle when provided with an appropriate diagram, and to state the function of each.

inner cell mass	trophoblast
amnion	allantois
yolk sac	trophoblast villi

8. To describe the process and timing of implantation in the human.

9. To define *decidua basalis* and *decidua capsularis.*

10. To state the germ-layer origin of several body organs and organ systems of the human: circulatory system, nervous system, skeleton, skeletal muscles, and lining of the digestive and respiratory tracts.

11. To describe developmental direction.

12. To describe the structure and function of the human placenta.

MATERIALS

Prepared slides of sea urchin development (zygote through larval stages)

Compound microscope

Human development models or plaques (if available)

Life Before Birth, Educational Reprint #27, available from Time-Life Educational Materials, Box 834, Radio City P.O., New York, N.Y. 10019

Pregnant cat, rat, or pig uterus (one per laboratory session) with uterine wall dissected to allow student examination

Dissecting instruments

Model of pregnant human torso

Preserved human embryos (if available)

Fresh or formalin-preserved placenta (obtained from a clinical agency)

Microscope slide of placenta tissue

EARLY EMBRYOLOGY OF THE SEA URCHIN AND THE HUMAN

Because reproduction is such a familiar event, we tend to lose sight of the wonder of the process. One part of that process, the development of the embryo, is the concern of embryologists who study the changes in structure that occur from the time of fertilization until the time of birth.

Early development in all animals involves three basic types of activities, which are integrated to ensure the formation of a viable offspring: (1) an increase in cell number and subsequent cell growth; (2) cellular specialization; and (3) morphogenesis, the formation of functioning organ systems.

This exercise provides a rather broad overview of the changes in structure that take place during embryonic development in the sea urchin. The pattern of changes in this marine animal

395

provides a basis for comparison with developmental events in the human.

Microscopic Study of Sea Urchin Development

1. Obtain a compound microscope and a set of slides depicting embryonic development of the sea urchin. Draw simple diagrams of your observations.

2. Observe the fertilized egg, or zygote, which appears as a single cell immediately surrounded by a fertilization membrane and a jelly-like membrane. After an egg is penetrated by a sperm, the egg and the sperm nuclei fuse to form a single nucleus. This process is called **fertilization.** Within 2 to 5 minutes after sperm penetration, a fertilization membrane forms beneath the jelly coat to prevent the entry of additional sperm. Label the fertilization and jelly membranes and the zygote.

2-cell stage   4-cell stage

Fertilized egg

8-cell stage   16-cell stage

3. Observe the cleavage stages. Once fertilization has occurred, the zygote begins to divide, forming a mass of successively smaller and smaller cells, called **blastomeres.** This series of mitotic divisions without intervening growth periods is referred to as **cleavage,** and it results in a multicellular embryonic body. As the division process continues, a solid ball of cells forms. (At the 32-cell stage, it is called **morula,** and the embryo resembles a raspberry in form.) Then the cell mass hollows out to become the embryonic form called the **blastula,** which is a ball of cells surrounding a central cavity. The blastula is the final product of cleavage.

The cleavage stage of embryonic development provides a large number of building blocks (cells) with which to fashion the forming body. (If this is a little difficult to understand, consider trying to build a structure with a huge block of granite rather than with small bricks.)

Diagram the 2-, 4-, 8-, and 16-cell stages as you observe them.

Identify and sketch the blastula stage of cleavage—a ball of cells with an apparently lighter center, owing to the presence of the central cavity.

Blastula

4. Identify the early **gastrula** form (which follows the blastula in the developmental sequence). The gastrula looks as if one end of the blastula has been indented or pushed into the central cavity, forming a two-layered embryo. In time, a third layer of cells appears between the initial two cell layers. Thus, as a result of gastrulation, a three-layered embryo forms, each layer corresponding to a primary germ layer from which all body tissues develop. The innermost layer, the **endoderm,** and the middle layer, the **mesoderm,** form the internal organs; the outermost layer, the **ectoderm,** forms the surface tissues of the body.

Draw a gastrula below. Label the ectoderm and endoderm. If you can see the third layer of cells, the mesoderm, budding off between the other two layers, label that also.

Gastrula

5. Gastrulation in the sea urchin if followed by the appearance of the free-swimming larval form, in which the three germ layers have differentiated into the various tissues and organs of the animal's body. The larvae exist for a few days in the unattached form and then settle to the ocean bottom to attach and develop into the sessile adult form. If time allows, observe the larval form on the prepared slides. Observations need not be recorded.

Developmental Stages of the Human

1. Go to the demonstration area where the models of human development are on display. (If these are not available, use Figure 46.1 for this study.) Observe the human development models and respond to the questions posed below.

Is the observed human cleavage process similar

to that in the sea urchin? _____

Why do you suppose this is so? _____

_____

_____

_____

_____

2. Observe the blastula, which is called the **blastocyst** or **chorionic vesicle** in the human. Unlike the sea urchin, only a portion of the blastula cells in the human contribute to the formation of the embryonic body—those seen on the top of the blastocyst forming the so-called **inner cell mass (ICM).** The rest of the blastocyst enclosing the central cavity and overriding the ICM is referred to as the **trophoblast.** The trophoblast becomes an extraembryonic membrane called the **chorion,** which forms the fetal portion of the **placenta.**

3. Observe the implanting blastocyst shown on the model or in the figure. By approximately the seventh day after fertilization, the developing embryo is in the blastocyst stage and is floating free in the uterine cavity. About that time, the blastocyst becomes attached to the uterine wall over the ICM area, and implantation begins. The trophoblast cells secrete enzymes that erode the uterine mucosa at the point of attachment to reach the vascular supply in the submucosa. By the fourteenth day after fertilization, implantation is completed and the uterine mucosa has grown over the burrowed-in embryo. The portion of the uterine wall beneath the ICM destined to take part in placenta formation is called the **decidua basalis** and that surrounding the rest of the blastocyst is called the **decidua capsularis.** Identify these regions. By the time implantation has been completed, embryonic development has progressed to the **gastrula** stage, and the three primary germ layers are present and are beginning to differentiate. Within the next 6 weeks, virtually all of the body organ systems will have been laid down at least in rudimentary form by the germ layers. The ectoderm gives rise to the epidermis of the skin and the nervous system; the endoderm forms the mucosa of the digestive tract and associated structures; and the mesoderm forms virtually everything lying between the two (skeleton, walls of the digestive organs, urinary system, muscular and circulatory systems, and others). By the ninth week of development, the embryo is referred to as a **fetus.** From this point on, the major activities are growth and tissue and organ specialization. All the groundwork has been completed by the eighth week.

4. Again observe the blastocyst or chorionic vesicle to follow the formation of the extraembryonic membranes and the placenta (see Figure 46.1). Note the villus extensions of the trophoblast. By the time implantation has been completed, the trophoblast

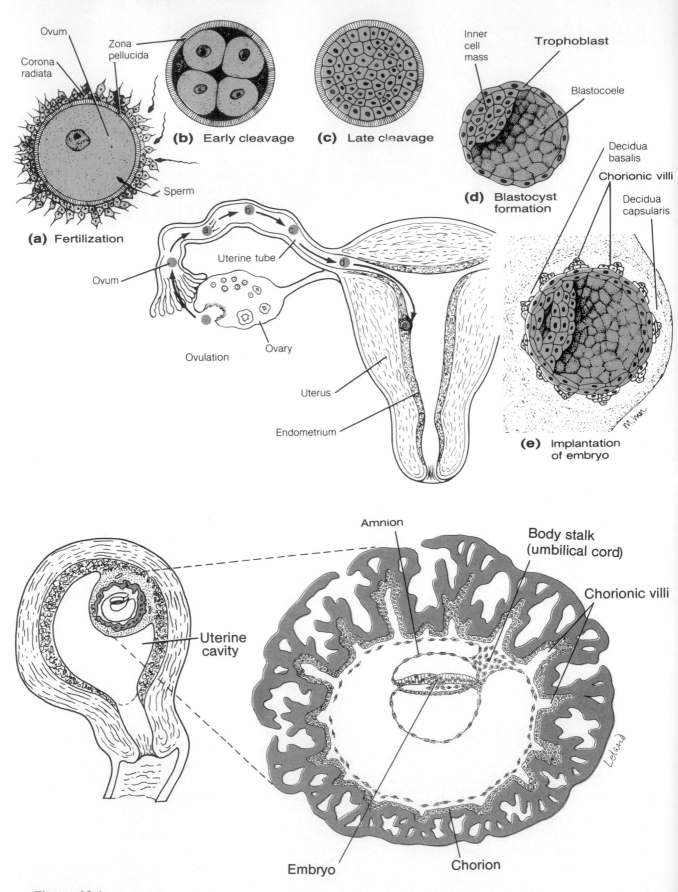

Figure 46.1

Early embryonic development of the huan: top, (a–e) from fertilization to blastocyst implantation in the uterus; below, embryo of approximately 22 days. Embryonic membranes and germ layers present.

has differentiated into the **chorion,** and the large elaborate villi extending from the chorion are lying in the blood-filled sinusoids in the uterine tissue. This composite of uterine tissue and chorionic villi is called the **placenta,** and all exchanges to and from the embryo occur through the chorionic membranes.

Three embryonic membranes (originating in the ICM) have also formed by this time: the amnion, the allantois, and the yolk sac. Identify each. The **amnion** encases the young embryonic body in a fluid-filled chamber that protects the embryo from mechanical trauma and prevents adhesions during rapid embryonic growth. The **yolk sac** in humans has lost its original function, which was to pass nutrients to the embryo after digesting the yolk mass, primarily because the placenta has taken over that task. (Also, the human egg has very little yolk.) However, the yolk sac is not totally useless since the embryo's first blood cells originate here, and the primordial germ cells migrate from it into the embryo's body to seed the gonadal tissue. The **allantois,** which protrudes from the posterior end of the yolk sac, is also a redundant structure in humans because of the placenta. In birds and reptiles, it is a repository for embryonic wastes; in the human, it is the structural basis on which the mesoderm migrates to form the body stalk, or **umbilical cord,** which attaches the embryo to the placenta.

5. Go to the demonstration area to view the photographic series, *Life Before Birth.* This series, by Lennart Nilsson, illustrates human development in a way you will long remember. After viewing it, respond to the following questions.

In your own words, what do the chorionic villi look like? _____

_____

_____

What organs or organ systems appear *very* early in

embryonic development? _____

_____

_____

_____

Does development occur in a rostral to caudal (head

to toe) direction, or vice versa? _____

_____

Does development occur in a distal-proximal direc-

tion, or vice versa? _____

Does spontaneous movement occur *in utero?* _____

_____

How does the mother recognize this? _____

_____

_____

The very young embryo has been described as resembling "an astronaut suspended and floating in space." Do you think this definition is appropriate?

_____

Why or why not? _____

_____

_____

What is vernix caseosa? _____

_____

_____

What is lanugo? _____

_____

_____

IN UTERO DEVELOPMENT

1. Go to the appropriate demonstration area and observe the fetuses in the Y-shaped animal uterus. Identify the following fetal or fetal-related structures:

Placenta. (a composite structure formed from the uterine mucosa and the fetal chor-

ion). Describe its appearance. _____

_____

_____

Umbilical cord. Note its relationship to the

placenta and fetus. _____

_____

_____

Amniotic sac. Identify the transparent amnion surrounding a fetus. Open one amniotic sac and note the amount, color, and consistency of the fluid. _____

_____

_____

Remove a fetus and observe the degree of development of the head, body, and extremities.

Is the skin thick or thin? _____

_____

Basis of your response? _____

_____

_____

2. Observe the model of a pregnant human torso. Identify the placenta. How does it differ in shape

from the animal placenta observed? _____

_____

_____

Identify the umbilical cord. In what region of the uterus does implantation usually occur, as indicated by the position of the placenta?

_____

_____

What might be the consequence if it occurred

lower? _____

_____

_____

Why would a feet-first position (breech presentation) be less desirable than the position-

ing in the model? _____

_____

_____

3. If human fetuses are available for demonstration, examine them, making mental comparisons with the appearance of the dissected animal fetuses and the illustrations in the *Life Before Birth* series. *Do not pick up the demonstration cases or jiggle the embryos around.* Embryonic tissues are exceedingly soft and tend to disintegrate easily, even when properly preserved.

GROSS AND MICROSCOPIC ANATOMY OF THE PLACENTA

The placenta is a remarkable temporary organ. It is composed of maternal and fetal tissues and is responsible for providing nutrients and oxygen to the embryo and fetus while removing carbon dioxide and metabolic wastes.

1. Note that the placenta on display has two very different-appearing surfaces—one smooth and the other spongy, roughened, and torn-looking.

Which is the fetal side? _____

Basis of your conclusion? _____

_____

Identify the umbilical cord. Within the cord, identify the umbilical vein and two umbilical arteries. What is the function of the umbilical

vein? _____

_____

The umbilical arteries? _____

_____

Are any of the fetal membranes still attached?

_____

If so, which? _____

_____

2. Obtain a microscope slide of placenta tissue. Observe the placental tissue carefully, comparing it to Figure 46.2. Identify the intervillus spaces (maternal sinusoids), which are blood-filled in life. Identify the villi, and notice their rich vascular supply. Draw a small representative diagram of your observations below and label appropriately.

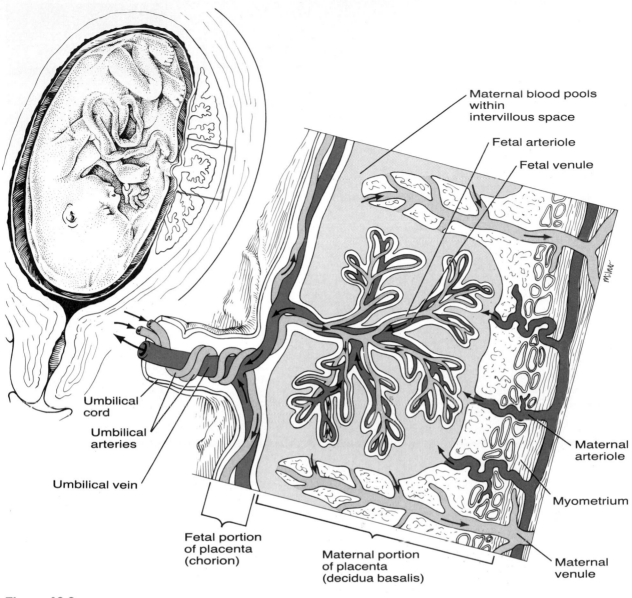

Figure 46.2

Diagrammatic representation of the structure of the placenta.

Principles of Heredity

OBJECTIVES

1. To define *allele, dominance, genotype, heterozygous, homozygous, incomplete dominance, phenotype,* and *recessiveness.*

2. To gain practice working out simple genetics problems, using a Punnet square.

3. To become familiar with the basic laws of probability.

4. To observe selected human phenotypes and determine their genotype basis.

MATERIALS

Pennies (for coin tossing)
PTC (phenylthiocarbamide) taste strips
Sodium benzoate taste strips
Chart drawn on chalkboard for tabulation of class results of human phenotype/genotype determinations
Blood typing sera: Anti-A and Anti-B; slides, toothpicks, wax pencils, sterile lancets, alcohol swabs

---

The field of genetics currently is bristling with excitement. Complex gene-splicing techniques have allowed researchers to precisely isolate genes coding for specific proteins and then to insert those genes into bacterial or tumor cells so that large amounts of particular proteins can be harvested. At present, growth hormone, insulin, and interferon, produced by these genetic engineering techniques, are undergoing clinical trials to determine their effectiveness (and possible side effects) in humans.

Comprehension of the science of genetics in relation to such studies is certainly complex and requires arduous training. However, a basic understanding and appreciation of how genes regulate our various traits (dimples and hair color, for example) can be gained by anyone. The thrust of this exercise is to provide a "genetics sampler" or relatively simple introduction to the principles of heredity.

INTRODUCTION TO THE LANGUAGE OF GENETICS

In humans all cells, except eggs and sperm, contain 46 chromosomes, that is, the diploid number. This number is established when fertilization occurs as a result of the fusion of the egg and sperm and the combination of the 23 chromosomes (or haploid complement) each is carrying. The diploid chromosomal number is maintained throughout life in nearly all cells of the body by the precise process of mitosis. As explained in Exercise 45, the diploid chromosomal number actually represents two complete (or nearly complete) sets of genetic instructions—one from the egg and the other from the sperm—or 23 pairs of *homologous chromosomes.*

Genes coding for the same traits on each pair of homologous chromosomes are called **alleles.** The alleles may be identical or different in their influence. For example, the members of the gene pair, or alleles, coding for hairline shape on your forehead may specify either straight across or widow's peak. When both alleles in a homologous chromosome pair have the same expression, the individual is said to be **homozygous** for that trait. When the alleles differ in their expression, the individual is **heterozygous** for the given trait and only one of the alleles, called the **dominant gene,** will exert its effects. The allele with less potency, the **recessive gene,** while still present, will be masked. While dominant genes, or alleles, will exert their effects in both homozygous and heterozygous conditions, recessive alleles *must* be present in double dose (homozygosity) to exert their influence. An individual's actual genetic makeup, that is, whether he is homozygous or heterozygous for the various alleles, is called **genotype.** The manner in which genotype is expressed (for example, the presence of a widow's peak or not, blue vs. brown eyes) is referred to as **phenotype.**

The complete story of heredity is much more complex than just outlined, and in actuality the expression of many traits (for example, eye color) is determined by the interaction of many allele pairs. However, our emphasis here will be to investigate only the less complex aspects of genetics.

DOMINANT-RECESSIVE INHERITANCE

One of the best ways to master the terminology and learn the principles of heredity is to work out the solutions to some genetic crosses much in the manner Mendel did in his classic experiments on pea plants. To work out the various simple monohybrid (one pair of alleles) crosses in this exercise, you will be given the genotype of the parents. You will then determine the possible genotypes of their offspring by using the Punnet square, and you will record both genotype and phenotype percentages. To illustrate the procedure, an example of one of Mendel's pea plant crosses is outlined next.

Alleles: T (determines *tallness;* dominant)
t (determines *dwarfness;* recessive)
Genotypes of parents: TT ($\male$) × tt ($\female$)
Phenotypes of parents: Tall × dwarf

To use the punnet or checkboard square, write the alleles (actually gametes) of one parent across the top and the gametes of the other parent down the left side. Then combine the gametes across and down to achieve all possible combinations as shown below:

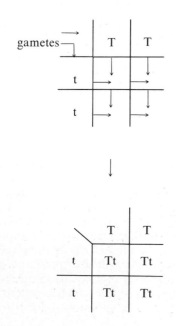

Results: Genotypes 100% Tt (all heterozygous)
Phenotypes 100% Tall (since T, which determines tallness, is dominant and all contain the T allele).

1. Using the technique outlined above, determine the genotypes and phenotypes of the offspring of the following crosses:

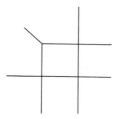

a. Genotypes of parents: Tt ($\male$) × tt ($\female$)

% of each genotype: _____

% of each phenotype: _____ % tall

_____ % dwarf

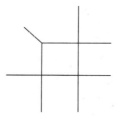

b. Genotypes of parents: Tt ($\male$) × Tt ($\female$)

% of each genotype: _____

% of each phenotype: _____ % tall

_____ % dwarf

2. In guinea pigs, rough coat (R) is dominant over smooth coat (r). What will be the genotypes and phenotypes of the following monohybrid crosses?

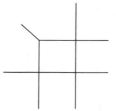

a. Genotypes of the parents: RR × rR

% of each genotype: _____

% of each phenotype: _____ % rough

_____ % smooth

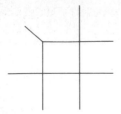

b. Genotypes of the parents: Rr × rr

% of each genotype: _____

% of each phenotype: _____ % rough

_____ % smooth

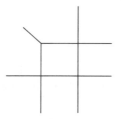

c. Genotypes of the parents: RR × rr

% of each genotype: _____

% of each phenotype: _____ % rough

_____ % smooth

INCOMPLETE DOMINANCE

In actuality, the concepts of dominance and recessiveness are somewhat arbitrary and artificial in some instances, since so-called dominant genes may be expressed differently in homozygous and heterozygous individuals. This gives rise to a condition called *incomplete dominance* or *intermediate inheritance.* In such cases, both alleles express themselves in their offspring. The crosses are worked out in the same manner as indicated previously, but the heterozygous offspring exhibit a phenotype intermediate between that of the homozygous individuals. Work the following problems:

1. The inheritance of flower color in snapdragons illustrates the principle of lack of dominance. The genotype RR is expressed as a red flower, Rr yields pink flowers, and rr produces white flowers. Work out the following problems to determine phenotypes seen and both genotype and phenotype percentages.

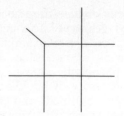

a. Genotypes of parents: RR × rr

Genotype %: _____

Phenotypes and %: _____

_____

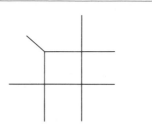

b. Genotypes of parents: Rr × rr

Genotype %: _____

Phenotypes and %: _____

_____

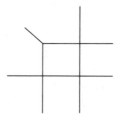

c. Genotypes of parents: Rr × Rr

Genotype %: _____

Phenotypes and %: _____

_____

2. In humans, the inheritance of sickle cell anemia/trait is determined by a single pair of alleles that exhibit incomplete dominance. Individuals homozygous for the sickling gene (s) are said to have *sickle cell anemia.* In double dose (ss) the sickling gene causes a very abnormal hemoglobin to be made, which crystallizes and becomes sharp and spiky under conditions of oxygen deficit. This, in turn, leads to the clumping and hemolysis of red blood cells in the circulation, which causes a great deal of pain and is often fatal. Heterozygous individuals (Ss) are said to have the *sickle cell trait* which is much less severe; however, they are carriers for the abnor-

mal gene and may pass it on to their offspring. Individuals with the genotype SS form normal hemoglobin. Work out the following crosses:

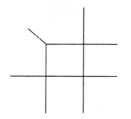

a. Parental genotypes: SS × ss

Genotype %: _____

Phenotypes and %: _____

_____

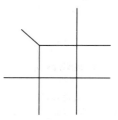

b. Parental genotypes: Ss × Ss

Genotype %: _____

Phenotypes and %: _____

_____

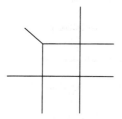

c. Parental genotypes: ss × Ss

Genotype %: _____

Phenotypes and %: _____

_____

SEX-LINKED INHERITANCE

Of the 23 pairs of homologous chromosomes, 22 pairs are referred to as *autosomes*. Autosomes contain genes that determine most body (somatic) characteristics. The 23rd pair, called the *sex chromosomes*, determine the sex of an individual, that is,

whether an individual will be male or female. Normal females possess two sex chromosomes that look alike, the X chromosomes. Males possess two dissimilar sex chromosomes, referred to as X and Y. Possession of the Y chromosome determines maleness. A photomicrograph of a male's chromosome complement (male karyotype) is shown in Figure 47.1. The Y sex chromosome is only about a third as large as the X sex chromosome, and lacks many of the genes (directing characteristics other than sex) that are found on the X. Genes present *only* on the X sex chromosome are called *sex-linked* (or X-linked) genes. Some examples of X-linked genes include those that determine normal color vision (or, conversely, color blindness), normal clotting ability (as opposed to hemophilia, or bleeder's disease), and retention of a full head of hair (as opposed to balding). The alleles that determine color blindness, hemophilia, and balding are all recessive alleles. In females, *both* X chromosomes must carry the recessive alleles for a woman to express any of these conditions, and thus they tend to be infrequently seen. However, should a male receive even one sex-linked recessive allele for these conditions, he will exhibit the recessive phenotype because his Y chromosome lacks any genes that might dominate or mask the recessive allele. The critical understanding of X-linked inheritance is the *absence* of male to male (that is, father to son) transmission of X-linked genes. The X of the father *will* pass to each of his daughters but to none of his sons. Since males always inherit sex-linked conditions from their mothers (through the X chromosome), a young man should look at his maternal grandfather (rather than at his father) to predict whether his hair will soon be thinning!

1. A heterozygous woman carrying the recessive gene for color blindness marries a man who is color-blind. Assume the dominant gene is X^C (allele for normal color vision) and the recessive gene is X^c (determines color blindness). The mother's genotype is $X^C X^c$ and the father's is $X^c Y$. Do a punnet square to determine the answers to the following questions.

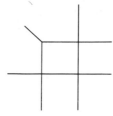

According to the laws of probability, what percent of their children will be color-blind?

_____ %

What is the proportion of color-blind individuals by sex? _____ male; _____ female

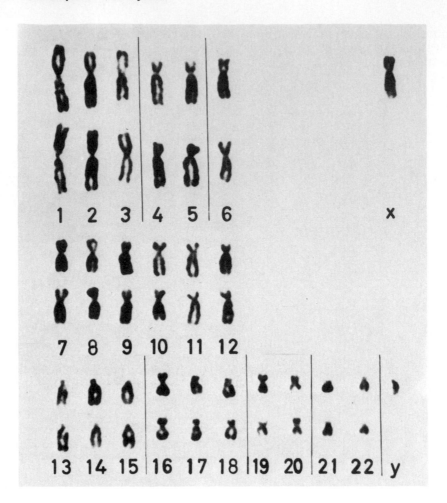

Figure 47.1

Karyotype of human male chromosomes. Each pair of homologous
chromosomes is numbered (1–22) except the sex chromosomes, which are identified by their letters, X and Y. (Courtesy of
T. T. Puck. 1972. *The Mammalian Cell as Microorganism.* San Francisco, CA: Holden-Day.)

What percentage will be carriers? _____ %

What is the sex of the carriers? _____

2. A heterozygous woman carrying the recessive
gene for hemophilia marries a man who is not
a hemophiliac. Assume the dominant gene is
X^H and the recessive gene is X^h. The woman's
genotype is $X^H X^h$ and her husband's genotype
is $X^H Y$. What is the potential percentage and
sex of their offspring that will be hemophiliacs?

_____ % males; _____ % females

What percentage can be expected to neither
exhibit nor carry the allele for hemophilia?

_____ %

What is the sex and percentage of individuals
(anticipated) that will be carriers for hemophilia?

_____ %; _____ sex

PROBABILITY

Since the segregation (or parceling out) of
chromosomes to daughter cells (gametes) during
meiosis and the combination of egg and sperm
are random or chance events, the possibility that
certain genomes will arise and be expressed is
based on the laws of probability. The randomness
of gene recombination from each parent determines individual uniqueness and explains why
siblings, however similar, never have totally corresponding traits (unless, of course, they are
identical twins). The Punnet square method that
you have been using to work out the genetics
problems actually provides information on the
probability of appearance of certain genotypes
considering all possible events.

Probability (P) is defined as:

$$P = \frac{\text{number of specific events/cases}}{\text{total number of events/cases}}$$

If an event is certain to happen, its probability is 1. If it happens one out of every two times, its probability is ½; if one out of 4 times, its probability is ¼, and so on.

When figuring the probability of separate events occurring together (or consecutively), the probabilities of each event must be multiplied together to get the final probability figure. For example, the probability of a penny coming up "heads" in each toss is ½ (since it has two sides—heads and tails). But the probability of a tossed penny coming up heads four times in a row is: ½ × ½ × ½ × ½ = 1/16.

1. Obtain two pennies and perform the following simple experiment to explore the laws of probability.
   a. Toss one penny into the air 10 times, and record the number of heads/tails observed.

   _____ heads _____ tails

   Probability: ____/10ths tails; ____/10ths heads

   b. Now toss two pennies into the air together for 24 tosses, and record the results of each toss below. In each case, report the probability in the lowest fractional terms.

   # HH _____ Probability _____

   # HT _____ Probability _____

   # TT _____ Probability _____

   Does the first toss have any influence on the

   second? _____

   Does the third toss have any influence on the

   fourth? _____

   c. Do a punnet square, using HT for one coin and HT for the alleles of the other.

   Probability of HH: _____

   Probability of HT: _____

   Probability of TT: _____

   How closely do your coin-tossing results correlate with the percentages obtained from the

   Punnet square results? _____

   _____

2. Determine the probability of having a boy or girl offspring for each conception.
   Parental genotypes: XY × XX

   Probability of males: _____ %

Probability of females: _____ %

3. Dad wants a baseball team! What are the chances of his having nine sons in a row?

   _____ (Sorry, Dad!)

GENETIC DETERMINATION OF SELECTED HUMAN CHARACTERISTICS

Many human traits are determined by a single pair of alleles easily identifiable by observation. For each of the characteristics described here, determine (as best you can) both your own phenotype and genotype, and record this information on Table 47.1. Since it is impossible to know whether you are homozygous or heterozygous for a trait when you exhibit its dominant expression, you are to record your genotype as A— (or B—, and so on, depending on the letter used to indicate the alleles) in such cases. If you exhibit the recessive trait, you are homozygous for the recessive allele and should record it accordingly as aa (bb, cc, and so on). When you have completed your observations, also record your data on the chart on the chalkboard, which has been set up for tabulation of class results.

Tongue rolling: extend your tongue and attempt to roll it into a U-shape longitudinally. People with this ability have the dominant allele for this trait. Use T for the dominant allele, and t for the recessive allele (see Figure 47.2).

Attached earlobes: have your lab partner examine your earlobes. If no portion of the lobe hangs free inferior to its point of attachment to the head, you are homozygous recessive (ee) for attached earlobes. If part of the lobe hangs free below the point of attachment, you possess at least one dominant gene (E) (see Figure 47.2).

Interlocking fingers: clasp your hands together by interlocking your fingers. Now observe your clasped hands. Which thumb is uppermost? If the left thumb is uppermost, you possess a dominant allele (I) for this trait. If you clasped your right over your left thumb, you are illustrating the homozygous recessive (ii) phenotype.

PTC taste: obtain a PTC taste strip. PTC or phenothiocarbamide is a harmless chemical that some people can taste and others find tasteless. Chew the strip. If it tastes slightly bitter, you are a "taster" and possess the dominant gene (P) for this trait. If you can not taste anything, you are a nontaster and are homozygous recessive (pp) for the trait. Approximately 70% of the United States' population are tasters.

Sodium benzoate taste: obtain a sodium benzoate taste strip and chew it. A different pair of alleles

TABLE 47.1 Record of Human Genotypes/Phenotypes

Characteristic	Phenotype	Genotype
Tongue rolling (T,t)		
Attached earlobes (E,e)		
Interlocking fingers (I,i)		
PTC taste (P,p)		
Sodium benzoate taste (S,s)		
Sex (X,Y)		
Dimples (D,d)		
Widow's peak (W,w)		
Bent little finger (L,l)		
Double-jointed thumb (J,j)		
Middigital hair (H,h)		
Freckles (F,f)		
Blaze (B,b)		
ABO blood type (I^A, I^B, i)		

determines the ability to taste sodium benzoate (as opposed to PTC taste). If you can taste it, you have at least one of the dominant alleles (S). If not, you are homozygous recessive (ss) for the trait. Also record whether sodium benzoate tastes salty, bitter, or sweet to you (if a taster). Even though PTC and sodium benzoate taste are inherited independently, they interact to determine a person's taste sensations. Individuals who find PTC bitter and sodium benzoate salty tend to be devotees of sauerkraut, buttermilk, spinach, and other slightly bitter or salty foods.

Sex: the genotype XX determines the female phenotype, whereas XY determines the male phenotype.

Dimpled cheeks: the presence of dimples in one or both cheeks is due to a dominant gene (D). Absence of dimples indicates the homozygous recessive condition (dd) (see Figure 47.2).

Widow's peak: a distinct downward V-shaped hairline at the middle of the forehead is referred to as a widow's peak. It is determined by a dominant allele (W), whereas the straight or continuous forehead hairline is determined by the homozygous recessive condition (ww) (see Figure 47.2).

Bent little finger: examine your little finger on each hand. If its terminal phalanx angles toward the ring finger, you are dominant for this trait. If one or both terminal digits are essentially

straight, you are homozygous recessive for the trait. Use L for the dominant allele and l for the recessive allele.

Double-jointed thumb: a dominant gene determines a condition of loose ligaments that allows one to throw the thumb out of joint. The homozygous recessive condition determines tight joints. Use J for the dominant allele and j for the recessive allele.

Middigital hair: critically examine the dorsum of the middle segment (phalanx) of your fingers. If *no* hair is obvious, you are recessive (hh) for this condition. If hair is seen, you have the dominant gene (H) for this trait (which, however, is determined by multigene inheritance) (see Figure 47.2).

Freckles: the appearance of freckles is a result of a dominant gene. Use F as the dominant allele and f as the recessive allele (see Figure 47.2).

Blaze: a lock of hair different in color from the rest of scalp hair is called a blaze; it is determined by a dominant gene. Use B for the dominant gene and b for the recessive gene.

Blood type: inheritance of ABO blood type is based on the existence of 3 alleles designated as I^A, I^B, and i. Both I^A and I^B are dominant over i, but neither is dominant over each other. Thus the possession of I^A and I^B will yield type AB blood, whereas the possession of the I^A and i alleles will yield type A blood, and so on as

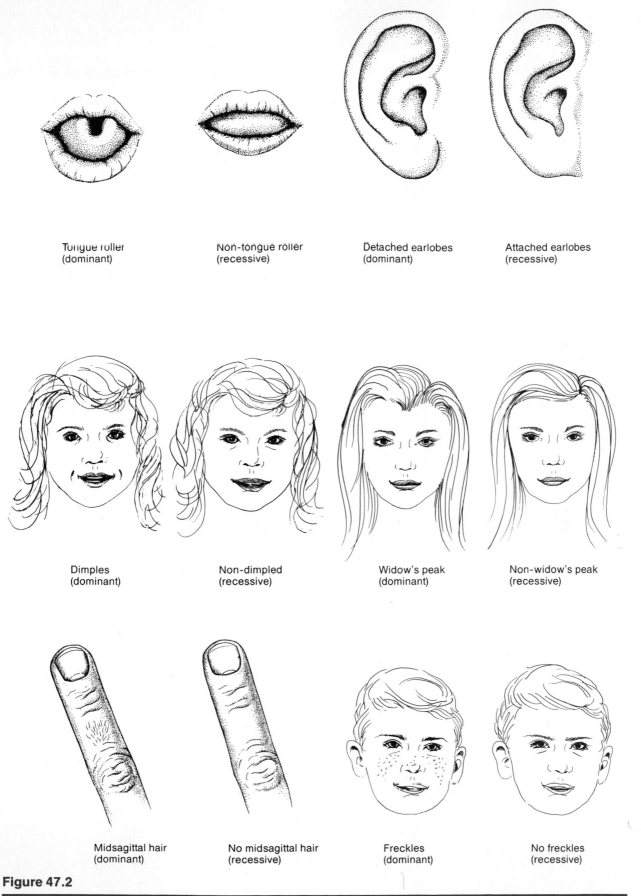

Tongue roller
(dominant)

Non-tongue roller
(recessive)

Detached earlobes
(dominant)

Attached earlobes
(recessive)

Dimples
(dominant)

Non-dimpled
(recessive)

Widow's peak
(dominant)

Non-widow's peak
(recessive)

Midsagittal hair
(dominant)

No midsagittal hair
(recessive)

Freckles
(dominant)

No freckles
(recessive)

Figure 47.2

Selected examples of human phenotypes.

explained in Exercise 29. There are four ABO blood groups or phenotypes, A, B, AB, and O, and their correlation to genotype is indicated as follows:

ABO blood group	Genotype
A	$I^A I^A$ or $I^A i$
B	$I^B I^B$ or $I^B i$
AB	$I^A I^B$
O	ii

Assuming you have previously typed your blood, record your phenotype and genotype on the chart. If not, type your blood following the instructions on pages 260–261, and then enter your results in the table. Once class data has been tabulated, scrutinize the results. Is there a single trait that is expressed in an identical manner by all members of the class?

Since all human beings have 23 pairs of homologues and each pair segregates independently at meiosis, the number of possible types of segregations includes over 5 million combinations! On the basis of this information, what would you guess are the chances of *any* two individuals in the class having identical phenotypes on all 14 traits investigated?

EXERCISE 1

The Language of Anatomy

Body Orientation, Direction, Planes, and Sections

1. Describe completely the standard human anatomic position? _____

_____

2. Define section: _____

3. Several incomplete statements are listed below. Correctly complete each statement by choosing the anatomic term from the key. Record the key letters in the same numbered blanks below.

Key: a. anterior   e. lateral   i. sagittal
           b. distal   f. medial   j. superior
           c. frontal   g. posterior   k. transverse
           d. inferior   h. proximal

In the anatomic position, the face and palms are on the _1_ body surface; the buttocks and shoulder blades are on the _2_ body surface; and the top of the head is the most _3_ part of the body. The ears are _4_ to the shoulders and _5_ to the nose. The heart is _6_ to the vertebral column (spine) and _7_ to the lungs. The elbow is _8_ to the fingers but _9_ to the shoulder.

The abdominopelvic cavity is _10_ to the thoracic cavity and _11_ to the spinal cavity. In humans, the dorsal surface can also be called the _12_ surface; however, in quadruped animals, the dorsal surface is the _13_ surface.

If an incision cuts the heart into right and left parts, the section is a _14_ section, but if the heart is cut so that the superior and inferior portions result, the section is a _15_ section. You are told to cut a dissection animal along two planes so that the kidneys are observable in both sections. The two sections that meet this requirement are the _16_ and _17_ sections.

1. _____   5. _____   9. _____   13. _____   17. _____

2. _____   6. _____   10. _____   14. _____

3. _____   7. _____   11. _____   15. _____

4. _____   8. _____   12. _____   16. _____

Surface Anatomy

1. In relation to the abdominal area, where would the following be located?

epigastric area _____

lumbar areas _____

hypochondriac areas _____

2. Indicate the following body areas on the accompanying diagram by placing the correct key letter at the end of each leader line.

Key:

a. abdominal
b. axillary
c. brachial
d. buccal
e. calf
f. cervical
g. cubital
h. femoral
i. gluteal
j. groin
k. lumbar
l. occipital
m. popliteal
n. pubic
o. thoracic
p. umbilical

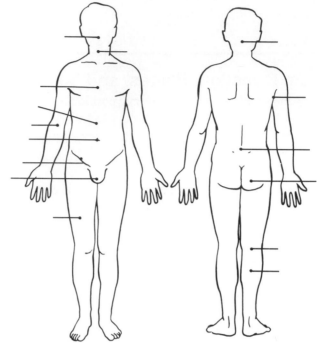

Body Cavities

1. Which body cavity would have to be opened for the following types of surgery? (Insert letter of key choice in same numbered blank.)

Key:
a. abdominopelvic   c. dorsal   e. thoracic
b. cranial   d. spinal   f. ventral

1. coronary bypass surgery
2. removal of the uterus or womb
3. removal of a brain tumor
4. appendectomy
5. stomach ulcer operation

The abdominopelvic and thoracic cavities are subdivisions of the _6_ body cavity, while the cranial and spinal cavities are subdivisions of the _7_ body cavity. The _8_ body cavity is totally surrounded by bone, and thus affords its contained structures very good protection.

2. *Name* the serous membranes covering the lungs (#9), the heart (#10), and the organs of the abdominopelvic cavity (#11), and insert your responses in the blanks on the right.

3. What muscle subdivides the ventral body cavity? (#12)

1. _____
2. _____
3. _____
4. _____
5. _____
6. _____
7. _____
8. _____
9. _____
10. _____
11. _____
12. _____

4. Which of the following organ systems are represented in all three subdivisions of the ventral body cavity? (Circle all appropriate responses.)

respiratory   circulatory   reproductive
nervous   excretory (urinary)   muscular

5. Which organ system would not be represented in any of the body cavities? _____

6. What are the bony landmarks of the abdominopelvic cavity? _____

_____

_____

7. Which body cavity affords the least protection to its internal structures? _____

8. What is the function of the serous membranes of the body? _____

_____

_____

_____

9. A nurse informs you that she is about to take blood from the cubital region. What portion of your body should you present to her? _____

10. What is the name given to the extension of the peritoneum holding the digestive organs in position?

_____

11. What do peritonitis, pleurisy, and pericarditis (pathologic conditions) have in common? _____

_____

12. Why are these conditions accompanied by a great deal of pain? _____

_____

_____

13. The mouth, or buccal cavity, and its extension, which stretches through the body inside the digestive system, is not listed as an internal body cavity. Why is this so? _____

_____

_____

EXERCISE 2

Organ Systems Overview

1. Use the key below to indicate the organ system(s) that perform the following functions for the body.

Key:
- a. circulatory
- b. digestive
- c. endocrine
- d. integumentary
- e. muscular
- f. nervous
- g. reproductive
- h. respiratory
- i. skeletal
- j. urinary

_____ rids the body of nitrogen-containing wastes

_____ is affected by removal of the thyroid gland

_____ provides support and levers on which the muscular system acts

_____ includes the heart

_____ causes the onset of the menstrual cycle

_____ protects underlying organs from drying out and from mechanical damage

_____ protects the body; destroys bacteria and tumor cells

_____ breaks down ingested food into its building blocks

_____ removes carbon dioxide from the blood

_____ delivers oxygen and nutrients to the tissues

_____ moves the limbs; facilitates facial expression

_____ conserves body water or eliminates excesses

_____ facilitates conception and childbearing

_____ controls the body with chemical molecules called hormones

_____ is damaged when you cut your finger or get a severe sunburn

2. *Name* the organ system to which each of the following sets of organs (or body structures) belongs:

blood vessels, spleen, heart _____

trachea, bronchi, alveoli _____

testis, vas deferens, urethra _____

adrenal glands, pancreas, pituitary _____

esophagus, large intestine, rectum _____

kidneys, bladder, ureters _____

bone, cartilages, tendons _____

3. Using the key below, place the following organs in their proper body cavity.

Key: a. abdominopelvic   b. cranial   c. spinal   d. thoracic

_____ 1. stomach   _____ 5. liver   _____ 9. lungs

_____ 2. small intestine   _____ 6. spinal cord   _____ 10. brain

_____ 3. large intestine   _____ 7. bladder   _____ 11. rectum

_____ 4. spleen   _____ 8. heart

4. Using the organs listed in item 3 above, record by number which would be found in the abdominal regions listed below.

_____ hypogastric region   _____ epigastric region

_____ right lumbar region   _____ left iliac region

_____ umbilical region   _____ left hypochondriac region

5. The five levels of organization of a living body are cell, _____, _____,

_____, and organism.

6. Define organ: _____

_____

_____

7. During the course of this laboratory exercise, a rat was dissected. What is the *value* of observing the

anatomy of a rat (or any other small mammal) when *human anatomy* is the actual topic of study? _____

_____

_____

EXERCISE 3   The Microscope

Care and Structure of the Compound Microscope

1. The following statements are true or false. If true, insert *T* on the answer blank. If false, correct the underlined word or phrase by inserting the correct answer.

_____ The microscope lens may be cleaned <u>with any soft tissue</u>.

_____ The coarse adjustment knob may be used in focusing <u>with all three objectives</u>.

_____ The microscope should be stored with the <u>oil immersion</u> lens in position over the stage.

_____ When beginning to focus, the <u>l.p.</u> lens should be used.

_____ Always focus <u>toward</u> the specimen in low power.

_____ A cover slip should always be used with <u>the high-dry and oil lenses</u>.

_____ The greater the amount of light delivered to the objective lens, the <u>less</u> the resolution.

2. Match the microscope structures given in column B with the statements that identify or describe them (column A).

Column A

_____ platform on which the slide rests for viewing

_____ lens located at the superior end of the body tube

_____ secure(s) the slide to the stage

_____ delivers a concentrated beam of light to the specimen

_____ used for precise focusing once initial focusing has been done

_____ carries the objective lenses; rotates so that the different objective lenses can be brought into position over the specimen

_____ used to increase or decrease the amount of light passing through the specimen

Column B

a. coarse adjustment knob

b. condenser

c. fine adjustment knob

d. iris diaphragm

e. mechanical stage or spring clips

f. movable nosepiece

g. objective lenses

h. ocular

i. stage

3. Explain the proper technique for transporting the microscope. _____

_____

_____

4. Define the following terms:

real image: _____

_____

virtual image: _____

_____

5. Define total magnification: _____

_____

6. Define resolution: _____

Viewing Objects Through the Microscope

1. Complete or respond to the following statements:

_____ The distance from the bottom of the objective lens in use to the specimen is called the _____ .

_____ The resolution of the human eye is _____ μm.

_____ The resolution of the optical microscope is _____ μm.

_____ The area of the specimen seen when looking through the microscope is the _____ .

_____ If a microscope has a 10× ocular and the total magnification at a particular time is 950×, the objective lens in use at that time is _____ ×.

_____ If after focusing in low power only the fine adjustment need be used to focus the specimen at the higher powers, the microscope is said to be _____ .

_____; _____ If the field size using a 10× ocular and a 15× objective is 1.5 mm, the approximate field size with a 30× objective is _____ mm or _____ μm.

_____ If the size of the high-power field is 1.2 mm, an object that occupies approximately a third of that field has an estimated size of _____ mm.

_____ Assume there is an object on the left side that you want to bring into the center of the field (that is, toward the apparent right). In what direction would you move your slide?

_____ If the object is in the top of the field and you want to move it downward into the center, you would move the slide _____ .

2. You have been asked to prepare a slide with the letter *k* on it. Draw its appearance in the l.p. field in the adjacent circle.

k

3. Say you are observing an object in the l.p. field. When you switch to h.p., it is no longer in your field of view.

Why might this occur? _____

_____

_____

What should be done initially to prevent this from happening? _____

_____

4. Do the following factors increase *or* decrease as one moves to higher magnifications with the microscope?

resolution _____   amount of light needed _____

working distance _____   depth of field _____

5. A student has the high-dry lens in position and appears to be intently observing the specimen. The instructor, noting a working distance of about 1 cm, knows the student isn't actually seeing the specimen.

How so? _____

6. Fill in the following chart with the appropriate information (relative to the microscope you used during the laboratory exercise):

	Low power	High power	Oil immersion
Magnification of the objective lenses			
Total magnification			
Detail observed			
Field size			
Working distance			

7. Why is it important to be able to use your microscope to perceive depth when studying tissue slides?

_____

_____

_____

8. If you are observing a slide of tissue two cell-layers thick, how can you determine which layer is superior?

_____

_____

_____

RS9

9. Describe the proper procedure for preparing a wet mount. _____

_____

_____

_____

_____

_____

10. Give two reasons why the light should be dimmed when viewing living or unstained material. _____

_____

_____

_____

EXERCISE 4

The Cell—
Anatomy and Division

Anatomy of the Composite Cell

1. Define the following:

organelle: _____

_____

cell: _____

_____

2. Although cells have differences that reflect their specific functions in the body, what functional capabilities

do all cells exhibit? _____

_____

3. Identify the following cell parts:

_____   external boundary of cell; confines cell contents, regulates entry
and exit of materials

_____   contains digestive enzymes of all varieties; "suicide sacs" of the cell

_____   scattered throughout the cell; controls release of energy from
foodstuffs

_____   slender projections of the cell membrane that increase its
surface area

_____   stored glycogen granules, proteins, crystals, pigments, and so on

_____   membranous system, consisting of flattened sacs and vesicles;
packages proteins for export

_____   control center of the cell; necessary for cell division and cell life

_____   two rod-shaped bodies near the nucleus; "spin" the mitotic spindle

_____   dense, darkly-staining nuclear body; possible packaging site for
ribosomes

_____   membranous system involved with synthesis of lipid-based
hormones

_____   membranous system; involved in intracellular transport of
proteins

_____   attached to membrane systems or scattered in the cytoplasm;
synthesize proteins

_____   threadlike structures in the nucleus; contain genetic material (DNA)

Observing Similarities and Differences in Cell Structure

1. List *one* important *structural* characteristic (a) of each of the following cell types observed in the laboratory, then give the *function* (b) that structure complements or ensures.

squamous epithelium   1a. _____

                             b. _____

sperm   2a. _____

                             b. _____

smooth muscle   3a. _____

                             b. _____

red blood cell   4a. _____

                             b. _____

2. What is the significance of the red blood cell being anucleate (without a nucleus)? _____

_____

Did it ever have a nucleus? _____ When? _____

_____

3. What are selective stains, and what is the basis of their action? _____

_____

_____

4. What does Janus green stain for? _____

Why can't you see these structures without Janus green? _____

Cell Division: Mitosis and Cytokinesis

1. Complete or respond to the following statements:

Division of the __1__ is referred to as mitosis. Cytokinesis is division of the __2__. The major structural difference between chromatin and chromosomes is that the latter is __3__. Chromosomes attach to the spindle fibers by undivided structures called __4__. If a cell undergoes mitosis but not cytokinesis, the product is __5__. The structure that acts as a scaffolding for chromosomal attachment and movement is called the __6__. __7__ is the period of cell life when the cell is not involved in division. Two cell populations in the body that do not undergo cell division are __8__ and __9__. The implication of an inability of a cell population to divide is that when some of its members die, they are replaced by __10__.

1. _____
2. _____
3. _____
4. _____
5. _____
6. _____
7. _____
8. _____
9. _____
10. _____

2. Using the key, categorize each of the events described below according to the phase in which it occurs.

Key: a. prophase   b. anaphase   c. telophase   d. metaphase   e. none of these

_____ Chromatin coils and condenses to form deeply-staining bodies.

_____ Centromeres break and chromosomes begin migration toward opposite poles of the cell.

_____ The nuclear membrane and nucleoli become reestablished.

_____ Chromosomes cease their poleward movement.

_____ Chromosomes align on the equator of the spindle.

_____ Nucleoli and nuclear membrane disappear.

_____ The spindle forms through the migration of the centrioles.

_____ Chromosomal material replicates.

_____ Centrioles replicate.

_____ Chromosomes first appear to be duplex structures.

_____ Chromosomes attach to the spindle fibers.

_____ Cleavage furrow forms.

_____ The nuclear membrane(s) is absent.

3. Identify the phases of mitosis depicted in the following diagrams:

_____   _____

4. What is the importance of mitosis? _____ _____

_____

_____

5. What is the physical advantage of the chromatin coiling and condensing to form short chromosomes at the onset of mitosis? _____

_____

_____

STUDENT NAME _____

LAB TIME/DATE _____

EXERCISE 5

The Cell—Transport Mechanisms and Cell Permeability

Choose all answers that apply to items 1 and 2, and place their letters in the appropriately numbered response blanks to the right.

1. Brownian motion
   a. reflects the kinetic energy of the smaller solvent molecules
   b. reflects the kinetic energy of the larger solute molecules
   c. is ordered and predictable
   d. is random and erratic

1. _____

2. Kinetic energy
   a. is higher in larger molecules
   b. is lower in larger molecules
   c. increases with increasing temperature
   d. decreases with increasing temperature
   e. is reflected in the speed of molecular movement

2. _____

3. Referring to the laboratory experiment using dialysis sacs 1 through 4 to study diffusion through nonliving membranes:

Sac 1: 40% glucose suspended in distilled water

Did glucose pass out of the sac? _____

Test used to determine presence of glucose was _____

Did the sac weight change? _____

If so, explain the reason for its weight change: _____

_____

_____

Sac 2: 40% glucose suspended in 40% glucose

Was there net movement of glucose in either direction? _____

Explanation: _____

Sac weight change? _____ Explanation: _____

_____

Sac 3: 10% NaCl in distilled water

Net movement of NaCl out of the sac? _____

Test used to determine the presence of NaCl: _____

Direction of net osmosis? _____

Sac 4: Boiled starch in distilled water

Net movement of starch out of the sac? _____

Test used to determine the presence of starch? _____

Direction of net osmosis? _____

4. What single characteristic of the semipermeable membranes used in the laboratory determines the

substances that can pass through them? _____

In addition to this characteristic, what other factors influence the passage of substances through living

membranes? _____

_____

5. A semipermeable sac containing 4% NaCl, 9% glucose, and 10% albumin is suspended in a solution
with the following composition: 10% NaCl, 10% glucose, and 40% albumin. Assume that the sac is permeable to
all substances except albumin. State whether each of the following will (a) move into the sac (b) move out of the
sac, or (c) not move.

glucose _____   water _____   albumin _____   NaCl _____

6. The diagrams below represent three microscope fields containing red blood cells. Arrows show the direction

of net osmosis. Which field contains a hypertonic solution? _____ The cells in this field are said to be

_____. Which field contains an isotonic bathing solution? _____

Which field contains a hypotonic solution? _____ What is happening to the cells in this field?

_____

(a)   (b)   (c)

7. What determines whether a transport process is active or passive? _____

_____

8. Use the key terms to characterize each of the statements below. (More than one key choice may apply.)

Key: a. diffusion, dialysis   c. permease system   e. pinocytosis
          b. diffusion, osmosis   d. phagocytosis   f. filtration

_____ require ATP (cellular energy)

_____ driven by kinetic energy of the molecules

_____ driven by hydrostatic (fluid) pressure

_____ follow a concentration gradient

_____ proceeds against a concentration gradient; requires a carrier

_____ engulf foreign substances

_____ moves water through a semipermeable membrane

_____ transports amino acids, some sugars, and Na$^+$ through the cell membrane

_____ provides for cellular uptake of solid or large particles from the cell exterior

_____ moves small or lipid-soluble solutes through the cell membrane

9. Define the following terms:

brownian motion: _____

_____

_____

diffusion: _____

_____

_____

osmosis: _____

_____

_____

dialysis: _____

_____

_____

filtration: _____

_____

_____

active transport: _____

_____

_____

phagocytosis: _____

_____

_____

pinocytosis: _____

_____

_____

EXERCISE 6

Classification of Tissues

Tissue Structure and Function—General Review

1. Define tissue: _____

_____

2. Use the key choices to identify the *major* tissue types described below:

Key: a. connective tissue   b. epithelium   c. muscle   d. nervous tissue

_____ forms membranes—mucous, serous, and epidermal

_____ allows for movement of limbs, and for organ movements within the body

_____ transmits electrochemical impulses

_____ supports body organs

_____ cells may absorb and/or secrete substances

_____ basis of the major controlling system of the body

_____ major function of the cells of this tissue type is to shorten

_____ forms hormones

_____ packages and protects body organs

_____ characterized by having large amounts of nonliving matrix

_____ most widely distributed in the body

_____ forms brain and spinal cord

Epithelial Tissue

1. Describe the general characteristics of epithelial tissue. _____

_____

_____

_____

_____

2. On what bases are epithelial tissues classified? _____

_____

3. What are the major functions of epithelium in the body? (Give examples.) _____

_____

_____

_____

4. How is the function of epithelium reflected in its arrangement? _____

_____

_____

5. Where is ciliated epithelium found? _____

_____

What role does it play? _____

_____

_____

6. Transitional epithelium is actually stratified squamous epithelium, but there is something special about it.

How does it differ structurally from other stratified squamous epithelia? _____

_____

How does this reflect its function in the body? _____

_____

_____

7. How do the endocrine and exocrine glands differ in structure and function? _____

_____

_____

_____

8. Respond to the following with the key choices:

Key: a. pseudostratified ciliated   c. simple cuboidal   e. stratified squamous
       b. simple columnar   d. simple squamous   f. transitional

_____ lining of the esophagus

_____ lining of the stomach and small intestine

_____ lung tissue, alveolar sacs

_____ collecting tubules of the kidney

_____ epidermis of the skin

_____ lining of bladder; peculiar cells that have the ability to slide over each other

_____ forms the thin serous membranes; a single layer of flattened cells

Connective Tissue

1. What are the general characteristics of connective tissues? _____
_____
_____
_____
_____
_____

2. What functions are performed by connective tissue? _____
_____
_____

3. How are the functions of connective tissue reflected in its structure? _____
_____
_____
_____

4. Using the key, choose the best response to identify the connective tissues described below:

Key: a. adipose connective tissue   e. fibrocartilage
           b. areolar connective tissue   f. hemopoietic tissue
           c. dense fibrous connective tissue   g. hyaline cartilage
           d. elastic cartilage   h. osseous tissue

_____ provides great strength through parallel bundles of collagenic fibers; found in tendons

_____ acts as a storage depot for fat

_____ composes the dermis of the skin

_____ makes up the intervertebral disks

_____ forms the bony skeleton

_____ composes the basement membrane and packages organs; includes a gellike matrix with all categories of fibers and diverse cell types

_____ forms the embryonic skeleton and the surfaces of bones at the joints; reinforces the trachea

_____ provides an elastic framework for the external ear

_____ structurally amorphous matrix heavily invaded with fibers; appears glassy and smooth

_____ contains cells arranged concentrically around a nutrient canal; matrix hard owing to calcium salts

_____ provides insulation for the body

5. Why are adipose cells called "signet ring" cells? _____
_____

RS21

Muscle Tissue

1. The three types of muscle tissue exhibit similarities as well as differences. Check the appropriate space in the chart below to indicate which muscle types exhibit each characteristic.

Characteristic	Skeletal	Cardiac	Smooth
Voluntarily controlled			
Involuntarily controlled			
Has a banded appearance			
Has a single nucleus in each cell			
Multinucleate			
Found attached to bones			
Allows you to direct your eyeballs			
Found in the walls of the stomach, uterus, and arteries			
Contains spindle-shaped cells			
Contains cylindrical cells with branching ends			
Contains long, nonbranching cylindrical cells			
Displays intercalated disks			
Concerned with locomotion of the body as a whole			
Changes the internal volume of an organ as it contracts			
Tissue of the circulatory pump			

Nervous Tissue

1. What two physiologic characteristics are highly developed in nervous tissue? _____

_____

2. In what ways are nerve cells similar to other cells? _____

_____

How are they different? _____

3. Sketch a neuron, recalling in your diagram the most important aspects of its structure. *Below* the diagram describe how its particular structure relates to its function in the body.

For Review

Attempt to draw the following tissue types from memory. Incorporate the important structural characteristics in your drawings.

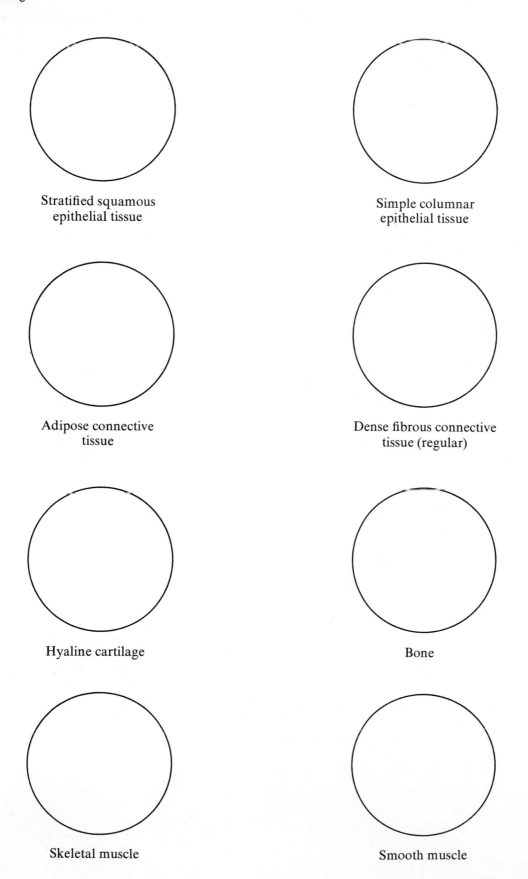

Stratified squamous
epithelial tissue

Simple columnar
epithelial tissue

Adipose connective
tissue

Dense fibrous connective
tissue (regular)

Hyaline cartilage

Bone

Skeletal muscle

Smooth muscle

EXERCISE 7

The Integumentary System

Basic Structure of the Skin

1. Complete the following statements. Insert responses in the blanks to the right.

a. The two basic tissues of which the skin is composed are dense irregular connective tissue, which makes up the dermis and _____, which forms the epidermis.

b. The waterproofing protein found in the epidermal cells is called _____.

c. Melanin and _____ contribute to skin color.

d. A localized concentration of melanin is referred to as a _____.

e–h. List four (4) protective functions of the skin.

a. _____

b. _____

c. _____

d. _____

e. _____

f. _____

g. _____

h. _____

2. Label the skin structures and areas on the accompanying diagram.

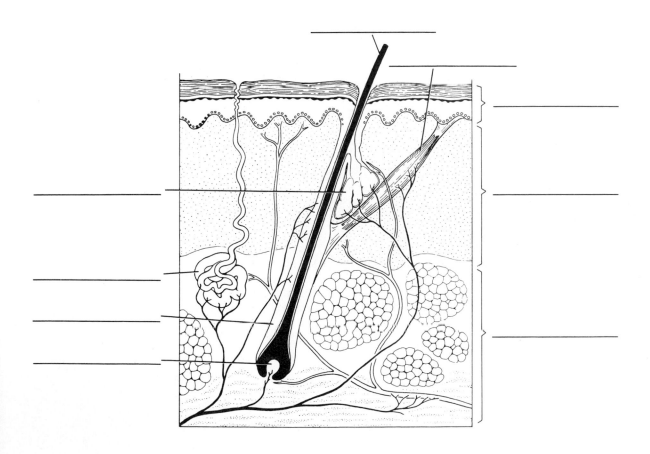

3. Using the key choices, choose all responses that apply to the following descriptions.

Key: a. stratum corneum   d. stratum lucidum   g. epidermis as a whole
        b. stratum germinativum   e. papillary layer   h. dermis as a whole
        c. stratum granulosum   f. reticular layer

_____ translucent cells containing eleidin

_____ dead cells

_____ dermis layer responsible for fingerprints

_____ vascular region

_____ major skin area where derivatives (nails, hair) arise

_____ epidermal region exhibiting rapid cell division; most inferior epidermal layer

_____ scalelike dead cells full of keratin that constantly slough off

_____ site of elastic and collagenic fibers

_____ site of melanin formation

4. What substance is manufactured in the skin (but is not a secretion) to play a role elsewhere in the body?

_____

5. What sensory receptors are found in the skin? _____

_____

_____

6. A nurse tells the doctor that a patient is cyanotic. What is cyanosis? _____

What does its presence imply? _____

7. What is the mechanism of a suntan? _____

_____

8. What is a decubitus ulcer? _____

Why does it occur? _____

_____

9. Some injections hurt more than others. On the basis of what you have learned about skin structure,

can you determine why this is so? _____

_____

Appendages of the Skin

1. Using key choices, respond to the following descriptions.

Key: a. arrector pili
   b. cutaneous receptors
   c. hair

   d. hair follicles
   e. sebaceous glands

   f. sudoriferous gland—apocrine
   g. sudoriferous gland—eccrine

_____ A blackhead is an accumulation of oily material that is produced by _____.

_____ Tiny muscles attached to hair follicles, which pull the hair upright during fright or cold.

_____ Most numerous variety of perspiration gland.

_____ Sheath formed of both epithelial and connective tissues.

_____ Less numerous variety of perspiration gland; secretion (often milky in appearance) contains proteins and other substances that promote bacterial growth.

_____ Found everywhere on body except palms of hands, soles of feet, and lips.

_____ Primarily dead/keratinized cells.

_____ Specialized nerve endings that respond to temperature, touch, and so on.

_____ Become more active at puberty.

_____ Part of the heat-liberating apparatus of the body.

2. How does the skin help in regulating body temperature? (Name two different mechanisms.) _____

_____

_____

_____

_____

STUDENT NAME ____________________

LAB TIME/DATE ____________________

EXERCISE 8

Classification of Membranes

1. Complete the following chart:

Membrane	Tissue type (epithelial/connective)	Common locations	Functions
mucous			
serous			
synovial			
epidermis			

2. Respond to the following statements by choosing an answer from the key.

Key: a. epidermis b. mucous c. serous d. synovial

_____ membrane type associated with skeletal system structures

_____ always formed of simple squamous epithelium

_____, _____ membrane types *not* found in the ventral body cavity

_____ the only membrane type in which goblet cells are found

_____ the only *external* membrane

EXERCISE 9

Bone Classification and Structure: An Overview

Bone Markings

1. Match the terms in column B with the appropriate description in column A and give one example of each.

Column A	Example	Column B
_____ sharp, slender process* _____		1. condyle
_____ small rounded projection* _____		2. crest
_____ narrow ridge of bone* _____		3. epicondyle
_____ large rounded projection* _____		4. fissure
_____ structure supported on neck† _____		5. foramen
_____ armlike projection† _____		6. fossa
_____ rounded, convex projection† _____		7. head
_____ narrow depression or opening ‡ _____		8. meatus
_____ canallike structure‡ _____		9. ramus
_____ opening through a bone‡ _____		10. sinus
_____ shallow depression† _____		11. spine
_____ air-filled cavity _____		12. trochanter
_____ large, irregularly shaped projection* _____		13. tubercle
_____ raised area of a condyle* _____		14. tuberosity

Classification of Bones

1. The four major anatomic classifications of bones are long, short, flat, and irregular. Which category

has the least amount of spongy bone relative to its total volume? _____

2. Classify each of the bones on p. RS 32 into one of the four major categories by checking the appropriate column. Use appropriate references as necessary.

---

*A site of muscle attachment.

†Takes part in joint formation.

‡A passageway for nerves or blood vessels.

	Long	Short	Flat	Irregular
humerus				
metacarpal				
frontal				
calcaneus				
rib				
vertebra				
radius				

Gross Anatomy of the Typical Long Bone

1. Using the terms to the right, characterize the following statements:

_____ site of spongy bone in the adult

_____ site of compact bone in the adult

_____ site of hemopoiesis in the adult

_____ major submembranous site of osteoclasts

_____ scientific name for bone shaft

_____ site of fat storage in the adult

_____ site of longitudinal growth in the child

a. diaphysis

b. endosteum

c. epiphyseal disk

d. epiphysis

e. periosteum

f. red marrow cavity

g. yellow marrow cavity

2. What differences between compact and spongy bone can be seen with the naked eye? _____

_____

3. What is the function of the periosteum? _____

_____

Microscopic Structure of Compact Bone

1. Trace the route taken by blood through a bone, starting with the periosteum and ending with an osteocyte

in a lacuna. Periosteum - - - → _____ - - - →

_____ - - - → _____ - - - → _____ osteocyte.

2. Several descriptions of bone structure are given in column B. Identify the structure involved by choosing the appropriate term from column A and placing the corresponding letter in the correct blank.

Column A

a. haversian canal

b. concentric lamellae

c. lacunae

d. canaliculi

e. matrix

Column B

_____ concentric layers of calcified matrix

_____ site of osteocytes

_____ longitudinal canal carrying blood vessels, lymphatics, and nerves

_____ nonliving, structural part of bone

_____ minute canals connecting lacunae

Chemical Composition of Bone

1. What is the function of the organic matrix in bone? _____

2. Name the important organic bone components. _____

3. Calcium salts form the bulk of the inorganic material in bone. What is the function of the calcium salts?

_____

4. Which is responsible for bone structure? (circle the appropriate response)

    inorganic portion   organic portion   both contribute

EXERCISE 10

The Axial Skeleton

The Skull

1. The skull is one of the major components of the axial skeleton. Name the other two.

_____ and _____

What structures do each of these areas protect? _____

_____

_____

2. Define suture: _____

3. With one exception, the skull bones are joined by sutures. Name the exception. _____

_____

4. What are the four major sutures of the skull, and what bones do they connect? _____

_____

_____

_____

_____

5. Name the eight bones composing the cranium.

_____ _____ _____ _____

_____ _____ _____ _____

6. Give two possible functions of the sinuses. _____

_____

7. What is the orbit? _____

What bones contribute to the formation of the orbit? _____

_____

_____

8. Why can the sphenoid bone be called the keystone of the cranial floor? _____

_____

9. What is a cleft palate? _____

10. Match the bone names in column B with the descriptions in column A.

Column A

_____ forehead bone

_____ cheekbone

_____ lower jaw

_____ bridge of nose

_____ posterior part of hard palate

_____ much of the lateral and superior cranium

_____ most posterior part of cranium

_____ single irregular, bat-shaped bone forming part of the cranial floor

_____ tiny bones bearing tear ducts

_____ anterior part of hard palate

_____ superior and medial nasal conchae formed from its projections

_____ site of mastoid process

_____ site of sella turcica

_____ site of cribriform plate

_____ site of mental foramen

_____ site of styloid processes

_____, _____, _____, _____ four bones containing paranasal sinuses

_____ its condyles articulate with the atlas

_____ foramen magnum contained here

_____ small U-shaped bone in neck where many tongue muscles attach

_____ middle ear found here

_____ nasal septum

_____ bears an upward protrusion, the "cock's comb," or crista galli

_____, _____ contain alveolar processes bearing teeth

Column B

a. ethmoid

b. frontal

c. hyoid

d. lacrimals

e. mandible

f. maxillae

g. nasals

h. occipital

i. palatines

j. parietals

k. sphenoid

l. temporals

m. vomer

n. zygomatic

The Vertebral Column

1. Using the key, correctly identify the vertebral parts/areas described below. (More than one choice may apply in some cases.)

Key: a. body
      b. intervertebral foramina
      c. spinous process
      d. superior articular process
      e. transverse process
      f. vertebral arch

_____ structure enclosing the nerve cord

_____ weight-bearing portion of the vertebra

_____ provides levers for the muscles to pull against

_____ provides an articulation point for the ribs

_____ openings providing for exit of spinal nerves

2. The distinguishing characteristics of the vertebrae composing the vertebral column are noted below. Correctly identify each described structure/region by choosing a response from the key.

Key: a. atlas
      b. axis
      c. cervical vertebra–typical
      d. coccyx
      e. lumbar vertebra
      f. sacrum
      g. thoracic vertebra

_____ vertebral type containing foramina in the transverse processes through which the vertebral arteries ascend to reach the brain

_____ its dens provides a pivot for rotation of the first cervical vertebra (C_1)

_____ transverse processes have facets for articulation with ribs; spinous process points sharply downward

_____ composite bone; articulates with the hipbone laterally

_____ massive vertebrae; weight-sustaining

_____ "tail-bone"; vestigeal fused vertebrae

_____ supports the head; allows a rocking motion in conjunction with the occipital condyles

_____ seven components; unfused

_____ twelve components; unfused

3. Name two factors/structures that allow for flexibility of the vertebral column.

_____ and _____

4. Describe how a spinal nerve exits from the vertebral column. _____

_____

5. Which two spinal curvatures are present at birth? _____ and _____

Under what conditions do the secondary curvatures develop? _____

_____

_____

6. Diagram the normal spinal curvatures and the abnormal spinal curvatures names below. (Use posterior or lateral views as necessary.)

       Normal curvature   Lordosis   Scoliosis   Kyphosis

7. Of what kind of tissue are the intervertebral disks composed? _____

_____

8. What is a herniated disk? _____

What problems might it cause? _____

_____

The Thorax

1. The major components of the thorax (excluding the vertebral column) are the _____ and the

_____ .

2. Differentiate between a true rib and a false rib. _____

_____

_____

_____

Is the floating rib a true or a false rib? _____

Why is it called this? _____

_____

3. What is the general shape of the thoracic cage? _____

EXERCISE 11　　The Appendicular Skeleton

Bones of the Shoulder Girdle and Upper Extremity

1. Match the bone names or markings in column B with the descriptions in column A.

Column A

_____ raised area on lateral surface of humerus to which deltoid muscle attaches

_____ upper arm bone

_____, _____ bones composing the shoulder girdle

_____, _____ forearm bones

_____ point where scapula and clavicle connect

_____ shoulder girdle bone that has no attachment to the axial skeleton

_____ shoulder girdle bone that articulates anteriorly with the sternum

_____ socket in the scapula for the upper arm bone

_____ process above the glenoid fossa that permits muscle attachment

_____ commonly called the collarbone

_____ distal medial process of the humerus; adjoins the ulna

_____ medial bone of forearm in anatomic position

_____ rounded knob on the humerus that articulates with the radius

_____ anterior depression, superior to the trochlea, which receives part of the ulna when the forearm is flexed

_____ forearm bone involved in formation of the elbow joint

_____ bones of the wrist

_____ bones of the fingers

_____ heads of these bones form the knuckles

_____, _____ bones that articulate with the clavicle

Column B

a. acromion process

b. capitulum

c. carpus

d. clavicle

e. coracoid process

f. coronoid fossa

g. deltoid tuberosity

h. glenoid fossa

i. humerus

j. metacarpus

k. olecranon fossa

l. olecranon process

m. phalanges

n. radial tuberosity

o. radius

p. scapula

q. sternum

r. styloid process

s. trochlea

t. ulna

2. Why does the clavicle often fracture when a person falls on his shoulder? _____

_____

3. Why is there generally no problem in the arm clearing the widest dimension of the thoracic cage?

_____

4. What is the total number of phalanges in the hand? _____

5. What is the total number of carpals in the wrist? _____

In the proximal row, the carpals are (medial to lateral) _____

In the distal row, they are (medial to lateral) _____

Bones of the Pelvic Girdle and Lower Extremity

1. Compare the pectoral and pelvic girdles by choosing appropriate descriptive terms from the key.

Key: a. flexibility most important   d. insecure axial and limb attachments
         b. massive   e. secure axial and limb attachments
         c. lightweight   f. weight-bearing most important

Pectoral: _____, _____, _____   Pelvic _____, _____, _____

2. What organs are protected, at least in part, by the pelvic girdle? _____

_____

3. Distinguish between the true pelvis and the false pelvis. _____

_____

_____

4. Name five differences between the male and female pelves. _____

_____

_____

_____

_____

_____

5. Why are the pelvic bones of a four-legged animal such as the cat or pig much less massive than those of

the human? _____

_____

6. A person instinctively curls over his abdominal area in times of danger. Why? _____

_____

7. For what anatomic reason do many women appear to be slightly knock-kneed? _____

_____

_____

8. What does *fallen arches* mean? _____

_____

9. Match the bone names and markings in column B with the descriptions in column A.

Column A

_____; _____; _____ fuse to form the os coxae

_____ receives the weight of the body when sitting

_____ point where the os coxae join anteriorly

_____ upper margin of iliac bones

_____ deep socket in the os coxa that receives the head of the thigh bone

_____ point where axial skeleton attaches to the pelvic girdle

_____ longest bone in body, articulates with the os coxa

_____ lateral bone of the lower leg

_____ medial bone of the lower leg

_____; _____ bones forming the knee joint

_____ point where the patellar tendon attaches

_____ kneecap

_____ shinbone

_____ distal process on medial tibial surface

_____ process forming the outer ankle

_____ heel bone

_____ bones of the ankle

_____ bones forming the instep of the foot

_____ opening in os coxa formed by the pubic and ischial rami

_____ sites of muscle attachment on the proximal end of the femur

_____ tarsal bone that articulates with the tibia

Column B

a. acetabulum

b. calcaneus

c. femur

d. fibula

e. gluteal tuberosity

f. greater sciatic notch

g. greater and lesser trochanters

h. iliac crest

i. ilium

j. ischial tuberosity

k. ischium

l. lateral malleolus

m. lesser sciatic notch

n. linea aspera

o. medial malleolus

p. obturator foramen

q. metatarsus

r. patella

s. pubic symphysis

t. pubis

u. sacroiliac joint

v. talus

w. tarsals

x. tibia

y. tibial tuberosity

10. Identify all bones (or groups of bones) indicated by leader lines in the diagram of the articulated skeleton.

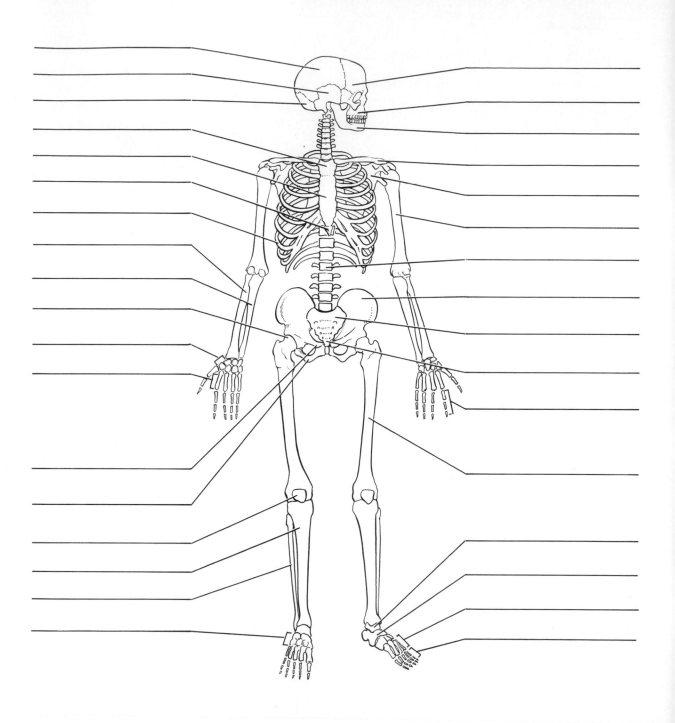

EXERCISE 12

The Fetal Skeleton

1. Are the same skull bones seen in the adult found in the fetal skull? _____

2. How does the relative size of the fetal face compare to its cranium? _____

_____

How does this compare to the adult skull? _____

_____

_____

3. Why are there outward conical projections in some of the fetal cranial bones? _____

_____

4. What is a fontanel? _____

What is its fate? _____

What is the function of the fontanels in the fetal skull? _____

_____

5. Describe how the fetal skeleton compares with the adult skeleton in the following areas:

vertebrae _____

_____

os coxae _____

_____

carpals and tarsals _____

sternum _____

frontal bone _____

patella _____

rib cage _____

_____

6. How does the size of the fetus's head compare to the size of its body? _____

_____

EXERCISE 13

Articulations and Body Movements

Types of Joints

1. Use key responses to identify the joint types described below.

Key: a. amphiarthroses   b. diarthroses   c. synarthroses

_____ allows a slight degree of movement

_____ essentially immovable

_____ characterized by cartilage connecting the bony portions

_____ all have a fibrous capsule lined with a synovial membrane surrounding a joint cavity

_____ freely movable

_____ bone regions united by fibrous connective tissue

2. Match the joint subcategories in column B with their descriptions in column A.

Column A	Column B
_____ joint between skull bones	a. ball and socket
_____ joint between the axis and atlas	b. ellipsoidal
_____ hip joint	c. gliding
_____ intervertebral joints	d. hinge
_____ joint between forearm bones and wrist	e. pivot
_____ elbow	f. saddle
_____ interphalangeal joints	g. suture
_____ intercarpal joints	h. symphysis
_____ joint between tarsus and tibia	i. synchondrosis
_____ joint between skull and vertebral column	j. syndesmosis
_____ joint between jaw and skull	k. symphysis
_____ joints between proximal phalanges and metacarpal bones	
_____ epiphyseal plate of a child's long bone	
_____ a multiaxial joint	
_____, _____ biaxial joints	
_____, _____ uniaxial joints	

3. What characteristics do all joints have in common? _____

_____

4. Describe the structure and function of the following tissues in relation to a synovial joint (use a simple diagram if you like):

ligament _____

_____

tendon _____

_____

hyaline cartilage _____

_____

synovial membrane _____

_____

bursa _____

5. What structural joint changes are common to the elderly? _____

_____

_____

Body Movements

1. Complete the following statements:

The movable attachment of a muscle is called its __1__, and its stationary attachment is called the __2__. Winding up for a pitch (as in baseball) can properly be called __3__. To keep your seat when riding a horse, the tendency is to __4__ your thighs. In running, the action at the hip joint is __5__ in reference to the leg moving forward and __6__ in reference to the leg in the posterior position. In kicking a football, the action at the knee is __7__. In climbing stairs, the hip and knee of the forward leg are both __8__. You have just touched your chin to your chest. This is __9__ of the neck. Using a screwdriver with a straight arm requires __10__ of the arm. Consider all the movements of which the arm is capable. One often used for strengthening all the upper arm and shoulder muscles is __11__. Movement of the head that signifies "no" is __12__. Standing on your toes as in ballet requires __13__ of the foot. Action that moves the distal end of the radius across the ulna is __14__. Raising the arms laterally away from the body is called __15__ of the arms. Walking on one's heels is __16__.

1. _____

2. _____

3. _____

4. _____

5. _____

6. _____

7. _____

8. _____

9. _____

10. _____

11. _____

12. _____

13. _____

14. _____

15. _____

16. _____

EXERCISE 14

Microscopic Anatomy, Organization, and Classification of Skeletal Muscles

Skeletal Muscle Cells and Their Packaging into Muscles

1. Use the letters of the items on the right to correctly identify the structures described on the left.

_____ connective tissue ensheathing a bundle of muscle cells

_____ another term for deep fascia

_____ contractile unit of muscle

_____ a muscle cell

_____ thin reticular connective tissue investing each muscle cell

_____ cell membrane of the muscle cell

_____ a long filamentous organelle found within muscle cells with a banded appearance

_____ actin or myosin-containing structure

_____ cordlike extension of connective tissue beyond the muscle, serving to attach it to a bone

a. endomysium

b. epimysium

c. myofiber

d. myofilament

e. myofibril

f. perimysium

g. sarcolemma

h. sarcomere

i. sarcoplasm

j. tendon

2. Why are the connective tissue wrappings of skeletal muscle important? (Give at least three reasons.)

_____

_____

_____

3. Why are indirect—that is, tendinous—muscle attachments to bone seen more often than direct attachments? _____

_____

_____

_____

4. How does an aponeurosis differ from a tendon? _____

_____

Skeletal, Smooth, and Cardiac Muscle—
A Comparison

1. The three types of muscle tissue exhibit similarities as well as differences. Check the appropriate space in the chart below to indicate which muscle types exhibit each characteristic.

Characteristic	Skeletal muscle	Cardiac muscle	Smooth muscle
Moves substances through internal organs			
Allow you to run, grasp, stand, smile, etc.			
Tissue of the heart			
Contains cells packaged in bundles called fascicles			
Contains cells arranged in a spiral pattern in the heart			
Contains cells that are loosely connected with fine connective tissue and arranged in layers running in opposing directions			

2. How would you distinguish skeletal muscle from cardiac muscle? _____

_____

_____

_____

3. What role does the unique structure of cardiac muscle play in its function? (Note: Before attempting a

response, *describe* the unique anatomy.) _____

_____

_____

_____

4. How are smooth muscle cells structurally different from cardiac and skeletal muscle cells? _____

_____

_____

5. What is meant by the statement "Smooth muscle is involuntary in action?" _____

_____

Which muscle type is voluntary in action? _____

6. What capability is most highly expressed in muscle tissue? _____

_____

The Myoneural Junction

1. Complete the following statements:

The junction between a motor neuron's axon and the muscle cell membrane is called a myoneural junction or a __1__ junction. A motor neuron and all of the skeletal muscle cells it stimulates is called a __2__. The end plates of each motor axon have numerous projections called __3__. The actual gap between the axon and the muscle cell is called a __4__. Within the axonal end plates are many small vesicles containing a neurotransmitter substance called __5__. When the __6__ reaches the ends of the axon, the neurotransmitter is released and diffuses to the muscle cell membrane to combine with receptors there. The combination of the neurotransmitter with the muscle membrane receptors causes the membrane to become permeable to sodium, which results in the influx of sodium ions and __7__ of the membrane. Then contraction of the muscle cell occurs. Before the muscle cell can be stimulated to contract again, __8__ must occur.

1. _____
2. _____
3. _____
4. _____
5. _____
6. _____
7. _____
8. _____

Classification of Skeletal Muscles

1. Several criteria were given relative to the naming of muscles. Match the criteria (column B) to the muscle cell names (column A). Note that more than one criterion may apply in some cases.

Column A	Column B
_____ gluteus maximus	a. action of the muscle
_____ adductor magnus	b. shape of the muscle
_____ biceps femoris	c. location of the muscle's origin and/or insertion
_____ abdominis transversus	d. number of origins
_____ extensor carpi ulnaris	e. location of muscle relative to a bone or body region
_____ trapezius	f. direction in which the muscle fibers run relative to some imaginary line
_____ rectus femoris	g. relative size of the muscle
_____ external oblique	

2. When muscles are discussed relative to the manner in which they interact with other muscles, the terms shown in the key are often used. Match the key terms with the appropriate definitions.

Key: a. antagonist   b. fixator   c. prime mover   d. synergist

_____ agonist

_____ postural muscles for the most part

_____ reverses and/or opposes the action of a prime mover

_____ stabilizes a joint so that the prime mover may act at more distal joints

_____ performs the same movement as the prime mover

_____ immobilizes the origin of a prime mover

EXERCISE 15

Gross Anatomy of the Muscular System

Muscles of the Head and Neck

1. Identify the major muscles described next:

_____ used in smiling

_____ used to suck in your cheeks

_____ used in winking

_____ used in pouting (pulls the corners of the mouth downward)

_____ raises your eyebrows

_____ used to form the vertical frown crease on the forehead

_____ "kissing muscle"

_____ prime mover of jaw closure

Muscles of the Trunk

1. Identify the major muscles described next:

_____ a major spine flexor

_____ prime mover for pulling the arm posteriorly

_____ prime mover for shoulder flexion

_____, _____ assume major responsibility for forming the abdominal girdle (three pairs of muscles)

_____

_____ pulls the shoulder backward and downward

_____ prime mover of shoulder abduction

_____, _____ important in shoulder adduction; antagonists of the shoulder abductor (two muscles)

_____ moves the scapula forward and downward

_____ small, inspiratory muscles between the ribs; elevate the ribs

_____ extends the head

_____ pulls scapulae medially

Arm Muscles

1. Identify the muscles described next:

_____ places the palm upward

_____ flexes the forearm and supinates the hand

_____, _____ forearm flexors; no role in supination (two muscles)

_____ elbow extensor

_____ power wrist flexor and hand adductor

_____ flexes wrist and distal phalanges

_____ pronate the hand (two muscles)

_____ flexes the thumb

_____, _____, extend and abduct the wrist (three muscles)

_____

_____ extend the wrist and digits

Muscles of the Lower Extremity

1. Identify the muscles described next:

_____ moves the thigh laterally to take the "at ease" stance

_____ used to extend the hip when climbing stairs

_____ "toe dancer's" muscles—calf pair

_____ inverts the foot

_____ allow you to draw your legs to the midline of your body, as when standing at attention

_____ "tailor's muscle"

_____, _____, extends thigh, and flexes knee (three muscles)

_____

_____ extends knee and flexes thigh

Muscle Descriptions: General Review

1. Completion: identify the muscles described below:

_____, _____, _____ are commonly used for intramuscular

injections (three muscles).

The insertion tendon of the _____ group contains a large sesamoid bone, the patella.

The triceps surae insert in common into the _____ tendon.

The bulk of the tissue of a muscle tends to lie _____ to the part of the body it causes to move.

The extrinsic muscles of the hand originate on the _____.

Most flexor muscles are located on the _____ aspect of the body; most extensors are located

_____. An exception to this generalization is the extensor-flexor musculature of the

_____.

Muscle Recognition: General Review

1. Identify the numbered muscles in the diagram of the human anterior superficial musculature by matching its number with one of the following muscle names.

_____ orbicularis oris

_____ pectoralis major

_____ external oblique

_____ sternocleidomastoid

_____ biceps brachii

_____ deltoid

_____ vastus lateralis

_____ brachioradialis

_____ frontalis

_____ rectus femoris

_____ pronator teres

_____ rectus abdominis

_____ sartorius

_____ gracilis

_____ flexor carpi ulnaris

_____ adductor longus

_____ palmaris longus

_____ flexor carpi radialis

_____ latissimus dorsi

_____ orbicularis oculi

_____ gastrocnemius

_____ masseter

_____ trapezius

_____ tibialis anterior

_____ extensor digitorum longus

_____ tensor fascia latae

_____ pectineus

_____ sternohyoid

_____ serratus anterior

_____ adductor magnus

_____ vastus medialis

_____ transversus abdominis

_____ peroneus longus

_____ temporalis

_____ zygomaticus

_____ coracobrachialis

_____ triceps brachii

_____ internal oblique

_____ iliopsoas

2. Identify each of the numbered muscles in this diagram of the human posterior superficial musculature by matching its number to one of the following muscle names.

_____ gluteus maximus

_____ semimembranosus

_____ gastrocnemius

_____ latissimus dorsi

_____ deltoid

_____ iliotibial tract (tendon)

_____ teres major

_____ semitendinosus

_____ trapezius

_____ biceps femoris

_____ triceps brachii

_____ external oblique

_____ gluteus medius

_____ gracilis

_____ flexor carpi ulnaris

_____ extensor carpi ulnaris

_____ extensor digitorum communis

_____ extensor carpi radialis longus

_____ occipitalis

_____ plantaris

_____ extensor carpi radialis brevis

_____ sternocleidomastoid

_____ adductor magnus

Doreen Davis

Dissection and Identification of Fetal Pig Muscles

1. Many human muscles are modified from those of the pig (or any quadruped) as a result of the requirements of an upright posture. The following questions refer to these differences:

How does the human trapezius muscle differ from the pig's? _____

_____

_____

How does the deltoid differ? _____

_____

_____

How does the extent and orientation of the human sartorius muscle differ from its relative position in the pig?

_____

_____

_____

Explain these differences in terms of differences in function. _____

_____

_____

The human rectus abdominis is definitely divided by four transverse tendons. These tendons are absent or difficult to identify in the pig's homologue. How do these tendons affect the human upright posture? _____

_____

_____

EXERCISE 16

Muscle Physiology

Muscle Activity

1. The following group of incomplete statements refers to a muscle cell in the resting or polarized state just before stimulation. Complete each statement by choosing the correct response from the key items below.

Key:
  a. Na^+ diffuses out of the cell
  b. K^+ diffuses out of the cell
  c. Na^+ diffuses into the cell
  d. K^+ diffuses into the cell
  e. inside the cell
  f. outside the cell
  g. relative ionic concentrations on the two sides of the membrane

  h. electrical conditions
  i. activation of the sodium-potassium pump, which moves K^+ into the cell and Na^+ out of the cell
  j. activation of the sodium-potassium pump, which moves Na^+ into the cell and K^+ out of the cell

There is a greater concentration of Na^+ _____; there is a greater concentration of K^+ _____. When the stimulus is delivered, the permeability of the membrane at that point is changed, and _____ initiating the depolarization of the membrane. Almost as soon as the depolarization wave has begun, a repolarization wave follows it across the membrane. This occurs as _____. Repolarization restores the _____ of the resting cell membrane. The _____ is (are) reestablished by _____.

2. Number the following statements in their proper sequence to describe the contraction mechanism in a skeletal muscle cell. Number 1 has already been designated.

__1__ Acetylcholine is released into the neuromuscular junction by the motor end plate of the axon.

_____ The action potential, carried deep into the cell, causes the sarcoplasmic reticulum to release calcium ions.

_____ The muscle cell relaxes and lengthens.

_____ Acetylcholine diffuses across the neuromuscular junction and binds to receptors on the sarcolemma.

_____ The calcium ion concentrations at the myofilaments increase; the myofilaments slide past one another, and the cell shortens.

_____ Depolarization occurs, and the action potential is generated.

_____ The concentration of the calcium ions at the myofilaments decreases as it is actively reabsorbed into the sarcoplasmic reticulum.

3. Muscle contraction is most commonly explained by the sliding filament hypothesis. What are the essential points of this hypothesis? _____

_____

_____

4. Relative to your observations of muscle fiber contraction (p. 135).

    a. What percentage of contraction was observed with the solution containing

        ATP,K^+,$-Mg^{++}$?_____%

        With *just* ATP? _____% With *just* Mg^{++} and K^+? _____%

    b. *Explain* your observations fully. _____

    _____

    _____

    _____

    c. What zones or bands disappear when the muscle cell contracts? _____

    d. *Draw* a relaxed and a contracted sarcomere below.

           Relaxed   Contracted

Induction of Contraction in the Frog Gastrocnemius Muscle

1. Why is it important to destroy the brain and spinal cord of a frog before conducting physiologic experiments on muscle contraction? _____

_____

_____

2. What sources of stimuli, other than electrical shocks, cause a muscle to contract? _____

_____

3. What is the most common stimulus for muscle contraction in the body? _____

4. Name the three phases of the muscle twitch, and state what happens during each phase:

_____, _____

_____, _____

_____, _____

5. Use the terms given on the right to identify the conditions described on the left:

_____ sustained contraction without any evidence of relaxation

_____ stimulus that results in no perceptible contraction

_____ stimulus at which the muscle first contracts perceptibly

_____ increasingly stronger contractions in the absence of increased stimulus intensity

_____ increasingly stronger contractions owing to stimulation at a rapid rate

_____ increasingly stronger contractions owing to increased stimulus strength

_____ weakest stimulus at which all muscle cells in the muscle are contracting

a. maximal stimulus

b. multiple motor unit summation

c. subthreshold or subliminal stimulus

d. tetanus

e. threshold stimulus

f. treppe

g. wave summation

6. With brackets and labels identify the portions of the tracing below that best correspond to three of the phenomena listed in the preceding key. Assume that only the timing of the stimulus has changed.

7. Complete the following statements. Insert your responses in the corresponding numbered blanks on the right.

The "all or none" law applies to skeletal muscle function at the __1__ level. When a weak but smooth muscle contraction is desired, a few motor units are stimulated at a __2__ rate. Treppe is referred to as the "warming up" process. It is believed that muscles contract more strongly after the first few contractions because the __3__ become more efficient. If blue litmus paper is pressed to the cut surface of a fatigued muscle, it changes color to red, indicating low pH. This situation is caused by the accumulation of __4__ in the muscle. Within limits, as the load on a muscle is increased, the muscle contracts __5__ strongly. The absolute refractory period is the time when the muscle cell will not respond to a stimulus because __6__ is occurring.

1. _____

2. _____

3. _____

4. _____

5. _____

6. _____

8. During the experiment on muscle fatigue, how did the muscle contraction pattern change as the muscle began to fatigue? _____

_____

How long was stimulation continued before fatigue was apparent? _____

If the sciatic nerve that stimulates the living frog's gastrocnemius muscle had been left attached to the muscle and the stimulus had been applied to the nerve rather than the muscle, would fatigue have become

apparent sooner or later? _____

Explain your answer. _____

_____

_____

9. Explain how the weak but sustained (smooth) muscle contractions of precision movements are produced.

_____

_____

_____

10. What do you think happens to a muscle in the body when its nerve supply is destroyed or badly

damaged? _____

_____

11. Explain the relationship between the load on a muscle and its strength of contraction. _____

_____

_____

12. The skeletal muscles are maintained in a slightly stretched condition for optimal contraction. How

is this accomplished? _____

_____

Why does overstretching a muscle drastically reduce its ability to contract? (Include an explanation

of the events at the level of the myofilaments.) _____

_____

_____

_____

13. If the length but not the tension of a muscle is changed, the contraction is called an isotonic con-traction. In an isometric contraction the tension is increased but the muscle does not shorten. Which type

of contraction did you observe most often during the laboratory experiments? _____

What is the role of isometric contractions in normal body functioning? _____

_____

_____

EXERCISE 17

Histology of Nervous Tissue

1. The cellular unit of the nervous system is the neuron. What is the major function of this cell type? _____

_____

2. Name four types of neuroglia and list at least four functions of these cells.

Types   **Functions**

_____   _____

_____   _____

_____   _____

_____   _____

3. Match each statement with a response chosen from the key.

Key: a. afferent neuron   f. ganglion   k. peripheral nervous
      b. associative neuron   g. neuroglia   system
      c. autonomic nervous   h. neurotransmitters   l. Schwann cells
         system   i. nerve   m. somatic nervous system
      d. central nervous system   j. nuclei   n. synapse
      e. efferent neuron   o. tract

_____ the brain and spinal cord collectively

_____ specialized cells that myelinate the fibers of neurons found in the peripheral nervous system

_____ specialized cells that myelinate the fibers of neurons found in the CNS

_____ junction or point of close contact between neurons

_____ a bundle of nerve processes inside the central nervous system

_____ neuron serving as part of the conduction pathway between sensory and motor neurons

_____ spinal and cranial nerves and ganglia

_____ collection of nerve cell bodies found outside the CNS

_____ neuron that conducts impulses away from the CNS to muscles and glands

_____ neuron that conducts impulses toward the CNS from the body periphery

_____ system that controls the involuntary activities of the body

_____ chemicals released by neurons that stimulate other neurons, muscles, or glands

_____ system involved in the voluntary activities of the body, such as activation of the skeletal muscles

Neuron Anatomy

1. Match the following anatomic terms (column B) with the appropriate description or function (column A).

Column A

_____ contains enlarged region of the nerve cell body from which the axon originates

_____ releases neurotransmitters

_____ conducts the impulse toward the nerve cell body

_____ increases the speed of impulse transmission and insulates the nerve fibers

_____ is site of the nucleus

_____ may be involved in the transport of substances within the neuron

_____ essentially rough endoplasmic reticulum, important metabolically

_____ conducts impulses away from the nerve cell body

Column B

a. axon

b. axonal end bulb

c. axon hillock

d. dendrite

e. myelin sheath

f. nerve cell body

g. neurofibril

h. Nissl body

2. Draw a "typical" neuron in the space below. Include and label the following structures on your diagram: cell body, nucleus, Nissl bodies, dendrites, axon, axon collaterals, myelin sheath, and nodes of Ranvier.

3. How is one-way conduction at synapses assured? _____

_____

4. What anatomic characteristic determines whether a particular neuron is classified as unipolar, bipolar, or multipolar? _____

Make a simple line drawing of each type here.

        Unipolar neuron   Bipolar neuron   Multipolar neuron

5. Describe how the Schwann cells form the myelin sheath and the neurilemma encasing the nerve processes. (You may want to diagram the process.) _____

_____

_____

_____

_____

Structure of a Nerve

1. What is a nerve? _____

_____

_____

2. State the location of each of the following connective tissue coverings:

endoneurium _____

perineurium _____

epineurium _____

3. What is the value of the connective tissue wrappings found in a nerve? _____

_____

4. Define mixed nerve: _____

_____

EXERCISE 18

Gross Anatomy of the Brain and Cranial Nerves

The Human Brain

1. Match the letters on the diagram of the human brain (right lateral view) to the proper terms listed at the left.

_____ frontal lobe

_____ parietal lobe

_____ temporal lobe

_____ precentral gyrus

_____ parieto-occipital fissure

_____ postcentral gyrus

_____ lateral fissure

_____ central fissure

_____ cerebellum

_____ medulla

_____ occipital lobe

_____ pons

2. In which of the cerebral lobes would the following functional areas be found?

auditory area _____   olfactory area _____

primary motor area _____   visual area _____

primary sensory area _____   Broca's area _____

3. Which of the following structures are *not* part of the brain stem? (Circle the appropriate response or responses.)

cerebral hemispheres   pons   midbrain   cerebellum   medulla   diencephalon

4. Completion: insert responses in corresponding blanks on the right.

A __1__ is an elevated ridge of cerebral tissue. The convolutions seen in the cerebrum are important because they increase the __2__. Gray matter is composed of __3__. White matter is composed of __4__. A fiber tract that provides for communication between different parts of the same cerebral hemisphere is called a(n) __5__, while one that carries impulses to and from the cerebrum from and to lower CNS areas is called a (n) __6__ tract. The lentiform nucleus along with the amygdaloid and caudate nuclei are collectively called the __7__.

1. _____

2. _____

3. _____

4. _____

5. _____

6. _____

7. _____

RS65

5. Identify the structures on the sagittal view of the human brain by matching the lettered areas to the proper terms at the left.

_____ cerebellum

_____ cerebral aqueduct

_____ cerebral hemisphere

_____ cerebral peduncle

_____ choroid plexus

_____ corpora quadrigemina

_____ corpus callosum

_____ fornix

_____ fourth ventricle

_____ hypothalamus

_____ mammillary bodies

_____ massa intermedia

_____ medulla oblongata

_____ optic chiasma

_____ pineal body

_____ pituitary gland   _____ septum pellucidum

_____ pons   _____ thalamus

6. Using the letters from the diagram in item 5, match the appropriate structure with the descriptions given below:

_____ site of regulation of body temperature and water balance; most important autonomic center

_____ consciousness depends on the function of this part of the brain

_____ located in the midbrain, contains reflex centers for vision and audition

_____ responsible for the regulation of posture and coordination of complex muscular movements

_____ important synapse site for afferent fibers traveling to the sensory cortex

_____ contains autonomic centers regulating blood pressure, heart rate, and respiratory rhythm, as well as coughing, sneezing, and swallowing centers

_____ large commisure connecting the cerebral hemispheres

_____ fiber tract involved with olfaction

_____ connects the third and fourth ventricles

_____ encloses the third ventricle

7. Embryologically, the brain arises from the rostral end of a tubelike structure that quickly becomes divided into three major regions. Groups of structures that develop from the embryonic brain are listed below. Designate the embryonic origin of each group as the hindbrain, midbrain, or forebrain.

_____ the diencephalon, including the thalamus, optic chiasma, and hypothalamus

_____ the medulla, pons, and cerebellum

_____ the cerebral hemispheres

8. What is the importance of the fact that the human cerebral hemispheres are highly convoluted? _____

_____

9. What is the function of the basal nuclei? _____

_____

_____

10. What is the corpus striatum, and how is it related to the fibers of the internal capsule? _____

_____

_____

11. A brain hemorrhage within the region of the right internal capsule results in paralysis of the left side

of the body. Explain why the left side (rather than the right side) is affected. _____

_____

12. Explain why trauma to the base of the brain is often much more dangerous than trauma to the frontal lobes. (Hint: Think about the relative functioning of the cerebral hemispheres and the brain stem structures.

Which contain centers more vital to life?) _____

_____

_____

13. In "split brain" experiments, the main commissure connecting the cerebral hemispheres is cut. First, *name*

this commissure: _____

Then, describe what results (in terms of behavior) can be anticipated in such experiments. (Use an appropriate reference if you need help with this one!)

_____

_____

_____

_____

Meninges of the Brain

1. Identify the meningeal (or associated) structures described below.

_____ outermost meninx covering the brain, composed of tough fibrous connective tissue

_____ innermost meninx covering the brain, delicate and highly vascular

_____ structure instrumental in returning cerebrospinal fluid to the venous blood in the dural sinuses

_____ structure that forms the cerebrospinal fluid

_____ middle meninx, like a cobweb in structure

_____ its outer layer forms the periosteum of the skull

_____ a dural fold that attaches the cerebrum to the crista galli of the skull

_____ a dural fold separating the cerebrum from the cerebellum

Cerebrospinal Fluid

1. Fill in the following flow sheet, which relates to the circulation of cerebrospinal fluid from its formation site (assume that this is one of the lateral ventricles) to the site of its reabsorption into the venous blood.

Lateral ventricle - - - - - - - - - - - - →   _____   - →

Third ventricle - - - - - - - - - - - - →   _____   - →

_____ - - - - - - - →   Foramen of _____   - →

- - - - →   Foramina of _____

_____ surrounding the brain and cord (and central canal of the cord)   - →

Arachnoid villi - - - - - - - - - - - - →   _____ containing venous blood

Cranial Nerves

1. Provide the name and number of the cranial nerves involved in each of the following activities, sensations, or disorders.

_____ shrugging the shoulders

_____ smelling a flower

_____ raising the eyelids; focusing the lens of the eye for accommodation; and pupillary constriction

_____ slows the heart; increases the mobility of the digestive tract

_____ involved in Bell's palsy (facial paralysis)

_____ chewing food

_____ listening to music; seasickness

_____ secretion of saliva; tasting well-seasoned food

_____ involved in "rolling" the eyes (three nerves—provide numbers only)

_____ feeling a toothache

_____ reading Playgirl or Playboy magazine

_____ purely sensory in function (three nerves—provide numbers only)

Dissection of the Sheep Brain

1. In your own words, describe the relative hardness of the sheep brain tissue as observed when cutting into it.

_____

_____

Since formalin hardens all tissue, what conclusions might you draw about the relative hardness and texture of

living brain tissue? _____

_____

2. How does the relative size of the cerebral hemispheres compare in sheep and human brains? _____

_____

What is the significance? _____

_____

3. Which cranial nerves are much larger in the sheep brain than in the human brain? _____

_____ What is the significance of this difference? _____

_____

Application of Knowledge

You have been given all of the information you need to identify the brain region involved in the situations described below. See how well your nervous system has integrated this information, and name the region most likely to be involved in each situation.

1. Following a train wreck, a man with an obvious head injury was observed stumbling about the scene. An inability to walk properly and a loss of balance were quite obvious. What brain region was injured?

_____

2. A young woman was brought into the emergency room with extremely dilated pupils. Her friends stated

that she had overdosed on cocaine. What cranial nerve was stimulated by the drug? _____

3. A young man is admitted to the hospital after receiving serious burns resulting from standing with his back too close to a bonfire. He is muttering that he never felt the pain—otherwise, he would have smothered the flames by rolling on the ground. What part of his CNS might be malfunctional? _____

_____

4. An elderly gentleman had just suffered a stroke. He is able to understand verbal and written language, but when he tries to respond, his words come out garbled. What cortical region has been damaged by the stroke?

_____

EXERCISE 19

The Spinal Cord, Spinal Nerves, and Autonomic Nervous System

Anatomy of the Spinal Cord

1. Match the descriptions given below to the proper anatomic term:

a. cauda equina   b. conus medullaris   c. filium terminale   d. foramen magnum

_____ most superior boundary of the spinal cord

_____ meningeal extension beyond the spinal cord terminus

_____ spinal cord terminus

_____ collection of spinal nerves traveling in the vertebral canal below the terminus of the spinal cord

2. Choose the proper answer from the following key to respond to the following descriptions relative to spinal cord anatomy.

Key: a. afferent   b. efferent   c. both afferent and efferent   d. internuncial

_____ neuron type found in dorsal horn   _____ fiber type in ventral root

_____ neuron type found in ventral horn   _____ fiber type in dorsal root

_____ neuron type in dorsal root ganglion   _____ fiber type in spinal nerve

3. Where in the vertebral column is a lumbar puncture generally done? _____

Why is this the site of choice? _____

_____

4. The spinal cord is enlarged in two regions, the _____ and the _____

region. What is the significance of these enlargements? _____

_____

5. How does the position of the gray and white matter differ in the spinal cord and the cerebral hemispheres?

_____

_____

_____

_____

6. Choose the name of one tract, from the following key, that might be damaged when the following conditions are observed. (More than one choice may apply.)

_____ uncoordinated movement

_____ lack of voluntary movement

_____ tremors, jerky movements

_____ diminished pain perception

_____ diminished sense of touch

Key:

a. fasciculus gracilis
b. fasciculus cuneatus
c. lateral corticospinal tract
d. ventral corticospinal tract
e. tectospinal tract
f. rubrospinal tract
g. lateral spinothalamic tract
h. ventral spinothalamic tract
i. dorsal spinocerebellar tract
j. vestibulospinal tract
k. olivospinal tract
l. ventral spinocerebellar tract

7. Use an appropriate reference to describe the functional significance of an upper motor neuron and a lower motor neuron:

upper motor neuron _____

_____

lower motor neuron _____

_____

Will contraction of a muscle occur if the lower motor neurons serving it have been destroyed? _____ If

the upper motor neurons serving it have been destroyed? _____ Using an appropriate reference, differen-

tiate between flaccid and spastic paralysis and note the possible causes of each. _____

_____

_____

_____

_____

_____

Spinal Nerves and Nerve Plexuses

1. In the human there are 31 pairs of spinal nerves named according to the region of the vertebral column from which they issue. The spinal nerves are named below; note the vertebral level at which they emerge by number.

cervical nerves _____   sacral nerves _____

lumbar nerves _____   thoracic nerves _____

2. The ventral rami of spinal nerves C_1 through T_1 and T_{12} through S_4 take part in forming _____,

which serve the _____ of the body. The ventral rami of T_1 through T_{12} run between the ribs to serve the _____. The posterior rami of the spinal nerves serve _____

_____.

3. What would happen if the following structures were damaged or transected? (Use key choices for responses.)

Key: a. loss of motor function   b. loss of sensory function   c. loss of both motor and
                                                                                                          sensory function

_____ dorsal root of a spinal nerve

_____ ventral root of a spinal nerve

_____ anterior ramus of a spinal nerve

4. Define plexus: _____

_____

5. Name the major nerves that serve the following body areas:

_____ head, neck, shoulders (name plexus only)

_____ diaphragm

_____ posterior thigh

_____ lower leg and foot (name two)

_____ most anterior forearm muscles

_____ upper arm muscles

_____ abdominal wall (name plexus only)

_____ anterior thigh

_____ medial side of the hand

The Autonomic Nervous System

1. For the most part, sympathetic and parasympathetic fibers serve the same organs and structures. How can they exert antagonistic effects? (After all, nerve impulses are nerve impulses—aren't they?)

_____

_____

2. You are alone in your home late in the evening, and you hear an unfamiliar sound in your backyard. List four physiologic events promoted by the sympathetic nervous system that would aid you in coping with this rather frightening situation.

_____

_____

_____

_____

3. The chart on p. RS 74 states a number of conditions. Use a check mark to show which division of the autonomic nervous system is involved in each.

Sympathetic division	Condition	Parasympathetic division
	secretes norepinephrine, adrenergic fibers	
	secretes acetylcholine, cholinergic fibers	
	long preganglionic axon, short postganglionic axon	
	short preganglionic axon, long postganglionic axon	
	arises from cranial and sacral nerves	
	arises from spinal nerves T_1 to L_3	
	normally in control	
	fight or flight system	
	has more specific control (look it up!)	

4. Often after surgery, people are temporarily unable to urinate, and bowel sounds are absent. What division of the ANS is affected by the anesthesia? _____

_____

5. Name three structures that receive sympathetic innervation but not parasympathetic innervation. _____

_____

6. The pelvic nerve contains (circle one):

    a. preganglionic sympathetic fibers   c. preganglionic parasympathetic fibers

    b. postganglionic sympathetic fibers   d. postganglionic parasympathetic fibers

EXERCISE 20

Neurophysiology of Nerve Impulses

The Nerve Impulse

1. Match each of the terms in column B to the appropriate definition in column A.

Column A

Column B

_____ period of repolarization of the neuron membrane during which it cannot respond to a second stimulus

a. action potential

_____ state of reversal of the resting potential owing to an influx of sodium ions

b. depolarization

c. refractory period

_____ period during which potassium ions diffuse out of the neuron owing to a change in membrane permeability

d. repolarization

_____ self-propagated transmission of the depolarization wave along the neuron membrane

e. sodium-potassium pump

_____ process during which ATP is used to move sodium out of the cell and potassium into the cell; restores the resting conditions of the membrane potential and intracellular ionic concentrations

2. Respond appropriately to each statement below either by completing the statement or by answering the question raised. Insert your responses in the corresponding numbered blanks on the right.

1. The cellular unit of the nervous system is the neuron. What is the major function of this cell type?

1. _____

2 and 3. What characteristics are highly developed to allow the neuron to perform this function?

2. _____

3. _____

4. Would a substance that decreases membrane permeability to sodium increase or decrease the probability of generating a nerve impulse?

4. _____

3. Why don't the terms *depolarization* and *action potential* mean the same thing? (Hint: Under which conditions will a local depolarization *not* lead to the action potential?) _____

_____

_____

_____

4. A nerve generally contains many thickly myelinated fibers that typically exhibit nodes of Ranvier. An action potential is generated along these fibers by "saltatory conduction." Use an appropriate reference

to explain how saltatory conduction differs from conduction along unmyelinated fibers. _____

_____

_____

Physiology of Nerve Fibers

1. Why was the frog pithed before experimentation was begun? _____

_____

2. Respond appropriately to each question posed below. Insert your responses in the corresponding numbered blanks to the right.

1–3. Name three types of stimuli that resulted in the generation of action potentials in the sciatic nerve of the frog during
the laboratory experiments.

1. _____

2. _____

4. Which of the stimuli resulted in the most effective nerve
stimulation?

3. _____

4. _____

5. Which of the stimuli employed in that experiment might
represent types of stimuli to which nerves in the human body
are subjected?

5. _____

6. _____

6. What is the usual mode of stimulus transfer in
neuron-to-neuron interactions?

7. _____

7. Since the action potentials themselves were not visualized
with an oscilloscope, how did you recognize that impulses
were being transmitted?

3. Describe the observed effects of ether on nerve-muscle interaction. _____

_____

Does ether exert its blocking effects on nerve or muscle? _____ What observations

made during the experiment support your conclusion? _____

_____

_____

4. At what site did the tubocurarine block the impulse transmission? _____ Provide

evidence from the experiment to substantiate your conclusion. _____

_____

_____

Why was one of the frog's legs ligated in this experiment? _____

_____

_____

Motor Points

1. Define motor point: _____

   _____

2. Using a specific example from the experiment, describe what happened when a motor point was localized

   and stimulated: _____

   _____

   _____

3. How do you explain the observation that in some cases stimulation of a motor point caused the contrac-

   tion of a muscle located a considerable distance away? _____

   _____

4. Why would stimulation of motor points be important therapy for a person with nerve damage? (Consider

   what the absence of nerve stimulation would mean to the skeletal muscles.) _____

   _____

   _____

   _____

EXERCISE 21　　　　General Sensation

Structure of Sensory Receptors

1. Differentiate between interoceptors and exteroceptors in terms of anatomic location and stimulus source:

Interoceptor: _____

Exteroceptor: _____

2. A number of activities or sensations are listed in the chart below. For each, check whether the receptors would be exteroceptors or interoceptors, and then name the specific receptor types. (Since visceral receptors were not described in detail in this exercise, you need only indicate that the receptor is a visceral receptor if it falls into that category.)

Activity or sensation	Exteroceptor	Interoceptor	Specific receptor type
walking on hot pavement			
feeling a pinch			
reading a book			
leaning on a shovel			
muscle sensations when rowing a boat			
the "too full" sensation			
seasickness			

Receptor Physiology

1. Explain how the sensory receptors act as transducers: _____

2. Define *stimulus:* _____

3. What was demonstrated by the two-point discrimination test? _____

How did the accuracy of the subject's tactile localization correlate with the results of the two-point discrimination test? _____

4. Define punctate distribution: _____

5. Several questions regarding general sensation are posed below. Answer each by placing your response in the appropriately numbered blanks to the right.

1. Which cutaneous receptors are the most numerous?

2–3. Which two body areas tested were most sensitive to touch?

4–5. Which two body areas tested were least sensitive to touch?

6. Which appears to be more numerous—cold or heat receptors?

7–9. Where would referred pain appear if the following organs were receiving painful stimuli—gallbladder (#7), kidneys (#8), and appendix (#9)? (Use your textbook if necessary.)

10. Where was referred pain felt when the elbow was immersed in ice water during the laboratory experiment?

11. What region of the cerebrum interprets the kind and intensity of stimuli that cause cutaneous sensations?

1. _____

2. _____

3. _____

4. _____

5. _____

6. _____

7. _____

8. _____

9. _____

10. _____

11. _____

6. Define adaptation: _____

_____

7. Why is it advantageous to have pain receptors that are sensitive to all vigorous stimuli, whether heat,

cold, or pressure? _____

_____ Why is the nonadaptability of pain receptors important? _____

_____

8. Imagine yourself without any cutaneous sense organs. Why might this be very dangerous? _____

_____

_____

9. Define referred pain: _____

_____

_____

What is the probable explanation for referred pain? (Consult your textbook or an appropriate reference

if necessary.) _____

_____

_____

RS80

EXERCISE 22

Human Reflex Physiology

The Reflex Arc

1. Define *reflex:* _____

_____

2. Name four essential components of a reflex arc:

_____ , _____ , _____ , _____

3. In general, what is the importance of reflex testing in a routine physical examination? _____

_____

_____

Somatic and Autonomic Reflexes

1. Use the key terms to complete the statements given below.

Key: a. abdominal reflex   d. ciliospinal reflex   g. gag reflex
                        b. Achilles jerk   e. corneal reflex   h. patellar reflex
                        c. Babinski reflex   f. crossed extensor reflex   i. pupillary light reflex

Reflexes classified as somatic reflexes include _____, _____, _____, _____, _____, _____, and _____.

Of these, the deep tendon reflexes are _____ and _____, and the superficial cord reflexes are _____

and _____. Reflexes classified as autonomic reflexes include _____, _____.

2. In what way do cord-mediated reflexes differ from those involving higher brain centers? _____

_____

Name two cord-mediated reflexes: _____ and _____

Name two somatic reflexes in which the higher brain centers participate: _____

and _____

3. Can the stretch reflex be elicited in a pithed animal? _____ Explain your answer. _____

_____

4. Trace the reflex arc, naming efferent and afferent nerves, receptors, effectors, and synapse centers, for the following reflexes:

patellar reflex: _____

_____

Achilles jerk: _____

_____

5. Three factors were investigated in conjunction with patellar reflex testing that influence the rapidity and effectiveness of reflex arcs; that is, mental distraction, effect of simultaneous muscle activity in another body area, and fatigue.

Which of these factors increases the excitatory level of the spinal cord? _____

Which factor decreases the excitatory level of the muscles? _____

When the subject was concentrating on an arithmetic problem, did the change noted in the patellar reflex indi-

cate that brain activity is necessary for the patellar reflex or only that it may modify it? _____

_____

6. Name the division of the autonomic nervous system responsible for each of the following reflexes:

ciliospinal reflex _____ salivary reflex _____

pupillary light reflex _____

7. The pupillary light reflex, the crossed extensor reflex, and the corneal reflex illustrate the purposeful nature of reflex activity. Describe the protective aspect of each.

pupillary light reflex _____

_____

_____

corneal reflex _____

_____

_____

crossed extensor reflex _____

_____

_____

8. Was the pupillary light response contralateral or ipsilateral? _____ Why would

such a response be of significant value in this particular reflex? _____

_____

_____

_____

9. Differentiate between the types of activities accomplished by somatic and automatic reflexes. _____

_____

_____

_____

_____

10. Several types of reflex activity were not investigated in this exercise. The most important of these are autonomic reflexes, which are difficult to illustrate in a laboratory situation. To rectify this omission, complete the following chart, using references as necessary.

Reflex	Organ involved	Receptors stimulated	Action
micturition (urination)			
Hering-Breuer			
defecation			
carotid sinus			

Reaction Time of Unlearned Responses

1. Name at least three factors that may modify reaction time to a stimulus. _____

_____

_____

2. In general, how did the response time for the unlearned activity performed in the laboratory compare to that for the simple patellar reflex? _____

3. Did the response time without verbal stimuli decrease with practice? _____ Explain the reason for this.

_____

_____

4. Explain in detail why response time increased when the subject had to react to a word stimulus. _____

_____

_____

_____

EXERCISE 23   Electroencephalography

Brain Wave Patterns and the Electroencephalogram

1. Define EEG: _____ _____

2. What are the four major types of brain wave patterns? _____

Match each statement below with a type of brain wave pattern.

_____ below 3 cps; slow, large waves; normally seen during deep sleep

_____ rhythm generally apparent when an individual is in a relaxed, nonattentive state with the eyes closed

_____ correlated to the alert state; usually about 15 to 30 cps

_____ large, irregular, low-frequency waves; uncommon in adults but common in children

3. What is meant by the term *alpha block?* _____

_____

_____

4. List at least four types of brain lesions that may be determined by EEG studies. _____

_____

5. What is the common result of hypoactivity or hyperactivity of the brain neurons? _____

Observing Brain Wave Patterns

1. How was alpha block demonstrated in the laboratory experiment? _____

_____

2. What was the effect of mental concentration on the brain wave pattern? _____

_____

_____

3. What effect did hyperventilation have on the brain wave pattern? _____

Why? _____ _____

_____

_____

EXERCISE 24

Special Senses: Vision

Anatomy of the Eye

1. Three accessory eye structures contribute to the formation of tears and/or aid in lubrication of the eyeball. Name each and then name its major secretory product. Indicate which has antibacterial properties by circling the correct secretory product.

Accessory structures	Product

2. The eyeball is wrapped in adipose tissue within the orbit. What is the function of the adipose tissue?

_____

What seven bones form the bony orbit? (Think! If you can't remember, check a skull or your text.)

_____   _____   _____

_____   _____

_____   _____

3. Why does one often have to blow one's nose after having a good cry? _____

_____

4. Identify the extrinsic eye muscle predominantly responsible for the actions described below.

_____ turns the eye laterally

_____ turns the eye medially

_____ turns the eye up and laterally

_____ turns the eye inferiorly

_____ turns the eye superiorly

_____ turns the eye down and laterally

5. What is a sty? _____

Conjunctivitis? _____

6. Match the key responses with the descriptive statements on the left.

_____ attaches the lens to the ciliary body

_____ fluid filling the anterior chamber of the eye

_____ the "white" of the eye

_____ retinal area devoid of photoreceptors

_____ modification of the choroid, which controls the shape of the crystalline lens

_____ nutritive (vascular) tunic of the eye

_____ drains the aqueous humor of the eye

_____ tunic containing the rods and cones

_____ substance occupying the posterior cavity of the eyeball

_____ heavily pigmented tunic that prevents light-scattering within the eye

_____, _____ smooth muscle structures (intrinsic eye muscles)

_____ area of acute or discriminatory vision

_____ form (by filtration) the aqueous humor

_____, _____, _____, _____ refractory media of the eye

_____ anteriormost portion of the sclera—your "window on the world"

_____ tunic composed of tough, white fibrous connective tissue

Key:

a. aqueous humor

b. canal of Schlemm

c. choroid coat

d. ciliary body

e. ciliary processes of the ciliary body

f. cornea

g. fovea centralis

h. iris

i. lens

j. optic disk

k. retina

l. sclera

m. suspensory ligaments

n. vitreous humor

7. The iris is composed primarily of two smooth muscle layers, one arranged radially and the other circularly. Which of these dilates the pupil? _____

8. You would expect the pupil to be dilated in which of the following circumstances? (Circle the correct response(s).)

    a. in brightly lighted surroundings   c. during focusing for near vision

    b. in dimly lighted surroundings   d. in observing distant objects

9. The intrinsic eye muscles are under the control of which of the following? (Circle the correct response.)

          autonomic nervous system   somatic nervous system

Dissection of the Cow (Sheep) Eye

1. What modification of the choroid that is not present in humans is found in the cow eye? _____

_____

What is its function? _____

_____

2.   What is the anatomic appearance of the retina? _____

_____

At what point is it attached to the posterior aspect of the eyeball? _____

Microscopic Anatomy of the Retina

1.   The two major layers of the retina are the epithelial and nervous layers. In the nervous layer, the neuron populations are arranged as follows from the epithelial layer to the vitreous humor. (Circle all proper responses.)

bipolar cells, ganglion cells, photoreceptors　　　　　photoreceptors, ganglion cells, bipolar cells

ganglion cells, bipolar cells, photoreceptors　　　　　photoreceptors, bipolar cells, ganglion cells

2.   The axons of the _____ cells form the optic nerve, which exits from the eyeball.

3.   The following statements may be completed by inserting either RODS or CONES. Complete the following sentences:

The dim light receptors are the _____. Only _____ are found in the fovea centralis, while only

_____ are found in the periphery of the retina. _____ are the photoreceptors that operate best in

bright light and allow for color vision.

Visual Pathways to the Brain

1.   The visual pathway to the occipital lobe of the brain consists most simply of a chain of five neurons. Beginning with the photoreceptor cell of the retina, name them and note their location in the pathway.

1   Photoreceptor cell, retina

2   _____

3   _____

4   _____

5   _____

2.   Visual field tests are done to reveal destruction along the visual pathway from the retina to the optic region of the brain. Note where the lesion is likely to be in the following cases:

normal vision in left eye visual field; absence of vision of right eye visual field: _____

normal vision in both eyes for right half of the visual field; absence of vision in both eyes for left half of the

visual field: _____

3.   How is the right optic *tract* anatomically different from the right optic *nerve*? _____

_____

_____

What does this difference result from? _____

Visual Tests and Experiments

1. Match the terms in column B with the descriptions in column A.

Column A	Column B
_____ light bending	a. accommodation
_____ ability to focus for close (under 20 ft.) vision	b. astigmatism
_____ normal vision	c. convergence
_____ inability to focus well on close objects (farsightedness)	d. emmetropia
_____ nearsightedness	e. hypermetropia
_____ blurred vision due to unequal curvatures of the lens or cornea	f. myopia
_____ medial movement of the eyes during focusing on close objects	g. refraction

2. Complete the following statements:

In farsightedness, the light is focused __1__ the retina. The lens required to treat myopia is a __2__ lens. The "near point" increases with age because the __3__ of the lens decreases as we get older. A convex lens, like that of the eye, produces an image that is upside down and reversed from left to right. Such an image is called a __4__ image.

1. _____
2. _____
3. _____
4. _____

3. Check the appropriate column under the vertical headings to characterize events occurring within the eye during close and distant vision.

	Ciliary muscle		Suspensory ligaments		Lens convexity		Degree of light refraction	
close vision	relaxed	contracted	relaxed	contracted	increased	decreased	increased	decreased
distant vision	relaxed	contracted	relaxed	contracted	increased	decreased	increased	decreased

4. Explain why vision is lost when light hits the blind spot. _____

5. What is meant by the expression "20/20 vision"? _____

6. Record your Snellen eye test results below:

Left eye (without glasses) _____ (with glasses) _____

Right eye (without glasses) _____ (with glasses) _____

Is your visual acuity normal, less than normal, or better than normal? _____

Explain. _____

_____

_____

RS90

Explain why each eye is tested separately when using the Snellen eye chart. _____

_____

Explain 20/40 vision: _____

_____

Explain 20/10 vision: _____

_____

7. Define astigmatism: _____

_____

How can it be corrected? _____

8. Record the distance of your near-point of accommodation as tested in the laboratory:

right eye _____   left eye _____

Is your near-point within the normal range for your age? _____

9. Define presbyopia: _____

_____

What causes it? _____

10. To which wavelengths of light do the three cone types of the retina respond maximally?

_____   _____   _____

11. Since only three cone types exist, how can you explain the fact that we see a much greater range of colors?

_____

_____

12. From what condition does color blindness result? _____

_____

13. Record the results of the demonstration of the relative positioning of rods and cones (page 215) in the circle below (use appropriately colored pencils).

14. Explain the difference between binocular and panoramic vision. _____

_____

_____

_____

What is the advantage of binocular vision? _____

What factor(s) are responsible for binocular vision? _____

_____

_____

_____

15. Why is the ophthalmoscopic examination an important diagnostic tool? _____

_____

16. In the experiment on the convergence reflex (page 218), what happened to the position of the eyeballs

as the object was moved closer to the subject's eyes? _____

_____

What extrinsic eye muscles control the movement of the eyes during this reflex? _____

_____

What is the value of this reflex? _____

_____

What would be the visual result of an inability of these muscles to function? _____

_____

17. In the experiment on the photopupillary reflex (page 218), what happened to the eye pupil exposed to

light? _____ What happened to the pupil of the nonilluminated eye? _____

_____ Explanation? _____

_____

_____

18. Many college students struggling through mountainous reading assignments are told that they need
glasses for "eyestrain." Why is it more of a strain on the extrinsic and intrinsic eye muscles to look at close

objects than at far objects? _____

_____

EXERCISE 25

Special Senses: Hearing and Equilibrium

Anatomy of the Ear

1. Select the terms from column B that apply to the column A descriptions. Some terms are used more than once.

Column A

_____, _____, _____ structures comprising the outer or external ear

_____, _____, _____ structures composing the bony or osseous labyrinth

_____, _____, _____ collectively called the ossicles

_____, _____ ear structures not involved with hearing

_____ allows pressure in the middle ear to be equalized with atmospheric pressure

_____ vibrates at the same frequency as sound waves hitting it; transmits the vibrations at the ossicles

_____ contains the organ of Corti

_____, _____ contain receptors for the sense of equilibrium

_____ transmits the vibratory motion of the stirrup to the fluid in the Scala vestibuli of the inner ear

_____ acts as a pressure relief valve for the increased fluid pressure in the Scala tympani; bulges into the tympanic cavity

_____ connects the nasopharynx and the middle ear

_____ fluid contained within the membranous labyrinth

_____ fluid contained within the osseous labyrinth and bathing the membranous labyrinth

Column B

a. anvil

b. cochlea

c. endolymph

d. eustachian tube

e. external auditory canal

f. hammer

g. oval window

h. perilymph

i. pinna

j. round window

k. semicircular canals

l. stirrup

m. tympanic membrane

n. vestibule

2. Sound waves hitting the eardrum initiate its vibratory motion. Trace the pathway through which vibrations and fluid currents are transmitted to finally stimulate the hair cells in the organ of Corti. (Name the appropriate ear structures in their correct sequence.) Eardrum - - - → _____ _____

_____

3. The oval and round windows are membranous partitions in the bony wall separating the middle and inner ear. Explain fully the role each plays in the auditory process. _____

_____

_____

4. Match the membranous labyrinth structures listed on the right with the descriptive statements on the left.

_____, _____ found within the vestibule

_____ contains the organ of Corti

_____, _____ sites of the maculae

_____ positioned in all spatial planes

_____ hair cells of organ of Corti rest on this membrane

_____ gelatinous membrane overlying the hair cells of the organ of Corti

_____ contains the crista ampullaris

_____, _____, _____, _____ function in static equilibrium

_____, _____, _____, _____ function in dynamic equilibrium

_____ carries auditory information to the brain

_____ gelatinous cap overlying hair cells of the crista ampullaris

_____ grains of calcium carbonate in the maculae

a. ampulla

b. basilar membrane

c. cochlear duct

d. cochlear nerve

e. cupula

f. membranous semicircular canals

g. saccule

h. statoconia

i. tectorial membrane

j. utricle

k. vestibular nerve

5. Describe how sounds of different frequency (pitch) are differentiated in the cochlea. _____

_____

_____

6. Explain the role of the endolymph of the semicircular canals in activating the receptors during angular

motion. _____

_____

_____

7. Explain the role of the statoconia in perception of static equilibrium (head position). _____

_____

_____

_____

Laboratory Tests

1. Was the auditory acuity measurement made during the experiment on page 224 the same or different

for both ears? _____ What factors might account for a difference in the

acuity of the two ears? _____

_____

2. During the sound localization experiment on page 224, in which position(s) was the sound least easily located? _____

How can this phenomenon be explained? _____

_____

3. In the experiment on page 225, which tuning fork was the most difficult to hear? _____ cps.

What conclusion can you draw? _____

_____

4. When the tuning fork handle was pressed to your forehead during the Weber test, where did the sound seem to originate? _____

Where did it seem to originate when one ear was plugged with cotton? _____

_____ How do sound waves reach the cochlea when conduction deafness is present?

_____

5. Indicate whether the following conditions described relate to conduction deafness (C) or sensineural (central) deafness (S):

_____ can result from the fusion of the ossicles

_____ can result from a lesion on the cochlear nerve

_____ sound heard in one ear but not in the other during bone and air conduction

_____ often improved by a hearing aid

_____ can result from otitis media

_____ can result from impacted cerumen or a perforated eardrum

_____ can result from a blood clot in the auditory cortex

6. The Rinne test evaluates an individual's ability to hear by air- or bone-conducted sound. Which is more indicative of normal hearing? _____

7. Define *nystagmus:* _____

_____

vertigo _____

_____

8. The Banary test investigated the effect of rotatory acceleration on the semicircular canals. Explain *why* the subject still had the sensation of rotation immediately after being stopped. _____

_____

9. The series I and II experiments on equilibrium tested the effect of acceleration on the semicircular canals. The head positioning determined which semicircular canals were activated. On the chart below, indicate the position of the head, note the semicircular canals affected, the type of nystagmus resulting from the experiment, and the direction of the fall under those conditions.

Head position	Rotation	Nystagmus	Canal	Direction of fall
Erect	Clockwise			
	Counterclockwise			
Forward on chest	Clockwise			
	Counterclockwise			
On left shoulder	Clockwise			
	Counterclockwise			
On right shoulder	Clockwise			
	Counterclockwise			

10. What is the usual reason for conducting the Romberg test? _____

_____

_____

Was the degree of sway observed greater with the eyes open or closed? _____

Why? _____

11. Normal balance or equilibrium is dependent on input from a number of sensory receptors. Name them.

_____

_____

EXERCISE 26

Special Senses: Taste and Olfaction

Localization and Anatomy of Taste Buds

1. Name three sites where receptors for taste are found, and circle the predominant site.

_____, _____ and _____

2. Describe the cellular makeup and arrangement of a taste bud. (Use a diagram if helpful.) _____

_____

_____

Localization and Anatomy of Olfactory Receptors

1. Describe the cellular composition and location of the olfactory epithelium. _____

_____

_____

2. How and why does sniffing improve your sense of smell? _____

_____

Laboratory Experiments

1. Taste and smell receptors are both classified as _____ because they both re-

spond to _____

2. Why is it impossible to taste substances with a dry tongue? _____

_____

3. State the most important sites of the taste-specific receptors as determined during the plotting exercise in the laboratory.

salt _____   sour _____

bitter _____   sweet _____

4. The basic taste sensations are elicited by specific chemical substances or groups. Name them.

salt _____   sour _____

bitter _____   sweet _____

5. Name three factors that influence our appreciation of foods. Substantiate each choice with an example from the laboratory experience.

_____   Substantiation _____

_____

_____

_____   Substantiation _____

_____

_____

_____   Substantiation _____

_____

_____

Which of the factors chosen is most important? _____

Substantiate your choice with an example from everyday life. _____

_____

Expand on your explanation and choices by explaining why a cold, greasy hamburger is unappetizing to most

people. _____

_____

6. Babies tend to favor bland foods, whereas adults tend to like highly seasoned foods. What is the basis for this

phenomenon? _____

_____

_____

7. How palatable is food when you have a cold? _____

Explain _____

_____

8. What is the mechanism of olfactory adaptation? _____

_____

_____

In your opinion, is olfactory adaptation desirable? _____ Explain your answer.

_____

_____

EXERCISE 27

Anatomy and Basic Function of the Endocrine Glands

Gross Anatomy and Basic Function of the Endocrine Glands

1. Both the endocrine and nervous systems are major regulating systems of the body; however, the nervous system has been compared to an airmail delivery system and the endocrine system to the pony express.

Briefly explain this comparison. _____

_____

_____

_____

2. Note several ways in which the endocrine and exocrine glands differ. Include in this discussion their mode

of formation, products, and mode of product export. _____

_____

_____

_____

3. Define hormone. _____

_____

_____

4. Chemically, hormones belong to two molecular groups, the _____

and the _____.

5. What do all hormones have in common? _____

_____

6. Define target organ: _____

_____

7. Why don't all tissues respond to all hormones? _____

_____

8. Identify the endocrine organ for each of the following descriptive statements:

_____ located in the throat; bilobed gland connected by an isthmus

_____ found close to the kidney

RS99

_____ a mixed gland, located in the mesentery close to the stomach and small intestine

_____ paired glands suspended in the scrotum

_____ ride "horseback" on the thyroid gland

_____ found in the pelvic cavity of the female, concerned with ova and female hormone production

_____ found in the upper thorax overlying the heart; short-lived endocrine function

_____ found in the roof of the third ventricle

9. For each statement describing hormonal effects, identify the hormone(s) involved by choosing a number from key A, and note the hormone's site of production with a letter from key B. More than one hormone may be involved in some cases.

Key A   Key B

1. ACTH   13. MSH   a. adrenal cortex
2. ADH   14. oxytocin   b. adrenal medulla
3. aldosterone   15. progesterone   c. anterior pituitary
4. cortisone   16. PTH   d. hypothalamus
5. epinephrine   17. serotonin   e. ovaries
6. estrogen   18. STH   f. pancreas
7. FSH   19. testosterone   g. parathyroid glands
8. glucagon   20. thymosin   h. pineal gland
9. insulin   21. thyrocalcitonin   i. posterior pituitary
10. LH   22. thyroxine   j. testes
11. LTH   23. TSH   k. thymus gland
12. melatonin   l. thyroid gland

_____, _____ basal metabolism hormone

_____, _____ programming of T-lymphocytes

_____, _____ and _____, _____ regulation of blood calcium levels

_____, _____ and _____, _____ released in response to stressors

_____, _____ and _____, _____ development of secondary sexual characteristics

_____, _____; _____, _____; _____, _____; and _____, _____ regulate the function of

another endocrine gland

_____, _____ mimics the sympathetic nervous system

_____, _____ and _____, _____ regulate blood glucose levels

_____, _____ and _____, _____ directly responsible for regulation of the menstrual cycle

_____, _____ and _____, _____ regulation of the ovarian cycle

_____, _____ and _____, _____ maintenance of salt and water balance in the ECF

_____, _____ and _____, _____ directly involved in milk production and ejection

_____, _____ questionable function; may stimulate the melanocytes of the skin

10. Although the pituitary gland is often referred to as the master gland of the body, recent studies show that the hypothalamus exerts some control over the pituitary gland. How does the hypothalamus control both anterior and posterior pituitary functioning? _____

_____

_____

_____

11. Indicate whether the release of the hormones listed below is stimulated by A, another hormone, B, the nervous system (neurotransmitters, or releasing factors); or C, humoral factors (the concentration of specific substances in the blood or extracellular fluid):

_____ thyroxin   _____ parathormone   _____ ADH

_____ insulin   _____ testosterone   _____ TSH, FSH

_____ estrogen   _____ epinephrine   _____ aldosterone

12. Name the hormone that would be produced in *inadequate* amounts under the following conditions:

_____ sexual immaturity

_____ tetany

_____ excessive diuresis without high blood glucose levels

_____ excessive thirst, high blood glucose levels

_____ abnormally small stature, normal proportions

_____ miscarriage

_____ lethargy, hair loss, low BMR, obesity

13. Name the hormone that is produced in *excessive* amounts in the following conditions:

_____ lantern jaw and large hands and feet in the adult

_____ bulging eyeballs, nervousness, increased pulse rate

_____ demineralization of bones, spontaneous fractures

Microscopic Anatomy of Selected Endocrine Glands

1. Choose a response from the key below to name the hormone(s) produced by the cell types listed:

Key: a. insulin   d. thyrocalcitonin   g. glucagon
     b. STH, LTH   e. TSH, ACTH, FSH, LH   h. PTH
     c. thyroxine   f. mineralocorticoids   i. glucocorticoids

_____ parafollicular cells of the thyroid   _____ zona reticularis cells

_____ follicular epithelial cells of the thyroid   _____ zona glomerulosa cells

_____ beta cells of the islets of Langerhans   _____ oxyphil cells

_____ alpha cells of the islets of Langerhans   _____ acidophil cells of the anterior pituitary

_____ basophil cells of the anterior pituitary

2. Five diagrams of the microscopic structures
of the endocrine glands are presented here. Identify
each and name all structures indicated with leader lines.

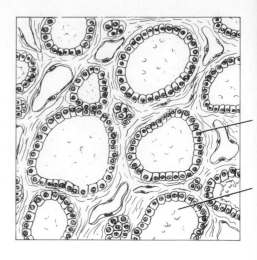

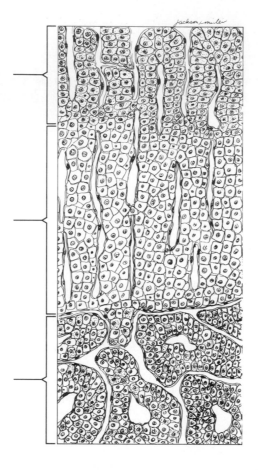

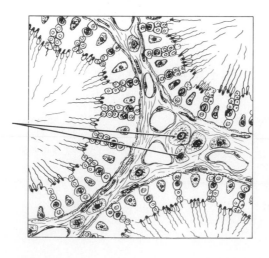

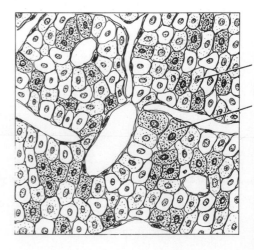

EXERCISE 28

Experiments on Hormonal Action

Effect of Pituitary Hormones on the Ovary

1. In the experiment on the effects of pituitary hormones, two anterior pituitary hormones caused ovulation to occur in the experimental animal. Which of these actually triggered ovulation or egg expulsion? _____ _____ The normal function of the second hormone involved, ____

   is to _____

2. Why was a second frog injected with saline? _____

   _____

Effects of Hyperinsulinism

1. Briefly explain what was happening within the fish's system when it was immersed in the insulin solution.

   _____

   _____

2. What is the mechanism of the recovery process observed? _____

   _____

3. What would you do to help a friend who had inadvertently taken an overdose of insulin? _____

   _____ Why? _____

4. What is a glucose tolerance test? (Use an appropriate reference as necessary to answer this question.)

   _____

   _____

5. How does diabetic coma differ from insulin shock? _____

   _____

   _____

Effect of Epinephrine on the Heart

1. Based on your observations, what is the effect of epinephrine on the force and rate of the heartbeat?

   _____

2. What is the role of this effect in the "flight or fight" response? _____

   _____

Effect of Thyroxine on Metabolic Rate

1. Relative to the measurement of oxygen consumption in rats, which group had the highest metabolic rate?

_____

Which group had the lowest metabolic rate? _____

Correlate these observations with the pretreatment these animals received. _____

_____

_____

Which group of rats was hyperthyroid? _____

Which euthyroid? _____ Which hypothroid? _____

2. Since oxygen usage = carbon dioxide evolved, how were you able to measure the oxygen consumption in

the experiments? _____

_____

3. What did changes in the fluid levels in the manometer arms indicate? _____

_____

4. The techniques used in this set of laboratory experiments probably allowed for several inaccuracies. One was the inability to control the activity of the rats. How would changes in their activity levels affect the results

observed? _____

_____

Another possible source of error is the fact that the amount of food consumed by the rats was not controlled in the 14-day period preceding the lab session. If each of the rats had been force-fed equivalent amounts of

food, which group (do you think) would have gained the most weight in that 14-day period? _____

Which the least? _____ Explain your answers: _____

_____

_____

5. TSH produced by the anterior pituitary prods the thyroid gland to release thyroxine to the blood. Which

group of rats can be assumed to have the *highest* blood levels of TSH? _____

Which the lowest? _____ Explain your reasoning. _____

_____

_____

6. Use an appropriate reference to determine how each of the following factors modifies metabolic rate. Indicate increase by ↑ and decrease by ↓ .

increased exercise _____   aging _____   infection/fever _____

small/slight stature _____   obesity _____   sex (♂ or ♀) _____

EXERCISE 29

Blood

Composition of Blood

1. What is the blood volume of an averaged-size adult? _____ liters

2. Explain why blood is classified as a connective tissue. _____
_____

_____

3. What determines whether blood is bright red or a dull brick-red? _____
_____

4. Use the key to identify the cell type(s) or blood elements that fit the following descriptive statements:

_____ most numerous leukocyte

_____, _____, _____ granular leukocytes

_____ also called an erythrocyte, anucleate

_____, _____ actively phagocytic leukocytes

_____, _____ agranular leukocytes

_____ fragments to form platelets

_____ (a) through (g) are all examples of these

_____ increases during allergy attacks

_____ releases histamine during inflammatory reactions

_____, _____ formed in lymphoid tissue

_____ contains hemoglobin; therefore involved in oxygen transport

_____ primarily water, noncellular; the fluid matrix of blood

_____ increases in number during prolonged infections

_____, _____, _____, _____, _____ also called white blood cells

Key: a. red blood cell

b. megakaryocyte

c. eosinophil

d. basophil

e. monocyte

f. neutrophil

g. lymphocyte

h. formed elements

i. plasma

5. Supply the following information based on what you know about substances normally found dissolved in plasma. Name three nutrient substances. _____ , _____ and

_____

Name two gases. _____ and _____

Name three ions. _____ , _____ , and _____

RS105

6. Describe the consistency and color of the plasma you observed in the laboratory. _____

_____

7. What is the average life span of a red blood cell? How does its anucleate condition affect this life span?

_____

_____

8. From memory, describe the structural characteristics of each of the following blood cell types as accurately as possible, and note the percentage of each in the total white blood cell population.

eosinophils _____

_____

neutrophils _____

_____

lymphocytes _____

_____

basophils _____

_____

monocytes _____

_____

Hematologic Tests

1. Broadly speaking, why are hematologic stuides of blood so important in the diagnosis of disease?

_____

_____

2. With information from the blood tests you conducted, fill in the following chart. (Record your values for healthy adults.)

Test	Normal values (healthy adults)	Significance	
		High values	Low values
total WBC count			
total RBC count (male)			

Test	Normal values (healthy adults)	Significance	
		High values	Low values
hematocrit (male)			
hemoglobin determination (male)			
sedimentation rate			
coagulation time			

3. Why is a differential WBC count more valuable when trying to pin down the specific source of pathology than a total WBC count? _____

_____

_____

4. Explain the reasons for using different diluents for the RBC and WBC counts. _____

_____

_____

5. What name is given to the process of RBC production? _____

What acts as a stimulus for this process? _____

What organ provides this stimulus and under what conditions? _____

_____

6. Discuss the effect of each of the following factors on RBC count. Consult an appropriate reference as necessary, and explain your reasoning.

athletic training (for example, running 4 to 5 miles per day over a period of 6 to 9 months) _____

_____

a permanent move from sea level to a high-altitude area _____

_____

_____

lack of iron containing foods in your diet _____

_____

_____

7. Define *hematocrit:* _____

8. If you had a high hematocrit, would you expect your hemoglobin determination to be high or low? _____ Why? _____

_____

9. What is an anticoagulant? _____

Name two anticoagulants used in conducting the hematologic tests. _____

and _____

What is the body's natural anticoagulant? _____

10. If your blood clumped with both anti-A and anti-B sera, your ABO blood type would be _____.

To what ABO blood groups could you give blood? _____ From which ABO donor types

could you receive blood? _____ Which ABO blood type is most common? _____

Least common? _____

11. Explain why an Rh-negative person does not have a transfusion reaction on the first exposure to Rh-positive blood, but *does* have a reaction on the second exposure. _____

_____

What happens when an ABO blood type is mismatched for the first time? _____

_____

12. Correctly identify the blood pathologies described in column A by choosing a response from column B.

Column A	Column B
_____ abnormal increase in the number of WBCs	a. anemia
_____ abnormal increase in the number of RBCs	b. leukocytosis
_____ condition of too few RBCs or RBCs with hemoglobin deficiencies	c. leukopenia
_____ abnormal decrease in the number of WBCs	d. polycythemia

RS108

EXERCISE 30

Anatomy of the Heart

Gross Anatomy of the Human Heart

1. Complete the following statements:

   The heart is a cone-shaped muscular organ located within the __1__. Its apex rests on the __2__ and its base is at the level of the __3__ rib.
   The coronary arteries, which nourish the myocardium, arise from the __4__. The coronary sinus empties into the __5__. Relative to the roles of the heart chambers, the __6__ are receiving chambers, while the __7__ are discharging chambers. The membrane that lines the heart and also forms the valve flaps is called the __8__. The outermost layer of the heart is called the __9__. The heart muscle itself is also called the __10__.

   1. _____
   2. _____
   3. _____
   4. _____
   5. _____
   6. _____
   7. _____
   8. _____
   9. _____
   10. _____

2. What is the function of the fluid that fills the pericardial sac? _____

3. Why might a thrombus (blood clot) in a branch of the left coronary artery cause sudden death? _____

4. An anterior view of the heart is shown here. Identify each numbered structure, and write its name on the corresponding numbered line.

   1. _____
   2. _____
   3. _____
   4. _____
   5. _____
   6. _____
   7. _____
   8. _____
   9. _____
   10. _____

   11. _____
   12. _____
   13. _____
   14. _____
   15. _____
   16. _____
   17. _____
   18. _____
   19. _____
   20. _____
   21. _____

5. What is the function of the valves found in the heart? _____

6. Can the heart function with leaky valves? (Think! Can a water pump function with leaky valves?) _____

7. What is the role of the chordae tendineae? _____

_____

_____

8. Define:

angina pectoris _____

_____

pericarditis _____

9. Differentiate clearly between the *roles* of the pulmonary and systemic circulations. _____

_____

_____

_____

10. Complete the following scheme of circulation in the human body:

Right atrium through the tricuspid valve to the _____ through the _____

_____ valve to the pulmonary trunk to the _____

to the capillary beds of the lungs to the _____ to the _____

of the heart through the _____ valve to the _____ through the

_____ valve to the _____ to the systemic arteries to the

_____ of the tissues to the systemic veins to the _____ and

_____ entering the right atrium of the heart.

11. If the mitral valve does not close properly, which circulation is affected? _____

_____

Dissection of the Sheep Heart

1. During the sheep heart dissection, you were asked initially to identify the right and left ventricles without cutting into the heart. During this procedure, what differences did you observe between the two

chambers? _____

_____

Since structure and function are related, how does this structural difference reflect the relative functions

of these two heart chambers? _____

_____

_____

2. Semilunar valves prevent backflow into the _____; AV valves prevent back-

flow into the _____. Using your own observations, explain how the operation of

the semilunar valves differs from that of the AV valves. _____

_____

_____

_____

3. What is the moderator band, where is it located, and what is its role? _____

_____

_____

4. Differentiate clearly between the location and appearance of pectinate muscle and trabeculae carneae.

_____

_____

_____

5. Two remnants of fetal structures are observable in the heart—the ligamentum arteriosum and the fossa ovalis. Where is each located, and what common purpose did they serve as functioning fetal structures?

_____

_____

_____

EXERCISE 31

Conduction System of the Heart and Electrocardiography

The Purkinje System

1. Is the Purkinje system considered part of the intrinsic or extrinsic regulation system of the heart? _____

_____

2. List the elements of the Purkinje system in order, starting from the SA node:

SA node ---→ _____ ---→ _____ ---→

_____ ---→ _____

Which of those structures is replaced when an artificial pacemaker is installed? _____

At what structure in the transmission sequence is the impulse temporarily delayed? _____

Why? _____

_____

3. Even though cardiac muscle has an inherent ability to beat, the Purkinje system plays a critical role in heart

physiology. What is that role? _____

_____

4. How does the "all or none" law apply to normal heart operation? _____

_____

_____

Electrocardiography

1. Define ECG: _____

_____

2. Draw an ECG wave form for one heart beat. Label the P, QRS, and T waves, the P-R interval, and the S-T segment.

3. What changes from *baseline* were noted in the ECG recorded during running? _____

Explain: _____

That recorded during breath holding? _____

_____

 Explain: _____

_____

4. Describe what happens in the cardiac cycle in the following situations.

during the P wave _____

immediately before the P wave _____

immediately after the P wave _____

during the QRS wave _____

immediately after the QRS wave (S-T interval) _____

_____

during the T wave _____

5. Define the following terms:

tachycardia: _____

bradycardia: _____

flutter: _____

fibrillation: _____/

myocardial infarction: _____

6. Which would be more serious, atrial or ventricular fibrillation? _____

Why? _____

_____

7. Abnormalities of heart valves can be detected more accurately by auscultation than by electrocardiography.

Why is this so? _____

_____

_____

_____

EXERCISE 32

Anatomy of Blood Vessels

Microscopic Structure of Blood Vessels

1. Use key choices to identify the blood vessel tunic described.

Key: a. tunica intima　　　　b. tunica media　　　　c. tunica externa

_____ single thin layer of endothelium

_____ bulky middle coat containing smooth muscle and elastin

_____ provides a smooth surface to decrease resistance to blood flow

_____ the only tunic of capillaries

_____ also called the adventitia

_____ the only tunic that plays an active role in blood pressure regulation

_____ supporting, protective coat

2. Servicing the capillaries is the essential function of the organs of the circulatory system. Explain this

statement. _____

_____

_____

3. Cross sectional views of an artery and a vein are shown here. Identify each and on the lines beneath, note the structural details that allowed you to make these identifications.

4. Why are valves present in veins but not in arteries? _____

_____

_____

5. Name two events *occurring within the body* that aid in venous return.

_____ and _____

6. Why are the walls of arteries proportionately thicker than those of the corresponding veins? _____

_____

Major Systemic Arteries and Veins of the Body

1. Use the key on the right to identify the arteries described on the left.

_____ and_____ two arteries formed by the bifurcation of the brachio-cephalic artery

_____ first branches off the ascending aorta

_____ and_____ two paired arteries serving the brain

_____ largest artery of the body

_____ arterial network on the dorsum of the foot

_____ serves the posterior thigh

_____ supplies the diaphragm

_____ bifurcates to form the radial and ulnar arteries

_____ artery generally auscultated to determine the blood pressure in the arm

_____ artery serving the kidney

_____ artery serving the liver

_____ artery that supplies the last half of the large intestine

_____ serves the pelvic organs

_____ the external iliac artery becomes the (what?) on entry into the thigh

_____ major artery serving the upper arm

_____ supplies most of the small intestine

_____ the terminal branches of the dorsal or descending aorta

_____ an arterial trunk that has three major branches, which run to the liver, spleen, and stomach

_____ major artery serving the tissues external to the skull

_____, _____, and _____ three arteries serving the leg inferior to the knee

_____ artery generally used to take the pulse at the wrist

a. anterior tibial

b. aorta

c. brachial

d. brachiocephalic

e. celiac trunk

f. common carotid

g. common iliac

h. coronary

i. deep femoral

j. dorsalis pedis

k. external carotid

l. femoral

m. hepatic

n. inferior mesenteric

o. intercostals

p. internal carotid

q. internal iliac

r. peroneal

s. phrenic

t. posterior tibial

u. radial

v. renal

w. subclavian

x. superior mesenteric

y. vertebral

z. ulnar

Special Circulations

Pulmonary circulation:

1. Trace the pathway of a carbon dioxide gas molecule in the blood in the inferior vena cava until it leaves the bloodstream. Name all structures (vessels, heart chambers, and others) passed through en route.

_____

_____

2. Trace the pathway of an oxygen gas molecule from an alveolus of the lung to the right atrium of the heart. Name all structures through which it passes. _____

_____

_____

3. Most arteries of the adult body carry oxygen-rich blood, and the veins carry oxygen-depleted, carbon dioxide–rich blood. What is different about the pulmonary arteries and veins? _____

_____

_____

4. How do the arteries of the pulmonary circulation differ structurally from the systemic arteries? What condition results from this anatomic difference? _____

_____

_____

Arterial supply of the brain and the circle of Willis:

1. What two paired arteries enter the skull to supply the brain?

_____ and _____

2. The paired arteries just named cooperate to form a ring of blood vessels encircling the pituitary gland at the base of the brain. What name is given to this communication network? _____

What is its function ? _____

_____

3. What portion of the brain is served by the anterior and middle cerebral arteries? _____

_____

Both the anterior and middle cerebral arteries arise from the _____

arteries.

4. Trace the pathway of a drop of blood from the aorta to the occipital lobe of the brain, noting all structures through which it flows. _____

_____

Hepatic portal circulation:

1. What is the source of blood in the hepatic portal system? _____

_____

2. Why is this blood carried to the liver before it enters the systemic circulation? _____

_____

3. The hepatic portal vein is formed by the union of the _____, which drains

the _____, _____, _____

_____, and the _____, which drains the _____

and _____ . The _____ vein, which drains the lesser

curvature of the stomach, empties directly into the hepatic portal vein.

4. Trace the flow of a drop of blood from the small intestine to the right atrium of the heart, noting all

structures encountered or passed through on the way. _____

_____

_____

_____

Fetal circulation:

1. You have learned that the fetal bypass structures become obliterated after birth. The failure of two of these to close could cause congenital heart disease, in which the youngster would have improperly oxygen-

ated blood. Which two structures are these? _____

and _____

2. For each of the following structures, first note its function in the fetus, and then note what happens to it or what it is converted to after birth. *Circle* the blood vessel that carries the most oxygen-rich blood.

Structure	Function in fetus	Fate
umbilical artery		
umbilical vein		
ductus venosus		
ductus arteriosus		
foramen ovale		

3. What organ serves as a respiratory/digestive/excretory organ for the fetus? _____

Dissection of the Blood Vessels of the Fetal Pig

1. What differences did you observe between the origin of the common carotid arteries in the pig and in

the human? _____

_____

_____

_____

2. How do the relative sizes of the external and internal jugular veins differ in the human and the pig?

_____

_____

3. How did the brachial veins of the pig differ from those of humans? _____

_____

_____

4. What difference was noted between the origin of the hepatic portal vein in the pig and in humans?

_____

_____

Between the origin of the internal and external iliac arteries? _____

_____

_____

_____

_____

5. In the pig, the inferior vena cava is called the _____

while the superior vena cava is called the _____.

EXERCISE 33

Human Cardiovascular Physiology—Blood Pressure and Pulse Determinations

Cardiac Cycle

1. Define the following terms:

systole _____

diastole _____

cardiac cycle _____

2. Using the key below, indicate the time interval occupied by the following events of the cardiac cycle:

a. 0.4 sec   b. 0.3 sec   c. 0.1 sec   d. 0.8 sec

_____ the length of the normal cardiac cycle   _____ the quiescent period, or pause

_____ time interval of atrial systole   _____ ventricular contraction period

3. If an individual's heart beats 80/min, what is the length of the cardiac cycle? _____ What portion of the

cardiac cycle is shortened by this more rapid heart beat? _____

4. Answer the following questions, which concern events of the cardiac cycle. When are the AV valves closed?

_____

Open? _____

What event within the heart causes the AV valves to open? _____

What causes them to close? _____

When are the semilunar valves closed? _____

_____

Open? _____

What event causes the semilunar valves to open? _____

_____

To close? _____

Are both sets of valves closed during any part of the cycle? _____

If so, when? _____

_____

Are both sets of valves open during any part of the cycle? _____

At what point in the cardiac cycle is the pressure highest in the heart? _____

Lowest? _____ What event results in the pressure deflection called the di-

crotic notch? _____

_____

5. What two factors promote the movement of blood through the heart? _____

_____ and

_____

Auscultation of Heart Sounds

1. Complete the following statements:

The monosyllables describing the heart sounds are __1__. The
first heart sound is a result of closure of the __2__ valves while
the second is a result of closure of the __3__ valves. The heart
chambers, which have just been filled when you hear the first
heart sound, are the __4__, and the chambers, which have just
emptied, are the __5__. Immediately after the second heart sound,
the __6__ are filling with blood and the __7__ are empty.

1. _____

2. _____

3. _____

4. _____

5. _____

6. _____

7. _____

2. As you listened to the heart sounds during the laboratory, what differences in pitch, length, and amplitude

(loudness) of the two sounds did you observe? _____

_____

_____

3. Indicate where you would place your stethoscope to auscultate the following most accurately:

closure of the tricuspid valve _____

closure of the aortic semilunar valve _____

apical heartbeat _____

Which valve is heard most clearly when the apical heartbeat is auscultated? _____

4. No one expects you to be a full-fledged physician on such short notice, but on the basis of what you have

learned about heart sounds, how might abnormal heart sounds be used to diagnose heart problems? _____

_____

_____

_____

Palpation of the Pulse

1. Define *pulse:* _____

_____

2. Describe the procedure used to take the pulse. _____

_____

_____

_____

3. Which artery is palpated at the following pressure points?

at the wrist _____   on the dorsum of the foot _____

in front of the ear _____   at the side of the neck _____

4. When you were palpating the various pulse or pressure points, which appeared to have the greatest

amplitude or tension? _____ Why do you think this was so? _____

_____

5. Assume someone has been injured in an auto accident and is hemorrhaging badly. What pressure point would you compress to help stop bleeding from the following areas?

the thigh _____   the calf _____

the forearm _____   the thumb _____

6. How could you tell by simple observation whether bleeding is arterial or venous? _____

_____

7. You may sometimes observe a slight difference between the values obtained from an apical pulse (beats/min)

and an arterial pulse taken elsewhere on the body. What is this difference called? _____

_____

Blood Pressure Determinations

1. Define _blood pressure:_ _____

_____

2. To what phase of the cardiac cycle do the following terms apply?

systolic pressure _____ diastolic pressure _____

3. What is the name of the instrument used to compress the artery and record pressures in the auscultatory

method of determining blood pressure? _____

4. What are the sounds of Korotkoff? _____

_____

What causes the systolic sound? _____

_____

The disappearance of sound? _____

_____

5. Interpret 145/85/82: _____

_____

RS123

6. Define *pulse pressure:* _____

Why is this measurement important? _____

_____

7. How do venous pressures compare to arterial pressures? _____

Why? _____

8. What maneuver to increase the thoracic pressure illustrates the effect of external factors on venous pres-

sure? _____ How was it performed? _____

_____

_____

9. What might an abnormal increase in venous pressure indicate? (Think!) _____

_____

Effect of Various Factors on Blood Pressure and Heart Rate

1. What is the effect of the following on blood pressure? (Indicate increase by I and decrease by D.)

_____ increased diameter of the arterioles _____ increased cardiac output _____ arteriosclerosis

_____ increased blood viscosity _____ hemorrhage _____ increased pulse rate

2. In what position—sitting, reclining, or standing—is the blood pressure normally the highest? _____

_____ The lowest? _____ What changes did you observe

immediately when the subject stood up after being in the reclining position? _____

_____

What changes in the blood vessels might account for the change? _____

_____

What changes in the subject's blood pressure were observed after standing for 3 minutes? _____

_____

How do you account for this change? _____

_____

3. What was the effect of exercise on blood pressure? _____

On pulse? _____ Do you think these effects reflect changes in cardiac output

or in peripheral resistance? _____

4. What effects of the following on blood pressure did you observe in the laboratory?

nicotine _____ cold temperature _____

What do you think the effect of heat would be? _____ Why? _____

_____

5. Why are there normally no significant increases in diastolic pressure after exercise? _____

_____

EXERCISE 34

Frog Cardiovascular Physiology

Intrinsic Properties of Cardiac Muscle: Automaticity and Rhythmicity

1. Define the following terms:

automaticity _____

_____

rhythmicity _____

_____

2. Discuss the anatomic differences you observed between frog and human hearts. _____

_____

_____

3. Which region of the dissected frog heart had the highest intrinsic rate of contraction? _____

The greatest automaticity? _____ The greatest regularity or rhythmicity?

_____ How do these properties correlate with the duties of a pacemaker?

_____

Is this region the pacemaker of the frog heart? _____

Which region had the lowest intrinsic rate of contraction? _____

Effect of Various Factors on Heart Activity

1. Define *extrasystole:* _____

2. Refer to the recordings you made during this exercise, and respond to the following questions.

What was the effect of stimulation of the heart during ventricular contraction? _____

_____

During ventricular relaxation (first portion)? _____

During the pause interval? _____

What does this indicate about the refractory period of cardiac muscle? _____

Can cardiac muscle be tetanized? _____ Why or why not? _____

_____

3. Describe the effect of thermal factors on the frog heart.

cold _____ heat _____

4. What was the effect of vagal stimulation on heart rate? _____

Which of the following factors (some tested and some not) would cause the same effect: epinephrine, acetyl-

choline, atropine sulfate, sympathetic nervous system activity, digitalis, potassium ions? _____

_____

Which of these factors would reverse or antagonize vagal effects? _____

_____

5. What is vagal escape? _____

_____

Why is vagal escape valuable in maintaining homeostasis? _____

_____

6. Once again refer to your recordings. Did the administration of the following produce any changes in force
of contraction (shown by peaks of increasing or decreasing height)? If so, explain the mechanism.

epinephrine _____

_____

acetylcholine _____

_____

calcium ions _____

_____

7. Excess concentration of each of the following ions would most likely interfere with normal heart activity.
Note the type of changes caused in each case.

K^+ _____

Ca^{++} _____

Na^+ _____

8. How does the Stannius ligature used in the laboratory produce heart block? _____

_____

9. Define *partial heart block*, and describe how it was recognized in the laboratory: _____

_____

10. Define *total heart block*, and describe how it was recognized in the laboratory: _____

_____

11. What do your heart block experiment results indicate about the spread of impulses from the atria to the ventricles? _____

_____

Observation of the Microcirculation and Investigation of Factors Affecting Local Blood Flow

1. In what way are the red blood cells of the frog different from those of the human? _____

On what basis of this one factor, would you expect their life spans to be longer or shorter? _____

2. The following statements refer to your observation of one or more of the vessel types observed in the microcirculation in the frog's web. Characterize each statement by choosing one or more responses from the key.

a. arteriole   b. venule   c. capillary

_____ smallest vessels observed

_____ vessel within which the blood flow is rapid, pulsating

_____ vessel in which blood flow is least rapid

_____ red blood cells pass through single file in these vessels

_____ blood flow smooth and steady

_____ most numerous vessels

_____ vessels that deliver blood to the capillary bed

_____ vessels that serve the needs of the tissues via exchanges

_____ vessels that drain the capillary beds

3. Which of the vessel diameters changed most? _____

What division of the nervous system controls the vessels? _____

4. Discuss the effects of the following on blood vessel diameter (state specifically the blood vessels involved) and rate of blood flow. Then explain the importance of the reaction observed to the general well-being of the body.

local application of cold _____

_____

_____

local application of heat _____

_____

_____

inflammation (or application of HCl) _____

_____

_____

histamine _____

_____

_____

EXERCISE 35   The Lymphatic System

Distribution of the Lymphatic Vessels and Lymph Nodes

1. Explain why the lymphatic system is a one-way system, whereas the blood vascular system is a two-way system. _____

_____

2. How do lymph vessels resemble veins? _____

How do lymph capillaries differ from blood capillaries? _____

_____

3. What is the function of the lymphatic vessels? _____

_____

4. What is lymph? _____

5. What factors are involved in the flow of lymphatic fluid? _____

_____

6. What name is given to the terminal duct draining most of the body? _____

What is the cisterna chyli? _____

_____

7. How does the composition of lymph differ in the cisterna chyli and in the general lymphatic stream?

_____

8. Which portion of the body is drained by the right lymphatic duct? _____

9. Note three areas where lymph nodes are densely clustered: _____ ,

_____ , _____

10. What are the two major functions of the lymph nodes? _____

_____

11. The radical mastectomy is an operation in which a cancerous breast, surrounding tissues, and the underlying muscles of the anterior thoracic wall plus the axillary lymph nodes are removed. After such an operation, the arm usually swells or becomes edematous and very uncomfortable—sometimes for months.

Why? _____

_____

RS129

Microscopic Anatomy of a Lymph Node

1. What structural characteristic ensures that the flow of lymph through a lymph node is slow? _____

_____

Why is this desirable? _____

EXERCISE 36

Anatomy of the Respiratory System

Upper and Lower Respiratory Structures

1. Two pairs of vocal folds are found in the larynx. Which pair are the true vocal cords? (Superior or inferior?) _____

2. What is the significance of the fact that the human trachea is reinforced with cartilage rings? _____

_____

Of the fact that the rings are incomplete posteriorly? _____

_____

3. Name the specific cartilages in the larynx that correspond to the following descriptions:

forms the Adam's apple _____ shaped like a signet ring _____

a "lid" for the larynx _____ vocal cord attachment _____

4. Why is the epiglottis called the guardian of the airways? _____

_____

_____

5. Trace a molecule of oxygen from the external nares to the pulmonary capillaries of the lungs: External

nares - - - → _____

_____

_____

_____

6. What is the function of the pleural membranes? _____

_____

7. Name two functions of the nasal cavity mucosa: _____

_____

8. The following questions refer to the primary bronchi:

Which is longer? _____ Larger in diameter? _____

More horizontal? _____

The most common site for lodging of a foreign object that had entered the respiratory passageways?

_____

RS131

9. Match the terms in column B to those in column A.

Column A	Column B
_____ nerve that serves the diaphragm during inspiration	a. alveoli
_____ separates the oral and nasal cavities	b. bronchioles
_____ food passageway posterior to the trachea	c. conchae
_____ flaps over the glottis during swallowing of food	d. epiglottis
_____ voice box	e. esophagus
_____ windpipe	f. glottis
_____ pleural layer covering the walls of the thorax	g. larynx
_____ actual site of gas exchange	h. palate
_____ autonomic nervous system nerve serving the thoracic region	i. parietal pleura
_____ opening between the vocal folds	j. phrenic nerve
_____ fleshy lobes in the nasal cavity	k. primary bronchi
	l. trachea
	m. vagus nerve
	n. visceral pleura

10. What portions of the respiratory system are referred to as anatomic dead space? Why? _____

_____

_____

11. Define *external respiration*: _____

   internal respiration: _____

Sheep Pluck Demonstration

1. Does the lung inflate part-by-part or as a whole, like a balloon? _____

What happened when the pressure was released? _____

What type of tissue insures this phenomenon? _____

Examination of Prepared Slides of Lung and Trachea Tissue

1. The tracheal epithelium is ciliated and has goblet cells. What is the function of each of these modifications? Cilia? _____

Goblet cells? _____

2. The tracheal epithelium is said to be pseudostratified. Why? _____

_____

_____

3. What structural characteristics of the alveoli make them an ideal site for the diffusion of gases? _____

_____

Why does oxygen move from the alveoli into the pulmonary capillary blood? _____

_____

_____

4. If you observed pathologic lung sections, what were the conditions responsible and how did the tissue differ from normal lung tissue?

Slide type	**Observations**

Dissection of the Respiratory System of the Fetal Pig

1. Are the cartilaginous rings in the pig trachea complete or incomplete? _____

2. How does the number of lung lobes in the pig compare with the number in humans? _____

_____

3. Describe the appearance of lung tissue under the dissection microscope. _____

_____

4. Why did the segment of lung tissue, cut from the fetal pig's lung, <u>sink</u> when placed in water? _____

_____

_____

_____

EXERCISE 37

Respiratory Physiology

Mechanics of Respiration

1. Relative to your observations on the operation of the model lung, for each of the following cases, check the appropriate column.

Change	Diaphragm pushed up		Diaphragm pulled down	
	increased	decreased	increased	decreased
in internal volume of the bell jar (thoracic cage)				
in internal pressure				
in the size of the balloons (lungs)				
in direction of air flow	into lungs	out of lungs	into lungs	out of lungs

2. Base your answers to the following on your observations in Question 1.

Under what internal conditions does air tend to flow into the lungs? Explain. _____

_____

Under what internal conditions does air tend to flow out of the lungs? Explain. _____

_____

_____

3. The activation of the diaphragm and the external intercostal muscles begins the inspiratory process. What

results from the contractions of these muscles, and how is this accomplished? _____

_____

_____

4. What was the approximate increase in diameter of chest circumference during a quiet inspiration?

_____ inches   During forced inspiration? _____ inches

What temporary physiologic advantage does the substantial increase in chest circumference during forced in-

spiration create? _____

_____

_____

5. The presence of a partial vacuum between the pleural membranes is integral to normal breathing movements. What would happen if an opening were made into the chest cavity, as with a puncture wound?

_____

_____

How is this condition treated medically? _____

_____

Respiratory Volumes and Capacities—Spirometry

1. Write the respiratory volume term and the normal value that is described by the following statements:

volume of air present in the lungs after a forceful expiration _____

volume of air that can be expired forcibly after a normal expiration _____

volume of air that is breathed in and out during a normal respiration _____

volume of air that can be inspired forcibly after a normal inspiration _____

volume of air corresponding to TV + IRV + ERV _____

2. Record experimental respiratory volumes as determined in the laboratory.

Average tidal volume _____ cc   Average ERV _____ cc

Average IRV _____ cc   Average VC _____ cc

3. Does your separate vital capacity measurement equal the figure gained by the addition of the other

volumes? _____ Why? _____

_____

4. Would your vital capacity measurement differ if you performed the test standing? _____

Lying down? _____ Explain. _____

_____

_____

5. Which respiratory ailments can respiratory volume tests be used to detect? Explain your reasoning.

_____

_____

_____

6. Using an appropriate reference, complete the chart below:

		O_2	CO_2	N_2
% composition of air	Inspired			
	Expired			

Use of the Pneumograph to Determine Factors Influencing Rate and Depth of Respiration

1. Where are the neural control centers of respiratory rhythm? _____

and _____

2. Based on pneumograph readings of respiratory variation, what was the rate of quiet breathing?

Initial testing _____ breaths/min

Record observations of how the initial pneumograph recording was modified during the variouys testing procedures listed below. Indicate the respiratory rate and include comments on the relative depth of the respiratory peaks observed.

Test performed	Observations
talking	
yawning	
laughing	
standing	
concentrating	
swallowing water	
coughing	
lying down	
running in place	

3. Breath-holding interval after a deep inhalation _____ sec   Length of recovery period _____ sec

Breath-holding after a forceful expiration _____ sec   Length of recovery period _____ sec

After breathing quietly and taking a deep breath (which you held), was the urge to inspire *or* expire?

_____

After exhaling and then breath-holding, was the desire for inspiration or expiration? _____

Explain these results? (Hint: what reflex is involved here?) _____

_____

_____

4. Observations after hyperventilation: _____

_____

5. Length of breath-holding after hyperventilation: _____ sec Why does hyperventilation produce apnea or a

reduced respiratory rate? _____

_____

_____

6. Observations after rebreathing breathed air: _____

_____

Why does rebreathing breathed air produce an increased respiratory rate? _____

_____

7. What was the effect of running in place (exercise) on the duration of breath holding? _____

Explain: _____

_____

8. Relative to the test illustrating the effect of respiration on circulation:

Radial pulse before beginning test _____/min   Radial pulse after testing _____/min

Relative pulse force before beginning _____   Relative force of radial pulse after testing _____

Condition of neck and facial veins after testing _____

Explain: _____

9. Do the following factors generally increase (indicate with I) or decrease (indicate with D) the respiratory rate and depth?

Increase in blood CO_2 _____   Increase in blood pH _____

Decrease in blood O_2 _____   Decrease in blood pH _____

Did it appear that CO_2 or O_2 had a more marked effect on modifying the respiratory rate? _____

10. Sensory receptors sensitive to changes in blood pressure are located where? _____

_____

11. Where are sensory receptors sensitive to changes in O_2 levels in the blood located? _____

_____

12. What is the primary factor that initiates breathing in a newborn infant? _____

_____

13. Blood CO_2 levels and blood pH are related. When blood CO_2 levels increase, the pH _____.
                                                                                                                                              (increases/decreases)

Explain why. _____

_____

_____

Respiratory Sounds

1. Which of the respiratory sounds is heard during both inspiration and expiration? _____

Which is heard primarily during inspiration? _____

2. Where did you best hear the vesicular respiratory sounds? _____

_____

Role of the Respiratory System
in Acid-Base Balance of Blood

1. Define buffer: _____

_____

2. How successful was the laboratory buffer (pH 7) in resisting changes in pH when the acid was added?

_____

_____

When the base was added? _____

_____

How successful was the buffer in resisting changes in pH when the additional aliquots (3 more drops) of

the acid and base were added to the original samples? _____

_____

3. Explain how the carbonic acid–bicarbonate buffer system of the blood operates. _____

_____

_____

_____

_____

Ciliary Action of the Respiratory Mucosa

1. Do cilia "beat" in one or more directions? _____

What evidence do you have for your response? _____

_____

_____

2. How long did it take for the ciliary action to move the cork when it was initially flushed with room-

temperature saline? _____

3. What was the effect of the cold saline on the speed of cork transport? _____

Explain. _____

_____

4. Was the ciliary action strong enough to move the cork uphill against gravity? _____

Could you have predicted this? _____

On what basis? _____

_____

5. Specifically, what is the purpose of the ciliary action in the body? _____

_____

_____

EXERCISE 38

Anatomy of the Digestive System

Gross Anatomy of the Human Digestive System

1. Match the key items on the right with the descriptive statements on the left.

_____ structure that suspends the small intestine from the posterior body wall

_____ fingerlike extensions of the intestinal mucosa that increase the surface area for absorption

_____ large collections of lymphatic tissue found in the submucosa of the small intestine

_____ deep folds of the mucosa and submucosa that extend completely or partially around the small intestine circumference

_____, _____ two anatomic regions involved in the mechanical breakdown of foodstuffs

_____ organ that manipulates food in the mouth and initiates swallowing

_____ common passageway for air and food

_____, _____, _____ three structures continuous with and representing modifications of the peritoneum

_____ literally a food chute; no digestive/absorptive function

_____ folds of the gastric mucosa

_____ sacculations of the large intestine

_____ projections of the plasma membrane of a mucosal epithelial cell

_____ prevents food from moving back into the small intestine once it has entered the large intestine

_____ primary region of food and water absorption

_____ membrane securing the tongue to the floor of the mouth

_____ region of the digestive tract primarily involved in water absorption and the formation of feces

_____ area between the teeth and lips/cheeks

_____ blind sac that outpockets from the cecum

_____ region in which protein digestion begins

_____ structure attached to the lesser curvature of the stomach

a. anus

b. appendix

c. esophagus

d. frenulum

e. greater omentum

f. hard palate

g. haustra

h. ileocecal valve

i. large intestine

j. lesser omentum

k. mesentery

l. microvilli

m. oral cavity

n. parietal peritoneum

o. Peyer's patches

p. pharynx

q. plicae circulares

r. pyloric valve

s. rugae

t. small intestine

u. soft palate

v. stomach

w. tongue

x. vestibule

(continued)

RS141

_____ organ into which the stomach empties

_____ sphincter controlling the movement of food from the stomach into the duodenum

_____ posterior superior boundary of the oral cavity; uvula forms its terminus

_____ organ that receives pancreatic and bile secretions

_____ serous lining of the abdominal cavity wall

_____ principal site for the synthesis of vitamin K by microorganisms

_____ region containing two sphincters through which feces are expelled from the body

_____ anteriosuperior boundary of the oral cavity, underlain by bone

y. villi

z. visceral peritoneum

2. The tubelike digestive system canal that extends from the mouth to the anus is the _____

canal.

3. The general anatomic features of the digestive tube have been presented. Fill in the table below to complete the information listed.

Wall layer	Subdivisions of the layer	Major functions
mucosa		
submucosa	(not applicable)	
muscularis externa		
serosa or adventitia	(not applicable)	

4. How is the muscularis externa of the stomach modified? _____

_____

How does this modification relate to the function of the stomach? _____

_____

RS142

Accessory Digestive Organs

1. Various types of glands form a part of the alimentary tube wall or duct their secretions into it. Match the glands listed in column B with the function/locations described in column A.

Column A

Column B

_____ produce mucus; found in the submucosa of the small intestine

a. Brunner's glands

_____ produce a product containing amylase that begins starch breakdown in the mouth

b. crypts of Lieberkühn

c. gastric glands

_____ produces a whole spectrum of enzymes and an alkaline fluid that is secreted into the duodenum

d. liver

e. pancreas

_____ produces bile that it secretes into the duodenum via the common bile duct

f. salivary glands

_____ produces HCl and pepsinogen

_____ found in the mucosa of the small intestine; produce digestive enzymes

2. Which of the salivary glands produces a secretion that is mainly serous? _____

3. What is the role of the gall bladder? _____
_____

4. Use the key below to identify each tooth area described.

Key: a. anatomic crown

_____ visible portion of the tooth in situ

b. cementum

_____ material covering the tooth root

c. clinical crown

_____ hardest substance in the body

d. dentin

_____ attaches the tooth to bone and surrounding alveolar structures

e. enamel

_____ portion of the tooth embedded in bone

f. gingiva

_____ forms the major portion of tooth structure; similar to bone

g. odontoblasts

_____ form the dentin

h. periodontal membrane

_____ site of blood vessels, nerves, and lymphatics

i. pulp

_____ entire portion of the tooth covered with enamel

j. root

5. In the human, the number of deciduous teeth is _____; the number of permanent teeth is _____.

6. The dental formula for deciduous teeth is: $\dfrac{2,1,0,2}{2,1,0,2}$

Explain what this means: _____
_____

What is the dental formula for the permanent teeth? _____

7. What teeth are the "wisdom teeth"? _____

Microscopic Anatomy of the Digestive System and Accessory Organs

1. You have studied the histologic structure of a number of organs in this laboratory. Four of these are diagrammed below. Identify each.

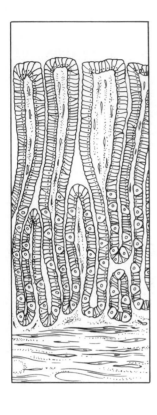

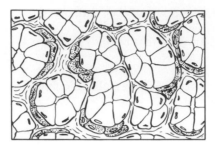

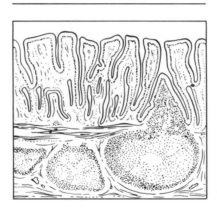

_____

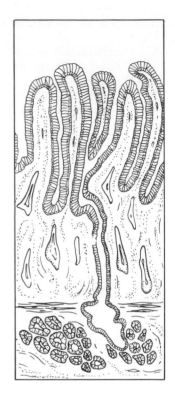

_____   _____   _____

2. What transition in epithelium type exists at the cardiac-esophageal junction? _____

_____

How do the epithelia of these two organs relate to their specific function? _____

_____   _____

_____

3. What cells of the stomach produce HCI? _____

Pepsinogen? _____

4. Name three structures always found in the triad regions of the liver. _____ ,

_____ , and _____ .

5. Where would you expect to find the van Kupffer cells of the liver? What is their function? _____

_____

6. Why is the liver so dark red in the living animal? _____

_____

Dissection of the Digestive System of the Fetal Pig

1.   Several differences between pig and human digestive anatomy should have become apparent during the dissection experience. Note the pertinent differences between the human and the pig relative to the following structures:

Structure	Pig	Human
Number of liver lobes		
Appendix		
Appearance and distribution of colon		
Presence of round ligament		

EXERCISE 39

Mechanisms of Food Propulsion and the Physiology of Smooth Muscle

Food Propulsion Mechanisms

1. Complete the following statements:

Swallowing occurs in three phases—the __1__, __2__, and __3__. Only one of these phases is voluntary; this is the __4__ phase. During the voluntary phase, the __5__ is used to push the food into the back of the throat. During swallowing the __6__ rises to insure that its passageway is covered by the epiglottis so that ingested substances don't enter the respiratory passageways. It is possible to swallow water while standing on your head because the water is carried along the esophagus involuntarily by the process of __7__. The pressure exerted by the foodstuffs on the __8__ sphincter causes it to open, allowing the foodstuffs to enter the stomach.

The three major types of movements that occur in the small intestine are __9__, __10__, and __11__. One of these movements, the __12__, acts to continually mix the foods but has no role in moving the foods along the digestive tract. Another type of movement seen in the large intestine, __13__, occurs infrequently and acts to move the feces over relatively long distances toward the anus. Three factors that influence the motility of the small intestine are __14__, __15__, and __16__.

1. _____
2. _____
3. _____
4. _____
5. _____
6. _____
7. _____
8. _____
9. _____
10. _____
11. _____
12. _____
13. _____
14. _____
15. _____
16. _____

Physiology of Smooth Muscle

1. Define the following terms:

automaticity _____

_____

plasticity _____

_____

2. Which type of smooth muscle is found in the walls of the body's hollow organs—multiunit or unitary?

_____ What are the major differences between these two types of smooth muscle

regarding the precision of their contraction and controlling factors? _____

_____

_____

_____

3. How does increasing the load on smooth muscle (unitary) modify its contractile activity? _____ _____ Explain the physiologic value of this response. _____

_____

_____

4. At what temperature (as determined during experimentation) was smooth muscle contractility the greatest? _____ °C How do you explain this? _____

5. You used several test solutions to investigate the effect of various ions on smooth muscle activity. Complete the table below by noting your observations and explaining each.

Ion added	Effect observed with increasing concentration	Explanation
OH^- (NaOH)		
H^+ (HCl)		
K^+ (KCl)		
Ca^{++} (CaCl$_2$)		

6. Generally speaking, what is the effect of the parasympathetic nervous system on the secretory activity and mobility of the digestive tract? _____ Of the sympathetic nervous system? _____ Which of the chemicals used, acetylcholine or epinephrine, elicited effects typical of the sympathetic nervous system? _____ Which elicited the effects of the parasympathetic nervous system? _____ Did you anticipate the effects? _____ Explain. _____

_____

_____

_____

EXERCISE 40

Chemical Breakdown of Foodstuffs: Enzymatic Action

1. Match the following definitions with the proper key letters.

Key: a. catalyst   b. control   c. enzyme   d. hydrolase   e. substrate

_____ increases the rate of a chemical reaction without becoming part of the product

_____ provides a standard of comparison for test results

_____ an enzyme that cleaves molecules by adding water to the bonds

_____ biologic catalyst; protein in nature

_____ substance on which a catalyst works

2. List three characteristics of enzymes. _____

_____

_____

3. The enzymes of the digestive system are classed as hydrolases. What does this mean? _____

_____

4. Fill in the following chart relative to the various digestive system enzymes encountered in this exercise.

Enzyme	Organ producing it	Site of action	Substrate(s)	Optimal pH
salivary amylase				
pepsin				
pancreatin				

5. Name the *end products* of digestion for the following types of foods:

proteins _____   fats _____   carbohydrates _____

6. You used several indicators or tests in the laboratory to determine the presence or absence of certain substances. Choose the correct test or indicator from the key to correspond to the conditions described below:

Key: a. IKI (Lugol's iodine)   c. litmus indicator   e. Ninhydrin test
         b. Benedict's solution   d. biuret test

_____ used to test for the presence of protein, which was indicated by a violet color

_____ used to test for the presence of starch, which was indicated by a blue-black color

_____ used to test for the presence of fatty acids, which was evidenced by a color change from blue to red

(*continued*)

_____ used to test for the presence of reducing sugars (maltose, sucrose, glucose) as indicated by a blue to red color change

_____ used to test for the presence of free alpha-amino groups as indicated by a blue color

7. In the procedure concerning starch digestion by salivary amylase, how do you explain the fact that neither 0 C incubation nor conditions that involved a preliminary boiling of the enzyme preparation resulted in positive test results for the digestion of starch? _____

_____

What conclusions can you draw when an experimental sample gives both a positive starch test and a positive Benedict's test after incubation? _____

_____

Why was paraffin chewed during saliva collection? _____

_____

Why was 37 C the optimal incubation temperature? _____

Why did very little, if any, starch digestion occur in test tubes 9 and 10? _____

_____

_____

8. In the procedure concerning pepsin digestion of protein, in which test tube did more protein hydrolysis occur? _____ Why? _____

_____

Why did test tubes 6 and 7 yield negative results for digestion? _____

_____

What functional relationship exists between HCl and pepsin? _____

_____

Pepsin is a protein-digesting enzyme, and the structural material of cells is largely protein. Why doesn't the stomach digest itself? _____

_____

9. In the procedure concerning pancreatic lipase digestion of fats and the action of bile salts, how did appearance of test tubes A and B differ? _____

_____

_____

How can you explain this difference? _____

_____

RS150

Why did litmus cream change from blue to red during the process of fat hydrolysis? _____

_____

Why is bile not considered an enzyme? _____

_____

_____

What role does bile play in fat digestion? _____

_____

10.   The three-dimensional structure of a functional protein is altered by intense heat or excesses of pH even though peptide bonds may not break. Such inactivation is called denaturation, and denatured enzymes are

nonfunctional. Explain why. _____

_____

_____

What specific experimental conditions in the various procedures resulted in the denaturation of the enzymes?

_____

_____

11.   Pancreatic and intestinal enzymes operate optimally at a pH that is slightly alkaline, yet the chyme entering the duodenum from the stomach is very acid. How is the proper pH for the functioning of the

pancreatic-intestinal enzymes assured? _____

_____

12.   Assume you have been chewing a piece of bread for 5 or 6 minutes. How would you expect its taste to

change during this interval? _____

Why? _____

13.   Note the mechanism of absorption (passive or active transport) of the following food breakdown products, and indicate (✔) whether the absorption would result in their movement into the blood capillaries or the lymph capillaries (lacteals).

Substance	Mechanism of absorption	Blood	Lymph
monosaccharides			
fatty acids and glycerol			
amino acids			
water			
Na^+, Cl^-, Ca^{++}			

14. People on a strict diet to lose weight begin to metabolize stored fats at an accelerated rate. How would this condition affect blood pH? _____

_____

15. Trace the pathway of a ham sandwich (ham = protein and fat; bread = starch) from the mouth to the site of absorption of its breakdown products, noting where digestion occurs and what specific enzymes are involved.

16. Some of the digestive organs have groups of secretory cells that liberate hormones (parahormones) into the blood. These exert an effect on the digestive process by acting on other cells or structures and causing them to release digestive enzymes, expel bile, or increase the mobility of the digestive tract. For each hormone below, note the organ producing the hormone and its effects on the digestive process. Include the target organs affected.

Hormone	Target organ(s) and effects
secretin	
gastrin	
cholecystokinin	

RS152

EXERCISE 41

Anatomy of the Urinary System

Gross Anatomy of the Human Urinary System

1. Complete the following statements:

The kidney is referred to as an excretory organ because it excretes __1__ wastes. It is also a major homeostatic organ because it maintains the electrolyte, __2__, and __3__ balance of the blood.

Urine is continuously formed by the __4__ and is routed down the __5__ by the mechanism of __6__ to a storage organ called the __7__. Eventually, the urine is conducted to the body __8__ by the urethra. In the male, the urethra is __9__ inches long and transports both urine and __10__. The female urethra is __11__ inches long and transports only urine.

Voiding or emptying the bladder is called __12__. Voiding has both voluntary and involuntary components. The voluntary sphincter is the __13__ sphincter. An inability to control this sphincter is referred to as __14__.

1. _____
2. _____
3. _____
4. _____
5. _____
6. _____
7. _____
8. _____
9. _____
10. _____
11. _____
12. _____
13. _____
14. _____

2. Completely describe the anatomic location of human kidneys. _____

_____

3. What is the function of the fat cushion that surrounds the kidneys in life? _____

_____

4. Define ptosis. _____

_____

5. Why is incontinence a normal phenomenon in the child under 1½ to 2 years old? _____

_____

What events may lead to its occurrence in the adult? _____

_____

Gross Internal Anatomy of the Pig or Sheep Kidney

1. Match the appropriate structure in column B to its description in column A.

Column A

_____ smooth membrane, tightly adherent to the kidney surface

_____ portion of the kidney containing collecting ducts

_____ portion of the kidney containing the bulk of the nephron structures

_____ outermost region of kidney tissue

_____ basinlike area of the kidney, continuous with the ureter

_____ a cup-shaped extension of the pelvis that encircles the apex of a pyramid

_____ areas of cortical tissue found between the medullary pyramids

Column B

a. cortex

b. medulla

c. minor calyces

d. renal capsule

e. renal columns

f. renal pelvis

Microscopic Anatomy of the Kidney and Bladder

1. Match each of the lettered structures on the diagram of the nephron (and associated renal blood supply) on the left with the terms on the right.

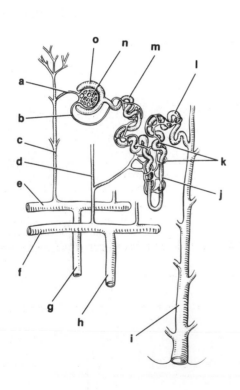

_____ collecting tubule

_____ glomerulus

_____ peritubular capillaries

_____ distal convoluted tubule

_____ proximal convoluted tubule

_____ interlobar artery

_____ interlobular artery

_____ arcuate artery

_____ interlobular vein

_____ efferent arteriole

_____ arcuate vein

_____ loop of Henle

_____ afferent arteriole

_____ interlobar vein

_____ Bowman's capsule

2. Using the terms provided in item 1, identify the following:

_____ site of filtrate formation

_____ primary site of tubular reabsorption

_____ secondarily important site of tubular reabsorption

_____ structure that conveys the processed filtrate (urine) to the renal pelvis

_____ blood supply that directly receives substances from the tubular cells

_____ its inner (visceral) membrane forms part of the filtration membrane

3. Complete the following statements:

The glomerulus is a unique high-pressure capillary bed because the __1__ arteriole feeding the glomerulus is larger in diameter than the __2__ arteriole draining the bed. Glomerular filtrate is very similar to __3__, but it has fewer proteins.

Three reabsorption mechanisms that occur in the renal tubule are __4__, __5__, and __6__. As an aid for the reabsorption process, the cells of the proximal convoluted tubule have dense __7__ on their luminal surface, which increases the surface area dramatically. Other than reabsorption, an important tubular function is __8__.

1. _____
2. _____
3. _____
4. _____
5. _____
6. _____
7. _____
8. _____

4. Trace a drop of blood from the time it enters the kidney in the renal artery until it leaves the kidney through the renal vein.   renal artery. - - - ➤ _____

_____

_____

_____   - - - ➤ renal vein.

5. Trace the anatomic pathway of a molecule of creatinine (metabolic waste) from Bowman's capsule to the urethra. Note each microscopic and/or gross structure it passes through in its travels. Name the subdivisions of the renal tubule.   Bowman's capsule - - - ➤ _____

_____

_____

_____   - - - ➤ urethra

6. What is important functionally about the specialized epithelium (transitional epithelium) in the bladder?

_____

_____

Dissection of the Urinary System of the Fetal Pig

1. How does the structure and distribution of the allantoic bladder of the fetal pig differ from the urinary

bladder of the human (or that of the adult pig for that matter)? _____

_____

2. What differences in fetal elimination of nitrogenous wastes accounts for the structural differences de-

scribed above? _____

_____

_____

3. How does the site of urethral emptying in the female pig differ from its termination point in the human

female? _____

_____

EXERCISE 42

Urinalysis

1. What is the normal volume of urine excreted in a 24-hour period?

a. 0.1–0.5 liters   b. 0.5–1.2 liters   c. 1.0–1.8 liters

2. Assuming normal conditions, note whether each of the following substances would be (a) in greater relative concentration in the urine than in the glomerular filtrate, (b) in lesser concentration in the urine than in the glomerular filtrate, or (c) absent in both the urine and the glomerular filtrate.

_____ water   _____ amino acids   _____ urea

_____ phosphate ions   _____ glucose   _____ uric acid

_____ sulfate ions   _____ albumin   _____ creatinine

_____ potassium ions   _____ red blood cells   _____ pus (WBC)

_____ sodium ions

3. Explain why urinalysis is a routine part of any good physical examination. _____

_____

4. What substance is responsible for the normal yellow color of urine? _____

5. Which has a greater specific gravity: 1 ml of urine or 1 ml of distilled water? _____

Explain. _____

_____

6. Explain the relationship between the color, specific gravity, and volume of urine. _____

_____

_____

_____

_____

7. A microscopic examination of urine may reveal the presence of certain abnormal urinary constituents.

Name three constituents that might be present if a urinary tract infection exists. _____,

_____ , and _____

8. How does a urinary tract infection influence urine pH? _____

How does starvation influence urine pH? _____

9. Several specific terms have been used to indicate the presence of abnormal urine constituents. Identify each of the abnormalities described below by inserting the term that names the condition.

_____ presence of erythrocytes in the urine

_____ presence of hemoglobin in the urine

_____ presence of glucose in the urine

_____ presence of albumin in the urine

_____ presence of ketone bodies (acetone and others) in the urine

_____ presence of pus (white blood cells) in the urine

10. What are renal calculi and what conditions favor their formation? _____

_____

11. All urine specimens become alkaline and cloudy on standing at room temperature. Explain. _____

_____

12. Glucose and albumin are both normally absent in the urine, but the reason for their exclusion differs.

Explain the reason for the absence of glucose. _____

_____

The reason for the absence of albumin. _____

_____

13. Several conditions (both pathologic and nonpathologic) are named below. Using the key provided, characterize the probable abnormal constituents or conditions of the urinary product of each. More than one choice may be necessary to fully characterize the condition in most cases.

a. albumin   d. glucose   g. pus   j. casts
b. hemoglobin   e. ketone bodies   h. high sp. gravity
c. blood cells   f. bilirubin   i. low sp. gravity

_____ glomerulonephritis   _____ pyelonephritis   _____ kidney stones

_____ diabetes mellitus   _____ gonorrhea   _____ eating a 5-lb box of
                                                                          candy at one sitting

_____ pregnancy, exertion   _____ starvation
                                                                _____ hemolytic anemias

_____ hepatitis, cirrhosis of _____ diabetes insipidus
          the liver   _____ cystitis (inflammation
                                                                          of the bladder)

14. Name the three major nitrogenous wastes found in the urine. _____,

_____ , and _____

EXERCISE 43

Kidney Regulation of Fluid and Electrolyte Balance

1. The following questions are based on the ingestion of 700 to 1000 ml of distilled water by student 1:

How did the volume collected during the experiment compare with the volume of distilled water drunk?

_____ How did the color and the specific gravity of the consecutive sam-

ples collected change during the course of this experiment? _____

_____

How did the specific gravity of the initial experimental urine samples compare with that of the control?

_____

Were the kidneys conserving water *or* salt during this experiment? _____

What hormone is responsible for the kidney's ability to conserve this substance? What elicits its release?

_____

_____

What hormone was being inhibited during this experiment? _____

Why? _____

_____

2. The following questions are based on the ingestion of 700 to 1000 ml 0.9% saline by student 2 and of 150 to 250 ml 2.0% saline by student 3:

How did the specific gravity of the experimental samples in each case compare with that of the control?

Student 2 _____

Student 3 _____

How did the total urinary volume collected during the 2½-hour period compare with the volumes ingested

in each case? Student 2 _____ Student 3 _____

In the case of Student 3, what hormone was inhibited and why? _____

_____

Were any sensations of thirst noted? _____ If so, which student had the greater subjective feelings of thirst? _____ Explain. _____

_____

_____

Which of the two salt solutions ingested was isotonic? _____ % Hypertonic _____ %

What hormone was exerting its effects on the kidneys of student 3 during these tests? _____

What activated this hormonal mechanism? _____

_____

3.   The following questions are based on the ingestion of 500 ml of 0.2% sodium bicarbonate by student 4:

How did the volume of urine excreted during the 2½-hour period compare to the volume ingested at the onset of the experiment? _____

How did the pH of the urine collected during the experiment compare with that of the control sample?

_____

What ions were the kidneys excreting? _____ Conserving? _____

Explain. _____

_____

4.   In which of the experimental procedures was renin released by the kidneys? _____

5.   Using your text or an appropriate reference, define the following:

polyuria _____

dysuria _____

anuria _____

oliguria _____

diuresis _____

Specifically, how do polyuria and diuresis differ? _____

_____

Which of the terms defined above apply to the conditions noted in the experiment conducted by

Student 1? _____   Student 2? _____

Student 3? _____   Student 4? _____

EXERCISE 44

Anatomy of
the Reproductive System

Gross Anatomy of the Human Male
Reproductive System

1. List the two principal functions of the testis:

_____   _____

2. Identify all structures or portions of structures provided with leader lines on the diagrammatic view of the male reproductive system below.

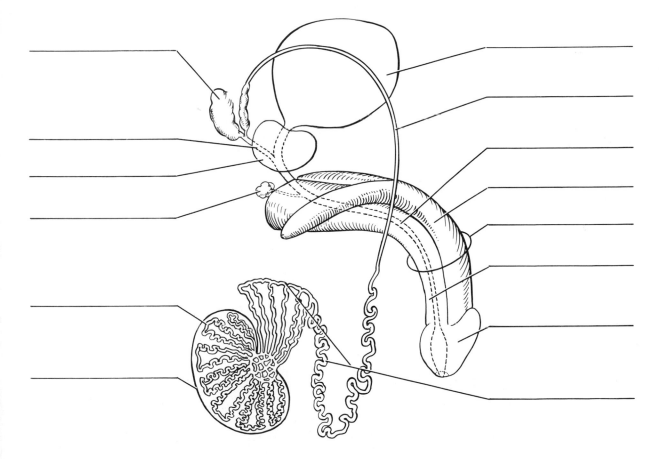

3. A common part of any physical examination of the male is palpation of the prostate gland. How is this accomplished? (Think!) _____

_____

4. How might enlargement of the prostate gland interfere with urination or the reproductive ability of the male? _____

5. Match the terms in column B to the descriptive statements in column A.

Column A

_____ copulatory organ/penetrating device

_____ site of sperm/androgen production

_____ passageway conveying sperm from the epididymis to the ejaculatory duct; in the spermatic cord

_____ conveys both sperm and urine down the length of the penis

_____ tubular storage and maturation site for sperm; hugs the lateral aspect of the testis

_____ extra-abdominal sac, which houses the testis

_____ cuff of skin encircling the glans penis

_____ portion of the urethra between the prostate gland and the penis

_____ empties a secretion into the prostatic urethra

_____ empties a secretion into the membranous urethra

Column B

a. Cowper's glands

b. epididymis

c. glans penis

d. membranous urethra

e. penile urethra

f. penis

g. prepuce

h. prostate gland

i. prostatic urethra

j. seminal vesicles

k. scrotum

l. testes

m. vas deferens

6. Why are the testes located in the scrotum? _____

_____

7. Describe the composition of semen and name all structures contributing to its formation. _____

_____

_____

8. Of what importance is the fact that semen is alkaline? _____

_____

9. What structures comprise the spermatic cord? _____

_____

Where is it located? _____

_____

10. Using the following terms, trace the pathway of sperm from the testes to the urethra: rete testis, epididymis, seminiferous tubule, ductus deferens.

_____ - - - → _____ - - - → _____ - - - → _____

11. Using an appropriate reference, define cryptorchidism and discuss its significance. _____

_____

_____

RS162

Gross Anatomy of the Human Female Reproductive System

1. On the diagram of a frontal section of a portion of the female reproductive system seen below, identify all structures provided with leader lines

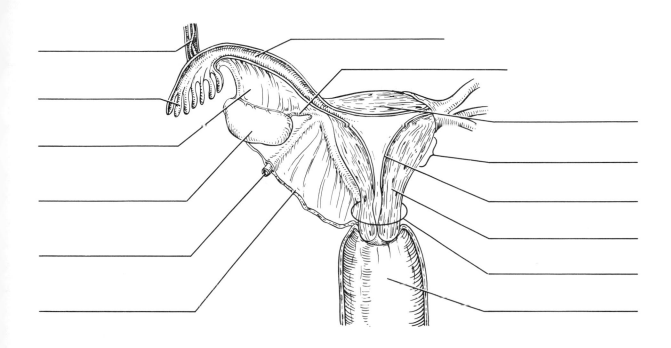

2. Identify the female reproductive system structures described below:

_____ chamber that houses the developing fetus

_____ copulatory canal

_____ usual site of fertilization

_____ becomes erectile during sexual stimulation

_____ duct conveying the ovum to the uterus

_____ partially closes the vaginal canal; a membrane

_____ primary reproductive organ

_____ undulate to create fluid currents to draw the ovulated egg into the fallopian tube

3. Do any sperm enter the pelvic cavity of the female? Why or why not? _____

_____

4. What is an ectopic pregnancy, and how can it happen? _____

_____

_____

5. Name the structures composing the external genitalia, or vulva, of the female. _____

_____

6. Put the following vestibular-perineal structures in their proper order from the anterior to the posterior aspect: vaginal orifice, anus, urethral opening, and clitoris.

Anterior limit: _____ ---→ _____ ---→ _____ ---→ _____

7. Name the male structure that is homologous to the female structures named below.

labia majora _____   clitoris _____

8. Assume a couple has just consummated the sex act and the male's sperm have been deposited in the woman's vagina. Trace the pathway of the sperm through the female reproductive tract.

_____

_____

9. Define *ovulation:* _____

_____

10. To describe breast function, complete the following sentences: Milk is formed by _____

within the _____ of the breast. Milk is then excreted into enlarged storage regions called

_____ and then finally through the _____.

11. Describe the procedure for self-examination of the breasts. (Men are not exempt from breast cancer, you

know!) _____

_____

_____

_____

_____

_____

_____

Microscopic Anatomy of Selected Reproductive Organs

1. The testis is divided into a number of lobes by connective tissue. Each of these lobes contains one to four

_____, which converge on a tubular region at the testis hilus called the

_____.

2. What is the function of the cavernous bodies seen in the male penis? _____

_____

3. Name the three layers of the uterine wall from the inside out.

_____, _____, _____,

Dissection of the Reproductive System of the Fetal Pig

1. The female pig has a _____ uterus; that of the human female is _____.

    Explain the difference in structure of these two uterine types. _____

    _____

    _____

2. What reproductive advantage is conferred by the pig's uterine type? _____

    _____

3. Cite differences noted between the pig and the human relative to the following structures:

    uterine tubes or oviducts _____

    _____

    urethral and vaginal openings in the female _____

    _____

    _____

EXERCISE 45   # Physiology of Reproduction: Gametogenesis and the Female Cycles

Meiosis

1. The following statements refer to events occurring during mitosis and/or meiosis. For each statement, decide if the event occurs in (a) mitosis only, (b) meiosis only, or (c) both mitosis and meiosis.

_____ dyads are visible

_____ tetrads are visible

_____ product is two diploid daughter cells

_____ product is four haploid daughter cells

_____ involves the phases prophase, metaphase, anaphase, and telophase

_____ occurs in all body tissues

_____ occurs only in the gonads

_____ increases the cell number for growth and repair

_____ synapsis and crossover of the homologous chromosomes occur

_____ daughter cells are quantitatively and qualitatively different from the mother cell

_____ daughter cells have the same number and types of chromosomes as the mother cell

_____ chromosomes are replicated before the division process begins

_____ provides cells for reproduction of the species

_____ consists of two consecutive nuclear divisions, without chromosomal replication occurring before the second division

2. Describe the process of synapsis. _____

_____

3. How does crossover introduce variability in the daughter cells? _____

_____

_____

4. Define homologous chromosomes. _____

_____

_____

Spermatogenesis

1. The cell types seen in the seminiferous tubules are listed in the key. Match the correct cell type(s) with the descriptions given below.

Key: a. primary spermatocytes   c. spermatogonia   e. spermatid
         b. secondary spermatocytes   d. Sertoli cell   f. sperm

_____ primitive stem cell   _____ product of meiosis II

_____ haploid   _____ product of spermiogenesis

_____ provides nutrients to developing sperm _____ product of meiosis I

2. Why are spermatids not considered functional gametes? _____

_____

3. Define *spermatogenesis:* _____

_____

Define *spermiogenesis:* _____

_____

4. Draw a sperm below and identify the *acrosome, head, midpiece,* and *tail.* Then beside each label, note the composition and function of each of these sperm structures.

5. The life span of a sperm is very short. What anatomic characteristics might lead you to suspect this even

if you didn't know its life span? _____

_____

Oogenesis, the Ovarian Cycle, and the Menstrual Cycle

1. The sequence of events leading to germ cell formation in the female begins during fetal development. By

the time the child is born, all the oogonia have been converted to_____.

In view of this fact, how does the total germ cell potential of the female compare to that of the male?

_____

2. The female gametes develop in structures called *follicles.* What is a follicle? _____

_____

How are primary and graafian follicles anatomically different? _____

_____

_____

What is a corpus luteum? _____

3. What hormone is produced by the graafian follicle? _____

By the corpus luteum? _____

4. Use the key to identify the cell type you would expect to find in the following structures.

Key: a. oogonium   b. primary oocyte   c. secondary oocyte   d. ovum

_____ forming part of the primary follicle in the ovary

_____ in the uterine tube before fertilization

_____ in the mature or graafian follicle of the ovary

_____ in the uterine tube shortly after sperm penetration

5. The cellular product of spermatogenesis is four _____; the final product of oogenesis

is one _____ and three _____. What is the function of this unequal

cytoplasmic division seen during oogenesis in the female? _____

_____

What is the fate of the three tiny cells produced during oogenesis? _____

Why? _____

6. The following statements deal with anterior pituitary and ovarian hormonal interrelationships. Name the
hormone(s) described in each statement.

_____ stimulates the growth of the ovarian follicle and its production of
                          estrogen

_____ triggers ovulation

_____ and _____ inhibit FSH release by the
                          anterior pituitary

_____ stimulates LH release by the anterior pituitary

_____ converts the ruptured follicle to a corpus luteum and stimulates it
                          to produce progesterone and estrogen

_____ maintains the hormonal production of the corpus luteum

7. Why does the corpus luteum deteriorate toward the end of the ovarian cycle? _____

_____

8. For each statement below dealing with hormonal blood levels during the female ovarian and menstrual cycles, decide whether the condition in column A is usually (a) greater than, (b) less than, or (c) essentially equal to the condition in column B.

Column A	**Column B**
_____ amount of estrogen in the blood during menses	amount of estrogen in the blood at ovulation
_____ amount of progesterone in the blood on the fourteenth day	amount of progesterone in the blood on the twenty-third day
_____ amount of LH in the blood during menses	amount of LH in the blood at ovulation
_____ amount of FSH in the blood on day 6 of the cycle	amount of FSH in the blood on day 20 of the cycle
_____ amount of estrogen in the blood on the tenth day	amount of progesterone in the blood on the tenth day

9. Ovulation and menstruation usually cease by the age of _____.

10. What uterine tissue undergoes dramatic changes during the menstrual cycle? _____

11. When during the female menstrual cycle would fertilization be unlikely? _____
_____

12. Assume that a woman could be an "on demand" ovulator like the rabbit, in which copulation stimulates the hypothalmic-anterior pituitary axis and causes LH release, and an oocyte was ovulated and fertilized on day 26 of her 28-day cycle. Why would a successful pregnancy be unlikely at this time? _____
_____
_____

13. The menstrual cycle depends on events within the female ovary. The stages of the menstrual cycle are listed below. For each, note its approximate time span and the events in the uterus, and then to the right, record the ovarian events going on at the same time. Pay particular attention to hormonal events.

Menstrual cycle stage	Uterine events	Ovarian events
Menstruation		
Proliferative		
Secretory		

EXERCISE 46

Survey of Embryonic Development

Early Embryology of the Sea Urchin and the Human

1. Use the key choices to identify the embryonic stage or process described below.

Key: a. cleavage   c. zygote   e. blastula
         b. morula   d. fertilization   f. gastrulation

_____ fusion of egg and sperm nuclei

_____ solid ball of embryonic cells

_____ process of rapid mitotic cell division without intervening growth periods

_____ the fertilized egg

_____ process involving cell migrations and rearrangements to form the three embryonic germ layers

_____ embryonic stage in which the embryo consists of a hollow ball of cells

2. What is the importance of cleavage in embryonic development? _____

_____

How is cleavage different from mitotic cell division, which occurs later in life? _____

_____

_____

3. Explain the process of gastrulation. _____

_____

_____

4. Name the three primary germ layers, and describe their relative positions in the embryo.

_____ , _____

_____ , _____

_____ , _____

5. The cells of the human blastula (more commonly called the blastocyst or chorionic vesicle) have various fates. Which blastocyst structures have the following fates?

_____ produces the embryonic body

_____ becomes the chorion and cooperates with uterine tissues to form the placenta

_____ produces the amnion, yolk sac, and allantois

_____ produces the primordial germ cells (an embryonic membrane)

_____ an embryonic membrane that provides the structural basis for the body stalk or umbilical cord

6. What is the function of the amnion and the amniotic fluid? _____

_____

_____

_____

7. Describe the process of implantation, noting the role of the trophoblast cells. _____

_____

_____

_____

8. How many days after fertilization is implantation generally completed? _____ What event in the female menstrual cycle ordinarily occurs just about this time if implantation does not occur? _____

_____

9. What name is given to the part of the uterine wall directly under the implanting embryo? _____

_____ That surrounding the rest of the embryonic structure? _____

_____

10. Using an appropriate reference, find out what decidua means and state the definition. _____

_____

How is this terminology applicable to the deciduas of pregnancy? _____

_____

11. Referring to the illustrations and text of "Life Before Birth," answer the following:

Which two organ systems are extensively developed in the *very young* embryo?

_____ and _____

Describe the direction of development by circling the correct descriptions below:

proximal-distal   distal-proximal   caudal-rostral   rostral-caudal

Does bodily control during infancy develop in the same directions? Think! Can an infant pick up a common pin (pincer grasp) or wave his arms earlier? Is arm-hand or leg-foot control achieved earlier? _____

_____

12. Note whether each of the following organs or organ systems develop from the (a) ectoderm, (b) endoderm, or (c) mesoderm. Use an appropriate reference as necessary.

_____ skeletal muscle   _____ respiratory mucosa   _____ nervous system

_____ skeleton   _____ circulatory system   _____ serosa membrane

_____ lining of gut   _____ epidermis of skin   _____ liver, pancreas

In Utero Development

1. Make the following comparisons between a human and the pregnant dissected animal structures.

Comparison object	Human	Dissected animal
shape of the placenta		
shape of the uterus		

2. Where in the human uterus do implantation and placentation ordinarily occur? _____

_____

3. Describe the function(s) of the placenta. _____

_____

_____

What embryonic membranes has it more or less "put out of business"? _____

_____

4. When does the human embryo come to be called a fetus? _____

5. What is the usual and most desirable fetal position in utero? _____

Why is this the most desirable position? _____

_____

_____

Gross and Microscopic Anatomy of the Placenta

1. Describe fully the gross structure of the human placenta as observed in the laboratory. _____

_____

_____

_____

_____

2. What is the tissue origin of the placenta: fetal, maternal, or both? _____

3. What are the placental barriers that must be crossed to exchange materials? _____

_____

EXERCISE 47

Principles of Heredity

Introduction to the Language of Genetics

1. Match the key choices with the definitions given below.

Key: a. allele   d. genotype   g. phenotype
      b. autosomes   e. heterozygous   h. recessive
      c. dominant   f. homozygous   i sex chromosomes

_____ actual genetic makeup

_____ chromosomes determining maleness/femaleness

_____ situation in which an individual has identical alleles for a particular trait

_____ genes not expressed unless they are present in homozygous condition

_____ expression of a genetic trait

_____ situation in which an individual has different alleles making up his
         genotype for a particular trait

_____ genes for the same trait that have different expressions

_____ chromosomes regulating most body characteristics

_____ the more potent gene allele; masks the expression of the less potent allele

Dominant-Recessive Inheritance

1. In humans, farsightedness is inherited by possession of a dominant gene. If a man who is homozygous for normal vision (aa) marries a woman who is heterozygous for farsightedness, what proportion of their children

would be expected to be farsighted? _____ %

2. A metabolic disorder called PKU is due to an abnormal recessive gene (p). Only homozygous recessive individuals exhibit this disorder. What percentage of the offspring will be anticipated to have PKU if the

parents are Pp and pp? _____ %

3. A man obtained 32 spotted and 10 solid color rabbits from a mating of two spotted rabbits.

Which trait is dominant? _____ Recessive? _____

What is the probable genotype of the rabbit parents? _____ × _____

4. Assume that the allele controlling brown eyes (B) is dominant over that controlling blue eyes (b) in human beings. (In actuality, eye color in humans is an example of multigene inheritance, which is much more complex than this.) A blue-eyed man marries a brown-eyed woman and they have six children, all brown-eyed. What are

the most likely genotypes of: the father? _____ the mother? _____

If the 7th child had *blue* eyes, what can you conclude about the parents' genotypes? _____

Incomplete Dominance

1. Tail length of a bobcat is controlled by incomplete dominance. The alleles are T for normal tail length and t for tail-less. What name could/would you give to the tails of heterozygous (Tt) cats? _____

_____ How would their tail length compare with that of TT or tt bob cats? _____

_____

2. If curly haired individuals are genotypically CC, straight-haired individuals are cc, and hetero-zygotes (Cc) have wavy hair, what percentage of the various phenotypes would be anticipated from a cross between a woman with CC and a man with cc?

_____ % curly _____ % wavy _____ % straight

Sex-Linked Inheritance

1. What does it mean when someone says a particular characteristic is sex-linked? _____

_____

2. You are a male, and you have been told that hemophilia "runs in your genes." Whose ancestors should you investigate—your mother's or your father's? _____ Why? _____

_____

3. An X^CX^c female marries an X^CY man. Do a punnet square for this match.

What is the probability of producing a color-blind son? _____ A color-blind daughter?

_____ A daughter that is a carrier for the color-blind gene? _____

4. Why are consanguinous marriages (marriages between blood relatives) prohibited in most cultures? _____

_____

_____

_____

Probability

1. What is the probability of having three daughters in a row? _____

2. A man and a woman, each of seemingly normal intellect, marry. While neither is aware of the fact, each is a heterozygote for the recessive allele. Is the allele for feeblemindedness dominant or recessive?

_____

What are the chances of them having one feebleminded child? _____

What are the chances that all their children (they plan a family of four) will be feebleminded?

_____

Genetic Determination of
Selected Human Characteristics

1. Look back at your data to complete this section. For each of the situations described here, determine if an offspring with the characteristics noted is possible with the parental genotypes listed. Check ($\checkmark$) the appropriate column.

Parental genotypes	Phenotype of child	Possibility	
		yes	no
Jj $\times$ jj	double-jointed thumbs		
FF $\times$ Ff	straight little finger		
EE $\times$ ee	detached ear lobes		
HH $\times$ Hh	mid-digital hair		
$I^A i \times I^B i$	Type O blood		
$I^A I^B \times ii$	Type B blood		

2. You have dimples, and you would like to know if you are homozygous or heterozygous for this trait. You have six brothers and sisters. How could you tell with some degree of certainty that you are a heterozygote by

observing your siblings? _____

Index